Turing's Legacy: Developments from Turing's Ideas in Logic

Alan Turing was an inspirational figure who is now recognized as a genius of modern mathematics. In addition to being a leading participant in the Allied forces' codebreaking effort at Bletchley Park in World War II, he proposed the theoretical foundations of modern computing, and anticipated developments in areas from information theory to computer chess. His ideas have been extraordinarily influential in modern mathematics and this book traces such developments by bringing together essays by leading experts in logic, artificial intelligence, computability theory, and related areas. Together, they give insight into this fascinating man, the development of modern logic, and the history of ideas. The articles within cover a diverse selection of topics, such as the development of formal proof, differing views on the Church–Turing thesis, the development of combinatorial group theory, and Turing's work on randomness which foresaw the ideas of algorithmic randomness that would emerge many years later.

ROD DOWNEY is Professor of Mathematics at Victoria University of Wellington, New Zealand. His main research interests lie in algebra, logic and complexity theory. Downey has received many professional accolades throughout his career, including the Schoenfield Prize of the Association for Symbolic Logic and the Hector Medal of the Royal Society of New Zealand, along with numerous fellowships to learned societies and institutes such as the Isaac Newton Institute (Cambridge) and the American Mathematical Society.

LECTURE NOTES IN LOGIC

**A Publication for
The Association for Symbolic Logic**

This series serves researchers, teachers, and students in the field of symbolic logic, broadly interpreted. The aim of the series is to bring publications to the logic community with the least possible delay and to provide rapid dissemination of the latest research. Scientific quality is the overriding criterion by which submissions are evaluated.

Editorial Board

H. Dugald Macpherson, Managing Editor
School of Mathematics, University of Leeds

Jeremy Avigad,
Department of Philosophy, Carnegie Mellon University

Volker Halbach,
New College, University of Oxford

Vladimir Kanovei,
Institute for Information Transmission Problems, Moscow

Manuel Lerman,
Department of Mathematics, University of Connecticut

Heinrich Wansing,
Department of Philosophy, Ruhr-Universität Bochum

Thomas Wilke,
Institut für Informatik, Christian-Albrechts-Universität zu Kiel

More information, including a list of the books in the series, can be found at http://www.aslonline.org/books-lnl.html

Lecture Notes in Logic 42

Turing's Legacy: Developments from Turing's Ideas in Logic

Edited by
ROD DOWNEY
Victoria University of Wellington

ASSOCIATION FOR SYMBOLIC LOGIC

CAMBRIDGE
UNIVERSITY PRESS

University Printing House, Cambridge CB2 8BS, United Kingdom

Cambridge University Press is part of the University of Cambridge.

It furthers the University's mission by disseminating knowledge in the pursuit of education, learning and research at the highest international levels of excellence.

www.cambridge.org
Information on this title: www.cambridge.org/9781107043480

© Association for Symbolic Logic 2014

This publication is in copyright. Subject to statutory exception and to the provisions of relevant collective licensing agreements, no reproduction of any part may take place without the written permission of Cambridge University Press.

First published 2014

Printed in the United Kingdom by CPI Group Ltd, Croydon CR0 4yy

A catalogue record for this publication is available from the British Library

Library of Congress Cataloging-in-Publication Data
Turing's legacy : developments from Turing's ideas in logic / edited by Rod Downey, Victoria University of Wellington.
 pages cm. – (Lecture notes in logic ; 42)
 Includes bibliographical references and index.
 ISBN 978-1-107-04348-0 (hardback)
1. Computational complexity. 2. Machine theory. 3. Turing, Alan Mathison, 1912-1954. I. Downey, R. G. (Rod G.), editor of compilation.
 QA267.7.T87 2014
 510.92–dc23   2014000240

ISBN 978-1-107-04348-0 Hardback

Cambridge University Press has no responsibility for the persistence or accuracy of URLs for external or third-party internet websites referred to in this publication, and does not guarantee that any content on such websites is, or will remain, accurate or appropriate.

CONTENTS

Rod Downey, editor
    Turing's legacy: developments from Turing's ideas in logic vii

Jeremy Avigad and Vasco Brattka
    Computability and analysis: the legacy of Alan Turing 1

Lenore Blum
    Alan Turing and the other theory of computation (expanded) 48

Harry Buhrman
    Turing in Quantumland 70

Rod Downey
    Computability theory, algorithmic randomness and Turing's
    anticipation 90

Ekaterina B. Fokina, Valentina Harizanov, and Alexander Melnikov
    Computable model theory... 124

Cameron E. Freer, Daniel M. Roy, and Joshua B. Tenenbaum
    Towards common-sense reasoning via conditional simulation:
    legacies of Turing in Artificial Intelligence........................ 195

Thomas C. Hales
    Mathematics in the age of the Turing machine..................... 253

Steven Homer and Alan L. Selman
    Turing and the development of computational complexity 299

Charles F. Miller III
    Turing machines to word problems 329

Anil Nerode
    Musings on Turing's Thesis....................................... 386

Dag Normann
    Higher generalizations of the Turing Model 397

Wilfried Sieg
    Step by recursive step: Church's analysis of effective calculability.... 434

Robert Irving Soare
    Turing and the discovery of computability 467
P. D. Welch
    Transfinite machine models 493

TURING'S LEGACY:
DEVELOPMENTS FROM TURING'S IDEAS IN LOGIC

§1. Introduction. The year 2012 was the centenary of the birth of one of the most brilliant mathematicians of the 20th century. There were many celebrations of this fact, and many conferences based around Turing's work and life during 2012. In particular, there was a half year program (Syntax and Semantics) at the Newton Institute in Cambridge, and many "Turing 100/Centenary" conferences throughout the year. These events included truly major meetings featuring many of the world's best mathematicians and computer scientists (and even Gary Kasparov) around his actual birth day of June 23, including *The Incomputable*, *ACM A. M. Turing Centenary Celebration*, *How the World Computes* (CiE 2012), and *The Turing Centenary Conference*. There are also a number of publications devoted to Turing's life, work and legacy.

To the general public, Turing is probably best known for his part in Bletchley Park and the war-winning efforts of the code-breakers at Hut 8. To biologists, Turing is best known for his work on morphogenesis, the paper "A Chemical Basis for Morphogenesis" being his most highly cited work.

To logicians, and computer scientists, Alan Turing is best known for his work in computation, arguably leading to the development of the digital computer. This development has caused almost certainly the most profound change in human history in the last century. Turing's work in computation grew from philosophical questions in logic. Thus it seems fitting that the *Association for Symbolic Logic* sponsored this volume.

The idea for this volume is to (mainly) look at the various ways Turing's ideas in logic have developed into major programs in the landscape of mathematics and philosophy in th early 21st Century. That is, where did these ideas go? A number of leading experts were invited to participate in this enterprise. All of the papers were reviewed both for readability by non-experts and for content by other experts.

§2. Turing's work. There is an excellent archive of Turing's work in
http://www.turing.org.uk/sources/biblio.html.
Jack Copeland (sometimes with Diane Proudfoot) have historical articles and books such as [1, 2, 3]. Below we give a few brief comments and refer the

---

Research supported by the Marsden Fund of New Zealand.

reader to these and the books Davis [4] and to Herken [5] for more historical comments, as well as the articles in this volume by Nerode, Sieg and Soare.

Turing [7] worked famously on the *Entscheidungsproblem*, the question of the decision problem for validity of first order predicate calculus. His work and that of Church, Kleene, Post and others solved the problem. Turing's paper of 1936 laid the foundations for the advent of stored program computers. His 1936 paper had the key idea of stored program computers via universal machines. Turing knew of the possibilities of large scale electronic computers following the groundbreaking ideas of Fred Flowers with his work on *Colossus* in the second world war. Turing's analysis of computation and his introduction of universal machines are discussed in both Sieg's and Soare's articles.

In a lecture of 1947, Turing said of his design of ACE (automated computing engine)

> "The special machine may be called the universal machine; it works in the following quite simple manner. When we have decided what machine we wish to imitate we punch a description of it on the tape of the universal machine .... The universal machine has only to keep looking at this description in order to find out what it should do at each stage. Thus the complexity of the machine to be imitated is concentrated in the tape and does not appear in the universal machine proper in any way.... [D]igital computing machines such as the ACE ... are in fact practical versions of the universal machine."

In 1943, McCulloch and Pitt used Turing ideas to show the control mechanism for a TM could be simulated by a finite collection of gates with delays. These ideas were later developed by von Neumann and others which lead to ENIAC in 1943. A friend of von Newmann who worked with him on the Atomic bomb project was Stanley Frankel (see [3]) who is quoted as saying the following:

> "von Neumann was well aware of the fundamental importance of Turing's paper of 1936 'On computable numbers ... ', which describes in principle the 'Universal Computer' ... Many people have acclaimed von Neumann as the 'father of the computer' (in a modern sense of the term) but I am sure that he would never have made that mistake himself. He might well be called the midwife, perhaps, but he firmly emphasized to me, and to others I am sure, that the fundamental conception is owing to Turing."

Turing designed the ACE. Whilst never built, Turing's design was the basis of the architecture of several computers. For example, Huxley's G15 computer, the first PC (about the size of a fridge) was based on it, with about 400 sold worldwide, and remaining in use until 1970(!).

The world's first programmable computer was built in Manchester by Turing's lifelong friend Max Newman (who Turing met in 1935). Turing was involved in this project and wrote the world's first programming manual. Turing also proposed methods of symbolic program verification, and logically constructing programs. His thesis, "Systems of logic based on ordinals," looked at transfinite methods of verification. In this thesis [8], Turing also introduces the notion of an oracle Turing machine which is essential for our understanding of relative computability and computational complexity.

Remarkably, Turing wrote the world's first computer chess program *before there were programmable computers*. The reader interested in this should look at Kasparov's talk video-ed at the Turing Centenary Conference at Manchester (Kasparov and Friedel [6]). Turing's ideas of using optimization of functions as a control method for artificial intelligence are at the vanguard of all such work.

Turing thought deeply about artificial intelligence with articles such as [9]. This is reflected by the well-known Turing test and the article he wrote whilst on sabbatical in Cambridge, infamously judged as a "schoolboy effort" by Charles Darwin, his government boss. For more on this see Copeland–Proudfoot [3].

The present volume has an article by Freer, Roy and Tennenbaum around AI. As well, we have a fascinating article on automated theorem proving by Hales. This article also is concerned with practical aspects of computation, something of great interest to Turing as witnessed by his famous article on ill-posedness of matrix operations. Lenore Blum has contributed a article to this volume about such questions and the "other theory of computation."

Other generalizations of the notion of algorithm are discussed in articles by Welch and by Normann. These are "higher" generalizations of the notion of computation to computations on ordinals and the transfinite. Finally, the generalization of the notion of computation to the quantum is given in Buhrman's article.

Turing's ideas of computation are critical to our understanding of randomness. The article by Downey looks at the development of pure computability theory, and its use in the theory algorithmic randomness. Downey's article also looks at Turing's anticipation of the Martin-Löf idea of using computation to bound the theory of measure as seen in his unpublished work on normality (Turing [10]).

Turing's original paper [7] was concerned with computation of the *real numbers and functions*. Thus, he wrote the first paper on computable analysis. Computable analysis and its developments are discussed in the article by Avigad and Brattka.

The article by Homer and Selman discusses how modern computational complexity theory has developed from Turing's ideas.

Turing had many technical contributions in mathematical logic. His early articles in the *Journal of Symbolic Logic* showed the equivalences of various models of computation. Turing also proved the undecidability of the word problem for cancellation semigroups. Charles Miller III contributes a long article concerning how these ideas have panned out in the area of combinatorial group theory, and where this important subject has gone in the last 50 years. In the same spirit, the article by Fokina, Harizanov and Melnikov looks at how Turing's ideas have developed in the computability theory of structures, such as logical models and algebraic structures.

We have not concerned ourselves with Turing's work on morphogenesis, as we are concentrating on *logical* developments. For this reason we have chosen not to follow his ideas on code-breaking and cryptography. These are very well chronicled in many books on the work of Turing and others at Bletchley park. It is widely reported that the work of this group of 1,200 workers and the efforts of the powerful group of mathematicians in "Hut 8" shortened the war by at least 2 years, and saved millions of lives. It is also clear that we could have added several other articles in other areas.

Nevertheless, we believe that this volume here represents an important collection of articles giving an insight into both a great mind and into how mathematics and logic (in and about computation) have developed in the last 70 years. We hope you enjoy the result.

The Editor
Rod Downey

REFERENCES

[1] JACK COPELAND, *The essential Turing*, Oxford University Press, Oxford and New York, September 2004.

[2] ———, *Alan Turing's automatic computing engine: The master codebreaker's struggle to build the modern computer*, Oxford University Press, Oxford and New York, June 2005.

[3] JACK COPELAND and DIANE PROUDFOOT, *Alan Turing father of the modern computer*, **The Rutherford Journal**, vol. 4 (2011–2012), http://www.rutherfordjournal.org/article040101.html.

[4] MARTIN DAVIS, *The undecidable: Basic papers on undecidable propositions, unsolvable problems and computable functions*, Dover, 1965.

[5] ROLF HERKEN, *The universal Turing machine: A half-century survey*, Springer-Verlag, 1995.

[6] GARY KASPAROV and FREDERIC FRIEDEL, *The reconstruction of Turing's "Paper Machine"*.

[7] ALAN TURING, *On computable numbers with an application to the Entscheidungsproblem*, **Proceedings of the London Mathematical Society**, vol. 42 (1936), pp. 230–265, correction in **Proceedings of the London Mathematical Society** vol. 43 (1937), pp. 544–546.

[8] ———, *Systems of logic based on ordinals*, **Proceedings of the London Mathematical Society**, vol. 45 (1939), no. 2, pp. 161–228.

[9] ———, *Computing machinery and intelligence*, **Mind**, vol. 59 (1950), pp. 433–460.

[10] ———, *A note on normal numbers*, **Collected works of A. M. Turing: Pure mathematics** (J. L. Britton, editor), North Holland, Amsterdam, 1992, pp. 117–119, with notes of the editor in 263–265.

COMPUTABILITY AND ANALYSIS: THE LEGACY OF ALAN TURING

JEREMY AVIGAD AND VASCO BRATTKA

§1. Introduction. For most of its history, mathematics was algorithmic in nature. The geometric claims in Euclid's *Elements* fall into two distinct categories: "problems," which assert that a construction can be carried out to meet a given specification, and "theorems," which assert that some property holds of a particular geometric configuration. For example, Proposition 10 of Book I reads "To bisect a given straight line." Euclid's "proof" gives the construction, and ends with the (Greek equivalent of) Q.E.F., for *quod erat faciendum*, or "that which was to be done." Proofs of theorems, in contrast, end with Q.E.D., for *quod erat demonstrandum*, or "that which was to be shown"; but even these typically involve the construction of auxiliary geometric objects in order to verify the claim.

Similarly, algebra was devoted to developing algorithms for solving equations. This outlook characterized the subject from its origins in ancient Egypt and Babylon, through the ninth century work of al-Khwarizmi, to the solutions to the quadratic and cubic equations in Cardano's *Ars Magna* of 1545, and to Lagrange's study of the quintic in his *Réflexions sur la résolution algébrique des équations* of 1770.

The theory of probability, which was born in an exchange of letters between Blaise Pascal and Pierre de Fermat in 1654 and developed further by Christian Huygens and Jakob Bernoulli, provided methods for calculating odds related to games of chance. Abraham de Moivre's 1718 monograph on the subject was entitled *The Doctrine of Chances: or, a Method for Calculating the Probabilities of Events in Play*. Pierre de Laplace's monumental *Théorie analytique des probabilités* expanded the scope of the subject dramatically, addressing statistical problems related to everything from astronomical measurement to the measurement in the social sciences and the reliability of testimony. Even so, the emphasis remained fixed on explicit calculation.

Analysis had an algorithmic flavor as well. In the early seventeenth century, Cavalieri, Fermat, Pascal, and Wallis developed methods of computing "quadratures," or areas of regions bounded by curves, as well as volumes. In the hands of Newton, the calculus became a method of explaining and predicting the motion of heavenly and sublunary objects. Euler's *Introductio in Analysis Infinitorum* of 1748 was the first work to base the calculus explicitly on the notion of a *function*; but, for Euler, functions were given by piecewise analytic expressions, and once again, his focus was on methods of calculation.

All this is not to say that all the functions and operations considered by mathematicians were computable in the modern sense. Some of Euclid's constructions involve a case split on whether two points are equal or not, and, similarly, Euler's piecewise analytic functions were not always continuous. In contrast, we will see below that functions on the reals that are computable in the modern sense are necessarily continuous. And even though Euler's work is sensitive to the rates of convergence of analytic expressions, these rates were not made explicit. But these are quibbles, and, generally speaking, mathematical arguments through the eighteenth century provided informal algorithms for finding objects asserted to exist.

The situation changed dramatically in the nineteenth century. Galois' theory of equations implicitly assumed that all the roots of a polynomial exist *somewhere*, but Gauss' 1799 proof of the fundamental theorem of algebra, for example, did not show how to compute them. In 1837, Dirichlet considered the example of a "function" from the real numbers to the real numbers which is equal to 1 on the rationals and 0 on the irrationals, without pausing to consider whether such a function is calculable in any sense. The Bolzano–Weierstraß Theorem, first proved by Bolzano in 1817, asserts that any bounded sequence of real numbers has a convergent subsequence; in general, there will be no way of computing such a subsequence. Riemann's proof of the open mapping theorem was based on the Dirichlet principle, an existence principle that is not computationally valid. Cantor's work on the convergence of Fourier series led him to consider transfinite iterations of point-set operations, and, ultimately, to develop the abstract notion of set.

Although the tensions between conceptual and computational points of view were most salient in analysis, other branches of mathematics were not immune. For example, in 1871, Richard Dedekind defined the modern notion of an *ideal* in a ring of algebraic integers, and defined operations on ideals in a purely extensional way. In other words, the operations were defined in such a way that they do not presuppose any particular representation of the ideals, and definitions do not indicate how to compute the operations in terms of such representations. More dramatically, in 1890, Hilbert proved what is now known as the Hilbert Basis Theorem. This asserts that, given any sequence $f_1, f_2, f_3, \ldots$ of multivariate polynomials over a Noetherian ring, there is some n such that for every $m \geq n$, f_m is in the ideal generated by $f_1, \ldots, f_n$. Such an m cannot be computed by surveying elements of the sequence, since it is not even a continuous function on the space of sequences; even if a sequence $x, x, x, \ldots$ starts out looking like a constant sequence, one cannot rule out the possibility that the element 1 will eventually appear.

Such shifts were controversial, and raised questions as to whether the new, abstract, set-theoretic methods were appropriate to mathematics. Set-theoretic paradoxes in the early twentieth century raised the additional question as to whether they are even consistent. Brouwer's attempt, in the 1910s, to found mathematics on an "intuitionistic" conception raised a further challenge to modern methods, and in 1921, Hermann Weyl, Hilbert's best student, announced that he was joining the Brouwerian revolution. The twentieth century *Grundlagenstreit*, or "crisis of foundations," was born.

At that point, two radically different paths were open to the mathematical community:

- Restrict the methods of mathematics so that mathematical theorems have direct computational validity. In particular, restrict methods so that sets and functions asserted to exist are computable, as well as infinitary mathematical objects and structures more generally; and also ensure that quantifier dependences are also constructive, so that a "forall-exists" statement asserts the existence of a computable transformation.
- Expand the methods of mathematics to allow idealized and abstract operations on infinite objects and structures, without concern as to how these objects are represented, and without concern as to whether the operations have a direct computational interpretation.

Mainstream contemporary mathematics has chosen decisively in favor of the latter. But computation is important to mathematics, and faced with a nonconstructive development, there are options available to those specifically interested in computation. For example, one can look for computationally valid versions of nonconstructive mathematical theorems, as one does in computable and computational mathematics, constructive mathematics, and numerical analysis. There are, in addition, various ways of measuring the extent to which ordinary theorems fail to be computable, and characterizing the data needed to make them so.

With Turing's analysis of computability, we now have precise ways of saying what it means for various types of mathematical objects to be computable, stating mathematical theorems in computational terms, and specifying the data relative to which operations of interest are computable. Section 2 thus discusses *computable analysis*, whereby mathematical theorems are made computationally significant by stating the computational content explicitly.

There are still communities of mathematicians, however, who are committed to developing mathematics in such a way that every concept and assertion has an *implicit* computational meaning. Turing's analysis of computability is useful here, too, in a different way: by representing such styles of mathematics in formal axiomatic terms, we can make this implicit computational interpretation mathematically explicit. Section 3 thus discusses different styles of constructive mathematics, and the computational semantics thereof.

§2. Computable analysis.

2.1. From Leibniz to Turing. An interest in the nature of computation can be found in the seventeenth century work of Leibniz (see [47]). For example, his *stepped reckoner* improved on earlier mechanical calculating devices like the one of Pascal. It was the first calculating machine that was able to perform all four basic arithmetic operations, and it earned Leibniz an external membership of the British Royal Society at the age of 24. Leibniz's development of calculus is better known, as is the corresponding priority dispute with Newton. Leibniz paid considerable attention to choosing notations and symbols carefully in order to facilitate calculation, and his use of the integral symbol $\int$ and the d symbol for derivatives have survived to the present day. Leibniz's work on the binary number system, long before the advent of digital computers, is also worth mentioning. A more important contribution to the study of computation was his notion of a *calculus ratiocinator*, that is, a calculus of reasoning. Such a calculus, Leibniz held, would allow one to resolve disputes in a purely mathematical fashion:[1]

> The only way to rectify our reasonings is to make them as tangible as those of the Mathematicians, so that we can find our error at a glance, and when there are disputes among persons, we can simply say: Let us calculate, without further ado, to see who is right.

His attempts to develop such a calculus amount to an early form of symbolic logic.

With this perspective, it is not farfetched to see Leibniz as initiating a series of developments that culminate in Turing's work. Norbert Wiener has described the relationship in the following way [231]:

> The history of the modern computing machine goes back to Leibniz and Pascal. Indeed, the general idea of a computing machine is nothing but a mechanization of Leibniz's *calculus ratiocinator*. It is, therefore, not at all remarkable that the theory of the present computing machine has come to meet the later developments of the algebra of logic anticipated by Leibniz. Turing has even suggested that the problem of decision, for any mathematical situation, can always be reduced to the construction of an appropriate computing machine.

2.2. From Borel to Turing. Perhaps the first serious attempt to express the mathematical concepts of a computable real number and a computable function on the real numbers were made by Émil Borel around 1912, the year that Alan Turing was born. Borel defined computable real numbers as follows:[2]

---
[1] Leibniz, *The Art of Discovery*, 1685 [232].
[2] All citations of Borel are from [19], which is a reprint of [18]. The translations here are by the authors of this article; obvious mistakes in the original have been corrected.

We say that a number α is computable if, given a natural number n, we can obtain a rational number that differs from α by at most $\frac{1}{n}$.

Of course, before the advent of Turing machines or any other formal notion of computability, the meaning of the phrase "we can obtain" remained vague. But Borel provided the following additional information in a footnote to that phrase:

> I intentionally leave aside the practical length of operations, which can be shorter or longer; the essential point is that each operation can be executed in finite time with a safe method that is unambiguous.

This makes it clear that Borel had an intuitive notion of an algorithm in mind. Borel then indicated the importance of number representations, and argued that decimal expansions have no special theoretical value, whereas continued fraction expansions are not invariant under arithmetic operations and hence of no practical value. He went on to discuss the problem of determining whether two real numbers are equal:

> The first problem in the theory of computable numbers is the problem of equality of two such numbers. If two computable numbers are unequal, this can obviously be noticed by computing both with sufficient precision, but in general it will not be known *a priori*. One can make clear progress in determining a lower bound on the difference of two computable numbers, whose definitions satisfy known conditions.

In modern terms, what Borel seems to recognize here is that although there is no algorithm that decides whether two computable real numbers are equal, the inequality relation between computable reals is, at least, computably enumerable. He then discussed a notion of the *height* of a number, which is based on counting the number of steps needed to construct that number, in a certain way. This concept can be seen as an early forerunner of the concept of Kolmogorov complexity. Borel considered ways that this concept might be utilized in addressing the equality problem.

In another section of his paper, Borel discussed the concept of a computable real number function, which he defined as follows:

> We say that a function is computable if its value is computable for any computable value of the variable. In other words, if α is a computable number, one has to know how to compute the value of $f(\alpha)$ with precision $\frac{1}{n}$ for any n. One should not forget that, by definition, to be given a computable number α just means to be given a method to obtain an arbitrary approximation to α.

It is worth noting that Borel only demanded computability at computable inputs in his definition. His definition is vague in the sense that he did not indicate whether he had in mind an algorithm that transfers a *method* to compute α into a *method* to compute $f(\alpha)$ (which would later become known as *Markov computability* or the *Russian approach*) or whether he had in mind an algorithm that transfers an *approximation* of α into an *approximation* of $f(\alpha)$ (which is closer to what we now call a computable function on the real numbers, under the *Polish approach*). He also did not say explicitly that his algorithm to compute f is meant to be uniform, but this seems to be implied by his subsequent observation:

> A function cannot be computable, if it is not continuous at all computable values of the variable.

A footnote to this observation then indicates that he had the Polish approach in mind:[3]

> In order to make the computation of a function effectively possible with a given precision, one additionally needs to know the modulus of continuity of the function, which is the [...] relation [...] between the variation of the function values with respect to the variation of the variable.

Borel went on to discuss different types of discontinuous functions, including those that we now call Borel measurable. In fact, the entire discussion of computable real numbers and computable real functions is preliminary to Borel's development of measure theory, and the discussion was meant to motivate aspects of that development.

2.3. Turing on computable analysis. Turing's landmark 1936 paper [207] is titled "On computable numbers, with an application to the Entscheidungsproblem." It begins as follows:

> The "computable" numbers may be described briefly as the real numbers whose expressions as a decimal are calculable by finite means. Although the subject of this paper is ostensibly the computable *numbers*, it is almost equally easy to define and investigate computable functions of an integrable variable or a real or computable variable, computable predicates, and so forth. The fundamental problems involved are, however, the same in each case, and I have chosen the computable numbers for explicit treatment as involving the least cumbrous technique. I hope shortly to give an account of the relations of the computable numbers, functions, and

---

[3] Hence Borel's result that computability implies continuity *cannot* be seen as an early version of the famous theorem of Ceĭtin, in contrast to what is suggested in [214]. The aforementioned theorem states that any Markov computable function is already effectively continuous on computable inputs and hence computable in Borel's sense, see sections 2.5 and 3, Figure 5. While this is a deep result, the observation that computable functions in Borel's sense are continuous is obvious.

so forth to one another. This will include a development of the theory of functions of a real variable expressed in terms of computable numbers. According to my definition, a number is computable if its decimal can be written down by a machine.

At least two things are striking about these opening words. The first is that Turing chose not to motivate his notion of computation in terms of the ability to characterize the notion of a computable function from $\mathbb{N}$ to $\mathbb{N}$, or the notion of a computable *set* of natural numbers, as in most contemporary presentations; but, rather, in terms of the ability to characterize the notion of a computable real number. In fact, Section 10 is titled "Examples of large classes of numbers which are computable." There, he introduced the notion of a computable function on computable real numbers, and the notion of computable convergence for a sequence of computable numbers, and so on; and argued that, for example, e and π and the real zeros of the Bessel functions are computable. The second striking fact is that he also flagged his intention of developing a full-blown theory of computable real analysis. As it turns out, this was a task that ultimately fell to his successors, as we explain below.

The precise definition of a computable real number given by Turing in the original paper can be expressed as follows: a real number r is *computable* if there is a computable sequence of 0s and 1s with the property that the fractional part of r is equal to the real number obtained by prefixing that sequence with a binary point. There is a slight problem with this definition, however, which Turing discussed in a later correction [208], published in 1937. Suppose we have a procedure that, for every i, outputs a rational number q_i with the property that

$$|r - q_i| < 2^{-i}.$$

Intuitively, in that case, we would also want to consider r to be a computable real number, because we can compute it to any desired accuracy. In fact, it is not hard to show that this second definition coincides with the first: a real number has a computable binary expansion if and only if it is possible to compute it in the second sense. In other words, the two definitions are extensionally equivalent.

The problem, however, is that it is not possible to pass *uniformly* between these two representations, in a computable way. For example, suppose a procedure of the second type begins to output the sequence of approximations $\frac{1}{2}, \frac{1}{2}, \frac{1}{2}, \ldots$. Then it is difficult to determine what the first binary digit is, because at some point the output could jump just above or just below $\frac{1}{2}$. This intuition can be made precise: allowing the sequence in general to depend on the halting behavior of a Turing machine, one can show that there is no algorithmic procedure which, given a description of a Turing machine describing a real number by a sequence of rational approximations, computes the digits of r. On the other hand, for any fixed description of the first sort, there is a computable description of the second sort: either r is a dyadic rational, which is to say, it has a finite binary expansion; or waiting long enough will always provide enough information to determine the digits. So the difference only shows up when one wishes to talk about computations which take descriptions of real numbers as input. In that case, as Turing noted, the second type of definition is more natural; and, as we will see below, these are the *descriptions* of the computable reals that form the basis for computable analysis.

Turing's second representation of computable reals, presented in his correction [208], is given by the formula

$$(2i-1)n + \sum_{r=1}^{\infty}(2c_r - 1)\left(\frac{2}{3}\right)^r,$$

where i and n provide the integer part of the represented number and the binary sequence c_r the fractional part. This representation is essentially what has later been called a *signed-digit representation* with base $\frac{2}{3}$. It is interesting that Turing acknowledged Brouwer's influence (see also [72]):

> This use of overlapping intervals for the definition of real numbers is due originally to Brouwer.

In the 1936 paper, in addition to discussing individual computable real numbers, Turing also defined the notion of a computable function on real numbers. Like Borel, he adopted the standpoint that the input to such a function needs to be computable itself:

> We cannot define general computable functions of a real variable, since there is no general method of describing a real number, but we can define a computable function of a computable variable.

A few years later Turing introduced oracle machines [210], which would have allowed him to handle computable functions on *arbitrary* real inputs, simply by considering the input as an oracle given from outside and not as being itself computed in some specific way. But the 1936 definition ran as follows. First, Turing extended his definition of computable real numbers γ_n from the unit interval to all real numbers using the formula $\alpha_n = \tan(\pi(\gamma_n - \frac{1}{2}))$. He went on:

> Now let $\varphi(n)$ be a computable function which can be shown to be such that for any satisfactory[4] argument its value is satisfactory. Then the function f, defined by $f(\alpha_n) = \alpha_{\varphi(n)}$, is a computable function and all computable functions of a computable variable are expressible in this form.

---

[4]Turing calls a natural number n *satisfactory* if, in modern terms, n is a Gödel index of a total computable function.

Hence, Turing's definition of a computable function on the real numbers essentially coincides with Markov's definition, which would later form the basis of the Russian approach to computable analysis (see sections 2.6 and 3.5). Computability of f is not defined in terms of approximations, but in terms of functions φ that transfer algorithms that describe the input into algorithms that describe the output. There are at least three objections against the technical details of Turing's definition, some of which have already been addressed by Guido Gherardi in [72], which is the first careful discussion of Turing's contributions to computable analysis:

1. Turing used the binary representation of real numbers in order to define γ_n, and hence the derived notion of a computable real function is somewhat unnatural and restrictive. However, we can consider this problem as being repaired by Turing's second approach to computable reals in [208], which we have described above.
2. Turing should have mentioned that the function f is only well-defined by $f(\alpha_n) = \alpha_{\varphi(n)}$ if φ is extensional in the sense that $\alpha_n = \alpha_k$ implies $\alpha_{\varphi(n)} = \alpha_{\varphi(k)}$. However, we can safely assume that this is what he had in mind, since otherwise his equation does not even define a single-valued function.
3. Turing did not allow arbitrary suitable computable functions φ, but he restricted his definition to total functions φ. However, it is an easy exercise to show that for any partial computable φ that maps all satisfactory numbers to satisfactory numbers there is a total computable function ψ such that $\psi(n)$ and $\varphi(n)$ are numbers of the same total computable function for all satisfactory inputs n. Hence the restriction to total φ is not an actual restriction.

These three considerations support our claim that Turing's definition of a computable real function is essentially the same as Markov's definition.

It is worth pointing out that Turing introduced another important theme in computable analysis, namely, developing computable versions of theorems of analysis. For instance, he pointed out:

> ... we cannot say that a computable bounded increasing sequence of computable numbers has a computable limit.

This shows that Turing was aware of the fact that the Monotone Convergence Theorem cannot be proved computably. He did, however, provide a (weak) computable version of the Intermediate Value Theorem:

> (vi) If α and β are computable and $\alpha < \beta$ and $\varphi(\alpha) < 0 < \varphi(\beta)$, where $\varphi(a)$ is a computable increasing continuous function, then there is a unique computable number γ, satisfying $\alpha < \gamma < \beta$ and $\varphi(\gamma) = 0$.

The fact that the limit of a computable sequence of computable real numbers need not be computable motivates the introduction of the concept of computable convergence:

> We shall say that a sequence β_n of computable numbers *converges computably* if there is a computable integral valued function $N(\varepsilon)$ of the computable variable ε, such that we can show that, if $\varepsilon > 0$ and $n > N(\varepsilon)$ and $m > N(\varepsilon)$, then $|\beta_n - \beta_m| < \varepsilon$. We can then show that:
> (vii) A power series whose coefficients form a computable sequence of computable numbers is computably convergent at all computable points in the interior of its interval of convergence.
> (viii) The limit of a computably convergent sequence is computable.
> And with the obvious definition of "uniformly computably convergent":
> (ix) The limit of a uniformly computably convergent computable sequence of computable functions is a computable function.
> Hence:
> (x) The sum of a power series whose coefficients form a computable sequence is a computable function in the interior of its interval of convergence.
> From (viii) and $\pi = 4(1 - \frac{1}{3} + \frac{1}{5} - \cdots)$ we deduce that π is computable. From $e = 1 + 1 + \frac{1}{2!} + \frac{1}{3!} + \cdots$ we deduce that e is computable. From (vi) we deduce that all real algebraic numbers are computable. From (vi) and (x) we deduce that the real zeros of the Bessel functions are computable.

Here we need to warn the reader that Turing erred in (x): the mere computability of the sequences of coefficients of a power series does not guarantee computable convergence on the whole interior of its interval of convergence, but only on each compact subinterval thereof (see Theorem 7 in [172] and Exercise 6.5.2 in [223]).

Before we leave Turing behind, some of his other, related work is worth mentioning. For example, in 1938, Turing published an article, "Finite Approximations to Lie Groups" [209], in the *Annals of Mathematics*, addressing the question as to which infinite Lie groups can be approximated by finite groups. The idea of approximating infinitary mathematical objects by finite ones is central to computable analysis. The problem that Turing considered here is more restrictive in that he required that the approximants be groups themselves, and the paper is not in any way directly related to computability; but the parallel is interesting. What is more directly relevant is the fact that Turing was not only interested in computing real numbers and functions in theory, but in practice. The Turing archive[5] contains a sketch of a proposal, in 1939, to build an analog computer that would calculate approximate values for the Riemann zeta-function on the critical line. His ingenious method was published in 1943 [211], and in a paper published in 1953 [213] he described calculations that were carried out in 1950 on the Manchester University Mark 1 Electronic Computer. Finally, we mention that, in a paper of 1948 [212], Turing studied linear equations from the perspective of algebraic complexity. There he introduced the concepts of a *condition number* and of *ill-conditioned* equations, which are widely used in numerical mathematics nowadays. (For a discussion of this work, see Lenore Blum's contribution to this collection, as well as [72].) In some very interesting unpublished work Turing provided an algorithm to compute an absolutely normal real number (see the discussion by Becher et al. in [13]) and in this way he even made an early contribution to algorithmic randomness.

2.4. From Turing to Specker. Turing's ideas on computable real numbers where taken up in 1949 by Ernst Specker [200], who completed his Ph.D. in Zürich under Hopf in the same year. Specker was probably attracted to logic by Paul Bernays, whom he thanks in that paper. Specker constructed a computable monotone bounded sequence (x_n) whose limit is not a computable real number, thus establishing Turing's claim. In modern terms one can easily describe such a sequence: given an injective computable enumeration $f: \mathbb{N} \to \mathbb{N}$ of a computably enumerable but non-computable set $K \subseteq \mathbb{N}$, the partial sums

$$x_n = \sum_{i=0}^{n} 2^{-f(i)}$$

form a computable sequence without a computable limit. Such sequences are now sometimes called *Specker sequences*.

It is less well-known that Specker also studied different representations of real numbers in that paper, including the representation by computably converging Cauchy sequences, the decimal representation, and the Dedekind cut representation (via characteristic functions). In particular, he considered both primitive recursive and general computable representations in all three instances. His main results include the fact that the class of real numbers with primitive recursive Dedekind cuts is strictly included in the class of real numbers with primitive recursive decimal expansions, and the fact that the latter class is strictly included in the class of real numbers with primitive recursive Cauchy approximation. He also studied arithmetic properties of these classes of real numbers.

Soon after Specker published his paper, Rosza Péter included Specker's discussion of the different classes of primitive recursive real numbers in her

---

[5]See AMT/C/2 in http://www.turingarchive.org/browse.php/C.

book, *Rekursive Funktionen* [171]. In a review of that book [182], Raphael M. Robinson noted that all the representations of real numbers mentioned above yield the same class of numbers if one uses computability in place of primitive recursiveness. However, this is no longer true if one moves from single real numbers to sequences of real numbers. Mostowski [159] proved that for computable sequences of real numbers the analogous strict inclusions hold, as they were proved by Specker for single primitive recursive real numbers: the class of sequences of real numbers with uniformly decidable Dedekind cuts is strictly included in the class of sequences of real numbers with uniformly computable decimal expansion, which in turn is strictly included in the class of sequences of real numbers with computable Cauchy approximations. Mostowski pointed out that the striking coincidence between the behavior of classes of single primitive recursive real numbers and computable sequences of real numbers is not very well understood:

> It remains an open problem whether this is a coincidence or a special case of a general phenomenon whose cause ought to be discovered.

A few years after Specker and Robinson, the logician H. Gordon Rice [181] proved what Borel had already observed informally, namely that equality is not decidable for computable real numbers. In contrast, whether $a < b$ or $a > b$ holds can be decided for two non-equal computable real numbers a, b. Rice also proved that the set $\mathbb{R}_c$ of computable real numbers forms a real algebraically closed field, and that the Bolzano–Weierstraß Theorem does not hold computably, in the following sense: there is a bounded c.e. set of computable numbers without a computable accumulation point. Kreisel wrote a review of Rice's paper for the *Mathematical Reviews* and pointed out that he had already proved results that are more general than Rice's; namely, he proved that any power series with a computable sequence of coefficients (that are not all identical to zero) has only computable zeros in its interval of convergence [121], and that there exists a computable bounded set of rational numbers which contains no computable subsequence with a computable modulus of convergence [120].

In a second paper on computable analysis [201], Specker constructed a computable real function that attains its maximum in the unit interval at a non-computable number. In a footnote he mentioned that this solves an open problem posed by Grzegorczyk [82], a problem that Lacombe had already solved independently [138]. Specker mentioned that a similar result can be derived from a theorem of Zaslavskiĭ [238]. He also mentioned that the Intermediate Value Theorem has a (non-uniform) computable version, but that it fails for sequences of functions.

2.5. Computable analysis in Poland and France. In the 1950s, computable analysis received a substantial number of contributions from Andrzej Grzegorczyk [82, 83, 84, 85] in Poland and Daniel Lacombe [82, 83, 84, 85, 122, 123, 124, 136, 137, 138, 139, 140, 141, 142, 143] in France. Grzegorczyk is the best known representative of the Polish school of computable analysis, a school of research that came into being soon after Turing published his seminal paper. Apparently, members of the Polish school of functional analysis became interested in questions of computability, and Stefan Banach and Stanisław Mazur gave a seminar talk on this topic in Lvov (now in Ukraine, at that time Polish) on January 23, 1937 [11]. Unfortunately, the second world war got in the way, and Mazur's course notes on computable analysis (edited by Grzegorczyk and Rasiowa) where not published until much later [152].

Banach and Mazur's notion of computability for functions on the real numbers was defined using computable sequences. Accordingly, we call a function $f: \mathbb{R}_c \to \mathbb{R}_c$ *Banach–Mazur computable* if it maps computable sequences to computable sequences. While this concept works well for well-behaved functions (see, for example, the discussion of Pour-El and Richards' work on linear functions in Section 2.6), it yields a weaker concept of computability of real number functions in general, as proved later by Aberth (again, see Section 2.6). One of Mazur's theorems was that every Banach–Mazur computable function is already continuous on the computable reals. This theorem can be seen as a forerunner and even as a stronger version of a theorem of Kreisel, Lacombe, and Shoenfield [124], which was independently proved in slightly different versions by Ceĭtin [36] and later by Moschovakis [158]. Essentially, the theorem says that every real number function that is computable in the sense of Markov is continuous (on the computable reals). (In this particular case one even has the stronger conclusion that the function is effectively continuous. See Figure 5 in Section 2.9.)

Grzegorczyk's work [84] was mostly concerned with computable functions $f: [0, 1] \to \mathbb{R}$ on all real numbers of the unit interval and he proved a number of equivalent characterizations of this concept. For instance f is computable if and only if each of the following conditions hold:

1. f maps computable sequences to computable sequences and it has a computable modulus of uniform convergence, and
2. f can be represented by a computable function h on rational intervals that is monotone with respect to inclusion and that approximates f in the sense that $\{f(x)\} = \bigcap \{h(I) : x \in I\}$ for all $x \in [0, 1]$.

Later on, the first characterization became the basic definition of computability used by Pour-El and Richards (see section 2.6), whereas the second characterization is the starting point for domain theoretic characterizations of computability [195, 54].

Lacombe's pioneering work opened up two research directions in computable analysis which were to play an important role. First, he initiated the study of c.e. open sets of reals, which are open sets that can be effectively obtained as unions of rational intervals [139]. Together with Kreisel, Lacombe proved that there are c.e. open sets that contain all computable reals and that have arbitrarily small Lebesgue measure [122]. Secondly, Lacombe anticipated the study of computability on more abstract spaces than Euclidean space, and was one of the first to define the notion of a computable metric space [143].

2.6. Computable analysis in North America. More comprehensive approaches to computable analysis were developed in the United States during the 1970s and 1980s. One stream of results is due to Oliver Aberth, and is accumulated in his book, *Computable Analysis* [2]. His work is built on Markov's notion of computability for real number functions. We call a function $f\colon \mathbb{R}_c \to \mathbb{R}_c$ on computable reals *Markov computable* if there is an algorithm that transforms algorithms that describe inputs x into algorithms that describe outputs $f(x)$. Among other results, Aberth provided an example of a Markov computable function $f\colon \mathbb{R}_c \to \mathbb{R}_c$ that cannot be extended to a computable function on all real numbers [2, Theorem 7.3]. Such a function f can, for instance, be constructed with a Specker sequence (x_n), which is a computable monotone increasing bounded sequence of real numbers that converges to a non-computable real x. Using effectively inseparable sets, such a Specker sequence can even be constructed so that it is effectively bounded away from any computable point (see, for instance, the Effective Modulus Lemma in [177]). One can use such a special Specker sequence and place increasingly steeper peaks on each x_n, as in Figure 1. If adjusted appropriately, this construction leads

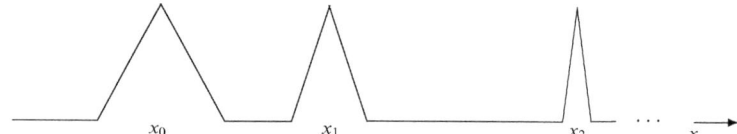

FIGURE 1. Construction of a function $f\colon \mathbb{R} \to \mathbb{R}$.

to a Markov computable function $f\colon \mathbb{R}_c \to \mathbb{R}_c$ that cannot be extended to a computable function on the whole of $\mathbb{R}$. In fact, it cannot even be extended to a continuous function on $\mathbb{R}$, since the peaks accumulate at x. Pour-El and Richards [177, Theorem 6, page 67] later refined this counter-example to obtain a differentiable, non-computable function $f\colon \mathbb{R} \to \mathbb{R}$ whose restriction to the computable reals $\mathbb{R}_c$ is Markov computable. This can be achieved by replacing the peaks in Aberth's construction by smoother peaks of decreasing height.

Another well-known counter-example due to Aberth [1] is the construction of a function $F\colon \mathbb{R}_c^2 \to \mathbb{R}_c$ that is Markov computable and uniformly continuous on a rectangle centered around the origin, with the property that the differential equation
$$y'(x) = f(x, y(x))$$
with initial condition $y(0) = 0$ has no computable solution y, not even on an arbitrarily small interval around the origin. This result can be extended to computable functions $f\colon \mathbb{R}^2 \to \mathbb{R}$, which was proved somewhat later by Pour-El and Richards [174]. It can be interpreted as saying that the Peano Existence Theorem does not hold computably.

Marian Pour-El and J. Ian Richards' approach to computable analysis was the dominant approach in the 1980s, and is nicely presented in their book [177]. The basic idea is to axiomatize computability structures on Banach spaces using computable sequences and a characterization of computability that is derived from Grzegorczyk's first characterization, which was mentioned in section 2.5. This approach is tailor-made for well-behaved maps such as linear ones. One of their central results is their First Main Theorem [176], which states that a linear map $T\colon X \to Y$ on computable Banach spaces is computable if and only if it is continuous and has a c.e. closed graph, i.e.,

T computable $\iff$ T continuous and graph(T) c.e. closed.

Moreover, the theorem states that a linear function T with a c.e. closed graph satisfies

T computable $\iff$ T maps computable inputs to computable outputs.

Here a closed set is called *c.e. closed* if it contains a computable dense sequence. The crucial assumption that the graph(T) is c.e. closed has sometimes been ignored tacitly in presentations of this theorem and this has created the wrong impression that the theorem identifies computability with continuity. In any case, the theorem is an important tool and provides many interesting counter-examples. This is because the contrapositive version of the theorem implies that every linear discontinuous T with a c.e. closed graph maps some computable input to a non-computable output. For instance, the operator of differentiation
$$d :\subseteq C[0,1] \to C[0,1], \ f \mapsto f'$$
is known to be linear and discontinuous and it is easily seen to have a c.e. closed graph (for instance the rational polynomials are easily differentiated) and hence it follows that there is a computable and continuously differentiable real number function $f\colon [0,1] \to \mathbb{R}$ whose derivative f' is not computable. This is a fact that has also been proved directly by Myhill [161].

The First Main Theorem is applicable to many other operators that arise naturally in physics, such as the wave operator [175, 173]. In particular, it follows that there is a computable wave $(x, y, z) \mapsto u(x, y, z, 0)$ that transform into a non-computable wave $(x, y, z) \mapsto u(x, y, z, 1)$ if it evolves for one time unit according to the wave equation

$$\frac{\partial^2 u}{\partial x^2} + \frac{\partial^2 u}{\partial y^2} + \frac{\partial^2 u}{\partial z^2} - \frac{\partial^2 u}{\partial t^2} = 0$$

from initial condition $\frac{\partial u}{\partial t} = 0$. This result has occasionally been interpreted as suggesting that it might be possible to build a wave computer that would violate Church's Thesis. However, a physically more appropriate analysis shows that this is not plausible (see Section 2.7 and also [170]). Pour-El and Richards' definitions and concepts regarding Banach spaces have been transferred to the more general setting of metric spaces by Yasugi, Mori and Tsujii [234] in Japan.

Another stream of research on computable analysis in the United States came from Anil Nerode and his students and collaborators. For instance, Metakides, Nerode and Shore proved that the Hahn–Banach Theorem is not computably valid, but admits a non-uniform computable solution in the finite dimensional case [155, 156]. Joseph S. Miller studied sets that appear as the set of fixed points of a computable self map of the unit ball in Euclidean space [157]. Among other things he proved that a closed set $A \subseteq [0, 1]^n$ is the set of fixed points of a computable function $f : [0, 1]^n \to [0, 1]^n$ if and only if it is co-c.e. closed and has a co-c.e. closed connectedness component. Zhou also studied computable open and closed sets on Euclidean space [241]. In a series of papers [103, 104, 102], Kalantari and Welch provided results that shed further light on the relation between Markov computable functions and computable functions. Douglas Cenzer and Jeffrey B. Remmel have studied the complexity of index sets for problems in analysis [37, 38, 39]. McNicholl and his co-authors have started a study of computable complex analysis [154, 153, 151].

It is possible to study computable functions of analysis from the point of view of computational complexity. We cannot discuss such results in detail here, but in passing we mention the important work of Harvey Friedman and Ker-I Ko [113, 114, 115, 117, 64, 118], which is well summarized in [116]; and, more recently, interesting work of Mark Braverman, Stephen Cook and Akitoshi Kawamura [28, 27, 29, 105, 106]. Braverman and Michael Yampolsky have written a nice monograph on computability properties of Julia sets [30].

2.7. Computable analysis in Germany. In Eastern Germany, Jürgen Hauck produced a sequence of interesting papers on computable analysis that were only published in German [86, 87, 88], building on the work of Dieter Klaua [107, 108]. The novel aspect of Hauck's work is that he studied the notion of a representation (of real numbers or other objects) in its own right. This aspect has been taken up in the work of Weihrauch's school of computable analysis [223], which has its origins in the work of Christoph Kreitz and Klaus Weihrauch [125, 126, 225, 221]. Kreitz and Weihrauch began to develop a systematic theory of representations that is centered around the notion of an *admissible* representation. Representations $\delta_X :\subseteq \mathbb{N}^\mathbb{N} \to X$ are (potentially partial) maps that are surjective. Admissible representations δ_X are particularly well-behaved with respect to some topology given on the represented set X. If δ_X and δ_Y are admissible representations of topological spaces X and Y, respectively, then a (partial) function $f :\subseteq X \to Y$ is continuous if and only if there exists a continuous $F :\subseteq \mathbb{N}^\mathbb{N} \to \mathbb{N}^\mathbb{N}$ on Baire space $\mathbb{N}^\mathbb{N}$ such that the diagram in Figure 2 commutes.

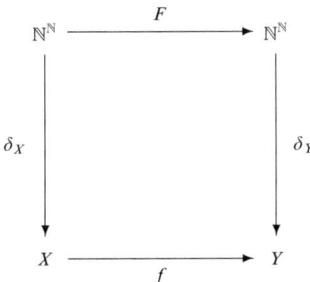

FIGURE 2. Notions of computable real number functions.

The perspective arising from the concept of a representation as a map sheds new light on representations of real numbers as they were studied earlier by Specker, Mostowski and others. Their results can be developed in a more uniform way, using the concept of computable reducibility for representations, which yields a lattice of real number representations. The position of a representation in this lattice characterizes the finitely accessible information content of the respective representation, see Figure 3. In particular, the behavior of representations that was mysterious to Mostowski can be explained naturally in this way.

The concept of an admissible representation has been extended to a larger classes of topological spaces by the work of Matthias Schröder [192]. In his definition, an admissible representation with respect to a certain topology on the represented space is just one that is maximal among all continuous representations with respect to computable reducibility. Hence, admissibility can be seen as a completeness property.

Many other interesting results have been developed using the concept of a representation (see, for instance, the tutorial [25] for a more detailed survey). For instance, Peter Hertling [91] proved that the real number structure is computably categorical in the sense that up to computable equivalence there is

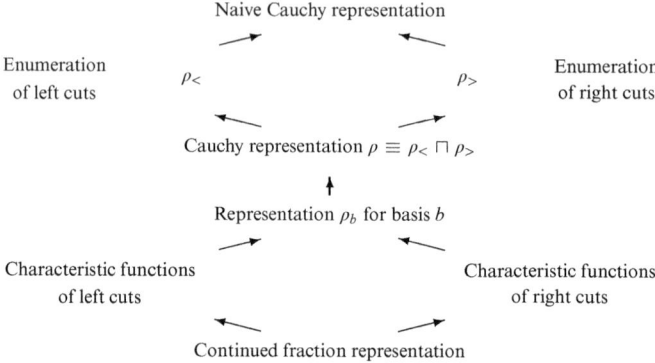

FIGURE 3. The lattice of real number representations.

one and only one representation that makes all the ingredients of the structure computable. He also studied a computable version of Riemann's Mapping Theorem [90], and proved that the Hyperbolicity Conjecture implies that the Mandelbrot set $M \subseteq \mathbb{R}^2$ is computable in the sense that its distance function is computable. Finally, he also proved that there is a Banach–Mazur computable function $f: \mathbb{R}_c \to \mathbb{R}_c$ that is not Markov computable [92].

Matthias Schröder and Klaus Weihrauch have studied a very general setting for computational complexity in analysis using proper representations [189, 224, 193] and Schröder has also studied online complexity of real number functions [190, 191], following earlier work of Schönhage [188]. Robert Rettinger and Klaus Weihrauch studied the computational complexity of Julia sets [178]. In another interesting stream of papers weaker notions of computability on real numbers were studied [239, 179, 180] by Xizhong Zheng, Klaus Weihrauch and Robert Rettinger.

Ning Zhong and Klaus Weihrauch studied computability of distributions [240] and proved computability of the solution operator of several partial differential equations [226, 227, 228, 229, 230]. In particular, they revised the the wave equation and proved that if this equation is treated in physically realistic spaces, then it turns out to be computable [226].

Martin Ziegler has studied models of hypercomputation and related notions of computability for discontinuous functions [242]. Martin Ziegler and Stéphane Le Roux have revisited co-c.e. closed sets. Among other things, they proved that the Bolzano–Weierstraß Theorem is not limit computable, in the sense that there is a computable sequence in the unit cube without a limit computable cluster point [183]. Baigger [10] proved that the Brouwer Fixed Point

Theorem has a computable counter-example (essentially following Orevkov's ideas [164], who proved the corresponding result for Markov computability much earlier).

Just as the computational content of representations of real numbers can be analyzed in a lattice, there is another lattice that allows one to classify the computational content of (forall-exists) theorems. This lattice is based on a computable reduction for functions that has been introduced by Klaus Weihrauch and that has recently been studied more intensively by Arno Pauly, Guido Gherardi, Alberto Marcone, Brattka, and others [169, 23, 21]. Theorems such as the Hahn–Banach Theorem [73], the Baire Category Theorem, the Intermediate Value Theorem [22], the Bolzano–Weierstraß Theorem [24] and the Radon–Nikodym Theorem [100] have been classified in this lattice as well as finding Nash equilibria or solving linear equations [168]. Figure 4 shows the respective relative positions of these theorems in the Weihrauch lattice. This classification yields a uniform perspective on the computational content of all these theorems, and the position of a theorem in this lattice fully characterizes certain properties of the respective theorem regarding computability. In particular, it implies all aforementioned computability properties of these theorems.

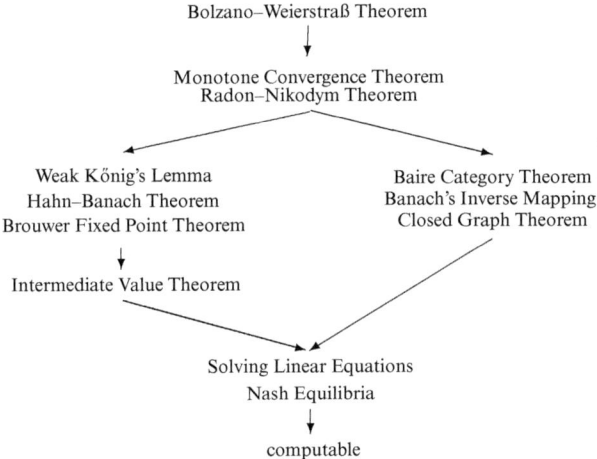

FIGURE 4. The computational content of theorems in the Weihrauch lattice.

2.8. Computable measure theory. The field of *algorithmic randomness* relies on effective notions of measure and measurable set. Historically, debates in the 1930s as to how to make sense of the notion of a "random" sequence of coin flips gave rise to the Kolmogorov axioms for a probability space, and the measure-theoretic interpretation of probability. On that interpretation, the statement that a "random" element of a space has a certain property should be interpreted as the statement that the set of counter-examples has measure 0. In that framework, talk of random elements is only a manner of speaking; there is no truly "random" sequence of coin flips x, since each such element is contained in the null set $\{x\}$.

In the 1960s, Per Martin-Löf, a student of Kolmogorov, revived the notion of a random sequence of coin flips [148, 149]. The idea is that, from a computability-theoretic perspective, it makes sense to call a sequence random if it avoids all suitably effective null sets. For example, a *Martin-Löf test* is an effectively null G_δ set, which is to say, an effective sequence of computably open sets G_i with the property that for each i, the Lebesgue measure of G_i is less than 2^{-i}. An sequence of coin flips *fails* such a test if it is in the intersection of the sets G_i. An element of the unit interval is said to be *Martin-Löf random* if it does not fail any such test. Osvald Demuth [48] defined an equivalent notion, independently, in 1975.

Martin-Löf randomness can be given alternate characterizations in terms of computable betting strategies, or in terms of information content. There are by now many other notions of randomness in the literature. One, based on the influential work of C.-P. Schnorr [187, 186], is similar to Martin-Löf randomness but requires that, moreover, the measures of the open sets G_i are uniformly computable. It is impossible to even begin to survey this very active area here, but detailed overviews can be found in the monographs [163, 50]. Kučera and Nies [133] provides a nice discussion of Demuth's contributions to the subject.

Measure theory, more generally, has been studied from a computability-theoretic perspective. Most work in algorithmic randomness has focused on random elements of $2^\mathbb{N}$, or, equivalently, random elements of the unit interval $[0, 1]$. But there are a number of ways of developing the notion of a computable measure on more general spaces; see the work of Gács [66], Hertling and Weihrauch [93] and Hoyrup and Rojas [98]. Weihrauch [222] and Schröder [194] have discussed different representations of probability measures and Weihrauch and Wu [233] have further developed the foundations of computable measure theory. Edalat [53] has studied computable measure and integration theory from the perspective of domain theory. Bosserhoff [20] and Hoyrup and Rojas [98] have defined various notions of computability for measurable functions. This makes it possible to consider results in measure theory, probability, and dynamical systems in computability-theoretic terms.

For example, computable aspects of the ergodic theorems have been studied extensively. The pointwise and mean ergodic theorems assert that if T is a measure-preserving transformation of a finite measure space and f is any measurable function, then the ergodic averages $A_n f = \frac{1}{n}\sum_{i<n} f \circ T^i$ converge pointwise, and in the L^2 norm, respectively. V'yugin [219, 220] and, independently, Avigad and Simic [9, 196, 6] have shown that the mean and pointwise ergodic theorems do not hold computably, but Avigad, Gerhardy, and Townser [8] have shown that a rate of convergence can be computed from T, f, and the norm of the limit. In particular, the rate of convergence can be computed from T and f when the transformation is ergodic. V'yugin [219, 220] has shown that if T is a measurable transformation of the unit interval and f is a computable function, then the sequence of ergodic averages $(A_n f)$ converges on every Martin-Löf random real. Franklin and Towsner [60] have recently shown that this is sharp. Galatolo, Hoyrup, and Rojas [69] have generalized V'yugin's result to computable measurable functions. Indeed, the work of Gács, Galatolo, Hoyrup, and Rojas [68, 67], together with the results of [8], shows that the Schnorr random reals are exactly the ones at which the ergodic theorem holds with respect to every ergodic measure and computable measure-preserving transformation (see also the discussion in [167]). Bienvenu, Day, Hoyrup, and Shen [15] and, independently, Franklin, Greenberg, Miller, and Ng [59] have extended V'yugin's result to the characteristic function of a computably open set, in the case where the measure is ergodic.

Other theorems of measure-theory and measure-theoretic probability have been considered in this vein. (Such work is, to some extent, foreshadowed by the study of measure theory in the context of reverse mathematics, by Simpson, Yu, and others; see, for example, [237, 235, 236].) The Lebesgue differentiation theorem states, roughly speaking, that for almost every point of $\mathbb{R}^n$, the value of an integrable function at that point is equal to the limit of increasingly small averages taken around that point. Pathak [166] showed that, when the integrable function is computable in an appropriate sense, the theorem is true at every Martin-Löf random point. Pathak, Rojas, and Simpson [167] later strengthened the conclusion to the Schnorr-random real numbers, and showed that this is sharp; this same result was obtained independently by Rute [185]. Another theorem of Lebesgue's states that a function of bounded variation is differentiable almost everywhere. Recasting results of Demuth [49], Brattka, Miller, and Nies [26] have shown that when the function in question is computable, the result holds at Martin-Löf random real numbers, and that this is sharp. Moreover, they obtained clean characterizations of other notions of randomness in terms of differentiability. Freer, Kjos-Hansen, Nies, and Stephan [61] have obtained similar results for the class of Lipschitz functions. Rute [185] has considered a number of differentiability and martingale convergence theorems, characterizing the points at which the theorems hold, and determining when rates of convergence are and are not computable. Ackerman, Freer, and Roy have studied the computability of conditional distributions [3], Hoyrup, Rojas, and Weihrauch have studied the computability of Radon–Nikodym derivatives [99], and Freer and Roy have provided a computable version of de Finetti's theorem [62].

2.9. Appendix: Notions of computability for real number functions. In this appendix, we summarize the various notions of computability for real number functions that were discussed in previous sections, and we briefly indicate the logical relations between these notions. For simplicity we restrict the discussion to functions that are total on all real numbers or on all computable real numbers.

A *rapidly converging Cauchy name* for a real number x is a sequence $(x_i)_{i \in \mathbb{N}}$ of rationals that converges to x rapidly, i.e., in such a way that for every i and $j \geq i$, $|x_j - x_i| < 2^{-i}$. (Any other computable rate of convergence would work equally well.)

1. $f : \mathbb{R} \to \mathbb{R}$ is called *computable* if there is an algorithm that transforms each given rapidly converging Cauchy name of an arbitrary x into such a name for $f(x)$ (on a Turing machine with one-way output tape).
2. $f : \mathbb{R}_c \to \mathbb{R}_c$ is called *Borel computable* if f is computable in the previously described sense, but restricted to computable reals.
3. $f : \mathbb{R}_c \to \mathbb{R}_c$ is called *Markov computable* if there is an algorithm that converts each given algorithm for a computable real x into an algorithm for $f(x)$. Here an algorithm for a computable real number x produces a rapidly converging Cauchy name for x as output.
4. $f : \mathbb{R}_c \to \mathbb{R}_c$ is called *Banach–Mazur computable* if f maps any given computable sequence (x_n) of real numbers into a computable sequence $(f(x_n))$ of real numbers.

The diagram in Figure 5 summarizes the logical relations between different notions of computability. Some of these equivalences can be generalized to partial functions with well-behaved domains, and to other computable metric spaces (see [89, 92]).

§3. Constructive analysis. As noted in Section 1, the growing divergence of conceptual and computational concerns in the early twentieth century left the mathematical community with two choices: either restrict mathematical language and methods to preserve direct computational meaning, or allow more expansive language and methods and find a place for computational mathematics within it. We have seen how the theory of computability has supported the latter option: given broad notions of real number, function, space, and so on, we can now say, in precise mathematical terms, which real numbers, functions, and spaces are *computable*.

There is still, however, a community of mathematicians that favors the first option, namely, adopting a "constructive" style of mathematics, whereby theorems maintain a direct computational interpretation. This simplifies language:

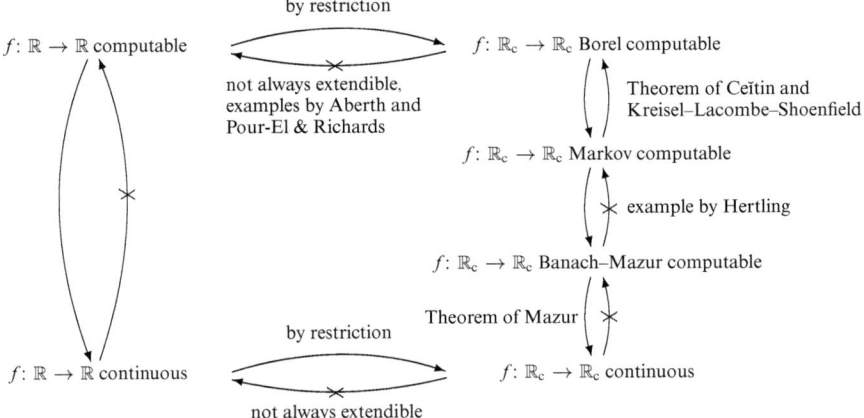

FIGURE 5. Notions of computable real number functions.

rather than say "there exists a computable function f" one can say "there exists a function f," with the general understanding that, from a constructive standpoint, proving the existence of such a function requires providing an algorithm to compute it. Similarly, rather than say "there is an algorithm which, given x, computes y such that ... ," one can say "given x, there exists y such that ... " with the general understanding that a constructive proof, by its very nature, provides such an algorithm.

In sum, from a constructive point of view, all existence claims are expected to have computational significance. We will see, however, that this general stricture is subject to interpretation, especially when it comes to infinitary mathematical objects. Today there are many different "flavors" of constructive mathematics, with varying norms and computational interpretations. Here the methods of logic and computability theory prove to be useful yet again: once we design a formal deductive system that reasonably captures a style of constructive reasoning, we can provide a precise computational semantics. This serves to clarify the sense in which a style of constructive mathematics has computational significance, and helps bridge the two approaches to preserving computational meaning.

This section will provide a brief historical overview of a some of the different approaches to constructive mathematics, as well as some associated formal axiomatic systems and their computational semantics. Troelstra and van Dalen [206] and Beeson [14] provide more thorough introductions to these topics.

3.1. Kroneckerian mathematics. In 1887, facing the rising Dedekind–Cantor revolution, the mathematician Leopold Kronecker wrote an essay in which he urged the mathematical community to maintain focus on symbolic and algorithmic aspects of mathematics.

> [A]ll the results of the profoundest mathematical research must in the end be expressible in the simple forms of the properties of integers. But to let these forms appear simply, one needs above all a suitable, surveyable manner of expression and representation for the numbers themselves. The human spirit has been working on this project persistently and laboriously since the greyest prehistory .... ([130, page 955])

One gets a better sense of Kronecker's views by considering the way they played out in his work, such as his treatment of algebraic numbers in his landmark *Grundzüge einer arithmetischen Theorie der algebraischen Grössen* [128], or in his treatment of the fundamental theorem of algebra in *Ein Fundamentalsatz der allgemeinen Arithmetik* [129]. In general, Kronecker avoided speaking of "arbitrary" real numbers and functions, and, rather, dealt with algebraic *systems* of such things, wherein the objects are given by symbolic expressions, and operations on the objects are described in terms of symbolic calculations. Of course, it is still an important task to understand how rational approximations can be obtained from an algebraic description of a real number. But even when dealing with convergent sequences and limits (as in, for example, the proof of Dirichlet's theorem on primes in an arithmetic progression in his lectures on number theory), Kronecker always provided explicit information as to how quickly a sequence converges. The approach is nicely described by his student Hensel:

> He believed that one can and must in this domain formulate each definition in such a way that its applicability to a given quantity can be assessed by means of a finite number of tests. Likewise that an existence proof for a quantity is to be regarded as entirely rigorous only if it contains a method by which that quantity can really be found. ([131], quoted in Stein [202, p. 250].)

A number of articles by Harold Edwards (including [55, 56, 57]) provide an excellent overview of Kronecker's outlook and contributions to mathematics.

3.2. Brouwerian intuitionism. During the 1910s, the Dutch mathematician L. E. J. Brouwer advanced a new philosophy of mathematics, *intuitionism*, which made strong pronouncements as to the appropriate methods of mathematical reasoning. Brouwer held that mathematical objects are constructions of the mind, and that we come to know a mathematical theorem only by carrying out a mental construction that enables us to see that it is true. In particular, seeing that a statement of the form $A \wedge B$ is true involves seeing that A is true, and that B is true; seeing that $A \to B$ is true involves having a mental procedure that transforms any construction witnessing the truth of A to a construction witnessing the truth of B; seeing that $\neg A$ is true involves having a procedure that transforms a construction witnessing the truth of A to a contradiction (that is, something which, evidently, cannot be the case); and seeing that $A \vee B$ is true requires carrying out a mental construction enabling one to see that A is true, or carrying out a mental construction to see that B is true; and similarly for statements involving the universal and existential quantifiers. This account, developed further by Heyting and Kolmogorov, has come to be known as the BHK interpretation. It is not hard to see that, according to this interpretation, the law of the excluded middle, $A \vee \neg A$, does not generally hold. For example, if A is the statement of the Goldbach conjecture, then we cannot presently assert $A \vee \neg A$, because we do not currently know that the Goldbach conjecture is true, nor do we know that it is false.

There is a strong solopsistic strand to Brouwer's philosophy, in that the mathematical knowledge one has is a reflection of one's inner mental life, independent of an external world or other thinkers. He also held that mathematical knowledge and thought are independent of language, which he took to be a deeply limited and flawed means of communicating mathematical ideas. Intuitionism may seem to have little to do with computation *per se*, but, as we will see below, if one replaces Brouwer's mental constructions and procedures with symbolic representations and algorithms, one is left with an essentially computational view of mathematics. Thus, Brouwer's mathematical ideas have proved to be influential in computer science, despite the different motivations.

Brouwer was also an influential topologist, and an important part of his intuitionist program was to develop an intuitionistically appropriate foundation for reasoning about the continuum [32, 33]. To that end, he introduced the notion of a *choice sequence*, that is, an "arbitrary" sequence of natural numbers. Some choice sequences are generated by a law, such as the sequence $0, 0, 0, \ldots$. At the other extreme, some sequences are "lawless," that is, generated by events that are not predetermined. Brouwer introduced the *continuity principle*, which, roughly speaking, asserts that any function from choice sequences to the natural numbers is continuous; in other words, the value of the function at a given choice sequence depends only on a finite initial segment of that sequence. (See Troelstra and van Dalen [206, I.4.6] for a precise formulation.)

Brouwer went on to develop an intuitionistic set theory based on such sequences. A *spread*, in his terminology, is a tree on $\mathbb{N}$ such that each node has at least one successor. Brouwer saw this data as a way of specifying a collection of choice sequences; that is, a choice sequence is in the spread if and only if every initial segment is in the tree. A *fan* is a spread with the property that every node has only finitely many successors. At the risk of clouding some intuitions, we can translate these notions to classical terms: a choice sequence is an arbitrary element of Baire space (the space of functions from $\mathbb{N}$ to $\mathbb{N}$ under the product topology), a spread corresponds to a closed subset of Baire space, and a fan corresponds to a compact subset of Baire space.

In 1927, Brouwer [34] introduced the *Bar Theorem*, which, despite the name, is essentially an axiomatic principle that provides transfinite induction on well-founded trees on $\mathbb{N}$. Roughly speaking, it asserts that any property that holds outside and at the leaves of such a well-founded tree, and is preserved in passing from all the children of a node to the node itself, holds of every finite sequence. An immediate corollary is the *Fan Theorem*, which asserts that a well-founded fan is finite. Using the fan theorem and the continuity principle, Brouwer showed that every function from [0, 1] to $\mathbb{R}$ is uniformly continuous. When properties of choice sequences are understood in terms of computable predicates on Baire space, this formulation of analysis accords quite well with Grzegorczyk's notion of computability, as discussed in Section 2.5.

For more information on intuitionistic mathematics, see [52, 95, 206, 31]. For more on Brouwer and his philosophical views, see [215, 216].

3.3. Early formal systems for constructive mathematics. Now let us consider some formal axiomatic systems that capture such constructive styles of reasoning. A theory now known as *primitive recursive arithmetic* was designed by Thoralf Skolem [198] to capture Hilbert's notion of "finitary" mathematical reasoning. In its strictest form, the theory has variables ranging over the natural numbers, but no quantifiers. One starts with some basic functions on the natural numbers, and is allowed to define new functions using a schema of *primitive recursion*:

- $f(0, z_1, \ldots, z_n) = g(z_1, \ldots, z_n)$, and
- $f(x + 1, z_1, \ldots, z_n) = h(x, f(x, z_1, \ldots, z_n), z_1, \ldots, z_n)$,

where g and h have previously been defined. Although this schema may seem limited, with ingenuity one can show that almost any reasonable function (and hence relation) on the natural numbers is primitive recursive. In particular, the primitive recursive relations are closed under bounded quantification. As a result, primitive recursive arithmetic can be viewed as a reasonable framework in which to carry out a Kroneckarian constructivism.

Indeed, primitive recursive arithmetic is often taken as a starting point for formalizing basic mathematical notions in many presentations of formal axiomatic foundations for mathematics, classical or constructive. It plays a central role in Hilbert and Bernays' landmark *Grundlagen der Mathematik* [96], in Kleene's *Introduction to Metamathematics* [111], and Goodstein's *Recursive Number Theory* [79]. Goodstein's later *Recursive Analysis* [80] develops topics like integration and differentiation on the basis of primitive recursive arithmetic.

Today, finitism is viewed as an extreme form of intuitionism. Foundational writings in the 1920s, however, do not show a clear distinction between the two. Indeed, as late as 1930, von Neumann used the two terms interchangeably in his exposition of formalism [218] at the Second Conference for Epistemology and the Exact Sciences in Königsberg. (This was, incidentally, the meeting where Gödel announced the first incompleteness theorem.)

The situation was clarified when Arend Heyting, a student of Brouwer's, provided formal axiomatizations of intuitionistic mathematics, despite Brouwer's antipathy towards formalization. Specifically, he provided formal treatments of intuitionistic logic, arithmetic, and Brouwer's theory of choice sequences, which appeared in a series of three papers published in 1930 [94]. In 1933, Gödel [76], and later Gentzen [70], showed that classical first-order arithmetic could be interpreted in intuitionistic first-order arithmetic via what is now known as the *double-negation translation*. At the time, classical first-order arithmetic, which extends primitive recursive arithmetic with quantifiers ranging over the natural numbers and induction for all formulas in the language, was viewed as strictly stronger than finitism. The interpretation of classical arithmetic in intuitionistic arithmetic thereby prompted the realization that intuitionism and finitism diverge as well.

3.4. Realizability. The developments described in this section so far all predate Turing's analysis of computability. Once notions of computability were in place, however, it was not long before Stephen Kleene showed that they can be used to provide a precise sense in which intuitionistic mathematics has a computational interpretation. Towards the end of a paper [109] published in 1943, but first presented to the American Mathematical Society in 1940, he put forth the following:

> Thesis III. A proposition of the form $(x)(Ey)A(x, y)$ containing no free variables is provable constructively, only if there is a general recursive function $\phi(x)$ such that $(x)A(x, \phi(x))$. ([109, Section 16, page 69])

One can turn this thesis into *theorems* by showing that the statement holds when we replace "constructive provability" by provability in particular axiomatic systems. Kleene carried out this program over the next few years, together with his student, David Nelson [110, 162]. For that purpose, Kleene introduced the notion of "realizability," which is, roughly speaking, an analysis of the Brouwer–Heyting–Kolmogorov interpretation in computability-theoretic terms. Specifically, Kleene defined a relation e *realizes* φ inductively, where e is a natural number and φ is a sentence in the language of first-order arithmetic, with the following clauses:

- If φ is atomic, then e realizes φ if and only if it is true.
- e realizes $\theta \wedge \eta$ if and only if e is of the form $2^a 3^b$, where a realizes θ and b realizes η.
- e realizes $\theta \vee \eta$ if and only if e is of the form $2^0 3^a$ and a realizes θ, or e is of the form $2^1 3^b$ and b realizes η.

- e realizes $\theta \to \eta$ if and only if e is the Gödel number of a partial recursive function f which, given any a realizing θ, returns a number $f(a)$ realizing η.
- e realizes $\exists x\, \theta(x)$ if and only if e is of the form $2^x 3^a$ and a realizes $\theta(\bar{x})$, where $\bar{x}$ is the numeral that denotes x.
- e realizes $\forall x\, \theta(x)$ if and only if e is the Gödel number of a partial recursive function f which, given any x, returns a number $f(x)$ realizing $\theta(\bar{x})$.

More generally, if φ has free variables, e realizes φ if and only if it realizes its universal closure. A formula φ is said to be *realizable* if there is some number realizing it.

Kleene emphasized that this does not provide a reductive analysis of intuitionistic truth, insofar as quantifiers and logical connectives themselves appear in the definition. In particular, if φ is a true, purely universal sentence, then anything realizes φ; and a realizer for a negated sentence carries no useful information at all. However, for any sentence φ in the language of arithmetic, the statement "e realizes φ" can also be expressed in the language of arithmetic, and one can inquire as to how these two assertions are related to one another. Nelson showed that if intuitionistic logic proves φ, then it proves that φ is realizable. Unfortunately, it is not the case that for any φ, intuitionistic arithmetic proves that φ is equivalent to its own realizability; but Nelson showed that intuitionistic arithmetic *does* prove that this equivalence is realizable. Moreover, one can strengthen the clause for implication:

- e realizes $\theta \to \eta$ if and only if θ implies η *and* e is the Gödel number of a partial recursive function f which, given any a realizing θ, returns a number $f(a)$ realizing η.

In that case, intuitionistic arithmetic proves that a formula φ is true if and only if it is realizable.

The idea that a constructive proof provides explicit "evidence" for the truth of a theorem in question, and that such evidence can often be described in computability-theoretic terms, is a powerful one. It can help illuminate the "meaning" or "computational content" of a formal axiomatic system; and, as a purely technical device, it can be used to obtain metamathematical properties of such systems, such as provability and unprovability results. In 1967, Kleene and Vesley [112] presented a formalization of Brouwerian analysis together with a suitable realizability interpretation.

By now, there is a bewildering array of realizability relations in the literature. Variations can depend on any of the following features:

- the language expressing the mathematical assertions (first-order, second-order, or higher-order, etc.);
- the kinds of realizers (arbitrary computable functions, computable functions in a particular class, or a broader class of functions);

- the descriptions of the realizers (e.g., whether they are represented by natural numbers, or terms in a formal language);
- whether or not the realizability relation itself is expressed in a formal system; or
- the particular clauses of the realizability relation itself (see, for example, the variant of the clause for implication above).

All these decisions have bearing on the axioms and rules that are realized, and the metamathematical consequences once can draw. See Troelstra [205] for a definitive reference, as well as [14, 204] for more information. Realizability theory can also be used to translate results from constructive analysis into computable analysis; see [12, 144].

3.5. The Russian school of constructive mathematics. The post-Turing era brought a new approach to constructive mathematics, the principal tenets of which were set forth by A. A. Markov in the late 1940s and developed through the 1950s [146, 147]. The result is what has come to be known as the "Markov school" or "Russian school" of constructive mathematics, with contributions by Nikolai Shanin, I. Zaslavskiĭ, Gregory Ceĭtin, Osvald Demuth, and Boris Kushner, among many others.[6] Aspects of their work have already been discussed above in connection with computable analysis. Indeed, a hallmark of this style of constructivity is that it relies explicitly on notions of computability, which is to say, the real numbers are explicitly defined to be computable reals; the notion of a function from the natural numbers to the natural numbers is explicitly defined to be a computable function; and so on. In contrast to contemporary computable analysis, however, the Russian school insisted that proofs also have a constructive character, so that, for example, a proof of a statement of the form $\forall x \, \exists y \, \varphi(x, y)$ statement can be seen to yield a computable dependence of y on x. This style of constructivity can be viewed as "constructive computable mathematics," or "constructive recursive mathematics," and, indeed, is often referred to as such.

Note that this style of constructivity stands in stark contrast to Brouwerian intuitionism: even if Brouwer could have identified his "constructions" with formal notions of computability, he would have been unlikely to do so. The Russian school also adopted a principle that is not found in Brouwerian intuitionism, namely, Markov's principle. This states that if P is a decidable property of natural numbers (that is, $P(x) \vee \neg P(x)$ holds for every x), and it is contradictory that no x satisfies P, then some x satisfies P. In symbols:

$$\neg \forall x \, \neg P(x) \to \exists x P(x).$$

---

[6]A detailed bibliography can be found on the *Computability and Complexity in Analysis Network*, http://cca-net.de/publications/.

The intuition is that one can find an x satisfying $P(x)$ by simply searching for it systematically, since the hypothesis guarantees that the search cannot fail to turn up such an x.

One can find good overviews of this style of constructivity in books by Aberth [2] and Kushner [134], as well as Kushner's survey [135].

3.6. Theories of finite types. In constructive recursive mathematics, one can interpret functions as Turing-machine indices, which can, in turn, be thought of as descriptions of a computer program. Thus, in constructive recursive mathematics, sets and functions can be "coded," or represented, by natural numbers. But ordinary mathematics deals not just with sets and functions of natural numbers, but also with sets of sets, sets of functions, functionals defined on spaces of functions, and so on. Rather than represent all of these using indices for computable objects, it is more natural, for some purposes, to keep the computational interpretation implicit, and take mathematical objects at face value.

To that end, it is convenient to adopt the language of *finite types*. The idea traces back to foundational frameworks designed by Frege, Russell and Whitehead, and Church, which will be discussed in Section 3.8. Roughly, a "type" can be thought of as a syntactic classification of mathematical objects. To obtain the finite types, start with the type N, intended to denote the natural numbers, so that an object of type N is a natural number. Add the rule that whenever A and B are finite types, so is A $\to$ B. Intuitively, an object of type A $\to$ B is a function from A to B. Thus, we can form the type N $\to$ N of functions from N to N, the type (N $\to$ N) $\to$ N of functionals from N $\to$ N to N, and so on. It is sometimes also convenient to add a base type Bool for the Boolean values "true" and "false," and product types A $\times$ B, but these are inessential.

Following Gödel [77], we can extend the set of primitive recursive functions to the set of *primitive recursive functionals of finite type* by extending the schema of primitive recursion to the higher types. This provides a syntactic calculus for defining objects of the various types, and Gödel's theory T provides a calculus for reasoning about these objects.

There are then two ways of giving T a computational interpretation. The first is to remain at the level of syntax: one provides an explicit procedure to "reduce" any term in the calculus to a canonical normal form (see Tait [203] and Girard [75]). Thus if F is a term of type A $\to$ B, one can view F as denoting the function which takes any term t, in normal form, and returns the normal form corresponding to $F(t)$.

A second approach, however, provides a more natural computational understanding of the finite types. For each type σ, define the set of *hereditarily recursive functions of type* σ, HRO$_\sigma$, inductively as follows: the hereditarily recursive functions of type N are simply the natural numbers, and the hereditarily recursive functions of type A to B are those indices e such that for every hereditarily recursive function x of type A, $\varphi_e(x)$ is defined, and is a hereditarily recursive function of type B. Thus the hereditarily recursive functions of type N → N are (indices of) total computable functions; hereditarily recursive functions of type (N → N) → N are computable functions which, for each total computable function, return a number; and so on. It is then easy to interpret each term of T as a hereditarily recursive functional.

In general, hereditarily recursive functions are not extensional: because functions act on indices, it can happen that two indices e and e' compute the same function from N to N, and yet $F(e) \neq F(e')$ for some hereditarily recursive functional F. One can repair this by inductively defining extensional equality between functionals, and insisting that the interpretations of the function types preserve this equality. The resulting set of functionals is then called the *hereditarily effective operations*, HEO. This is an instance of what computer scientists refer to as a *PER model*, since for each σ, HEO$_\sigma$ is given by a partial equivalence relation, which is to say, an equivalence relation on a subset of the natural numbers.

One obtains the finite type extensions HA$^\omega$ of Heyting arithmetic by extending T with quantifiers and induction over arbitrary finite types. Gödel's *Dialectica interpretation* provides an interpretation of HA$^\omega$ in T. Alternatively, one can show that provable formulas in HA$^\omega$ are realized by terms in T. Thus, whenever HA$^\omega$ proves $\forall x \, \exists y \, \varphi(x, y)$, where x and y are variables of any finite type, there is a hereditarily effective operation which computes y from x. The models HRO and HEO can be formalized in intuitionistic arithmetic, HA, showing that HA$^\omega$ is also interpretable in first-order intuitionistic arithmetic. (For discussions of HA$^\omega$, T, and models thereof, see [7, 205, 206].)

3.7. Bishop-style constructive mathematics. In 1967, the American analyst, Errett Bishop, published a book, *Foundations of Constructive Analysis* [16], which launched a new era in constructivity. In the preface, he wrote:

> It appears ... that there are certain mathematical statements that are merely evocative, which make assertions without empirical validity. There are also mathematical statements of immediate empirical validity, which say that certain performable operations will produce certain observable results, for instance, the theorem that every positive integer is the sum of four squares. Mathematics is a mixture of the real and the ideal, sometimes one, sometimes the other, often so presented that it is hard to tell which is which. The realistic component of mathematics—the desire for pragmatic interpretation—supplies the control which determines the course of development and keeps mathematics from lapsing into meaningless formalism. The idealistic component permits simplifications and opens possibilities which would otherwise be closed. The methods of proof and objects of investigation have been idealized to form a

> game, but the actual conduct of the game is ultimately motivated by pragmatic considerations.
>
> For 50 years now there have been no significant changes in the rules of this game. Mathematicians unanimously agree on how mathematics should be played ...
>
> This book is a piece of constructivist propaganda, designed to show that there does exist a satisfactory alternative.

This was a landmark in the history of constructive mathematics. Brouwerian intuitionism relied not only on a vocabulary that is foreign to most working mathematicians, but also on principles, such as the continuity of every function defined on $[0, 1]$, that are not classically valid. Similarly, the Russian school's explicit restriction to computable objects sets it apart from contemporary mathematics. A central feature of Bishop's constructive mathematics is that the theorems are classically valid, and, indeed, look like ordinary mathematical theorems. The point, however, is that they are established in such a way that every theorem has computational significance. This is achieved by stating definitions and theorems carefully, restricting generality ("avoiding pseudogenerality," in Bishop's words), and adhering to methods that retain computational meaning.

Foundations thus began with an informal statement of constructive set-theoretic principles. To start with, "a *sequence* is a rule which associates to each positive integer n a mathematical object a_n." Then:

> The totality of all mathematical objects constructed in accord with certain requirements is called a *set*. The requirements of the construction, which vary with the set under consideration, determine that set. Thus the integers are a set, the rational numbers are a set, and the collection of all sequences each of whose terms is an integer is a set. Each set A will be endowed with a relation $=$ of equality. This relation is a matter of convention, except that it must be an *equivalence relation*.

And, as far as functions are concerned:

> The dependence of one quantity on another is expressed in the basic notion of an operation. An *operation* from a set A to a set B is a rule f which assigns an element $f(a)$ of B to each element a of A. The rule must afford an explicit, finite, mechanical reduction of the procedure for constructing $f(a)$ to the procedure for constructing a ... The most important case occurs when
>
> $$f(a_1) = f(a_2)$$
>
> whenever a_1 and a_2 are equal elements of A. Such an operation f is called a *function*.

Although Bishop did not provide a formal axiomatic foundation, it is reasonable to seek a formal interpretation of this framework. One option is to use a system like HA$^\omega$, and interpret each set as a definable subset of a type with a definable equivalence relation; that is, take the membership relation $x \in A$ to be given by a formula $\varphi_A(x)$ and take equality $x =_A y$ to be given by another formula $\psi_{=_A}(x, y)$. The interpretations of HA$^\omega$ discussed in the last section then give Bishop-style constructive mathematics a direct computational interpretation. Bishop-style mathematics can also be developed in constructive type theory (which will be discussed below) viewing sets as types equipped with an equivalence relation (a.k.a. "setoids").

Bishop-style constructive mathematics is, in a sense, the most "pure" (or, at least, minimal) constructive approach discussed so far, in that it does not rely on bar induction or the continuity principle from intuitionistic mathematics, nor Markov's principle from constructive recursive mathematics. This makes the framework more appealing to classical mathematicians. For example, we have seen that the Russian school defines a real number to be a computable real number, and a function from reals to reals to be computable in an appropriate sense. This means that classical mathematicians cannot view their theorems about real numbers as such; they have to be interpreted as saying something more restrictive. Similarly, the fact that, in a Brouwerian framework, all functions on the real numbers are continuous shows that the Brouwerian notion of function departs from the classical one. In contrast, even though it is consistent with Bishop-style mathematics to think of all real numbers and functions as being computable, the framework is fully consistent with ordinary classical mathematics. In other words, the theorems in the framework can be viewed as contemporary mathematical theorems, proved in such a way that the results have additional computational significance.

Bishop's work was continued by a number of people, including Douglas Bridges, Ray Mines, Fred Richman, and many others, and remains a mainstay of modern constructivity (see, for example, [17, 31]).

3.8. Intuitionistic higher-order arithmetic. The finite types, discussed in Section 3.6, trace back to Gottlob's Frege foundational system [63], which was built on a single base type of individuals. They made their way, in modified form, into the ramified type theory of Russell and Whitehead's *Principia Mathematica* [184], and ultimately into Alonzo Church's formulation of higher order logic as *simple type theory* [40]. All of these systems include some form of a scheme of *comprehension*,

$$\exists X \, \forall y \, (y \in X \leftrightarrow \varphi(y)),$$

which asserts that any formula φ with a free variable y of type σ gives rise to a set, or predicate, X, of type $\sigma \to$ Bool. (More precisely, Frege's "extensions" of formulas stood as proxy for such objects.)

All of the systems just described are based on classical logic, but one can just as well consider intuitionistic versions, for example, adding comprehension axioms to HA^ω. The result is *intuitionistic higher-order logic*, or IHOL. From a logical perspective, such a system is much stronger than HA^ω. One reason to be interested in such a system is that it represents the internal logic of a *topos*, the algebraic structure that Alexander Grothendieck used to study sheaves over a space (see, for example, [145]).

It is perhaps a matter of debate whether intuitionistic higher-order logic deserves to be called "constructive." But one thing that speaks in favor of this is that one can give IHOL a computational interpretation. In fact, this can be done in various ways, paralleling the various ways of giving a computational interpretation to Gödel's T. For example, one can define an explicit normalization procedure which reduces proofs of IHOL to a canonical normal form; methods based on Girard's *candidats de reducibilité* [74, 75] show the reduction procedure always terminates. Alternatively, one can give a realizability interpretation by interpreting IHOL in Martin Hyland's *effective topos* [101].

3.9. Constructive type theory. At present, the predominant foundational frameworks for constructive mathematics take the form of *constructive type theory*. Such frameworks unify two of the foundational trends we have seen so far:

- type theory, in the Frege–Russell–Church–Gödel tradition; and
- the notion of explicit "evidence" for a constructive claim, in the tradition of the BHK interpretation and realizability.

It is the use of *dependent types* that makes this unification possible.

In simple type theory, types cannot depend on parameters. For example, given a type A and a fixed natural number n, one can form the type A^n of n-tuples of elements of A, but one cannot view these as a *family* of related types that depend on the parameter n. In other words, one cannot view A^n as a type that depends on a variable n of type N. This is exactly the sort of thing that dependent type theory is designed to support. To start with, the type A → B of functions which take an argument in A and return a value in B can be generalized to a dependent product $\prod_{x:A} B(x)$, where $B(x)$ is a type that can depend on x. Intuitively, elements of this type are functions that map an element a of A to an element of $B(a)$. When B does not depend on x, the result is just A → B. Similarly, product types A × B can be generalized to dependent sums $\sum_{x:A} B(x)$. Intuitively, elements of this type are pairs (a, b), where a is an element of A and b is an element of $B(a)$. When B does not depend on x, this is just A × B.

The second conceptual innovation in constructive type theory is to internalize the notion of constructive evidence. According to the BHK interpretation, knowing a mathematical theorem amounts to having a construction that realizes the claim. Thus we can view any mathematical proposition as specifying a special type of data, namely, the type of construction that is appropriate to realizing it. This has come to be known as the "propositions as types correspondence" or the "Curry–Howard correspondence" [45, 46, 97], developed by Curry, Howard, Tait, Martin-Löf, and Girard, among others. The point is that logical operations look a lot like operations on datatypes. For example, in propositional logic, from A and B one can conclude $A \wedge B$. One can read this as saying that given a proof a of A and a proof b of B of B one can "pair" them to obtain a proof (a, b) of $A \wedge B$. In other words, $A \wedge B$ and $A \times B$ are governed by the same rules. Similarly, the interpretation of implication $A \to B$ mirrors the rules for function types: giving a proof of $A \to B$ amounts to constructing a function from A to B. In the same way, a proof of $\forall x : A\, B(x)$ is a function which, given any a in A, return a proof of $B(a)$.

Thus, in constructive type theory, some types are naturally viewed as types of data and some are naturally viewed as propositions, but the two interact and are governed by similar rules. A single calculus gives the rules for defining mathematical objects and proving propositions; that is, the calculus provides a set of rules for carrying out mathematical constructions of both sorts. Two of the most commonly used systems today are *Martin-Löf type theory* [150], and the *calculus of inductive constructions* [43], which extends the original *calculus of constructions* due to Coquand and Huet [42]. The relationship between the Martin-Löf type theory and the calculus of constructions is similar to the relationship between HA^ω and intuitionistic higher-order logic: the calculus of constructions has "impredicative" comprehension principles that render it much stronger than Martin-Löf's predicative counterpart.

As was the case with HA^ω and intuitionistic higher-order logic, one can give these systems a computational interpretation in various ways. Indeed, semantics for constructive type theory draws on the full range of the theory of programming language semantics, making use of realizability interpretations, strong normalization proofs, domain theory, and more. The literature on this subject is extensive; see, for example, [4, 44].

3.10. Computational interpretations of classical theories. We characterized the Russian school of constructive recursive mathematics as reasoning about computable objects in a constructive way. One can maintain the first requirement axiomatically while giving up the second: for example, the subsystem of second-order arithmetic known as RCA_0 is a classical system for which first-order quantifiers can be interpreted as ranging over the natural numbers and second-order quantifiers can be interpreted as ranging over computable sets and functions. Thus RCA_0 is a reasonable setting for formalizing classical computable analysis (see [197]).

To what extent can one preserve a computational interpretation of quantifier dependences in a formal system that includes the law of the excluded middle? There is a trivial sense in which any reasonable classical theory has a computational interpretation. Let φ be a Π_2 statement, that is, an assertion of the form $\forall x \, \exists y \, R(x, y)$ where x and y range over natural numbers and R is a primitive recursive relation. Suppose some classical theory T proves φ. Then, assuming T can be trusted in this regard, φ is *true*, which means that a simple-minded computer program that, on input x, searches systematically for a y satsifying $R(x, y)$ is guaranteed to succeed.

There is also a fundamental sense in which such an interpretation cannot be extended to Π_3 sentences. Let $T(e, x, s)$ be Kleene's primitive recursive relation that expresses that s is a halting computation sequence for Turing machine e on input x. Then the classically valid statement that any given Turing machine e either halts on input 0 or doesn't can be expressed as follows:

$$\forall e \, \exists s \, \forall s' \, (T(e, 0, s) \vee \neg T(e, 0, s')).$$

But any function mapping an e to an s satisfying the conclusion provides a solution to the halting problem, and thus there is no computable function witnessing the $\forall e \, \exists s$ dependence.

Nonetheless, one can often give classical logic an *indirect* computational interpretation. One way to do this is to interpret a classical theory in a constructive one, using the double-negation translation and tricks such as the Friedman–Dragalin translation [51, 65] or the Dialectica interpretation [77, 7] to recover Π_2 theorems (see also [5, 41]). One can also provide more direct computational interpretations, such as realizability relations of various sorts. Griffin [81] has shown that classical logic can be understood in terms of a standard semantics for programming languages with control operators, such as exceptions. This computational interpretation is captured by the $\lambda\mu$-calculus designed by Parigot [165] (see also [199]). Chetan Murthy [160] has shown that this interpretation can be seen as the result of combining a double-negation translation with the Friedman–Dragalin trick, and then using intuitionistic realizability. Jean-Louis Krivine [127] has provided a realizability interpretation for full classical set theory.

Sometimes what one wants from a classical proof is not a computational interpretation *per se*, but useful computational or quantitative information that is hidden by classical methods. In the 1950s, Kreisel spoke of "unwinding" classical proofs to obtain such information, a program which has developed under the heading of "proof mining," by Kohlenbach and others [119].

Acknowledgments. We are grateful to Martín Escardó, Guido Gherardi, André Nies, Klaus Weihrauch, and an anonymous referee for helpful comments and suggestions. Avigad's work has been partially supported by National Science Foundation grant DMS-1068829 and Air Force Office of Scientific Research grant FA9550-12-1-0370. Brattka's research was supported by a Marie Curie International Research Staff Exchange Scheme Fellowship within the 7th European Community Framework Programme and by the National Research Foundation of South Africa.

REFERENCES

[1] OLIVER ABERTH, *The failure in computable analysis of a classical existence theorem for differential equations*, **Proceedings of the American Mathematical Society**, vol. 30 (1971), pp. 151–156.
[2] ———, *Computable analysis*, McGraw-Hill, New York, 1980.
[3] N. L. ACKERMAN, C. E. FREER, and D. M. ROY, *Noncomputable conditional distributions*, **Logic in Computer Science (LICS), 2011**, IEEE Conference Presentations, Los Alamitos, California, 2011, pp. 107–116.
[4] THORSTEN ALTENKIRCH, *Proving strong normalization of CC by modifying realizability semantics*, **Types for proofs and programs (Nijmegen, 1993)**, Springer, Berlin, 1994, pp. 3–18.
[5] JEREMY AVIGAD, *Interpreting classical theories in constructive ones*, **The Journal of Symbolic Logic**, vol. 65 (2000), pp. 1785–1812.
[6] ———, *Uncomputably noisy ergodic limits*, **Notre Dame Journal of Formal Logic**, vol. 53 (2012), pp. 347–350.
[7] JEREMY AVIGAD and SOLOMON FEFERMAN, *Gödel's functional ("Dialectica") interpretation*, **Handbook of proof theory** (Samuel R. Buss, editor), North-Holland, Amsterdam, 1998, pp. 337–405.
[8] JEREMY AVIGAD, PHILIPP GERHARDY, and HENRY TOWSNER, *Local stability of ergodic averages*, **Transactions of the American Mathematical Society**, vol. 362 (2010), pp. 261–288.
[9] JEREMY AVIGAD and KSENIJA SIMIC, *Fundamental notions of analysis in subsystems of second-order arithmetic*, **Annals of Pure and Applied Logic**, vol. 139 (2006), pp. 138–184.
[10] G. BAIGGER, *Die Nichtkonstruktivität des Brouwerschen Fixpunktsatzes*, **Archiv für Mathematische Logik und Grundlagenforschung**, vol. 25 (1985), pp. 183–188.
[11] STEFAN BANACH and STANISŁAW MAZUR, *Sur les fonctions calculables*, **Annales de la Société Polonaise de Mathématique**, vol. 16 (1937), p. 223.
[12] ANDREJ BAUER, *The realizability approach to computable analysis and topology*, Ph.D. thesis, School of Computer Science, Carnegie Mellon University, Pittsburgh, 2000.
[13] VERÓNICA BECHER, SANTIAGO FIGUEIRA, and RAFAEL PICCHI, *Turing's unpublished algorithm for normal numbers*, **Theoretical Computer Science**, vol. 377 (2007), pp. 126–138.
[14] MICHAEL J. BEESON, *Foundations of constructive mathematics*, Springer, Berlin, 1985.
[15] LAURENT BIENVENU, ADAM R. DAY, MATHIEU HOYRUP, ILYA MEZHIROV, and ALEXANDER SHEN, *A constructive version of Birkhoff's ergodic theorem for Martin-Löf random points*, **Information and Computation**, vol. 210 (2012), pp. 21–30.
[16] ERRETT BISHOP, *Foundations of constructive analysis*, McGraw-Hill, New York, 1967.
[17] ERRETT BISHOP and DOUGLAS S. BRIDGES, *Constructive analysis*, Springer, Berlin, 1985.
[18] ÉMILE BOREL, *Le calcul des intégrales définies*, **Journal de Mathématiques Pures et Appliquées. Série 6**, vol. 8 (1912), pp. 159–210.
[19] ———, *La théorie de la mesure et al théorie de l'integration*, **Leçons sur la théorie des fonctions**, Gauthier-Villars, Paris, 1950, pp. 214–256.
[20] VOLKER BOSSERHOFF, *Notions of probabilistic computability on represented spaces*, **Journal of Universal Computer Science**, vol. 14 (2008), pp. 956–995.
[21] VASCO BRATTKA, *Effective Borel measurability and reducibility of functions*, **Mathematical Logic Quarterly**, vol. 51 (2005), pp. 19–44.
[22] VASCO BRATTKA and GUIDO GHERARDI, *Effective choice and boundedness principles in computable analysis*, **The Bulletin of Symbolic Logic**, vol. 17 (2011), pp. 73–117.

[23] ———, *Weihrauch degrees, omniscience principles and weak computability*, **The Journal of Symbolic Logic**, vol. 76 (2011), pp. 143–176.

[24] VASCO BRATTKA, GUIDO GHERARDI, and ALBERTO MARCONE, *The Bolzano–Weierstrass theorem is the jump of weak Kőnig's lemma*, **Annals of Pure and Applied Logic**, vol. 163 (2012), pp. 623–655.

[25] VASCO BRATTKA, PETER HERTLING, and KLAUS WEIHRAUCH, *A tutorial on computable analysis*, **New computational paradigms: Changing conceptions of what is computable** (S. Barry Cooper, Benedikt Löwe, and Andrea Sorbi, editors), Springer, New York, 2008, pp. 425–491.

[26] VASCO BRATTKA, JOSEPH S. MILLER, and ANDRÉ NIES, *Randomness and differentiability*, submitted; preliminary version at http://arxiv.org/abs/1104.4465.

[27] MARK BRAVERMAN, *Parabolic Julia sets are polynomial time computable*, **Nonlinearity**, vol. 19 (2006), pp. 1383–1401.

[28] MARK BRAVERMAN and STEPHEN COOK, *Computing over the reals: Foundations for scientific computing*, **Notices of the American Mathematical Society**, vol. 53 (2006), pp. 318–329.

[29] MARK BRAVERMAN and M. YAMPOLSKY, *Non-computable Julia sets*, **Journal of the American Mathematical Society**, vol. 19 (2006), pp. 551–578.

[30] MARK BRAVERMAN and MICHAEL YAMPOLSKY, **Computability of Julia sets**, Springer, Berlin, 2008.

[31] DOUGLAS BRIDGES and FRED RICHMAN, **Varieties of constructive mathematics**, Cambridge University Press, Cambridge, 1987.

[32] L. E. J. BROUWER, *Begründung der Mengenlehre unabhängig vom logischen Satz vom ausgeschlossen Dritten. Erster Teil: Allgemeine Mengenlehre*, **Koninklijke Nederlandse Akademie van Wetenschappen te Amsterdam**, vol. 12 (1918), no. 5, reprinted in [35], pp. 150–190.

[33] ———, *Begründung der Mengenlehre unabhängig vom logischen Satz vom ausgeschlossen Dritten. Zweiter Teil: Theorie der Punkmengen*, **Koninklijke Nederlandse Akademie van Wetenschappen te Amsterdam**, vol. 12 (1919), no. 7, reprinted in [35], pp. 191–221.

[34] ———, *Über Definitionsbereiche von Funktionen*, **Mathematische Annalen**, vol. 97 (1927), no. 1, pp. 60–75.

[35] ———, *Collected works, volume I: Philosophy and foundations of mathematics*, (A. Heyting, editor), North-Holland Publishing Company, Amsterdam, 1975.

[36] G. S. CEĬTIN, *Algorithmic operators in constructive complete separable metric spaces*, **Doklady Akademii Nauk**, vol. 128 (1959), pp. 49–52, (in Russian).

[37] DOUGLAS CENZER and JEFFREY B. REMMEL, *Index sets in computable analysis*, **Theoretical Computer Science**, vol. 219 (1999), pp. 111–150.

[38] ———, *Effectively closed sets and graphs of computable real functions*, **Theoretical Computer Science**, vol. 284 (2002), pp. 279–318.

[39] ———, *Index sets for computable differential equations*, **Mathematical Logic Quarterly**, vol. 50 (2004), pp. 329–344.

[40] ALONZO CHURCH, *A formulation of the simple theory of types*, **The Journal of Symbolic Logic**, vol. 5 (1940), pp. 56–68.

[41] THIERRY COQUAND and MARTIN HOFMANN, *A new method of establishing conservativity of classical systems over their intuitionistic version*, **Mathematical Structures in Computer Science**, vol. 9 (1999), pp. 323–333.

[42] THIERRY COQUAND and GÉRARD HUET, *The calculus of constructions*, **Information and Computation**, vol. 76 (1988), pp. 95–120.

[43] THIERRY COQUAND and CHRISTINE PAULIN, *Inductively defined types*, **COLOG-88 (Tallinn, 1988)**, Springer, Berlin, 1990, pp. 50–66.

[44] THIERRY COQUAND and ARNAUD SPIWACK, *A proof of strong normalisation using domain theory*, **Logical Methods in Computer Science**, vol. 3 (2007), no. 4.

[45] HASKELL B. CURRY, *Functionality in combinatory logic*, **Proceedings of the National Academy of Sciences**, vol. 20 (1934), pp. 584–590.

[46] HASKELL B. CURRY and ROBERT FEYS, *Combinatory logic*, vol. I, North-Holland, Amsterdam, 1958.

[47] MARTIN DAVIS, *The universal computer: The road from Leibniz to Turing*, W. W. Norton, New York, 2000.

[48] OSVALD DEMUTH, *Constructive pseudonumbers*, **Commentationes Mathematicae Universitatis Carolinae**, vol. 16 (1975), pp. 315–331, (in Russian).

[49] ———, *The differentiability of constructive functions of weakly bounded variation on pseudo numbers*, **Commentationes Mathematicae Universitatis Carolinae**, vol. 16 (1975), pp. 583–599, (in Russian).

[50] RODNEY G. DOWNEY and DENIS R. HIRSCHFELDT, *Algorithmic randomness and complexity*, Springer, New York, 2010.

[51] ALBERT DRAGALIN, *Mathematical intuitionism: Introduction to proof theory*, American Mathematical Society, Providence, Rhode Island, 1988.

[52] MICHAEL DUMMETT, *Elements of intuitionism*, second ed., Oxford University Press, New York, 2000.

[53] ABBAS EDALAT, *A computable approach to measure and integration theory*, **Information and Computation**, vol. 207 (2009), pp. 642–659.

[54] ABBAS EDALAT and PHILIPP SÜNDERHAUF, *A domain-theoretic approach to computability on the real line*, **Theoretical Computer Science**, vol. 210 (1999), pp. 73–98.

[55] HAROLD M. EDWARDS, *Kronecker's views on the foundations of mathematics*, **The history of modern mathematics** (D. E. Rowe and J. McCleary, editors), Academic Press, San Diego, 1989, pp. 67–77.

[56] ———, *Kronecker's fundamental theorem of general arithmetic*, **Episodes in the history of modern algebra (1800–1950)**, American Mathematical Society, Providence, Rhode Island, 2007, pp. 107–116.

[57] ———, *Kronecker's algorithmic mathematics*, **Mathematical Intelligencer**, vol. 31 (2009), pp. 11–14.

[58] William Ewald (editor), *From Kant to Hilbert: A source book in the foundations of mathematics*, vol. 1 and 2, Oxford University Press, Oxford, 1996.

[59] JOHANNA FRANKLIN, NOAM GREENBERG, JOSEPH S. MILLER, and KENG MENG NG, *Martin-Löf random points satisfy Birkhoff's ergodic theorem for effectively closed sets*, **Proceedings of the American Mathematical Society**, vol. 140 (2012), pp. 3623–3628.

[60] JOHANNA FRANKLIN and HENRY TOWSNER, *Randomness and non-ergodic systems*, arXiv:1206.2682.

[61] CAMERON FREER, BJØRN KJOS-HANSSEN, ANDRÉ NIES, and FRANK STEPHAN, *Algorithmic aspects of Lipschitz functions*, preprint.

[62] CAMERON E. FREER and DANIEL M. ROY, *Computable de Finetti measures*, **Annals of Pure and Applied Logic**, vol. 163 (2012), pp. 530–546.

[63] GOTTLOB FREGE, *Grundgesetze der Arithmetik, Band I*, Hermann Pohle, Jena, 1893.

[64] HARVEY FRIEDMAN, *On the computational complexity of maximization and integration*, **Advances in Mathematics**, vol. 53 (1984), pp. 80–98.

[65] HARVEY M. FRIEDMAN, *Classically and intuitionistically provable functions*, **Higher set theory** (H. Müller and D. Scott, editors), Springer, Berlin, 1978, pp. 21–27.

[66] PETER GÁCS, *Uniform test of algorithmic randomness over a general space*, **Theoretical Computer Science**, vol. 341 (2005), pp. 91–137.

[67] PETER GÁCS, MATHIEU HOYRUP, and CRISTÓBAL ROJAS, *Randomness on computable probability spaces—a dynamical point of view*, **Theory of Computing Systems**, vol. 48 (2011), pp. 465–485.

[68] STEFANO GALATOLO, MATHIEU HOYRUP, and CRISTÓBAL ROJAS, *A constructive Borel–Cantelli lemma: constructing orbits with required statistical properties*, **Theoretical Computer Science**, vol. 410 (2009), pp. 2207–2222.

[69] STEFANO GALATOLO, MATHIEU HOYRUP, and CRISTÓBAL ROJAS, *Effective symbolic dynamics, random points, statistical behavior, complexity and entropy*, **Information and Computation**, vol. 208 (2010), pp. 23–41.

[70] GERHARD GENTZEN, *Die Widerspruchsfreiheit der reinen Zahlentheorie*, **Mathematische Annalen**, vol. 112 (1936), pp. 493–465, translated as *The consistency of elementary number theory* in [71], pp. 132–213.

[71] ———, *Collected works*, (M. E. Szabo, editor), North-Holland, Amsterdam, 1969.

[72] GUIDO GHERARDI, *Alan Turing and the foundations of computable analysis*, **The Bulletin of Symbolic Logic**, vol. 17 (2011), pp. 394–430.

[73] GUIDO GHERARDI and ALBERTO MARCONE, *How incomputable is the separable Hahn–Banach theorem?*, **Notre Dame Journal of Formal Logic**, vol. 50 (2009), pp. 393–425.

[74] JEAN-YVES GIRARD, *Une extension de l'interprétation de Gödel à l'analyse, et son application à l'élimination des coupures dans l'analyse et la théorie des types*, **Proceedings of the second Scandinavian Logic Symposium**, North-Holland, Amsterdam, 1971, pp. 63–92.

[75] JEAN-YVES GIRARD, YVES LAFONT, and PAUL TAYLOR, *Proofs and types*, Cambridge University Press, 1989.

[76] KURT GÖDEL, *Zur intuitionistischen Arithmetik und Zahlentheorie*, **Ergebnisse eines mathematischen Kolloquiums**, vol. 4 (1933), pp. 34–38, translated by Stefan Bauer-Mengelberg and Jean van Heijenoort as *On intuitionistic arithmetic and number theory* in [217], reprinted in [78] (1933e), pp. 287–295.

[77] ———, *Über eine bisher noch nicht benützte Erweiterung des finiten Standpunktes*, **Dialectica**, vol. 12 (1958), pp. 280–287, reprinted with English translation in Feferman et al., (editors), **Kurt Gödel: Collected Works**, volume 2, Oxford University Press, New York, 1990, pp. 241–251.

[78] ———, *Collected works*, (Solomon Feferman et al., editors), vol. I, Oxford University Press, New York, 1986.

[79] R. L. GOODSTEIN, *Recursive number theory: A development of recursive arithmetic in a logic-free equation calculus*, North-Holland, Amsterdam, 1957.

[80] ———, *Recursive analysis*, North-Holland, Amsterdam, 1961.

[81] TIMOTHY G. GRIFFIN, *A formulae-as-type notion of control*, **Proceedings of the 17th ACM SIGPLAN–SIGACT symposium on Principles of Programming Languages** (*POPL '90*), Association for Computing Machinery, New York, 1990, pp. 47–58.

[82] ANDRZEJ GRZEGORCZYK, *Computable functionals*, **Fundamenta Mathematicae**, vol. 42 (1955), pp. 168–202.

[83] ———, *On the definition of computable functionals*, **Fundamenta Mathematicae**, vol. 42 (1955), pp. 232–239.

[84] ———, *On the definitions of computable real continuous functions*, **Fundamenta Mathematicae**, vol. 44 (1957), pp. 61–71.

[85] ———, *Some approaches to constructive analysis*, **Constructivity in mathematics** (A. Heyting, editor), North-Holland, Amsterdam, 1959, pp. 43–61.

[86] JÜRGEN HAUCK, *Ein Kriterium für die Annahme des Maximums in der Berechenbaren Analysis*, **Zeitschrift für Mathematische Logik und Grundlagen der Mathematik**, vol. 17 (1971), pp. 193–196.

[87] ———, *Konstruktive Darstellungen reeller Zahlen und Folgen*, **Zeitschrift für Mathematische Logik und Grundlagen der Mathematik**, vol. 24 (1978), pp. 365–374.

[88] ———, *Konstruktive Darstellungen in topologischen Räumen mit rekursiver Basis*, **Zeitschrift für Mathematische Logik und Grundlagen der Mathematik**, vol. 26 (1980), pp. 565–576.

[89] PETER HERTLING, *Effectivity and effective continuity of functions between computable metric spaces*, **Combinatorics, complexity, and logic** (Douglas S. Bridges, Cristian S. Calude, Jeremy Gibbons, Steve Reeves, and Ian H. Witten, editors), Springer, Singapore, 1997.

[90] ———, *An effective Riemann Mapping Theorem*, **Theoretical Computer Science**, vol. 219 (1999), pp. 225–265.

[91] ——, *A real number structure that is effectively categorical*, **Mathematical Logic Quarterly**, vol. 45 (1999), pp. 147–182.

[92] ——, *A Banach–Mazur computable but not Markov computable function on the computable real numbers*, **Annals of Pure and Applied Logic**, vol. 132 (2005), pp. 227–246.

[93] PETER HERTLING and KLAUS WEIHRAUCH, *Random elements in effective topological spaces with measure*, **Information and Computation**, vol. 181 (2003), pp. 32–56.

[94] ARENDT HEYTING, *Die formalen Regeln der intuitionistischen Logik. I, II, III*, **Sitzungsberichte der königlich-preussischen Akademie der Wissenschaften. Berlin**, (1930), pp. 42–56, 57–71, and 158–169.

[95] ——, *Intuitionism: An introduction*, North-Holland, Amsterdam, 1956.

[96] DAVID HILBERT and PAUL BERNAYS, **Grundlagen der Mathematik**, vol. 1, Springer, Berlin, 1934, vol. 32, 1939.

[97] W. A. HOWARD, *The formulae-as-types notion of construction*, **To H. B. Curry: Essays on combinatory logic, lambda calculus and formalism**, Academic Press, London, 1980, pp. 480–490.

[98] MATHIEU HOYRUP and CRISTÓBAL ROJAS, *Computability of probability measures and Martin-Löf randomness over metric spaces*, **Information and Computation**, vol. 207 (2009), pp. 830–847.

[99] MATHIEU HOYRUP, CRISTÓBAL ROJAS, and KLAUS WEIHRAUCH, *Computability of the Radon–Nikodym derivative*, **Models of computation in context** (Benedikt Löwe, Dag Normann, Ivan Soskov, and Alexandra Soskova, editors), Springer, Berlin, 2011, pp. 132–141.

[100] ——, *Computability of the Radon–Nikodym derivative*, **Computability**, vol. 1 (2012), pp. 3–13.

[101] J. M. E. HYLAND, *The effective topos*, **The L.E.J. Brouwer Centenary Symposium** (*Noordwijkerhout, 1981*), North-Holland, Amsterdam, 1982, pp. 165–216.

[102] IRAJ KALANTARI and LARRY WELCH, *A blend of methods of recursion theory and topology*, **Annals of Pure and Applied Logic**, vol. 124 (2003), pp. 141–178.

[103] IRAJ KALANTARI and LAWRENCE WELCH, *Point-free topological spaces, functions and recursive points; filter foundation for recursive analysis. I*, **Annals of Pure and Applied Logic**, vol. 93 (1998), pp. 125–151.

[104] ——, *Recursive and nonextendible functions over the reals; filter foundation for recursive analysis, II*, **Annals of Pure and Applied Logic**, vol. 98 (1999), pp. 87–110.

[105] AKITOSHI KAWAMURA, *Lipschitz continuous ordinary differential equations are polynomial-space complete*, **Computational Complexity**, vol. 19 (2010), pp. 305–332.

[106] AKITOSHI KAWAMURA and STEPHEN COOK, *Complexity theory for operators in analysis*, **Proceedings of the 42nd ACM Symposium on Theory of Computing** (**STOC '10**), Association for Computing Machinery, New York, 2010, pp. 495–502.

[107] DIETER KLAUA, *Berechenbare Analysis*, **Zeitschrift für Mathematische Logik und Grundlagen der Mathematik**, vol. 2 (1956), pp. 265–303.

[108] ——, **Konstruktive Analysis**, Deutscher Verlag der Wissenschaften, Berlin, 1961.

[109] STEPHEN COLE KLEENE, *Recursive predicates and quantifiers*, **Transactions of the American Mathematical Society**, vol. 53 (1943), pp. 41–73.

[110] ——, *On the interpretation of intuitionistic number theory*, **The Journal of Symbolic Logic**, vol. 10 (1945), pp. 109–124.

[111] STEPHEN COLE KLEENE, **Introduction to metamathematics**, D. Van Nostrand Company, New York, 1952.

[112] STEPHEN COLE KLEENE and RICHARD EUGENE VESLEY, **The foundations of intuitionistic mathematics, especially in relation to recursive functions**, North-Holland, Amsterdam, 1965.

[113] KER-I KO, *The maximum value problem and NP real numbers*, **Journal of Computer and Systems Sciences**, vol. 24 (1982), pp. 15–35.

[114] ——, *Some negative results on the computational complexity of total variation and differentiation*, **Information and Control**, vol. 53 (1982), pp. 21–31.

[115] ———, *On the computational complexity of ordinary differential equations*, **Information and Control**, vol. 58 (1983), pp. 157–194.

[116] ———, *On the computational complexity of integral equations*, **Annals of Pure and Applied Logic**, vol. 58 (1992), pp. 201–228.

[117] KER-I KO and H. FRIEDMAN, *Computational complexity of real functions*, **Theoretical Computer Science**, vol. 20 (1982), pp. 323–352.

[118] ———, *Computing power series in polynomial time*, **Advances in Applied Mathematics**, vol. 9 (1988), pp. 40–50.

[119] ULRICH KOHLENBACH, *Applied proof theory: Proof interpretations and their use in mathematics*, Springer, Berlin, 2008.

[120] GEORG KREISEL, *On the interpretation of non-finitist proofs II. Interpretation of number theory. Applications*, **The Journal of Symbolic Logic**, vol. 17 (1952), pp. 43–58.

[121] ———, *Some elementary inequalities*, **Koninklijke Nederlandse Akademie van Wetenschappen, Proceedings, Series A**, vol. 55 (1952), pp. 334–338, **Indagationes Mathematicae**, vol. 14.

[122] GEORG KREISEL and DANIEL LACOMBE, *Ensembles récursivement mesurables et ensembles récursivement ouverts et fermés*, **Comptes Rendus de l'Académie des Sciences. Paris**, vol. 245 (1957), pp. 1106–1109.

[123] GEORG KREISEL, DANIEL LACOMBE, and J. R. SHOENFIELD, *Fonctionnelles récursivement définissables et fonctionnelles récursives*, **Comptes Rendus de l'Académie des Sciences. Paris**, vol. 245 (1957), pp. 399–402.

[124] ———, *Partial recursive functionals and effective operations*, **Constructivity in mathematics** (A. Heyting, editor), North-Holland, Amsterdam, 1959, pp. 290–297.

[125] CHRISTOPH KREITZ and KLAUS WEIHRAUCH, *Theory of representations*, **Theoretical Computer Science**, vol. 38 (1985), pp. 35–53.

[126] ———, *Compactness in constructive analysis revisited*, **Annals of Pure and Applied Logic**, vol. 36 (1987), pp. 29–38.

[127] JEAN-LOUIS KRIVINE, *Typed lambda-calculus in classical Zermelo–Frænkel set theory*, **Archive for Mathematical Logic**, vol. 40 (2001), pp. 189–205.

[128] LEOPOLD KRONECKER, **Grundzüge einer arithmetischen Theorie der algebraischen Grössen**, Riemer, Berlin, 1882, also published in **Journal für reine und angewandte Mathematik**, vol. 92 (1882), pp. 1–122, and [132], vol. II, pp. 237–387.

[129] ———, *Ein Fundamentalsatz der allgemeinen Arithmetik*, **Journal für die reine und angewandte Mathematik**, vol. 100 (1887), pp. 490–510, reprinted in [132], vol. IIIa, pp. 209–240.

[130] ———, *Über den Zahlbegriff*, **Philosophische Aufsätze, Eduard Zeller zu seinem fünfzigjährigen Doctorjubiläum gewidmet**, Fues, Leipzig, 1887, pp. 261–274, reprinted in [132], vol. IIIa, pp. 249–274. Translated as *On the concept of number* by William Ewald in [58], vol. 2, pp. 947–955.

[131] ———, **Vorlesungen über Zahlentheorie**, (Kurt Hensel, editor), Teubner, Leipzig, 1901, republished by Springer, Berlin, 1978.

[132] ———, **Werke**, (Kurt Hensel, editor), vol. 1–5, Chelsea Publishing Company, New York, 1968.

[133] ANTONÍN KUČERA and ANDRÉ NIES, *Demuth's path to randomness*, **Proceedings of the 2012 international Workshop on Theoretical Computer Science: Computation, physics and beyond (WTCS 2012)**, Springer, Berlin, 2012, pp. 159–173.

[134] BORIS A. KUSHNER, **Lectures on constructive mathematical analysis**, American Mathematical Society, Providence, RI, 1984, translated from the Russian by E. Mendelson.

[135] ———, *The constructive mathematics of A. A. Markov*, **American Mathematical Monthly**, vol. 113 (2006), pp. 559–566.

[136] DANIEL LACOMBE, *Classes récursivement fermés et fonctions majorantes*, **Comptes Rendus de l'Académie des Sciences. Paris**, vol. 240 (1955), pp. 716–718.

[137] ——, *Extension de la notion de fonction récursive aux fonctions d'une ou plusieurs variables réelles I–III*, **Comptes Rendus de l'Académie des Sciences. Paris**, vol. 240 and 241 (1955), pp. 2478–2480 and 13–14 and 151–153.

[138] ——, *Remarques sur les opérateurs récursifs et sur les fonctions récursives d'une variable réelle*, **Comptes Rendus de l'Académie des Sciences. Paris**, vol. 241 (1955), pp. 1250–1252.

[139] ——, *Les ensembles récursivement ouverts ou fermés, et leurs applications à l'Analyse récursive*, **Comptes Rendus de l'Académie des Sciences. Paris**, vol. 245 (1957), pp. 1040–1043.

[140] ——, *Quelques propriétés d'analyse récursive*, **Comptes Rendus de l'Académie des Sciences. Paris**, vol. 244 (1957), pp. 838–840 and 996–997.

[141] ——, *Les ensembles récursivement ouverts ou fermés, et leurs applications à l'Analyse récursive*, **Comptes Rendus de l'Académie des Sciences. Paris**, vol. 246 (1958), pp. 28–31.

[142] ——, *Sur les possibilites d'extension de la notion de fonction récursive aux fonctions d'une ou plusieurs variables réelles*, **Le raisonnement en mathematiques et en sciences**, Editions du Centre National de la Recherche Scientifique, Paris, 1958, pp. 67–75.

[143] ——, *Quelques procédés de définition en topologie récursive*, **Constructivity in mathematics** (A. Heyting, editor), North-Holland, Amsterdam, 1959, pp. 129–158.

[144] PETER LIETZ, **From constructive mathematics to computable analysis via the realizability interpretation**, Ph.D. thesis, Fachbereich Mathematik, TU Darmstadt, Darmstadt, 2004.

[145] SAUNDERS MAC LANE and IEKE MOERDIJK, **Sheaves in geometry and logic**, Springer, New York, 1994.

[146] A. A. MARKOV, *On the continuity of constructive functions*, **Uspekhi Matematicheskikh Nauk**, vol. 9 (1954), pp. 226–230, (in Russian).

[147] ——, *On constructive functions*, **Trudy Matematicheskogo Instituta Imeni V. A. Steklova**, vol. 52 (1958), pp. 315–348, (in Russian, English translation in **American Mathematical Society Translations** Series 2, vol. 29, 1963).

[148] PER MARTIN-LÖF, *The definition of random sequences*, **Information and Control**, vol. 9 (1966), pp. 602–619.

[149] ——, **Notes on constructive mathematics**, Almqvist and Wiksell, Stockholm, 1970.

[150] ——, *An intuitionistic theory of types: predicative part*, **Logic Colloquium '73** (H. E. Rose and J. C. Shepherdson, editors), North-Holland, Amsterdam, 1973.

[151] ALEC MATHESON and TIMOTHY H. MCNICHOLL, *Computable analysis and Blaschke products*, **Proceedings of the American Mathematical Society**, vol. 136 (2008), pp. 321–332.

[152] STANISŁAW MAZUR, **Computable analysis**, Razprawy Matematyczne, Warsaw, 1963.

[153] TIMOTHY H. MCNICHOLL, *Uniformly computable aspects of inner functions: estimation and factorization*, **Mathematical Logic Quarterly**, vol. 54 (2008), pp. 508–518.

[154] ——, *A uniformly computable implicit function theorem*, **Mathematical Logic Quarterly**, vol. 54 (2008), pp. 272–279.

[155] GEORGE METAKIDES and ANIL NERODE, *The introduction of non-recursive methods into mathematics*, **The L. E. J. Brouwer Centenary Symposium** (A. S. Troelstra and D. van Dalen, editors), North-Holland, Amsterdam, 1982, pp. 319–335.

[156] GEORGE METAKIDES, ANIL NERODE, and R. A. SHORE, *Recursive limits on the Hahn–Banach theorem*, **Errett Bishop: Reflections on him and his research** (Murray Rosenblatt, editor), American Mathematical Society, 1985.

[157] JOSEPH STEPHEN MILLER, **Pi-0-1 classes in computable analysis and topology**, Ph.D. thesis, Cornell University, Ithaca, USA, 2002.

[158] YIANNIS NICHOLAS MOSCHOVAKIS, *Recursive metric spaces*, **Fundamenta Mathematicae**, vol. 55 (1964), pp. 215–238.

[159] ANDRZEJ MOSTOWSKI, *On computable sequences*, **Fundamenta Mathematicae**, vol. 44 (1957), pp. 37–51.

[160] CHETAN R. MURTHY, *An evaluation semantics for classical proofs*, **Proceedings, Sixth Annual IEEE Symposium on Logic in Computer Science**, Amsterdam, 1991, pp. 96–107.

[161] JOHN MYHILL, *A recursive function defined on a compact interval and having a continuous derivative that is not recursive*, **Michigan Mathematical Journal**, vol. 18 (1971), pp. 97–98.

[162] DAVID NELSON, *Recursive functions and intuitionistic number theory*, **Transactions of the American Mathematical Society**, vol. 61 (1947), pp. 307–368.

[163] ANDRÉ NIES, **Computability and randomness**, Oxford University Press, New York, 2009.

[164] V. P. OREVKOV, *A constructive mapping of the square onto itself displacing every constructive point*, **Doklady Akademii Nauk**, vol. 152 (1963), pp. 55–58, (in Russian) Translated in: **Soviet Mathematics Doklady**, vol. 4 (1963), pp. 1253–1256.

[165] MICHEL PARIGOT, *$\lambda\mu$-calculus: an algorithmic interpretation of classical natural deduction*, **Logic programming and automated reasoning** (**St. Petersburg, 1992**), Springer, Berlin, 1992.

[166] NOOPUR PATHAK, *A computational aspect of the lebesgue differentiation theorem*, **Journal of Logic and Analysis**, vol. 1 (2009), pp. 1–15.

[167] NOOPUR PATHAK, CRISTÓBAL ROJAS, and STEPHEN G. SIMPSON, *Schnorr randomness and the Lebesgue differentiation theorem*, **Proceedings of the American Mathematical Society**, (to appear).

[168] ARNO PAULY, *How incomputable is finding Nash equilibria?*, **Journal of Universal Computer Science**, vol. 16 (2010), pp. 2686–2710.

[169] ———, *On the (semi)lattices induced by continuous reducibilities*, **Mathematical Logic Quarterly**, vol. 56 (2010), pp. 488–502.

[170] ROGER PENROSE, **The emperor's new mind. concerning computers, minds and the laws of physics**, Oxford University Press, New York, 1989.

[171] RÓZSA PÉTER, **Rekursive Funktionen**, Akademischer Verlag, Budapest, 1951.

[172] MARIAN POUR-EL and J. CALDWELL, *On a simple definition of computable functions of a real variable*, **Zeitschrift für Mathematische Logik und Grundlagen der Mathematik**, vol. 21 (1975), pp. 1–19.

[173] MARIAN POUR-EL and NING ZHONG, *The wave equation with computable initial data whose unique solution is nowhere computable*, **Mathematical Logic Quarterly**, vol. 43 (1997), pp. 499–509.

[174] MARIAN BOYKAN POUR-EL and J. IAN RICHARDS, *A computable ordinary differential equation which possesses no computable solution*, **Annals of Mathematical Logic**, vol. 17 (1979), pp. 61–90.

[175] ———, *The wave equation with computable initial data such that its unique solution is not computable*, **Advances in Mathematics**, vol. 39 (1981), pp. 215–239.

[176] ———, *Noncomputability in analysis and physics: a complete determination of the class of noncomputable linear operators*, **Advances in Mathematics**, vol. 48 (1983), pp. 44–74.

[177] ———, **Computability in analysis and physics**, Springer, Berlin, 1989.

[178] ROBERT RETTINGER and KLAUS WEIHRAUCH, *The computational complexity of some Julia sets*, **Proceedings of the 35th Annual ACM Symposium on Theory of Computing** (Michel X. Goemans, editor), Association for Computing Machinery, New York, 2003, pp. 177–185.

[179] ROBERT RETTINGER and XIZHONG ZHENG, *On the hierarchy and extension of monotonically computable real numbers*, **Journal of Complexity**, vol. 19 (2003), pp. 672–691.

[180] ———, *A hierarchy of Turing degrees of divergence bounded computable real numbers*, **Journal of Complexity**, vol. 22 (2006), pp. 818–826.

[181] H. GORDON RICE, *Recursive real numbers*, **Proceedings of the American Mathematical Society**, vol. 5 (1954), pp. 784–791.

[182] R. M. ROBINSON, *Review of "Peter, R., Rekursive Funktionen"*, **The Journal of Symbolic Logic**, vol. 16 (1951), pp. 280–282.

[183] STÉPHANE LE ROUX and MARTIN ZIEGLER, *Singular coverings and non-uniform notions of closed set computability*, **Mathematical Logic Quarterly**, vol. 54 (2008), pp. 545–560.

[184] BERTRAND RUSSELL and ALFRED NORTH WHITEHEAD, **Principia mathematica**, vol. 1, Cambridge University Press, 1910, vol. 2, 1912; vol. 3, 1913.

[185] JASON RUTE, *Algorithmic randomness, martingales, and differentiability*, in preparation.
[186] CLAUS PETER SCHNORR, *Komplexität von Algorithmen mit Anwendung auf die Analysis*, **Archiv für Mathematische Logik und Grundlagenforschung**, vol. 14 (1971), pp. 54–68.
[187] ———, *Zufälligkeit und Wahrscheinlichkeit*, Lecture Notes in Mathematics, vol. 218, Springer, Berlin, 1971.
[188] ARNOLD SCHÖNHAGE, *Numerik analytischer Funktionen und Komplexität*, **Jahresbericht der Deutschen Mathematiker-Vereinigung**, vol. 92 (1990), pp. 1–20.
[189] MATTHIAS SCHRÖDER, *Topological spaces allowing type 2 complexity theory*, **Computability and complexity in analysis** (Ker-I Ko and Klaus Weihrauch, editors), FernUniversität, Hagen, September 1995, pp. 41–53.
[190] ———, *Fast online multiplication of real numbers*, **STACS 97** (Rüdiger Reischuk and Michel Morvan, editors), Springer, Berlin, 1997, pp. 81–92.
[191] ———, *Online computations of differentiable functions*, **Theoretical Computer Science**, vol. 219 (1999), pp. 331–345.
[192] ———, *Extended admissibility*, **Theoretical Computer Science**, vol. 284 (2002), pp. 519–538.
[193] ———, *Spaces allowing type-2 complexity theory revisited*, **Mathematical Logic Quarterly**, vol. 50 (2004), pp. 443–459.
[194] ———, *Admissible representations for probability measures*, **Mathematical Logic Quarterly**, vol. 53 (2007), pp. 431–445.
[195] DANA SCOTT, *Outline of a mathematical theory of computation*, **Technical monograph prg-2**, Oxford University, Oxford, November 1970.
[196] KSENIJA SIMIC, *The pointwise ergodic theorem in subsystems of second-order arithmetic*, **The Journal of Symbolic Logic**, vol. 72 (2007), pp. 45–66.
[197] STEPHEN G. SIMPSON, **Subsystems of second order arithmetic**, second ed., Cambridge University Press, Cambridge, 2009.
[198] THORALF SKOLEM, *Begründung der elementaren Arithmetik durch die rekurrierende Denkweise ohne Anwendung scheinbarer Veränderlichen mit unendlichem Ausdehnungsbereich*, **Skrifter, Norske Videnskaps-Akademi i Oslo, Matematisk-Naturvidenskapelig Klasse**, vol. 6, J. Dybwad, Oslo, 1923, pp. 1–38, translated in [217], pp. 302–333.
[199] MORTEN H. SØRENSEN and PAWEL URZYCZYN, *Lectures on the Curry–Howard isomorphism*, Elsevier, Amsterdam, 2006.
[200] ERNST SPECKER, *Nicht konstruktiv beweisbare Sätze der Analysis*, **The Journal of Symbolic Logic**, vol. 14 (1949), pp. 145–158.
[201] ———, *Der Satz vom Maximum in der rekursiven Analysis*, **Constructivity in mathematics** (A. Heyting, editor), North-Holland, Amsterdam, 1959, pp. 254–265.
[202] HOWARD STEIN, *Logos, Logic, and Logistiké*, **History and philosophy of modern mathematics** (William Aspray and Philip Kitcher, editors), University of Minnesota, Minneapolis, 1988, pp. 238–259.
[203] WILLIAM W. TAIT, *Intensional interpretations of functionals of finite type, I*, **The Journal of Symbolic Logic**, vol. 32 (1967), pp. 198–212.
[204] A. S. TROELSTRA, *Metamathematical investigation of intuitionistic arithmetic and analysis*, Springer, Berlin, 1973.
[205] ———, *Realizability*, **Handbook of proof theory** (Samuel R. Buss, editor), North-Holland, Amsterdam, 1998, pp. 407–473.
[206] A. S. TROELSTRA and DIRK VAN DALEN, **Constructivism in mathematics: An introduction**, North-Holland, Amsterdam, 1988.
[207] ALAN M. TURING, *On computable numbers, with an application to the "Entscheidungsproblem"*, **Proceedings of the London Mathematical Society**, vol. 42 (1936), pp. 230–265.
[208] ———, *On computable numbers, with an application to the "Entscheidungsproblem". A correction*, **Proceedings of the London Mathematical Society**, vol. 43 (1937), pp. 544–546.

[209] ———, *Finite approximations to Lie groups*, **Annals of Mathematics**, vol. 39 (1938), pp. 105–111.

[210] ———, *Systems of logic based on ordinals*, **Proceedings of the London Mathematical Society. Series 2**, vol. 45 (1939), pp. 161–228.

[211] ———, *A method for the calculation of the zeta-function*, **Proceedings of the London Mathematical Society. Series 2**, vol. 48 (1943), pp. 180–197.

[212] ———, *Rounding-off errors in matrix processes*, **Quarterly Journal of Mechanics and Applied Mathematics**, vol. 1 (1948), pp. 287–308.

[213] ———, *Some calculations of the Riemann zeta-function*, **Proceedings of the London Mathematical Society. Series 3**, vol. 3 (1953), pp. 99–117.

[214] V. A. USPENSKY and A. L. SEMENOV, *Basic developments connected with the concept of algorithm and with its applications in mathematics*, **Algorithms in modern mathematics and computer science** (Andrei P. Ershov and Donald E. Knuth, editors), Springer, Berlin, 1981.

[215] MARK VAN ATTEN, *On Brouwer*, Wadsworth/Thomson Learning, Belmont, CA, 2004.

[216] ———, *Luitzen Egbertus Jan Brouwer*, **The Stanford encyclopedia of philosophy** (Edward N. Zalta, editor), summer 2011 ed., 2011.

[217] JEAN VAN HEIJENOORT, **From Frege to Gödel: A sourcebook in mathematical logic, 1879–1931**, Harvard University Press, Cambridge, 1967.

[218] JOHN VON NEUMANN, *Die formalistische Grundlegung der Mathematik*, **Erkenntnis**, vol. 2 (1931), 116–121, translated by E. Putnam and G. J. Massey as *The formalist foundations of mathematics* (P. Benacerraf and H. Putnam, editors), **Philosophy of Mathematics: Selected Readings**, 2nd edition, Cambridge University Press, Cambridge, 1983, pp. 61–65.

[219] V. V. V'YUGIN, *Ergodic convergence in probability, and an ergodic theorem for individual random sequences*, **Teoriya Veroyatnostei i ee Primeneniya**, vol. 42 (1997), pp. 35–50.

[220] ———, *Ergodic theorems for individual random sequences*, **Theoretical Computer Science**, vol. 207 (1998), pp. 343–361.

[221] KLAUS WEIHRAUCH, **Computability**, Springer, Berlin, 1987.

[222] ———, *Computability on the probability measures on the Borel sets of the unit interval*, **Theoretical Computer Science**, vol. 219 (1999), pp. 421–437.

[223] ———, **Computable analysis**, Springer, Berlin, 2000.

[224] ———, *Computational complexity on computable metric spaces*, **Mathematical Logic Quarterly**, vol. 49 (2003), pp. 3–21.

[225] KLAUS WEIHRAUCH and CHRISTOPH KREITZ, *Representations of the real numbers and of the open subsets of the set of real numbers*, **Annals of Pure and Applied Logic**, vol. 35 (1987), pp. 247–260.

[226] KLAUS WEIHRAUCH and NING ZHONG, *Is wave propagation computable or can wave computers beat the Turing machine?*, **Proceedings of the London Mathematical Society**, vol. 85 (2002), pp. 312–332.

[227] ———, *Computing the solution of the Korteweg-de Vries equation with arbitrary precision on Turing machines*, **Theoretical Computer Science**, vol. 332 (2005), pp. 337–366.

[228] ———, *An algorithm for computing fundamental solutions*, **SIAM Journal on Computing**, vol. 35 (2006), pp. 1283–1294.

[229] ———, *Computing Schrödinger propagators on Type-2 Turing machines*, **Journal of Complexity**, vol. 22 (2006), pp. 918–935.

[230] ———, *Computable analysis of the abstract Cauchy problem in a Banach space and its applications I*, **Mathematical Logic Quarterly**, vol. 53 (2007), pp. 511–531.

[231] NORBERT WIENER, *Time, communication, and the nervous system*, **Annals of the New York Academy of Sciences**, vol. 50 (1948), pp. 197–220.

[232] Philip P. Wiener (editor), **Leibniz: Selections**, Charles Scribner's Sons, New York, 1951.

[233] YONGCHENG WU and KLAUS WEIHRAUCH, *A computable version of the Daniell-Stone theorem on integration and linear functionals*, **Theoretical Computer Science**, vol. 359 (2006),

pp. 28–42.

[234] Mariko Yasugi, Takakazu Mori, and Yoshiki Tsujii, *Effective properties of sets and functions in metric spaces with computability structure*, **Theoretical Computer Science**, vol. 219 (1999), pp. 467–486.

[235] Xiaokang Yu, *Riesz representation theorem, Borel measures and subsystems of second-order arithmetic*, **Annals of Pure and Applied Logic**, vol. 59 (1993), pp. 65–78.

[236] ———, *Lebesgue convergence theorems and reverse mathematics*, **Mathematical Logic Quarterly**, vol. 40 (1994), pp. 1–13.

[237] Xiaokang Yu and Stephen G. Simpson, *Measure theory and weak König's lemma*, **Archive for Mathematical Logic**, vol. 30 (1990), pp. 171–180.

[238] I. D. Zaslavskiĭ, *Disproof of some theorems of classical analysis in constructive analysis*, **Uspekhi Matematicheskikh Nauk**, vol. 10 (1955), pp. 209–210, (in Russian).

[239] Xizhong Zheng and Klaus Weihrauch, *The arithmetical hierarchy of real numbers*, **Mathematical Logic Quarterly**, vol. 47 (2001), pp. 51–65.

[240] Ning Zhong and Klaus Weihrauch, *Computability theory of generalized functions*, **Journal of the Association for Computing Machinery**, vol. 50 (2003), pp. 469–505.

[241] Qing Zhou, *Computable real-valued functions on recursive open and closed subsets of Euclidean space*, **Mathematical Logic Quarterly**, vol. 42 (1996), pp. 379–409.

[242] Martin Ziegler, *Real hypercomputation and continuity*, **Theory of Computing Systems**, vol. 41 (2007), pp. 177–206.

DEPARTMENTS OF PHILOSOPHY AND MATHEMATICAL SCIENCES
CARNEGIE MELLON UNIVERSITY, PITTSBURGH, PA, USA
E-mail: avigad@cmu.edu

FACULTY OF COMPUTER SCIENCE
UNIVERSITÄT DER BUNDESWEHR, MÜNCHEN, GERMANY
and
DEPARTMENT OF MATHEMATICS AND APPLIED MATHEMATICS
UNIVERSITY OF CAPE TOWN, CAPE TOWN, SOUTH AFRICA
and
ISAAC NEWTON INSTITUTE FOR MATHEMATICAL SCIENCES
CAMBRIDGE, UNITED KINGDOM
E-mail: Vasco.Brattka@cca-net.de

ALAN TURING AND THE OTHER THEORY OF COMPUTATION (EXPANDED)

LENORE BLUM

> **Abstract.** We recognize Alan Turing's work in the foundations of numerical computation (in particular, his 1948 paper "Rounding-Off Errors in Matrix Processes"), its influence in modern complexity theory, and how it helps provide a unifying concept for the two major traditions of the theory of computation.

§1. Introduction. The two major traditions of the theory of computation, each staking claim to similar motivations and aspirations, have for the most part run a parallel non-intersecting course. On one hand, we have the tradition arising from logic and computer science addressing problems with more recent origins, using tools of combinatorics and discrete mathematics. On the other hand, we have numerical analysis and scientific computation emanating from the classical tradition of equation solving and the continuous mathematics of calculus. Both traditions are motivated by a desire to understand the essence of computation, of algorithm; both aspire to discover useful, even profound, consequences.

While the logic and computer science communities are keenly aware of Alan Turing's seminal role in the former (discrete) tradition of the theory of computation, most remain unaware of Alan Turing's role in the latter (continuous) tradition, this notwithstanding the many references to Turing in the modern numerical analysis/computational mathematics literature, e.g., [Bur10, Hig02, Kah66, TB97, Wil71]. These references are not to recursive/computable analysis (suggested in Turing's seminal 1936 paper), usually cited by logicians and computer scientists, but rather to the fundamental role that the notion of "condition" (introduced in Turing's seminal 1948 paper) plays in real computation and complexity.

In 1948, in the first issue of the *Quarterly Journal of Mechanics and Applied Mathematics*, sandwiched between a paper on "Use of Relaxation Methods

---

This paper amplifies a shorter version [Blu13] contained in *Alan Turing: His Work and Impact*, Elsevier and follows the perspective presented in "Computing over the Reals: Where Turing Meets Newton," [Blu04].

Turing's Legacy: Developments from Turing's Ideas in Logic
Edited by Rod Downey
Lecture Notes in Logic, 42

and Fourier Transforms" and "The Position of the Shock-Wave in Certain Aerodynamic Problems," appears the article "Rounding-Off Errors in Matrix Processes." This paper introduces the notion of the *condition number* of a matrix, the chief factor limiting the accuracy in solving linear systems, a notion fundamental to numerical computation and analysis, and a notion with implications for complexity theory today. This paper was written by Alan Turing [Tur48].

ROUNDING-OFF ERRORS IN MATRIX PROCESSES
By A. M. TURING
(*National Physical Laboratory, Teddington, Middlesex*)

[Received 4 November 1947]

SUMMARY

A number of methods of solving sets of linear equations and inverting matrices are discussed. The theory of the rounding-off errors involved is investigated for some of the methods. In all cases examined, including the well-known 'Gauss elimination process', it is found that the errors are normally quite moderate: no exponential build-up need occur.

Included amongst the methods considered is a generalization of Choleski's method which appears to have advantages over other known methods both as regards accuracy and convenience. This method may also be regarded as a rearrangement of the elimination process.

THIS paper contains descriptions of a number of methods for solving sets of linear simultaneous equations and for inverting matrices, but its main concern is with the theoretical limits of accuracy that may be obtained in the application of these methods, due to rounding-off errors.

After the war, with the anticipation of a programmable digital computing device on the horizon, it was of great interest to understand the comparative merits of competing computational "processes" and how accurate such processes would be in the face of inevitable round-off errors. Solving linear systems is basic. Thus for Turing (as it was for John von Neumann [vNG47]), examining methods of solution with regard to the ensuing round-off errors presented a compelling intellectual challenge.[1]

In 1945, Turing had submitted an 86 page proposal to the British National Physical Laboratory (NPL) for the Automatic Computing Engine (ACE), *an automatic electronic digital computer with internal program storage* [Tur45]. This was to be an incarnation of the universal machine envisioned by his theoretical construct in "Computable Numbers" [Tur36], blurring the boundary

---

[1] It is clear that Turing and von Neumann were working on similar problems, for similar reasons, in similar ways at the same time, probably independently. However while Turing acknowledges von Neumann, as far as I know, von Neumann never cites Turing's work in this area.

between data and program. Thus, and in contrast to other proposed "computing machines" at the time, Turing's computer would have simplified hardware, with universality emanating from the power of programming.

Turing envisioned an intelligent machine that would learn from its teachers and from its experience and mistakes, and hence have the ability to create and modify its own programs. Turing also felt that the most conducive environment for realizing such a machine was to have mathematicians and engineers working together in close proximity, not in separate domains [Hod92].

Unfortunately, the ACE computer was never to see the light of day;[2] a less ambitious non-universal machine, the PILOT ACE, was constructed after Turing left the NPL for the University of Manchester in 1948. Although Turing's central passion during his time at the NPL was the promised realization of his universal computer, his only published paper to come out of this period (1945–1948) was "Round-Off Errors in Matrix Processes."[3]

"Rounding-Off Errors in Matrix Processes" was by no means an anomaly in Turing's creative pursuits. In his 1970 Turing Award Lecture, James Wilkinson writes [Wil71]:

> Turing's international reputation rests mainly on his work on computable numbers but I like to recall that he was a considerable numerical analyst, and a good part of his time from 1946 onwards was spent working in this field ....

Wilkinson attributes this interest, and Turing's decision to go to the NPL after the war, to the years he spent at Bletchley Park gaining knowledge of electronics and "one of the happiest times of his life."

Wilkinson also credits Turing for converting him from a classical to numerical analyst. From 1946 to 1948, Wilkinson worked for Turing at the NPL on the logical design of Turing's proposed Automatic Computing Engine and the problem of programming basic numerical algorithms:

> The period with Turing fired me with so much enthusiasm for the computer project and so heightened my interest in numerical analysis that gradually I abandoned [the idea of returning to Cambridge to take up research in classical analysis].

Here I would like to recognize Alan Turing's work in the foundations of numerical computation. Even more, I would like to indicate how this work has seeded a major direction in complexity theory of real computation and provides a unifying concept for the two major traditions in the theory of computation.

---

[2]An interrelated combination of personalities, rivalries, politics and bureaucracy seems to have been in play. For an in-depth chronicle of the saga, see Andrew Hodges' chapter, *Mercury Delayed* in [Hod92].

[3]The paper ends with the cryptic acknowledgement: "published with the permission of the Director of the National Physical Laboratory."

§2. Rounding-Off Errors in Matrix Processes.

This paper contains descriptions of a number of methods for solving sets of linear simultaneous equations and for inverting matrices, *but its main concern is with the theoretical limits of accuracy* that may be obtained in the application of these methods, due to round-off errors.

So begins Turing's paper [Tur48]. (Italics are mine, I'll return to this shortly.)

The basic problem at hand: Given the linear system, $Ax = b$ where A is a real non-singular $n \times n$ matrix and $b \in \mathbb{R}^n$. Solve for $x \in \mathbb{R}^n$.

Prompted by calculations [FHW48] challenging the arguments by Harold Hotelling [Hot43] that Gaussian elimination and other direct methods would lead to exponential round-off errors, Turing introduces quantities not considered earlier to bound the magnitude of errors, showing that for all "normal" cases, the exponential estimates are "far too pessimistic."[4]

In this paper, Turing introduced the notion of *condition number* of a matrix making explicit for the first time a measure that helps formalize the informal notion of ill and well-conditioned problems.[5]

§3. The matrix condition number: Where Turing meets Newton.

When we come to make estimates of errors in matrix processes, we shall find that the chief factor limiting the accuracy that can be obtained is 'ill-conditioning' of the matrices involved [Tur48].

Turing provides an illustrative example:

$$\left.\begin{aligned} 1.4x + 0.9y &= 2.7 \\ -0.8x + 1.7y &= -1.2 \end{aligned}\right\} \quad (8.1)$$

$$\left.\begin{aligned} -0.786x + 1.709y &= -1.173 \\ -0.800x + 1.700y &= -1.200 \end{aligned}\right\} \quad (8.2)$$

---

[4] In their 1946 paper, Valentine Bargmann, Deane Montgomery and von Neumann [BMvN63] also dismissed Gaussian elimination as likely being unstable due to magnification of errors at successive stages (pp. 430–431) and so turn to iterative methods for analysis. However, in 1947 von Neumann and Herman Goldstine reassess [vNG47] noting, as does Turing, that it is the computed solution, not the intermediate computed numbers, which should be the salient object of study. They re-investigated Gaussian elimination for computing matrix inversion and now give optimistic error bounds similar to those of Turing, but for the special case of positive definite symmetric matrices. Turing in his paper notes that von Neumann communicated these results to him at Princeton [during a short visit] in January 1947 before his own proofs were complete.

[5] In sections 3 and 4 of his paper, Turing also formulates the LU decomposition of a matrix (actually the LDU decomposition) and shows that Gaussian elimination computes such a decomposition.

The set of equations (8.1) is fully equivalent to (8.2)[6] but clearly if we attempt to solve (8.2) by numerical methods involving rounding-off errors we are almost certain to get much less accuracy than if we worked with equations (8.1)....

We should describe the equations (8.2) as an *ill-conditioned* set, or, at any rate, as ill-conditioned compared with (8.1). It is characteristic of ill-conditioned sets of equations that small percentage errors in the coefficients given may lead to large percentage errors in the solution.

Turing defines two condition numbers (he calls them N and M), which in essence measure the intrinsic potential for magnification of errors. He then analyzes various standard methods for solving linear systems, including Gaussian elimination, and gets error bounds proportional to his measures of condition. Turing is "also as much interested in statistical behaviour of the errors as in the maximum possible values" and presents probabilistic estimates; he also improves Hotelling's worst case bound from 4^{n-1} to 2^{n-1} [Tur48].

The following widely used (*spectral*) matrix condition number $\kappa(A)$, wedged somewhat between Turing's condition numbers, is often attributed to Turing, though it is unclear who first defined it. (See the discussion in Section 10, *Who invented the condition number?*) John Todd in his survey [Tod68] is vague on its genesis although he specifically credits Turing with recognizing "that a condition number should depend symmetrically on A and A^{-1}, specifically as a product of their norms."[7]

DEFINITION. Suppose A is a real non-singular $n \times n$ matrix. The (*spectral*) *matrix condition number* of A is given by

$$\kappa(A) = \|A\| \|A^{-1}\|$$

where

$$\|A\| = \sup_{y \neq 0} \frac{|Ay|}{|y|} = \sup_{|y|=1} |Ay|$$

is the *operator* (spectral) *norm* with respect to the Euclidean norm. For singular matrices, define $\kappa(A) = \infty$.

This definition can be generalized using other norms. In the case of the Euclidean norm, $\kappa(A) = \sigma_1/\sigma_n$, where σ_1 and σ_2 are the largest and smallest singular values of A, respectively. It follows that $\kappa(A) \geq 1$.[8]

---

[6] The third equation (in the set of four) is the second plus 0.01 times the first.

[7] Turing's N condition number is defined as $N(A)N(A^{-1})$ and the M condition number as $nM(A)M(A^{-1})$, where $N(A)$ is the Frobenius norm of A and $M(A)$ is the max norm. Turing also defines the spectral norm in this paper, so he could have easily defined the spectral condition number.

[8] For the case of computing the inverse of a positive definite symmetric matrix A by Gaussian elimination, von Neumann and Goldstine [vNG47] give an error estimate bounded

To see how natural a measure this is, consider a slightly more general situation. Let X and Y be normed vector spaces with associated map $\varphi : X \to Y$. A measure of the "condition" of *problem instance* (φ, x) should indicate how small perturbations of the *input x* (the *problem data*) can alter the *output* $\varphi(x)$ (the *problem solution*).

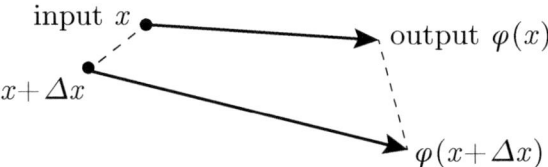

So let Δx be a small perturbation of input x and $\Delta \varphi = \varphi(x + \Delta x) - \varphi(x)$. The limit as $\|\Delta x\|$ goes to zero of the ratio

$$\frac{\|\Delta \varphi\|}{\|\Delta x\|},$$

or of the *relative* ratio

$$\frac{\|\Delta \varphi\| / \|\varphi(x)\|}{\|\Delta x\| / \|x\|}$$

(favored by numerical analysts), will be a measure of the condition of the problem instance.[9] If large, computing the output with small error will require increased precision, and hence from a computational complexity point of view, increased time/space resources.

DEFINITION. [10] The *condition number of problem instance* (φ, x) is defined by

$$\widehat{\kappa}(\varphi, x) = \lim_{\delta \to 0} \sup_{\|\Delta x\| \leq \delta} \frac{\|\Delta \varphi\|}{\|\Delta x\|},$$

and the *relative condition number* by

$$\kappa(\varphi, x) = \lim_{\delta \to 0} \sup_{\|\Delta x\| \leq \delta \|x\|} \frac{\|\Delta x\| / \|\varphi(x)\|}{\|\Delta x\| / \|x\|}.$$

If $\kappa(\varphi, x)$ is small, the problem instance is said to be *well-conditioned* and if large, *ill-conditioned*. If $\kappa(\varphi, x) = \infty$, the problem instance is *ill-posed*.

---

by $14.2n^2(\lambda_1 / \lambda_2)u$. Here λ_1 and λ_2 are the largest and smallest eigenvalues of A and u is the smallest number recognized by the machine. For the case of positive definite symmetric matrices, λ_1 / λ_2 is equal to $\kappa(A)$. Thus, the spectral condition number appears implicitly in the von Neumann–Goldstine paper for this case.

[9] All norms are assumed to be with respect to the relevant spaces.

[10] Here I follow the notation in [TB97], a book I highly recommend for background in numerical linear algebra.

As Turing envisaged it, the condition number measures the theoretical limits of accuracy in solving a problem. In particular, the logarithm of the condition number provides an *intrinsic* lower bound for the *loss of precision* in solving the problem instance, independent of algorithm.[11] Thus it also provides a key intrinsic parameter for specifying "input word size" for measuring computational complexity over the reals—and in connecting the two traditions of computation—as we shall see in Section 6.

If φ is differentiable, then

$$\widehat{\kappa}(\varphi, x) = \|D\varphi(x)\|$$

and

$$\kappa(\varphi, x) = \|D\varphi(x)\| \left(\|x\| / \|\varphi(x)\| \right),$$

where $D\varphi(x)$ is the Jacobian (derivative) matrix of φ at x and $\|D\varphi(x)\|$ is the operator norm of $D\varphi(x)$ with respect to the induced norms on X and Y.

Thus we have a conceptual connection between the condition number (Turing) and the derivative (Newton). Indeed, the following theorem says the matrix condition number $\kappa(A)$ is essentially the relative condition number for solving the linear system $Ax = b$. In other words, the condition number is essentially the (normed) derivative.[12]

THEOREM. 1. Fix A, a real non-singular $n \times n$ matrix, and consider the map $\varphi_A \colon \mathbb{R}^n \to \mathbb{R}^n$ where $\varphi_A(b) = A^{-1}(b)$. Then $\kappa(\varphi_A, b) \leq \kappa(A)$ and there exist $\bar{b}$ such that $\kappa(\varphi_A, \bar{b}) = \kappa(A)$. *Thus, with respect to perturbations in b, the matrix condition number is the worst case relative condition for solving the linear system $Ax = b$.*

2. Fix $b \in \mathbb{R}^n$ and consider the partial map $\varphi_b \colon \mathbb{R}^{n \times n} \to \mathbb{R}^n$ where, for A non-singular, $\varphi_b(A) = A^{-1}(b)$. Then for A non-singular, $\kappa(\varphi_b, A) = \kappa(A)$.

So the condition number $\kappa(A)$ indicates the number of digits that can be lost in solving the linear system. Trouble is, computing the condition number seems as hard as solving the problem itself. Probabilistic analysis can often be employed to glean information.

In the mid 1980s, in response to a challenge by Steve Smale, there was a flurry of work estimating the expected loss of precision in solving linear systems, e.g., by Adrian Ocneanu (unpublished), Eric Kostlan [Kos88], and [BS86]. The sharpest results here are due to Alan Edelman [Ede88] who showed that, with respect to the standard normal distribution, the average $\log \kappa(A) \sim \log(n)$, a result that Turing was clearly pursuing.

---

[11] Velvel Kahan points out that "pre-conditioning" can sometimes alter the given problem instance to a better conditioned one with the same solution. (Convert equations (8.2) to (8.1) in Turing's illustrative example.)

[12] This inspired in part the title of my paper, "Computing over the Reals: Where Turing meets Newton" [Blu04].

§4. Turing's evolving perspective on computing over the reals. The Turing Machine is the canonical abstract model of a general purpose computer, studied in almost every first course in theoretical computer science. What most students of theory are not aware of, however, is that Turing defined his "machine" in order to define a theory of real computation. The first paragraph of his seminal 1936 paper [Tur36] begins and ends as follows:

> The "computable" numbers may be described briefly as the real numbers whose expressions as a decimal are calculable by finite means. ... According to my definition, a number is computable if its decimal can be written down by a machine.

Of course, the machine thus developed becomes the basis for the classical theory of computation of logic and theoretical computer science.

In the same first paragraph Turing writes, "I hope shortly to give an account of the relation of the computable numbers, functions, and so forth to one another. This will include a development of the theory of functions of a real variable expressed in terms of computable numbers." As far as I know, Turing never returned to computing over the reals using this approach; *recursive analysis* (also known as *computable analysis*) was developed by others, [PER89, Wei00].

When Turing does return to computing over the reals, as in "Rounding-off Errors" written while he was preoccupied with the concrete problem of computing solutions to systems of equations, his implicit real number model is vastly different. Now real numbers are considered as individual entities and each basic algebraic operation is counted as one step. In the first section of this paper, Turing considers the "measures of work in a process:"

> It is convenient to have a measure of the amount of work involved in a computing process, even though it be a very crude one. ... We might, for instance, count the number of additions, subtractions, multiplications, divisions, recordings of numbers, ...

This is the basic approach taken by numerical analysts, qualified as Turing also implies, by condition and round-off errors. It is also the approach taken by Mike Shub, Steve Smale and myself in [BSS89], and later with Felipe Cucker in our book, *Complexity and Real Computation* [BCSS98]. See also [Blu90], [Sma97], and [Cuc02].

§5. Complexity and real computation in the spirit of Turing, 1948. From the late 1930s to the 1960s, a major focus for logicians was the classification of what was computable (by a Turing Machine, or one of its many equivalents) and what was not. In the 1960s, the emerging community of theoretical computer scientists embarked on a more down to earth line of inquiry—of the computable, what was feasible and what was not—leading to a formal theory of complexity with powerful applications and deep problems, viz., the famous/infamous $P = NP$? challenge.

Motivated to develop analogous foundations for numerical computation, [BSS89] present a model of computation over an arbitrary field R. For example, R could be the field of real or complex numbers, or $\mathbb{Z}_2$, the field of integers mod 2. In the spirit of Turing 1948, inputs to the so-called *BSS machines* are vectors over R and the basic algebraic operations, comparisons and admissible retrievals are *unit cost*. Algorithms are represented by directed graphs (or in more modern presentations, circuits) where interior nodes are labelled with basic operations, and computations flow from the input to output nodes. The *cost of a computation* is the number of nodes traversed from input to output.

As in the discrete case, *complexity* (or cost of a computation) is measured as a function of *input word size*. At the top level, input word size is defined as the vector length. When R is $\mathbb{Z}_2$, the input word size is just the bit length, as in the discrete case. Complexity classes over R, such as P, NP and EXP, are defined in natural ways. When R is $\mathbb{Z}_2$, the BSS theory of computation and complexity reduces to the classical discrete theory.

The problem of deciding whether or not a finite system of polynomial equations over R has a solution over R, so fundamental to mathematics, turns out to be a universal NP-complete problem [BSS89]. More precisely, for any field $(R, =)$, or real closed field $(R, <)$, instances of NP-complete problems over R can be coded up as polynomial systems such that an instance is a "yes" instance if and only if the corresponding polynomial system has a solution over R.[13] We call this problem the *Hilbert Nullstellensatz* over R, or HN_R.

There are many subtleties here. For example, the fact that $NP \subset EXP$ over $\mathbb{Z}_2$ is a simple counting argument on the number of possible witnesses. Over the reals or complexes, there are just too many witnesses. Indeed, it's not a priori even clear that in those cases, NP problems are decidable. Decidability in those cases follows from the decidability of HN_R by Alfred Tarski [Tar51], and membership in EXP from Jim Renegar's exponential-time decision algorithms [Ren88a].

New challenges arise: Does $P = NP$? over the reals or complex numbers, or equivalently, is $HN_R \in P$ over those fields? And what is the relation between these questions and the classical P vs. NP challenge?

In attempt to gain new insight or to access more tools, mathematicians often position hard problems within new domains. It is tempting thus to speculate if tools of algebraic geometry might have a role to play in studying classical

---

[13] The notation $(R, =)$ denotes that branching in *BSS* machines over R are decided by equality comparisons, while $(R, <)$ indicates that R is an ordered field and branching is now decided by order comparisons. When we talk about computing over the complex numbers or $\mathbb{Z}_2$, we are supposing our machines branch in the former sense, while over the real numbers, we mean the latter.

complexity problems. Salient transfer results: If $P = NP$ over the complex numbers, then $BPP \supseteq NP$ over $\mathbb{Z}_2$ [CSS94]. And for algebraically closed fields of characteristic 0, either $P = NP$ for all, or for none [BCSS96].

We shall return to this discussion in Section 9, but first we introduce condition into the model.

§6. Introducing condition into the model: Connecting the two traditions.
At the top level, the model of computation and complexity over the reals or complex numbers discussed above is an exact arithmetic model. But, like Turing, we note the cost of obtaining a solution (to a given accuracy) will depend on a notion of condition. Ill-conditioned problem instances will require additional resources to solve. Thus, it would seem natural to measure cost as a function of an *intrinsic* input word size which depends on the condition as well as the input dimension and desired accuracy of solution and not "on some perhaps whimsically given input precision" [Blu90, Blu91].

This perspective helps connect the two major traditions of computation.

To illustrate, consider the linear programming problem (LPP): The problem is to optimize a linear function on a polytope in Euclidean space defined by linear inequalities. The discovery of the first practical polynomial time algorithms for linear programming in the 1980s, the so-called *interior point methods*, crystalized for me and for others, flaws in the discrete theory's analysis of algorithms for problems whose natural domains are continuous.

For suppose some coefficient of an LPP instance is 1. If the instance is well-conditioned, the answer to a slightly perturbed instance should not vary wildly, nor should the cost of computation. However, when cost is measured as a function of input word size in bits, as the discrete theory prescribes, the complexity *analysis* allows much more time to solve a very slightly perturbed instance, e.g., when 1 is replaced by $1 + 10^{-10^{10}}$.

When the underlying space is the real numbers, this makes no sense.[14] What is clearly desired: a more intrinsic notion of input word size that takes into account condition.

But how to measure condition of a linear program? [15]

---

[14] The interior point methods are not even finite as real number algorithms. They are iterative processes. In the discrete case, a stopping rule halts the iteration in a timely manner and then a diophantine "jump" to an exact solution is performed. Such a jump does not work over the reals. For these algorithms over the real numbers, solutions to within a prescribed accuracy will have to suffice. This contrasts with George Dantzig's Simplex Method [Dan47] which, although exponential in the worst case [KM72], is a finite exact arithmetic algorithm. This conundrum yields the open question: Is linear programming *strongly polynomial* in the sense that, is there an algorithm for LPP that is polynomial in both the discrete and the exact arithmetic models?

[15] In 1990, with this question in mind, I posed the challenge [Blu90]: "For more general classes of problems over continuous domains, an important research direction is to develop analogous measures of condition, as well as other intrinsic parameters, and to study their relationship to computational complexity."

§7. The Condition Number Theorem sets the stage.

Let Σ_n be the variety of singular (*ill-posed*) $n \times n$ matrices over $\mathbb{R}$, i.e.,

$$\Sigma_n = \{A \in \mathbb{R}^{n \times n} \mid A \text{ is not invertible}\}.$$

We might expect that matrices close to Σ_n would be ill-conditioned while those at a distance, well-conditioned. That is what the *Condition Number Theorem* says. It provides a geometric characterization of the matrix condition number which suggests, for other computational problems, how condition could be measured.

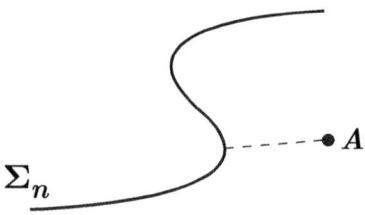

THE CONDITION NUMBER THEOREM.

$$\kappa(A) = \frac{\|A\|}{dist(A, \Sigma_n)}.$$

Here $dist(A, \Sigma_n) = \inf\{\|A - B\| \mid B \in \Sigma_n\}$ where *dist* is measured respect to the operator norm or the Frobenius norm. (The Frobenius norm is given by $\|A\|_F = \sqrt{\sum a_{ij}^2}$, where $A = [a_{ij}]$.)

The Condition Number Theorem is a re-interpretation of a classical theorem of Carl Eckart and Gale Young [EY36]. Although published in 1936, Turing and von Neumann seem not to have been aware of it. Velvel Kahan [Kah66], and later his student Jim Demmel [Dem87]), were the first to exploit it connecting condition with distance to ill-posedness.

Jim Renegar, inspired by the Condition Number Theorem, answers our query on how to measure condition of a linear program, and then uses this measure as a key parameter in the complexity analysis of his beautiful algorithm [Ren88b, Ren95a, Ren95b].[16]

Recall, the linear programming problem (A, b, c) is to maximize $c^T x$ such that $Ax \geq b$. Here A is a real $m \times n$ matrix, $b \in \mathbb{R}^m$ and $c \in \mathbb{R}^n$. Let (A, b) denote the above inequalities, which also define a polytope in $\mathbb{R}^n$. In the following, we assume that $m \geq n$. We call

$\Sigma_{m,n} = \{(A, b) \mid (A, b)$ is on the boundary between the feasible and infeasible$\}$

the space of *ill-posed* linear programs.

---

[16]Jim tells me that he started thinking about this after a talk I gave at at MSRI during the 1985–1986 Computational Complexity Year [Ren88b].

Let $C_P(A, b) = \|(A, b)\|_F / dist_F((A, b), \Sigma_{m,n})$. Here $\|\|_F$ is the Frobenius norm, and $dist_F$ is measured with respect to that norm. Similarly, define $C_D(A, c)$ for the dual program.

DEFINITION. The Renegar *condition number* [Ren95b] for the linear program (A, b, c) is given by

$$C(A, b, c) = \max[C_P(A, b), C_D(A, c)].$$

Renegar's algorithm for the LPP [Ren88b] imagines that each side of a (bounded, non-empty) polytope, given by inequalities, exerts a force inward, yielding the "center of gravity" of the polytope. Specifically, the center ξ is gotten by maximizing

$$\sum_{i=1}^{m} \ln(\alpha_i \cdot x - b_i)$$

where α_i is the ith row vector of the matrix A, and b_i is the ith entry in b.

In essence, the algorithm starts at this initial center of gravity ξ. It then follows a path of centers (approximated by Newton) generated by adding a sequence of new inequalities $c \cdot x \geq k^{(j)}$ ($k^{(j)}$, $j = 1, 2, \ldots$, chosen so that each successive new polytope is bounded and non-empty). Let $k^{(1)} = c \cdot \xi$. If the new center is $\xi^{(1)}$, then $k^{(2)}$ is chosen so that $k^{(1)} \leq k^{(2)} \leq c \cdot \xi^{(1)}$. And so on.

Conceptually, the hyperplane $c \cdot x = k^{(1)}$ is successively moved in the direction of the vector c, pushing the initial center ξ towards optimum.

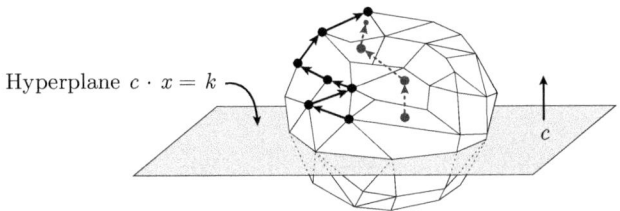

THEOREM (Renegar).

$$\sharp((A, b, c), \epsilon) = O(\sqrt{m}(\log m + \log C(A, b, c) + \log 1/\epsilon).$$

Here $\sharp(A, b, c), \epsilon)$ is the *number of iterates* for Renegar's algorithm to return a solution to within accuracy ϵ, or declare that the linear program is infeasible or unbounded.

The total number of *arithmetic operations* is

$$O(m^3 \log(mC(A, b, c)/\epsilon)).$$

For numerical analysis, it makes sense to define an algorithm to be *polynomial time* if there are positive constants k and c such that for all input instances x,

the number of steps $T(x)$ to output with accuracy ϵ satisfies,
$$T(x, \epsilon) \leq k(\dim(x) + \log \mu(x) + \log(1/\epsilon))^c.$$
Here $\dim(x)$ is its vector length and $\mu(x)$ is a number representing the condition of x. "Bit size" has been replaced with a more intrinsic input word size.

Renegar's algorithm is polynomial time in this sense.

John Dunagan, Dan Spielman and Shang-Hua Teng [DSTT02] showed that on average, $\log C(A, b, c) = O(\log m)$, with respect to the standard normal distribution.

This implies that the expected number of iterates of Renegar's algorithm is
$$O(\sqrt{m}(\log m/\epsilon)),$$
and the expected number of arithmetic operations,
$$O(m^3 \log m/\epsilon).$$
Thus in expectation, condition has been eliminated as a parameter.

§8. Condition numbers and complexity. The results above illustrate a two-part scheme for complexity of numerical analysis proposed by Smale [Sma97]:
1. Estimate the running time $T(x, \epsilon)$ of an algorithm as a function of $(\dim(x), \log \mu(x), \log 1/\epsilon)$.
2. Estimate $\text{Prob}\{x \mid \mu(x) \geq t\}$, assuming a given probability distribution on the input space.

Taken together, the two parts give a probability bound on the expected running time of the algorithm, eliminating the condition μ.

So, how to estimate the (tail) probability that the condition is large?

Suppose (by normalizing) that all problem instances live within the unit sphere. Suppose Σ is the space of ill-posed instances and that condition is inversely proportional to "distance" to ill-posedness. Then the ratio of the volume of a thin tube about Σ to the volume of the unit sphere provides an estimate that the condition is large. To calculate these volumes, techniques from integral geometry [San04] and geometric measure theory [Fed69] are often used as well as volume of tube formulas of Hermann Weyl [Wey39].

This approach to estimating statistical properties of condition, was pioneered by Smale [Sma81]. Mike Shub and I used these techniques to get log linear estimates for the average loss of precision in evaluating rational functions [BS86]. It is the approach employed today to get complexity estimates in numerical analysis, see e.g., [BCL08].

Many have observed that average analysis of algorithms may not necessarily reflect their typical behavior. In 2001, Spielman and Tang introduced the concept of *smoothed analysis* which, interpolating between worst and average case, suggests a more realistic scheme [ST01].

The idea is to first smooth the complexity measure locally. That is, rather than focus on running time at a problem instance, compute the average running time over all slightly perturbed instances. Then globally compute the worst case over all the local "smoothed" expectations.

More formally, assuming a normal distribution on perturbations, *smoothed running time* (or *cost*) is defined as

$$T_s(n, \epsilon) = \sup_{\bar{x} \in \mathbb{R}^n} \mathbf{E}_{x \sim \mathcal{N}(\bar{x}, \sigma^2)} T(x, \epsilon).$$

Here $\mathcal{N}(\bar{x}, \sigma^2)$ designates the distribution of $\bar{x}$ with variance σ^2, and $x \sim \mathcal{N}(\bar{x}, \sigma^2)$ means x is chosen according to this distribution.

If $\sigma = 0$, smoothed cost reduces to worst case cost; if $\sigma = \infty$ then we get the average cost.

Part 2 of Smale's scheme is now replaced by:

2*. Estimate $\sup_{\bar{x} \in \mathbb{R}^n} \text{Prob}_{x \sim \mathcal{N}(\bar{x}, \sigma^2)} \{x \mid \mu(x) \geq t\}$.

Now, 1 and 2* combine to give a *smoothed complexity analysis* eliminating μ. Estimating 2* employs techniques described above, now intersecting tubes about Σ with discs about $\bar{x}$ to get local probability estimates.

Dunagan, Spielman and Teng [DSTT02] also give a smoothed analysis of Renegar's condition number. Assuming $\sigma \leq 1/m$), the smoothed value of $\log C(A, b, c)$ is $O(\log m/\sigma)$. This in turn yields smoothed complexity analyses of Renegar's linear programming algorithm. For the number of iterates:

$$\sup_{\|(\bar{A}, \bar{b}, \bar{c})\| \leq 1} \mathbf{E}_{(A,b,c) \sim \mathcal{N}((\bar{A}, \bar{b}, \bar{c}), \sigma^2 I)} \sharp((A, b, c), \epsilon) = O(\sqrt{m}(\log m/\sigma\epsilon)).$$

And for the smoothed arithmetic cost:

$$T_s(m, \epsilon) = O(m^3 \log m/\sigma\epsilon).$$

§9. What does all this have to do with the classical *P* vs. *NP* challenge? This is (essentially) the question asked by Dick Karp in the Preface to our book, *Complexity and Real Computation* [BCSS98].

As noted in Section 5, the problem of deciding the solvability of finite polynomial systems, HN, is a universal *NP*-Complete problem. Deciding quadratic (degree 2) polynomial systems is also universal [BSS89]. If solvable in polynomial time over the complex numbers $\mathbb{C}$, then any classical *NP* problem is decidable in probabilistic polynomial time in the bit model [CSS94]. While premise and conclusion seem unlikely, understanding the complexity of $HN_\mathbb{C}$ is an important problem in its own right. Much progress has been made here, with condition playing an essential role.

During the 1990s, in a series of papers dubbed "Bezout I–V," Shub and Smale showed that the problem of finding *approximate zeros* to "square" complex polynomial systems can be solved probabilistically in polynomial time on average over $\mathbb{C}$ [SS94, Sma97].

The notion of an approximate zero means that Newton's method converges quadratically, immediately, to an actual zero. Achieving output accuracy to within ϵ requires only $\log \log 1/\epsilon$ additional steps.

For a system $f = (f_1, \ldots, f_n)$ of n polynomials in n variables over $\mathbb{C}$, the Shub–Smale homotopy algorithm outputs an approximate solution in cN^5 arithmetic steps. Here N is the number of coefficients in the system and c is a universal constant. For quadratic systems, this implies that the number of arithmetic operations is bounded by a polynomial in n (since in this case, $N \leq n^3$).

The trouble is, the algorithm is non-uniform. The starting point depends on the degree $d = (d_1, \ldots, d_n)$ of the systems, and has a probability of failure.

In *Math Problems for the Next Century*, Smale [Sma00] posed the following question (his 17th problem):

> Can a zero of n complex polynomial equations in n unknowns be found approximately, on the average, in polynomial time with a uniform algorithm?

Building on the Shub–Smale homotopy algorithm, there has been exciting progress here. As with Shub–Smale, the newer path-following algorithms approximate a path, starting from a given pair (g, ξ) where $\xi = (\xi_0, \ldots, \xi_n)$ and $g(\xi) = 0$, and output (f, ξ^*) where ξ^* is an approximate zero of f.[17]

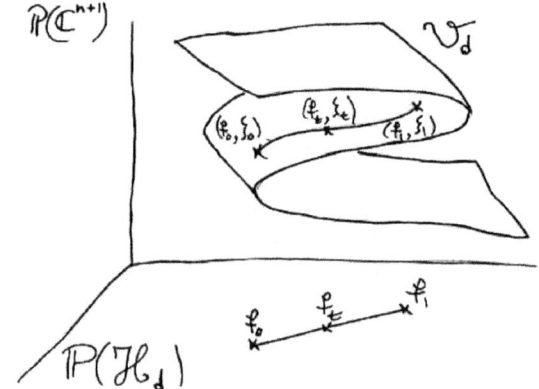

Sketch by Jean-Pierre Dedieu

The idea is to lift the "line" segment
$$f_t = (1-t)g + tf, \ t \in [0, 1]$$

---

[17]The polynomial systems considered are homogeneous (for good scaling properties) and the ambient spaces, projective (for compactness). Here g and f are systems of n homogeneous polynomials in $n + 1$ variables of degree $d = (d_1, \ldots, d_n)$.

to the variety $\mathcal{V}_d = \{(f, \xi) \mid f(\xi) = 0\}$. Note that $f_0 = g$, and that $f_1 = f$, the system to be solved.

By the *implicit function theorem*, this lift exits if the line does not intersect the *discriminant variety* of polynomial systems that have zeros with multiplicities. These algorithms approximate the lifting.

To steer clear of the *singular variety* of ill-posed pairs (i.e., pairs (f, ξ) where ξ is a multiple zero of f), they take into account the condition along the way. The condition will determine appropriate step size at each stage and hence running time.

Major considerations are: how to choose good initial pairs, how to construct good partitions (for approximating the path and steering clear of the singular variety), how to define measures of condition.

In two additional papers Bezout VI [Shu09] and Bezout VII [BS09], Shub and Carlos Beltrán present an Adaptive Linear Homotopy (ALH) algorithm with incremental time step dependent on the inverse of a normalized condition number squared.

Beltrán and Luis Miguel Pardo [BP11] show how to compute a random starting pair yielding a uniform Las Vegas algorithm, polynomial time on average. Utilizing the numerical algebraic geometry package Macaulay2, the randomized algorithm was implemented by Beltrán and Anton Leykin [BL12].

Bürgisser and Cucker give a hybrid deterministic algorithm which is almost polynomial time on average [BC11]. Let D be the maximum of the degrees, d_i. Then for $D \leq n$ the algorithm is essentially the ALH of Beltrán and Shub with initial pair:

$$g = (g_1, \ldots, g_n)$$

where

$$g_i(x_0, \ldots, x_n) = 1/\sqrt{2n}(x_0^{d_i} - x_i^{d_i}) \text{ and } \zeta = (1, \ldots, 1).$$

And for $D > n$, the algorithm calls on Renegar's symbolic algorithm [Ren89].

The algorithm takes $N^{(\log \log N)}$ arithmetic steps on average, coming close to answering Smale's question in the affirmative.

For a tutorial on the subject of this section, see [BS12].

§10. Postscript: Who invented the condition number? It is clear that Alan Turing was first to explicitly formalize a measure that would capture the informal notion of condition (of solving a linear system) and to call this measure a *condition number*. Formalizing what's "in the air" serves to illuminate essence and chart new direction. However, ideas in the air have many proprietors.

To find out more about the origins of the *spectral condition number*, I emailed a number of numerical analysts.

I also looked at many original papers. The responses I received, and related readings, uncover a debate concerning the origins of the (concept of) condition number not unlike the debate surrounding the origins of the general purpose computer—with Turing and von Neumann figuring central to both. (For an insightful assessment of the latter debate see Mike Shub's article, "Mysteries of Mathematics and Computation" [Shu94].)

Gauss himself [Gau03] is referenced for considering perturbations and preconditioning. Pete Stewart points to Helmut Wittmeyer [Wit36] in 1936 for some of the earliest perturbation bounds where products of norms appear explicitly. In 1949, John Todd [Tod50] explicitly focused on the notion of condition number, citing Turing's N and M condition numbers and the implicit von Neumann–Goldstine measure, which he called the P-condition number (P for Princeton).

Beresford Parlett tells me that "the notion was 'in the air' from the time of Turing and von Neumann et al.," that the concept was used by George Forsythe in a course he took from him at Stanford early in 1959 and that Wilkinson most surely "used the concept routinely in his lectures in Ann Arbor [summer, 1958]." The earliest explicit definition of the spectral condition number I could find in writing was in Alston Householder's 1958 SIAM article [Hou58] (where he cites Turing) and then in Wilkinson's book [Wil63], p. 91).

By far, the most informative and researched history can be found in Joe Grcar's 76 page article, "John von Neumann's Analysis of Gaussian Elimination and the Origins of Modern Numerical Analysis" [Grc11]. Here he uncovers a letter from von Neumann to Goldstine (dated January 11, 1947) that explicitly names the ratio of the extreme singular values as ℓ. Why this was not included in their paper [vNG47] or made explicit in their error bounds is a mystery to me.[18] Grcar chalks this up to von Neumann using his speeches (e.g., [vN89]) to expound on his ideas, particularly those given to drum up support for his computer project at the Institute for Advanced Study.[19]

Grcar's article definitely puts von Neumann at center stage. Of von Neumann's role as a key player in this area there is no doubt. However, Grcar also implies, that Turing's work on rounding error, and the condition number, was prompted by Turing's meeting with von Neumann in Princeton in January

---

[18] Joe was also kind enough to illuminate for me in detail how one could unravel von Neumann's and Goldstine's error analysis for the general case in their paper.

Many authors have cited the obtuseness and non-explicitness. For example, Edelman, in his PhD thesis [Ede89], recasts von Neumann's and Goldstine's ideas in modern notation given "the difficulty of extracting the various ideas from their work" and cites Wilkinson's referring to the paper's "indigestibility."

[19] An unpublished and undated paper by Goldstine and von Neumann [GvN63] containing material presented by von Neumann in various lectures going back to 1946, but clearly containing later perspectives as well, explicitly singles out (on page 14) ℓ as the "figure of merit." Also interesting to me, in the same paragraph, are the words "loss of precision" connected to the condition number, possibly for the first time.

1947. Indeed, prominently on page 630 he says, "No less than Alan Turing produced the first derivative work from the inversion paper."

This flies in the face of all we know about Alan Turing's singular individuality, both in personality and in research. In their personal remembrances of Turing, both Wilkinson [Wil71], who worked closely with him at the NPL in Teddington, and Max Newman [New55], earlier at Cambridge and Bletchley and later in Manchester, point to Turing's "strong predilection for working everything out from first principles, usually in the first instance without consulting any previous work on the subject, and no doubt it was this habit which gave his work that characteristically original flavor."

It also flies in the face of fact. As recounted by Wilkinson, Turing's experience with his team at the NPL, prior to meeting von Neumann in Princeton in 1947, was the stimulus for his paper:

> ... some time after my arrival [at the NPL in 1946], a system of 18 equations arrived in Mathematics Division and after talking around it for some time we finally decided to abandon theorizing and to solve it. ... The operation was manned by [Leslie] Fox, [Charles] Goodwin, Turing, and me, and we decided on Gaussian elimination with complete pivoting. Turing was not particularly enthusiastic, partly because he was not an experienced performer on a desk machine and partly because he was convinced that it would be a failure. ... the system was mildly ill-conditioned, the last equation had a coefficient of order 10^{-4} ... and the residuals were ... of order 10^{-10}, that is of the size corresponding to the exact solution rounded to ten decimals. ...
>
> ... I'm sure that this experience made quite an impression on him and set him thinking afresh on the problem of rounding errors in elimination processes. About a year later he produced his famous paper "Rounding-off errors in matrix process" ....[20]

Velvel Kahan (also a Turing Award recipient), in his 1966 paper [Kah66] and in a lengthy phone conversation (August 2011), asserts that von Neumann and Goldstine were misguided in their approach to matrix inversion (by computing A^{-1} from the formula $A^{-1} = (A^T A)^{-1} A^T$).

Kahan's assessment of Turing is a fitting conclusion to this paper:

> A more nearly modern error-analysis was provided by Turing (1948) in a paper whose last few paragraphs foreshadowed much of what was to come, but his paper lay unnoticed for several years until

---

[20]On June 23, 2012, at the Turing Centenary Conference in Cambridge, England, Andrew Hodges told me that in Turing's 1945 paper on the design of the ACE programmable computer (classified for many years) there is a discussion of the need for accurate numerical computations, and Gaussian Elimination is one of the examples given.

Wilkinson began to publish the papers which have since become a model of modern error-analysis.[21]

§11. **Acknowledgements.** Much of what I know about Alan Turing's life comes from Andrew Hodges' definitive biography, *Alan Turing: the Enigma* [Hod92]. I had the pleasure to discuss the material in this paper with Hodges at the Turing Centenary Conference, CiE 2012—*How the World Computes*, at the University of Cambridge, England. I thank the organizers of the conference, and in particular, Barry Cooper, for inviting me to speak.

Mike Shub and Steve Smale are the architects, *sans pareil*, of the most fundamental results in the area of complexity and real computation. My perspectives are clearly fashioned by my long time collaboration with them. Jim Renegar, who constantly amazes with his deep insights and beautiful algorithms, has been a constant sounding board for me.

I would like to thank the Mathematics Department of the University of Toronto for inviting me to be Dean's Distinguished Visitor during the Fields Institute Program on the Foundations of Computational Mathematics, Fall 2009. There I co-taught a course with my collaborator Felipe Cucker, and learned about new advances concerning condition from lectures given by Felipe and Peter Bürgisser whose influence is apparent here. Their forthcoming book, *Condition* [BC13], is certain to become the bible in this area. For an excellent survey paper see [Bur10].

Jean-Pierre Dedieu, who celebrated his 60th birthday during the Fields program, inspired us all with his beautiful lectures on *Complexity of Bezout's Theorem and the Condition Number*. We lost Jean-Pierre during the Centenary Year of Alan Turing. I dedicate this paper to him.

REFERENCES

[BMvN63] VALENTINE BARGMANN, DEANE MONTGOMERY, and JOHN VON NEUMANN, *Solution[s] of linear systems of high order*, **John von Neumann collected works** (A. H. Taub, editor), vol. 5, Macmillan, New York, 1963, report prepared for Navy Bureau of Ordnance, 1946, pp. 421–478.

[BL12] CARLOS BELTRÁN and ANTON LEYKIN, *Certified numerical homotopy tracking*, **Experimental Mathematics**, vol. 21 (2012), no. 1, pp. 69–83.

[BP11] CARLOS BELTRÁN and LUIS MIGUEL PARDO, *Fast linear homotopy to find approximate zeros of polynomial systems*, **Foundations of Computational Mathematics**, vol. 11 (2011), no. 1, pp. 95–129.

---

[21] Kahan is referring to *backward error analysis* which, rather than estimating the errors in a computed solution of a given problem instance (i.e., *forward error analysis*), estimates the closeness of a nearby problem instance whose exact solution is the same as the approximate computed solution of the original. Grcar [Grc11] also points to the von Neumann and Goldstine paper as a precursor to this notion as well.

[BS09] CARLOS BELTRÁN and MICHAEL SHUB, *Complexity of Bezout's theorem VII: Distance estimates in the condition metric*, **Foundations of Computational Mathematics**, vol. 9 (2009), no. 2, pp. 179–195.

[BS12] ———, *The complexity and geometry of numerically solving polynomial systems*, 2012, preprint submitted November 7, 2012, arXiv:1211.1528 [math.NA].

[Blu90] LENORE BLUM, *Lectures on a theory of computation and complexity over the reals (or an arbitrary ring)*, **1989 Lectures in the sciences of complexity II** (E. Jen, editor), Addison Wesley, 1990, pp. 1–47.

[Blu91] ———, *A theory of computation and complexity over the real numbers*, **Proceedings of the International Congress of Mathematicians (ICM 1990)** (I. Satake, editor), vol. 2, Springer-Verlag, 1991, pp. 1492–1507.

[Blu04] ———, *Computing over the reals: Where Turing meets Newton*, **Notices of the American Mathematical Society**, vol. 51 (2004), pp. 1024–1034, http://www.ams.org/notices/200409/fea-blum.pdf.

[Blu13] ———, *Alan Turing and the other theory of computation*, **Alan Turing: His work and impact** (S. Barry Cooper and Jan van Leeuwen, editors), Elsevier, 2013, pp. 377–384.

[BCSS96] LENORE BLUM, FELIPE CUCKER, MICHAEL SHUB, and STEPHEN SMALE, *Algebraic settings for the problem "$P \neq NP$?"*, **The mathematics of numerical analysis**, Lectures in Applied Mathematics, vol. 32, American Mathematical Society, 1996, pp. 125–144.

[BCSS98] ———, *Complexity and real computation*, Springer, New York, 1998.

[BS86] LENORE BLUM and MICHAEL SHUB, *Evaluating rational functions: Infinite precision is finite cost and tractable on average*, **SIAM Journal on Computing**, vol. 15 (1986), no. 2, pp. 384–398.

[BSS89] LENORE BLUM, MICHAEL SHUB, and STEPHEN SMALE, *On a theory of computation and complexity over the real numbers: NP-completeness, recursive functions and universal machines*, **Bulletin of the American Mathematical Society**, vol. 21 (1989), pp. 1–46.

[Bur10] PETER BÜRGISSER, *Smoothed analysis of condition numbers*, **Proceedings of the International Congress of Mathematicians (ICM 2010)**, vol. 4, World Scientific, 2010, pp. 2609–2633.

[BC11] PETER BÜRGISSER and FELIPE CUCKER, *On a problem posed by Steve Smale*, **Annals of Mathematics**, vol. 174 (2011), no. 3, pp. 1785–1836.

[BC13] ———, *Condition*, Grundlehren der mathematischen Wissenschaften, Springer-Verlag, 2013.

[BCL08] PETER BÜRGISSER, FELIPE CUCKER, and MARTIN LOTZ, *The probability that a slightly perturbed numerical analysis problem is difficult*, **Mathematics of Computation**, vol. 77 (2008), no. 263, pp. 1559–1583.

[Cuc02] FELIPE CUCKER, *Real computations with fake numbers*, **Journal of Complexity**, vol. 18 (2002), no. 1, pp. 104–134.

[CSS94] FELIPE CUCKER, MICHAEL SHUB, and STEPHEN SMALE, *Separation of complexity classes in Koiran's weak model*, **Theoretical Computer Science**, vol. 133 (1994), no. 1, pp. 3–14.

[Dan47] GEORGE B. DANTZIG, *Origins of the simplex method*, **A history of scientific computing** (S. G. Nash, editor), ACM Press, New York, 1990 (1947), pp. 141–151.

[Dem87] JAMES WELDON DEMMEL, *On condition numbers and the distance to the nearest ill-posed problem*, **Numerische Mathematik**, vol. 51 (1987), pp. 251–289, 10.1007/BF01400115.

[DSTT02] JOHN DUNAGAN, DANIEL A. SPIELMAN, and SHANG-HUA TENG, *Smoothed analysis of Renegar's condition number for linear programming*, **SIAM Conference on Optimization**, 2002.

[EY36] CARL ECKART and GALE YOUNG, *The approximation of one matrix by another of lower rank*, **Psychometrika**, vol. 1 (1936), no. 3, pp. 211–218.

[Ede88] ALAN EDELMAN, *Eigenvalues and condition numbers of random matrices*, **SIAM Journal on Matrix Analysis and Applications**, vol. 9 (1988), no. 4, pp. 543–560.

[Ede89] ———, *Eigenvalues and condition numbers of random matrices*, PhD Dissertation, MIT, 1989, http://math.mit.edu/~edelman/thesis/thesis.pdf.

[Fed69] HERBERT FEDERER, *Geometric measure theory*, Grundlehren der mathematischen Wissenschaften, Springer, 1969.

[FHW48] LESLIE FOX, HARRY D. HUSKEY, and JAMES HARDY WILKINSON, *Notes on the solution of algebraic linear simultaneous equations*, **Quarterly Journal of Mechanics and Applied Mathematics**, vol. 1 (1948), pp. 149–173.

[Gau03] CARL FRIEDRICH GAUSS, *Letter to Gerling, December 26, 1823*, **Werke**, vol. 9 (1903), pp. 278–281, English translation, by George E. Forsythe, **Mathematical tables and other aids to computation**, vol. 5 (1951) pp. 255–258.

[GvN63] HERMAN H. GOLDSTINE and JOHN VON NEUMANN, *On the principles of large scale computing machines* (unpublished), **John von Neumann collected works** (Abraham H. Taub, editor), vol. 5, Macmillan, New York, 1963, pp. 1–33.

[Grc11] JOSEPH F. GRCAR, *John von Neumann's analysis of Gaussian elimination and the origins of modern numerical analysis*, **SIAM Review**, vol. 53 (2011), pp. 607–682.

[Hig02] NICHOLAS J. HIGHAM, *Accuracy and stability of numerical algorithms*, second ed., Society for Industrial and Applied Mathematics, Philadelphia, PA, USA, 2002.

[Hod92] ANDREW HODGES, *Alan Turing: The enigma*, Vintage, 1992.

[Hot43] HAROLD HOTELLING, *Some new methods in matrix calculation*, **Annals of Mathematical Statistics**, vol. 14 (1943), no. 1, pp. 1–34.

[Hou58] ALSTON S. HOUSEHOLDER, *A class of methods for inverting matrices*, **Journal of the Society for Industrial and Applied Mathematics**, vol. 6 (1958), pp. 189–195.

[Kah66] WILLIAM M. KAHAN, *Numerical linear algebra*, **Canadian Mathematical Bulletin**, vol. 9 (1966), pp. 757–801.

[KM72] VICTOR KLEE and GEORGE J. MINTY, *How good is the simplex algorithm?*, **Inequalities III. Proceedings of the third symposium on inequalities** (New York) (O. Shisha, editor), Academic Press, 1972, pp. 159–175.

[Kos88] ERIC KOSTLAN, *Complexity theory of numerical linear algebra*, **Journal of Computational and Applied Mathematics**, vol. 22 (1988), no. 2–3, pp. 219–230.

[New55] MAXWELL H. A. NEWMAN, *Alan Mathison Turing. 1912–1954*, **Biographical Memoirs of Fellows of the Royal Society**, vol. 1 (1955), pp. 253–263.

[PER89] MARIAN. B. POUR-EL and J. IAN RICHARDS, *Computability in analysis and physics*, Perspectives in mathematical logic, Springer Verlag, 1989.

[Ren88a] JAMES RENEGAR, *A faster PSPACE algorithm for deciding the existential theory of the reals*, **Proceedings of the 29th annual symposium on Foundations of Computer Science**, IEEE Computer Society, 1988, pp. 291–295.

[Ren88b] ———, *A polynomial-time algorithm, based on Newton's method, for linear programming*, **Mathematical Programming**, vol. 40 (1988), pp. 59–93.

[Ren89] ———, *On the worst-case arithmetic complexity of approximating zeros of systems of polynomials*, **SIAM Journal on Computing**, vol. 18 (1989), no. 2, pp. 350–370.

[Ren95a] ———, *Incorporating condition measures into the complexity theory of linear programming*, **SIAM Journal on Optimization**, vol. 5 (1995), no. 3, pp. 506–524.

[Ren95b] ———, *Linear programming, complexity theory and elementary functional analysis*, **Mathematical Programming**, vol. 70 (1995), pp. 279–351.

[San04] LUIS ANTONIO SANTALÓ, *Integral geometry and geometric probability*, Cambridge Mathematical Library, Cambridge University Press, 2004.

[Shu94] MICHAEL SHUB, *Mysteries of mathematics and computation*, **The Mathematical Intelligencer**, vol. 16 (1994), pp. 10–15.

[Shu09] ———, *Complexity of Bezout's theorem VI: Geodesics in the condition (number) metric*, **Foundations of Computational Mathematics**, vol. 9 (2009), no. 2, pp. 171–178.

[SS94] MICHAEL SHUB and STEPHEN SMALE, *Complexity of Bezout's theorem V: Polynomial time*, **Theoretical Computer Science**, vol. 133 (1994), pp. 141–164.

[Sma81] STEPHEN SMALE, *The fundamental theorem of algebra and complexity theory*, **American Mathematical Society. Bulletin**, vol. 4 (1981), no. 1, pp. 1–36.

[Sma97] ——, *Complexity theory and numerical analysis*, **Acta Numerica**, vol. 6 (1997), pp. 523–551.

[Sma00] ——, *Mathematical problems for the next century*, **Mathematics: Frontiers and perspectives** (V. I. Arnold, M. Atiyah, P. Lax, and B. Mazur, editors), American Mathematical Society, 2000, pp. 271–294.

[ST01] DANIEL A. SPIELMAN and SHANG-HUA TENG, *Smoothed analysis of algorithms: Why the simplex algorithm usually takes polynomial time*, **Journal of the ACM**, (2001), pp. 296–305.

[Tar51] ALFRED TARSKI, *A decision method for elementary algebra and geometry*, University of California Press, 1951.

[Tod50] JOHN TODD, *The condition of a certain matrix*, **Proceedings of the Cambridge Philosophical Society**, vol. 46 (1950), pp. 116–118.

[Tod68] ——, *On condition numbers*, **Programmation en mathématiques numériques, Besançon, 1966**, vol. 7, Éditions du Centre National de la Recherche Scientifique, Paris, no. 165, 1968, pp. 141–159.

[TB97] LLOYD N. TREFETHEN and DAVID BAU, **Numerical linear algebra**, SIAM, 1997.

[Tur36] ALAN MATHISON TURING, *On computable numbers, with an application to the Entscheidungsproblem*, **Proceedings of the London Mathematical Society. Second Series**, vol. 42 (1936), no. 1, pp. 230–265.

[Tur45] ——, *Proposal for development in the Mathematics Division of an Automatic Computing Engine (ACE)*, **Report E.882, Executive Committee**, inst-NPL, 1945.

[Tur48] ——, *Rounding-off errors in matrix processes*, **Quarterly Journal of Mechanics and Applied Mathematics**, vol. 1 (1948), pp. 287–308.

[vN89] JOHN VON NEUMANN, *The principles of large-scale computing machines*, **Annals of the History of Computing**, vol. 10 (1989), no. 4, pp. 243–256, transcript of lecture delivered on May 15, 1946.

[vNG47] JOHN VON NEUMANN and HERMAN H. GOLDSTINE, *Numerical inverting of matrices of high order*, **Bulletin of the American Mathematical Society**, vol. 53 (1947), no. 11, pp. 1021–1099.

[Wei00] KLAUS WEIHRAUCH, **Computable analysis: An introduction**, Texts in Theoretical Computer Science, Springer, 2000.

[Wey39] HERMANN WEYL, *On the volume of tubes*, **American Journal of Mathematics**, vol. 61 (1939), pp. 461–472.

[Wil63] JAMES HARDY WILKINSON, **Rounding errors in algebraic processes**, Notes on Applied Science No. 32, Her Majesty's Stationery Office, London, 1963, also published by Prentice-Hall, Englewood Cliffs, NJ, USA and reprinted by Dover, New York, 1994.

[Wil71] ——, *Some comments from a numerical analyst*, **Journal of the ACM**, vol. 18 (1971), no. 2, pp. 137–147, 1970 Turing Lecture.

[Wit36] HELMUT WITTMEYER, *Einfluß der Änderung einer Matrix auf die Lösung des zugehörigen Gleichungssystems, sowie auf die charakteristischen Zahlen und die Eigenvektoren*, **Zeitschrift für Angewandte Mathematik und Mechanik**, vol. 16 (1936), pp. 287–300.

COMPUTER SCIENCE DEPARTMENT
CARNEGIE MELLON UNIVERSITY
PITTSBURGH, PA 15213, USA
E-mail: lblum@cs.cmu.edu

TURING IN QUANTUMLAND

HARRY BUHRMAN

Abstract. We revisit the notion of a *quantum Turing-machine*, whose design is based on the laws of quantum mechanics. It turns out that such a machine is not more powerful, in the sense of computability, than the machine originally constructed by Turing. Quantum Turing-machines do not violate the Church–Turing thesis. The benefit of quantum computing lies in *efficiency*. Quantum computers appear to be more efficient, in time, than classical Turing-machines, however its exact additional computational power is unclear, as this question ties in with deep open problems in complexity theory. We will sketch where BQP, the quantum analogue of the complexity class P, resides in the realm of complexity classes.

§1. **Introduction.** A decade before Turing developed his theory of computing, physicists struggled with the advent of quantum mechanics. During the famous 5th Solvay Conference in 1927 it was clear that a new era of physics had surfaced. Its strange features like superposition and entanglement still lead to heated discussions and much confusion. However strange and counterintuitive, the theory has never been refuted by experiments that are performed daily and in great numbers throughout laboratories around the world. Time after time the predictions of quantum mechanics are in full agreement with experiment.

Shortly after the advent of quantum mechanics, Church, Turing and Post developed the notion of computability [Chu36, Tur36, Pos36]. Less than 10 years later these formal ideas would be put to practice resulting in the ENIAC, the first general purpose machine. During that time Turing also specified an electromechanical machine that helped break Enigma-ciphers during the Second World War. It was not the first time that mechanical computing and cryptanalysis were developed side by side. Around 1820 Babbage worked on the blueprints of a mechanical computing device, which he called Difference Engine #1 and #2. The first prototype was never finished and funding was cut for the second prototype. Babbage was also a gifted cryptanalyst, in 1854 he broke the, until then unbreakable, Vigenère cypher [Sin00].

Partially supported by the EU 7th framework project SIQS.

Turing's Legacy: Developments from Turing's Ideas in Logic
Edited by Rod Downey
Lecture Notes in Logic, 42
© 2014, Association for Symbolic Logic

Today computers occupy an indispensable place in our every day life, leading to an everlasting demand on faster and bigger computing power. The mathematical notions developed in the the 1930s are now firmly rooted in our physical world. As Landauer [Lan61] put it: "computing is physical". Indeed, trying to satisfy Moores law [Moo65], which states that the number of transistors on integrated circuits doubles approximately every two years, the physical limitations of computing have become apparent. In particular quantum mechanical effects are starting to pop up. Incorporating quantum mechanics into our computational paradigm is therefore the next logical step.

In this chapter we will review quantum mechanics and show how this lead Deutsch [Deu85] to the definition of a quantum Turing-machine. The new field of quantum information processing gained a lot of momentum after Shor [Sho94] constructed an efficient quantum algorithm for the factorization problem. Since the security of most of modern public-key cryptography is based on our *inability* to factorize numbers efficiently, a quantum computer, once built, will severely compromise the security of these protocols. It is interesting to see that again a new computing paradigm is developed alongside advances in cryptanalysis. Quantum cryptography is the subfield that studies cryptography in a quantum world, with as cornerstone the quantum key distribution protocol by Bennet and Brassard [BB84]. The field has also developed new areas like quantum communication complexity [BCMdW10], quantum information theory [BS98], and quantum inspired proofs [DdW11]. We won't discuss these developments here, but work out the basic definitions and show how this new paradigm fits into the classical framework of complexity theory.

§2. Quantum mechanics. Quantum mechanics is the most complete description of nature to date, that governs the atomic and subatomic world. Some of its features are very counter intuitive, but have been corroborated by experiments time after time. The best way to highlight some of the properties of quantum mechanics is by means of an experiment, which we will discuss next.

2.1. An experiment with photons. In the first part of the experiment we have a light source that emits light that is polarized at an angle of $45°$. It is not necessary to understand in detail what polarized light exactly is. The only important issue at this point is that polarization is a *property* of light, which has a *direction*, that can be measured. The outcome of such a measurement can be expressed by an angle. In our experiment we shine light with a certain polarization through a calcite crystal. A calcite crystal is transparent and has the property of birefringence, a light beam polarized in a certain direction gets split into two beams (see Figure 1). The birefringence of calcite crystals is known for a long time and a recent article [RGF+12] argues that the Vikings used the calcite crystal, the mysterious "sunstone" according to the Norse sagas, to determine the position of the sun on cloudy days. Perhaps the first quantum computer avant la lettre!

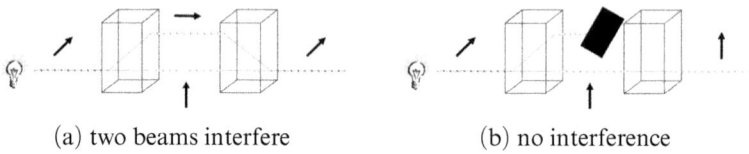

(a) two beams interfere   (b) no interference

FIGURE 1. Polarized light going through two calcite crystals.

The 45° polarized light is sent through the first calcite crystal and is split into two beams (Figure 1a). When measured right after exiting the first crystal, the light in the upper beam turns out to have horizontal polarization whereas the lower beam is vertical. The light in these two beams has only half the *intensity* of the original light coming directly out of the source. If we don't measure the polarization but let the two beams enter a second crystal, they merge and exit the second crystal as one beam, which has full intensity. The polarization after the second crystal is again 45°. This can be explained classically as an interference effect. By blocking the upper beam with a piece of black material (Figure 1b), interference is no longer possible and the light emitted by the second crystal is vertically polarized.

(a) single photons   (b) path A or B

FIGURE 2. Single photon experiment.

Let's now do the same experiment with a very very dim light source. We dim the light so much so that only *single* photons are produced each time we do the experiment. This second experiment is much harder than the first, but very good single photon sources have been accomplished [EFMP11]. Single photons still have a polarization and the photons that we use are again polarized at an angle of 45°.

The first observation (Figure 2a) to make is that in every run of the experiment the photons end up at two different spots, precisely where in the first experiment the two light beams were. Moreover, after doing the experiment many times, 50% of the time we find a photon at the top position and 50% of the time at the bottom. This, in itself is already a bit strange. What determines which path a photon takes? The ones coming from the source are all created in the same way, as much as possible.

When we measure the polarization of the photons that have gone through the calcite crystal it turns out that the top ones are always horizontally polarized and the bottom ones vertically. This is all in perfect agreement with the first experiment using a bright light source.

Next we place another calcite crystal and measure the photon when it exits the second one (Figure 2b). The photons always come out in the same place and when their polarization is measured it is again 45°. From all we observed so far it seems reasonable to assert that each photon either took the path labelled A in Figure 2b or path B. Let's examine this assertion by blocking

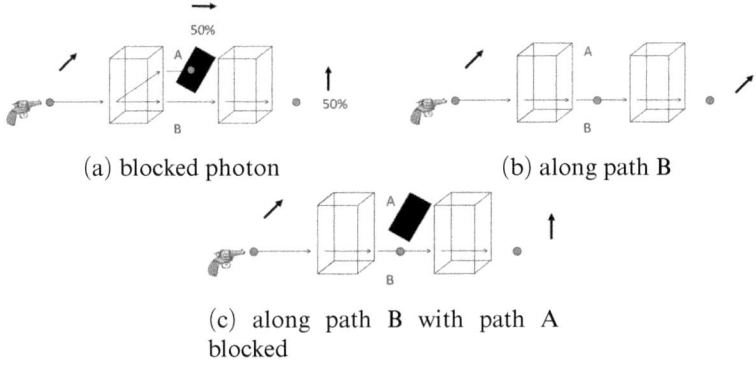

(a) blocked photon

(b) along path B

(c) along path B with path A blocked

FIGURE 3. Examining the paths.

path A (Figure 3a) and measuring the photons that end up there. We find that half of the time the photons travel along path A and their polarization is always horizontal, as it should be in order to be consistent with Figure 2a. Moreover whenever the photons don't end up at the blockage along path A, they end up behind the second crystal and have polarization vertical.

Our assertion says that the photons took either path A or path B. We see that half of the time the photons go along path A. Let's focus on the ones that take path B (Figure 3b). These photons exit the second crystal and have a polarization of 45° in total agreement with the situation in Figure 2b. But now we have a very strange situation when we consider these photons, going along path B, when we also block path A as in Figure 3c. In these cases the photon exits the second crystal with vertical polarization. It appears as if the photon that takes path B *knows* whether path A is blocked or not and changes polarization accordingly. The way out of this conundrum is given by quantum mechanics. The photon in Figure 2b does not travel along path A or B, it is in a *superposition* of going along A and B at the *same* time. Moreover it interferes with itself at the second calcite crystal. In the next Section we will formalize this behavior.

2.2. Qubits, superposition and measurement. Superposition is one of the main and counter intuitive ingredients of quantum mechanics. As we saw in the experiment above, a photon can be in a superposition of being in two different locations at the same time. But the a superposition principle also applies to larger systems. Famous is the example of Schroedinger's cat [Sch35] who is in a superposition of being dead and alive. In what follows we will apply these ideas to more familiar objects: bits. Classically a bit can be in any of two states: 0 or 1. Quantum mechanically a quantum bit or *qubit* may be in a superposition of both 0 and 1. It is useful to describe such systems as vectors in a finite dimensional Hilbert space, in this case a two dimensional one. We will identify the (basis) vector $\begin{pmatrix} 1 \\ 0 \end{pmatrix}$ with $|0\rangle$ to denote the classical bit 0 and (basis) vector $\begin{pmatrix} 0 \\ 1 \end{pmatrix}$ with $|1\rangle$ denoting the classical bit 1. This notation is called Dirac or bra-ket notation. Define for any vector $|a\rangle$ the complex conjugate transpose $\langle a|$, the expression $\langle a| \cdot |b\rangle$ also denoted as $\langle a \mid b \rangle$, then boils down to the inner product between $|a\rangle$ and $|b\rangle$. *Superposition* of two classical (basis) states/vectors, in this case the $|0\rangle$ and $|1\rangle$, is modelled as follows:

$$\alpha|0\rangle + \beta|1\rangle \qquad (1)$$

where α and β, called *amplitudes*, are complex numbers with the property that:

$$|\alpha|^2 + |\beta|^2 = 1. \qquad (2)$$

When one *observes* or *measures* a qubit $\alpha|0\rangle + \beta|1\rangle$ the outcome 0 is obtained with probability $|\alpha|^2$ and 1 with probability $|\beta|^2$. After the measurement has been performed and one of the outcomes has been observed, the state *collapses* to the outcome that has been observed. For example if outcome 0 was observed, the state has collapsed to the state $|0\rangle$. Note that Equation 2 guarantees that measurement of a qubit induces a probability distribution over the outcomes 0 and 1.

Let's try to plug in some values for α and β:

$$\frac{1}{\sqrt{2}}|0\rangle + \frac{1}{\sqrt{2}}|1\rangle. \qquad (3)$$

Measuring this qubit will yield with probability $\frac{1}{2}$ in observing a 0 and with probability $\frac{1}{2}$ a 1.

In general our system will consist of more than just one qubit. If we have two qubits $|x\rangle$ and $|y\rangle$ then $|x\rangle \otimes |y\rangle$ describes the two qubits together in a 4 dimensional Hilbert space. This construction is called the tensor or Kronecker product. If $|x\rangle = \alpha_0|0\rangle + \alpha_1|1\rangle$ and $|y\rangle = \beta_0|0\rangle + \beta_1|1\rangle$ then

$$|x\rangle \otimes |y\rangle = (\alpha_0|0\rangle + \alpha_1|1\rangle) \otimes (\beta_0|0\rangle + \beta_1|1\rangle)$$
$$= \alpha_0\beta_0|00\rangle + \alpha_0\beta_1|01\rangle + \alpha_1\beta_0|10\rangle + \alpha_1\beta_1|11\rangle.$$

By convention $|0\rangle \otimes |0\rangle$, $|0\rangle|0\rangle$, and $|00\rangle$ will denote the same state. The 4 dimensional Hilbert space obtained in this way has a natural basis, called the computational basis: $|00\rangle, |01\rangle, |10\rangle$, and $|11\rangle$, and any two qubit state can be expressed as a linear combination/superposition of these basis states. Any state on n qubits becomes:

$$\sum_{i \in \{0,1\}^n} \alpha_i |i\rangle \tag{4}$$

with the additional requirement that:

$$\sum_{i \in \{0,1\}^n} |\alpha_i|^2 = 1. \tag{5}$$

When measuring these n qubits we will observe i with probability $|\alpha_i|^2$.

In general not all the 2 qubit states that satisfy Equations 2 and 4 are obtained as the tensor of two single qubits. Such states are called *entangled*. For example the well known EPR-pair $\frac{1}{\sqrt{2}}(|00\rangle + |11\rangle)$ is entangled.

2.3. Partial measurement. In the previous Section we saw how to measure a quantum system

$$\sum_{x \in \{0,1\}^n} \alpha_x |x\rangle \tag{6}$$

in the computational basis. Such a measurement induces a probability distribution over the outcomes x

$$\Pr[\text{outcome} = x] = |\alpha_x|^2. \tag{7}$$

The state after measuring outcome x will collapse to $|x\rangle$.

For the definition of the quantum Turing-machine it will be important to also be able to *partially* measure a system. This is a measurement that only "looks" at part of the state. For example take the following two register state:

$$\sum_{x,y \in \{0,1\}^n} \alpha_{x,y} |x\rangle \otimes |y\rangle \tag{8}$$

on which we measure the first register in the computational basis. This will result in a probability distribution over outcomes of the first register:

$$P_x = \Pr[\text{outcome} = x] = \sum_y |\alpha_{x,y}|^2. \tag{9}$$

Again the state will collapse to x in the first register, but the second register stays in a superposition that is consistent with outcome x:

$$\frac{1}{\sqrt{P_x}} \sum_y \alpha_{x,y} |x\rangle \otimes |y\rangle. \tag{10}$$

The $1/\sqrt{P_x}$ factor renormalizes the partially collapsed state, so that it has norm 1.

2.4. Unitary operations. Next we need to model operations on qubits, such evolution of the system, according to quantum mechanics, is a *linear* map with the additional constraint that it preserves the probability interpretation, that is the squares of the amplitudes sum up to 1 (see Equations 2 and 5). Such norm-preserving linear transformations are called *unitary* and can be defined in mathematical terms:

$$UU^* = I \tag{11}$$

where U^* is the complex conjugate transpose of U and I is the identity matrix. In terms of computation the unitary constraint implies that the computation is *reversible*.

The following transformation on a single qubit is important and very useful. It is called the Hadamard transform.

$$H = \frac{1}{\sqrt{2}} \begin{pmatrix} 1 & 1 \\ 1 & -1 \end{pmatrix}. \tag{12}$$

It is a unitary operation since:

$$\frac{1}{\sqrt{2}} \begin{pmatrix} 1 & 1 \\ 1 & -1 \end{pmatrix} \times \frac{1}{\sqrt{2}} \begin{pmatrix} 1 & 1 \\ 1 & -1 \end{pmatrix} = \begin{pmatrix} 1 & 0 \\ 0 & 1 \end{pmatrix}.$$

Let's do a Hadamard operation on a qubit that is in the classical state $|0\rangle$:

$$\frac{1}{\sqrt{2}} \begin{pmatrix} 1 & 1 \\ 1 & -1 \end{pmatrix} \times \begin{pmatrix} 1 \\ 0 \end{pmatrix} = \frac{1}{\sqrt{2}} \begin{pmatrix} 1 \\ 1 \end{pmatrix}. \tag{13}$$

This state in ket notation, $\frac{1}{\sqrt{2}}|0\rangle + \frac{1}{\sqrt{2}}|1\rangle$, is the random qubit from Equation 3. Also the Hadamard transform models the behavior of the calcite crystal from Section 2.1, it puts the photon in an equal superposition of the two paths, and the probabilistic nature of where we found the photon (Figure 2a) is now explained by the measurement axiom!

When we apply the Hadamard transform again on this qubit:

$$\frac{1}{\sqrt{2}} \begin{pmatrix} 1 & 1 \\ 1 & -1 \end{pmatrix} \times \frac{1}{\sqrt{2}} \begin{pmatrix} 1 \\ 1 \end{pmatrix} = \begin{pmatrix} \frac{1}{2} + \frac{1}{2} \\ \frac{1}{2} - \frac{1}{2} \end{pmatrix} = \begin{pmatrix} 1 \\ 0 \end{pmatrix}. \tag{14}$$

We get the $|0\rangle$ again. Note the minus sign in the Hadamard transform. Its effect is illustrated in Equation 14. The minus sign caused the $\frac{1}{2} - \frac{1}{2}$ in the lower half of the vector to cancel out, or destructively interfere, while both terms in the upper half constructively interfered. It is both the superposition principle together with this interference behavior that gives quantum computing its power.

The tensor product is also defined on linear operations. In general if we have an $m \times n$ matrix A and an $n' \times m'$ matrix B then $A \otimes B$ is a $(m \cdot m') \times (n \cdot n')$ matrix defined as:

$$\begin{pmatrix} a_{1,1} \cdot B & a_{1,2} \cdot B & \cdots & a_{1,n} \cdot B \\ a_{2,1} \cdot B & a_{2,2} \cdot B & \cdots & a_{2,n} \cdot B \\ \vdots & \vdots & \ddots & \vdots \\ a_{m,1} \cdot B & a_{m,2} \cdot B & \cdots & a_{m,n} \cdot B \end{pmatrix}.$$

For example applying the Hadamard transform to n qubits ($H^{\otimes n}$) in the state $|0\rangle$ will generate a uniform superposition of all the basis states.

$$\overbrace{H|0\rangle \otimes \ldots \otimes H|0\rangle}^{n} = H^{\otimes n}|0^n\rangle = \frac{1}{\sqrt{2^n}} \sum_{y \in \{0,1\}^n} |y\rangle. \tag{15}$$

§3. **Quantum Turing-Machines.** In order to appreciate the definition of a quantum Turing-machine we will first remind the reader of the definition of a classical Turing-machine, see for example [AB09]. This is a device that has a two-way infinite tape of cells, which are ordered and labelled by an integer, and a read/write head that can move left and right ($\{L, R\} = D$) on the tape and write a symbol from a fixed finite alphabet Σ in the cell that it is currently reading. There is also a finite control that governs the behavior of the Turing-machine. This is modelled by a finite set of states Q, a dedicated starting state q_0, and a final halting state q_f. The transition function δ describes a single computation step.

$$\delta : Q \times \Sigma \to Q \times \Sigma \times D. \tag{16}$$

When the TM is in state $q \in Q$ and reading symbol $\sigma \in \Sigma$ at the current location of the head, $\delta(q, \sigma) = (q', \sigma', d)$ indicates to write σ' in the current tape cell, move the tape-head one cell in direction $d \in D$, and go to state q'. The computation starts with the input $x \in \Sigma^n$ written in the first n cells, the tape-head is at the cell with label 0 and the TM is in state q_0. All the cells, except for the first n have a special symbol[1] # written in them indicating that this cell has no content. During each step of the computation, the transition function δ is applied as described above. The computation halts when it enters the final state q_f, and the non-empty contents of the tape indicate the result of the computation.

A quantum Turing-machine [Deu85] is a generalization of the classical one, extended with quantum mechanical properties. As before we describe a QTM

---

[1] We assume that # is an element of Σ.

by means of its transition function

$$\delta: Q \times \Sigma \to \mathbb{C}^{Q \times \Sigma \times D}. \tag{17}$$

The way to interpret this is as follows. In the classical case δ changed the configuration (q, a, m) of a TM, where a is a description of the tape contents, q is the state of the TM, and m is the place where the head is located, to an new configuration (q', a', m'). In the quantum case δ defines a similar operation, but it now maps a state $|q, a, m\rangle$ to a superposition of states $\sum_{q',a',m'} \alpha_{q',a',m'} |q', a', m'\rangle$, where the a' are the tape contents that differ in at most one location from a, and $m' = m \pm 1$. Moreover the resulting state has to be norm one: $\sum_{q',a',m'} |\alpha_{q',a',m'}|^2 = 1$, which implies that δ is restricted to implement a unitary transformation on the configuration space of the QTM. Since δ specifies the behavior on basis states $|q, a, m\rangle$, by linearity it is also defined on super positions of basis states. One step of the QTM maps a superposition of basis configurations to a new superposition:

$$\sum_{q,a,m} \alpha_{q,a,m} |q, a, m\rangle \xrightarrow{\delta} \sum_{q,a,m} \alpha'_{q,a,m} |q, a, m\rangle. \tag{18}$$

Observe that at any point in time the sums in Equation 18 contain a *finite* number of terms with non-zero amplitude. The measurement axioms of quantum mechanics tell us what happens when we measure the QTM in configuration: $\sum_{q,a,m} \alpha_{q,a,m} |q, a, m\rangle$. We will observe the classical configuration $|q, a, m\rangle$ with probability $|\alpha_{q,a,m}|^2$. However after making this measurement, and observing, for example configuration $|q, a, m\rangle$, the original superposition has collapsed to the state $|q, a, m\rangle$.

As with classical TMs a QTM starts in the starting state q_0, the input $x \in \Sigma^n$ is written in the first n cells, and the head is at the cell with label 0, reading the first symbol of x. Each computation step is now applied according to Equation 17. With a classical Turing-machine it is clear when the computation is over, namely when the machine is in the final configuration q_f. How to define this with a QTM? Since the configuration of a QTM can be a superposition of states that are the final sate q_f and other states, this may not be well defined. Moreover how can one determine when a QTM has halted, since observing a machine that has not yet halted may collapse its superposition and thus disturb the computation. One way out is to add an additional register to the quantum state, that indicates whether the QTM has halted: $|q, a, f, m\rangle$, where $f = 1$ indicates that the QTM has halted and $f = 0$ means that it is still running. The transition function from Equation 17 needs to be extended to deal with this extra register:

$$\delta: Q \times F \times \Sigma \to \mathbb{C}^{Q \times F \times \Sigma \times D} \tag{19}$$

where $F = \{0, 1\}$. In terms of configuration space, an application of δ translates to:

$$\sum_{q,a,f,m} \alpha_{a,q,f,m} |q, a, f, m\rangle \xrightarrow{\delta} \sum_{q,a,f,m} \alpha'_{a,q,f,m} |q, a, f, m\rangle. \quad (20)$$

The computation starts in state q_0 and $f = 0$. After each computation step only register f is measured, leaving the remaining registers unmeasured. Such a partial measurement (see Section 2.3) can affect the total state, but when this flag register is not entangled with the other registers, such a measurement will not effect the remaining state. If this partial measurement yields $f = 0$ then δ is applied again to the state, over and over again, until $f = 1$ is measured in which case the computation has ended, and the remaining register is measured[2] in the computational basis and the output of the computation can be read off. Note that halting of such a computation is a probabilistic event. Just like for a classical TM it has to halt after a finite amount of steps, we require that the *expected* number of steps that the QTM takes is finite. Since the computation is probabilistic in nature we only demand that with probability $\frac{1}{2} + \epsilon$ the function value is computed, for $\epsilon > 0$.

A few more words about the description of the transition function δ are in place. In order to yield a finite description of δ it is necessary that the complex numbers in 19 come from a finite subset and have an efficient description. It turns out that this is not a problem since it can be proven that just a finite number of transformations are sufficient to approximate *any* unitary function arbitrary well by composing them using tensor products and products. This also forms the main ingredient for proving that there exists a universal quantum Turing-machine [Deu85, BV97]. For example the Hadamard transform (H) and the rotation over $\pi/4$ (R), and the phase flip (S) form a universal set of operations:

$$H = \frac{1}{\sqrt{2}} \begin{pmatrix} 1 & 1 \\ 1 & -1 \end{pmatrix} \quad R = \begin{pmatrix} 1 & 0 \\ 0 & e^{i\pi/4} \end{pmatrix} \quad S = \begin{pmatrix} 1 & 0 \\ 0 & i \end{pmatrix}. \quad (21)$$

As mentioned before the transition function δ corresponds to a unitary matrix. We can also see how for example the Hadamard matrix H corresponds to some transition function. For example it could be implemented as follows:

$$(q, 0, 0) \to \frac{1}{\sqrt{2}}(|q, 0, 0, L\rangle + |q, 0, 1, L\rangle) \quad (22)$$

$$(q, 0, 1) \to \frac{1}{\sqrt{2}}(|q, 0, 0, L\rangle - |q, 0, 1, L\rangle). \quad (23)$$

---

[2] It is not necessary to measure the remaining register. The output of a QTM could also be a quantum state, something that a classical TM simply can not output. Here we will only be concerned however with classical outputs.

The right-hand side of Equations 22 and 23 are a vector of complex numbers, in this example with entries $\frac{1}{\sqrt{2}}, -\frac{1}{\sqrt{2}}$, and 0, as is the right-hand side of Equation 19.

Another important point is that, due to the unitarity of the transition function, the computation of any QTM is *reversible*. Since the transition function δ is a unitary transformation U_δ, it follows that U_δ^*, its complex conjugate transpose, also models a transition function. Since $U_\delta U_\delta^* = I$, it follows that U_δ^* reverses a computation step. This seems to be a severe restriction since classical computations do not have to be reversible. However, it possible to transform any non-reversible computation into a reversible one with only a little overhead in time and space [Ben89].

Summarizing, we have the following definition.

DEFINITION 1. *A QTM is specified by its transition function*

$$\delta: Q \times F \times \Sigma \to \mathbb{C}^{Q \times F \times \Sigma \times D}$$

which represents a unitary transformation on the configuration space. One step of the machine is the application of δ to each of the basis states, followed by a partial measurement of the F register. A QTM computes a function f if the expected number of computation steps is finite for every input x and it outputs $f(x)$ with probability greather than $\frac{1}{2} + \epsilon$, for some $\epsilon > 0$.

§4. The Church–Turing thesis.
We will now examine the computational power of a QTM and prove that it does not exceed that of ordinary TMs.

THEOREM 2. [Deu85] *The class of languages accepted by QTMs is equal to the computable sets.*

PROOF. Since every TM can be transformed in one that is reversible it is clear that the class of languages accepted by QTMs contains the computable sets. We only need to show that it is not larger. We do this by simulating a QTM by means of a classical algorithm. Given a QTM M and its δ function, we need to establish what it outputs with probability $\frac{1}{2} + \epsilon$ conditioned on halting. We have already remarked that after t steps the number of terms with non-zero amplitude in 20 is finite. This is because in each step the number of non-zero configurations, can go in a super-position of at most a finite number of new ones, specified by δ. The idea now is for t to compute the configuration vector $|v_t\rangle$ after t steps.

$$|v_t\rangle = \sum_{a,q,f,m} \alpha_{a,q,f,m} |a, q, f, m\rangle. \qquad (24)$$

This involves doing the partial measurements also $t - 1$ times, and collapsing $|v_{t'}\rangle$ ($t' < t$) to the state corresponding to $f = 0$ as outcome of the partial measurement. This way we can compute the probability of halting after exactly t steps, which we call p_t. We know that the expected running time of M is finite:

$$\sum_t p_t \cdot t \leq c. \tag{25}$$

There could be runs of the machine that never halt, but they occur with vanishing probability. In particular, using Markov's inequality, there is a time $t_0 \leq c/\epsilon$ such that the probability that the machine runs for more than t_0 steps is bounded by ϵ, which implies that

$$\sum_{t \leq t_0} p_t \geq 1 - \epsilon. \tag{26}$$

Since we can compute p_t we can find t_0. Next we compute the probability of accepting conditioned on halting within t_0 steps. We accept the input if and only if this conditional accepting probability is $> \frac{1}{2}$. ⊣

It is easy to see that the above argument also works for functions instead of languages. Theorem 2 shows that the Church–Turing thesis, which states that effective computability is captured by Turing-machines, is not violated when we introduce quantum mechanical Turing-machines.

4.1. Efficient Church–Turing thesis. Although one can not compute more functions on a quantum computer, the proof sketched above shows that an exponential overhead in time is used in order to *simulate* a QTM by a classical one. Therefore Deutsch [Deu85] asked the question of whether it is possible that a quantum computer is more *efficient* than its classical counterpart. This question was related to Feynman [Fey82], who asked whether physics could be efficiently simulated on a Turing-machine. This touches upon an extension of the Church–Turing thesis that does not only ask about simulation of an algorithmic process by a Turing-machine but also requires that the simulation is efficient, with only polynomial overhead in time and space [vEB90].

Deutsch gave some indication that indeed QTMs can be more efficient, by exhibiting a quantum algorithm that determines for any $f: \{0,1\}^n \to \{0,1\}$, what the value of $f(x) \oplus f(y)$ is, querying f only one time. In particular Deutsch argues, if $f(x)$ and $f(y)$ are computed on a classical TM and each computation costs a day, then it would cost two days to compute $f(x) \oplus f(y)$, but on a quantum computer it would only cost one day. Deutsch's Definition of a quantum Turing-machine and his quantum algorithm mark the start of the field of quantum information processing. See also the excellent book [NC00]. These algorithms are best described in the black-box setting. See [BBC+01, BdW02] for precise definitions.

The mother of all quantum algorithms was generalized by Deutsch and Jozsa [DJ92] showing that there exists a problem that can be computed with a single query, but classically requires n queries when the solution has to be computed exactly. The drawback of this (super)exponential separation is that there does exist an efficient randomized algorithm that solves this problem also in a constant number of queries. Building upon this, Simon [Sim97] managed to construct a quantum algorithm, that truly establishes an exponential separation between quantum and randomized computations. All these separations are proven in the black-box or oracle setting, which essentially only counts queries to the input f.

A major advance was made by Shor [Sho94] who constructed, extending Simon's ideas, an efficient quantum algorithm to factorize numbers in their prime factors in time $O(n^3)$. Shor's quantum algorithm is special in that it is not a black-box algorithm, although at the heart of it resides a black-box procedure, called period-finding. The problem of factorization is well studied and it is believed that no fast (randomized) classical algorithms exists. The best known classical algorithm has expected running time $O(2^{n^{1/3} \log(n)^{2/3}})$.

The relevance of an efficient quantum algorithm for factorization is not just academic. The security of most public-key cryptography techniques, like RSA [RSA78], that are used frequently by numerous applications, rely on the absence of fast factorization algorithms. These cryptographic protocols become *insecure* when an efficient factorization method can be implemented. In particular they will be rendered useless when Shor's algorithm can be executed on a quantum computer that operates on a few thousand qubits.

After Shor's algorithm the field of quantum information processing got a tremendous boost and evolved into a flourishing community with active researchers from experimental and theoretical physics, computer science, and mathematics.

We need to mention one more important algorithm, which is due to Grover [Gro96]. This algorithm is able to search a marked item in an unordered list with n items, using only $O(\sqrt{n})$ look-ups or queries. Classically $\Omega(n)$ queries are necessary. Though not as impressive as Shor's speed-up, Grover's is only quadratic, search is a prominent primitive in many algorithms, and it is not surprising that variants of Grover's algorithm yield a quantum advantage in many computational settings.

Where does this leave us with respect to the extended Church–Turing thesis? In order to answer that question we need to have a better understanding of efficient (classical) computation, which is the realm of complexity theory.

4.2. Complexity theory. Complexity theory is the area of mathematics and theoretical computer science, that studies the question of the efficiency of optimal algorithms for a broad class of computational problems. Central is the class of problems that admit an efficient algorithm. An algorithm is efficient if its running time is upper-bounded by a polynomial in the size of the input.

DEFINITION 3. *A language or set $A \subseteq \Sigma^*$ is in the complexity class* P *if there is a polynomial $p(n)$ and a Turing-machine M, such that $M(x) = A(x)$ for all x and the running time of $M(x)$ is bounded by $p(|x|)$-many steps.*

Many problems that arise in practice are not known to be in P nor do we have a proof that no efficient algorithm exists. In order to study the complexity of these problems the complexity class NP was introduced by Cook and Levin [Coo71, Lev73].

DEFINITION 4. *A is in* NP *if there exists a* TM *M and a polynomial p such that*:

$$x \in A \Leftrightarrow \exists y \colon M(x, y) = 1$$

with $|y| \leq p(|x|)$ *and M runs in time* $p(|x| + |y|)$.

The name NP is somewhat esoteric, it is an abbreviation for Non-deterministic Polynomial-time. NP is the class of problems for which there is an efficient procedure to *verify* that a given solution to a problem instance is correct. Such correct solutions are also called witnesses. An example of an NP problem is the satisfiability problem SAT. Given a formula ϕ in conjunctive normal-form on n variables, $\phi \in$ SAT iff there exists an assignment $\alpha = \alpha_1 \ldots \alpha_n$ to the n variables such that $\phi(\alpha)$ evaluates to true. It is easy to see that SAT is in NP, since it is easy to check whether an assignment α satisfies ϕ, but the best known algorithm requires 2^n time steps, which tries all 2^n possible assignments.

Cook and Levin showed that SAT and a handful of other computational problems are in NP and in fact *characterize* NP in the sense that each of them is in P if and only if all of NP is in P. Such problems are called NP-complete. The handful of problems has grown to a long list of problems that come from many areas of science and operations research [GJ79], with new ones being added every year. The question of whether P = NP is one of the central open problems in mathematics and computer science. It is also one of the seven millennium prize problems.[3]

In order to study the power of efficient QTMs we need the quantum analogue of P, but before we do that we first define randomized polynomial-time.

DEFINITION 5. *A is in* BPP *if there exists a* TM *M*, $\epsilon > 0$, *and polynomial p such that*

- $x \in A \Rightarrow \Pr[M(x, y) = 1] > \frac{1}{2} + \epsilon$,
- $x \notin A \Rightarrow \Pr[M(x, y) = 0] > \frac{1}{2} + \epsilon$

where the probability is taken uniformly over the $y \in \{0, 1\}^{p(|x|)}$ *and the running time of M is bounded by* $p(|x| + |y|)$.

The complexity class BPP could be a first contender for violating the extend Church–Turing thesis. The best known deterministic simulation of a BPP computation is to loop over all the $y \in \{0, 1\}^{p(|x|)}$, simulate $M(x, y)$ and accept iff the majority of these y lead to an accepting computation.[4] There is

---

[3]See: http://www.claymath.org/millennium/.
[4]In fact a $\frac{1}{2} + \epsilon$ fraction of the y will give the correct answer.

evidence, starting from the beautiful work of Nisan and Wigderson [NW94] that efficient simulations exist, that only need to examine a polynomial number of *pseudo random* strings y. Under a fairly natural complexity-theoretical assumptions about the computational hardness of functions in exponential time, BPP can be simulated in polynomial time [AB09]. Indeed computational problems for which initially an efficient randomized algorithm was discovered get later on *derandomized*. The polynomial-time algorithm for primality testing [AKS04] being a prime example of this.

We next define, the quantum variant of BPP.

DEFINITION 6. *A is in* BQP *if there exists a* QTM $M, \epsilon > 0$, *and a polynomial* $p(n)$ *such that*:

- $x \in A \Rightarrow \Pr[M(x) = 1] > \frac{1}{2} + \epsilon$,
- $x \notin A \Rightarrow \Pr[M(x) = 0] > \frac{1}{2} + \epsilon$

and the running time of $M(x)$ is bounded by $p(|x|)$.

Using the same idea as in the proof of Theorem 2 one can show that if the expected running time of $M(x)$ is bounded by $p(|x|)$ then there exists another machine whose *total* running time is bounded by $p'(|x|)$ for some other polynomial p'. That is why we dropped the expectation in the running time in Definitions 5 and 6.

It is not hard to see that BPP $\subseteq$ BQP. Applying the Hadamard operation, like in Equation 2.4, to each of the qubits of the state $|0^{p(|x|)}\rangle$ gives a uniform superposition over all the strings $y \in \{0, 1\}^{p(|x|)}$, next running $M(x, y)$ in superposition and measuring the state:

$$\frac{1}{\sqrt{2^{p(|x|)}}} \sum_{y \in \{0,1\}^{p(|x|)}} |y\rangle \otimes |M(x, y)\rangle \qquad (27)$$

will give the same probability distribution of accepting and rejecting computations as the original BPP-machine.

Unlike for BPP, there is no evidence that BQP can be efficiently simulated. If this is not the case (BQP $\nsubseteq$ P) then the extended Church–Turing thesis is false. It turns out that it is quite hard to establish this because of the following theorem.

THEOREM 7. [ADH97] BQP $\subseteq$ PP.

PP is defined as follows.

DEFINITION 8. *A is in* PP *if there exists a* TM M, *polynomial p, such that*:

$$x \in A \Leftrightarrow \Pr[M(x, y) = 1] > \frac{1}{2}$$

where the probability is taken uniformly over the $y \in \{0, 1\}^{p(|x|)}$ and M runs in time $p(|x|)$.

PP is the class of sets recognized by probabilistic Turing-machines with success probability larger than $\frac{1}{2}$. Compare this to BPP which has correctness probability $\frac{1}{2} + \epsilon$. The derandomization techniques that are amenable for BPP do not work for PP, and in fact it is believed that PP is a much more powerful class that contains NP. The complexity class PP itself is included in PSPACE, the class of languages accepted by Turing-machines that use at most a polynomial amount of space.

DEFINITION 9. *A is in* PSPACE *if there exists a* TM *M and polynomial p such that M accepts A and for every input* x, $M(x)$ *writes in at most* $p(|x|)$ *many different tape cells.*

We now have the following inclusions:

$$P \subseteq BPP \subseteq BQP \subseteq PP \subseteq PSPACE \qquad (28)$$

and

$$P \subseteq NP \subseteq PP. \qquad (29)$$

From Equation 28 we see that if $P = PSPACE$ it follows that $BQP = P$. Hence showing that $P \neq BQP$ entails separating P from PP and PSPACE, which would be a major advance in complexity theory.

4.3. NP and BQP. What is the precise power of BQP in this landscape of complexity classes? In particular does BQP contain NP? The answer to this question is one of the main open problems in quantum complexity theory.

First we need some more notation. The polynomial-time hierarchy is defined as the complexity classes one gets by giving NP access to an oracle in NP. A Turing-machine equipped with an extra oracle tape, is called an oracle Turing-machine. It can write a string q, called the query, on this extra tape and enter the special query state. Then in one step it enters the YES or NO state, depending on whether q was in the oracle set O. It is as if it had a subroutine that computes membership in O for free.

DEFINITION 10. $\Sigma_1^p = NP$ *and* $\Sigma_k^p = NP^{\Sigma_{k-1}^p}$. $PH = \bigcup_k \Sigma_k^p$.

It is not hard to prove that $P = NP$ iff $P = PH$, and if $\Sigma_k^p = \Sigma_{k-1}^p$ then $PH = \Sigma_{k-1}^p$. It is believed that the PH is infinite.

The most compelling evidence for $P \neq BQP$ is the fact that factoring[5] is in BQP. But factoring is in $NP \cap co\text{-}NP$. It is therefore unlikely to be NP-complete since this would imply a collapse of the PH to $NP \cap co\text{-}NP \subseteq NP$.

Before we go on let's investigate this question for BPP, because here the situation is much clearer. Not only is there reasonable evidence that $P = BPP$, it can also be proven that if $NP = BPP$ then the polynomial-time hierarchy

---

[5]Strictly speaking factoring is a function and not a set. One can however define the set factoring $= \{\langle x, i, b \rangle \mid i^{\text{th}}$ bit of the prime factoriszation of x is $b\}$, which has the same complexity as factoring and is in BQP.

collapses to its second level. This follows from results of Adleman [Adl78] (BPP in P/poly) and Karp and Lipton [KL80]. We give a different proof below.

The following theorem, lists a few facts about BPP and its relation to PH and NP.

THEOREM 11. *The following are a few results about* BPP.
1. BPP $\subseteq \Sigma_2^p$ [Sip83],
2. BPP^{BPP} = BPP, *and*
3. NP$^{\text{BPP}} \subseteq$ BPP^{NP} [Zac86].

The following is a well known fact about BPP.

THEOREM 12. *if* NP $\subseteq$ BPP *then* PH = Σ_2^p.

PROOF. Assume that NP $\subseteq$ BPP. This implies that NP$^{\text{NP}} \subseteq$ NP$^{\text{BPP}} \subseteq$ BPP^{NP}. The first inclusion follows from the assumption and the second from Theorem 11 item 3. Using the assumption again we have that BPP$^{\text{NP}} \subseteq$ BPP^{BPP} = BPP. The equality is due to item 2. This way we have shown that Σ_2^p = NP$^{\text{NP}} \subseteq$ BPP. Employing the same chain of idea's again we get, using item 1 that $\Sigma_3^p \subseteq$ BPP $\subseteq \Sigma_2^p$ which implies PH = Σ_2^p. ⊣

It is an interesting open problem to prove a similar consequence from the assumption NP $\subseteq$ BQP. This would follow if one could show the same properties for BQP as in Theorem 11. Unfortunately only item 2 is known to hold: BQP$^{\text{BQP}} \subseteq$ BQP. With respect to item 1, the question of whether this is false with repect to an oracle is even a challenging open question. See the papers [Aar10, Aar11, FU10] for a possible approach to showing that relative to an oracle BQP $\not\subseteq$ PH.

§5. Conclusions and open problems. We have shown how the laws of quantum mechanics can be incorporated in the computing paradigm of Turing. We have seen how the resulting quantum Turing-machine makes use of superposition and interference. These quantum mechanical add-ons do not enrich the computing power, quantum Turing-machines do not go beyond the computable sets, and the Church–Turing thesis is not violated.

The quantum model does appear to be more efficient for certain computational problems, and a general simulation by classical means seems to require exponentially more time. Notably there is an efficient quantum algorithm for the factorization of numbers into their prime factors. No such efficient algorithms are known to exist for the classical setting. This fundamental result is the driving force behind the field of quantum information processing.

In order to understand the actual power of quantum computers we defined the class BQP and studied its relation to classical complexity classes like P, BPP, NP, PH, PP, and PSPACE. If P $\neq$ BQP then the extended Church–Turing thesis, which asks for an efficient simulation on classical

Turing-machines, is violated. However establishing such a feat will imply the separation of P from PP and PSPACE, which would be a major breakthrough.

Last we have tried to establish the relationship between BQP and NP. In particular we would like to see evidence that BQP can not efficiently simulate NP, showing the computational limits of quantum Turing-machines. Such evidence is currently missing. We showed how for BPP the picture is much clearer, NP $\subseteq$ BPP impplies PH collapses, and tried to adapt this reasoning to BQP. Specific open problems are the following:

- Is P $\neq$ BQP?
- Show that BQP $\subseteq$ PH or construct an oracle for which this is false.
- Show that $NP^{BQP} \subseteq BQP^{NP}$ or construct an oracle for which this is false.
- Does NP $\subseteq$ BQP imply that PH collapses?

REFERENCES

[Aar10] SCOTT AARONSON, *BQP and the polynomial hierarchy*, **STOC**, 2010, pp. 141–150.

[Aar11] ———, *A counterexample to the generalized linial-nisan conjecture*, CoRR abs/1110.6126, 2011.

[Adl78] LEONARD M. ADLEMAN, *Two theorems on random polynomial time*, **FOCS**, 1978, pp. 75–83.

[ADH97] LEONARD M. ADLEMAN, JONATHAN DEMARRAIS, and MING-DEH A. HUANG, *Quantum computability*, **SIAM Journal on Computing**, vol. 26 (1997), no. 5, pp. 1524–1540.

[AKS04] MANINDRA AGRAWAL, NEERAJ KAYAL, and NITIN SAXENA, *PRIMES Is in P*, **Annals of Mathematics. Second Series**, vol. 160 (2004), no. 2, pp. 781–793.

[AB09] SANJEEV ARORA and BOAZ BARAK, *Computational complexity—a modern approach*, Cambridge University Press, 2009.

[BBC+01] ROBERT BEALS, HARRY BUHRMAN, RICHARD CLEVE, MICHELE MOSCA, and RONALD DE WOLF, *Quantum lower bounds by polynomials*, **Journal of the ACM**, vol. 48 (2001), no. 4, pp. 778–797.

[BB84] C. H. BENNETT and G. BRASSARD, *Quantum cryptography: Public key distribution and coin tossing*, **Proceedings of IEEE international conference on Computers, Systems, and Signal Processing**, IEEE, 1984, pp. 175–179.

[Ben89] CHARLES H. BENNETT, *Time/space trade-offs for reversible computation*, **SIAM Journal on Computing**, vol. 18 (1989), no. 4, pp. 766–776.

[BS98] CHARLES H. BENNETT and PETER W. SHOR, *Quantum information theory*, **IEEE Transactions on Information Theory**, vol. 44 (1998), no. 6, pp. 2724–2742.

[BV97] ETHAN BERNSTEIN and UMESH V. VAZIRANI, *Quantum complexity theory*, **SIAM Journal on Computing**, vol. 26 (1997), no. 5, pp. 1411–1473.

[BCMdW10] HARRY BUHRMAN, RICHARD CLEVE, SERGE MASSAR, and RONALD DE WOLF, *Nonlocality and communication complexity*, **Reviews of Modern Physics**, vol. 82 (2010), no. 1, pp. 665–698.

[BdW02] HARRY BUHRMAN and RONALD DE WOLF, *Complexity measures and decision tree complexity: a survey*, **Theoretical Computer Science**, vol. 288 (2002), no. 1, pp. 21–43.

[Chu36] ALONZO CHURCH, *An unsolvable problem of elementary number theory*, **American Journal of Mathematics**, (1936).

[Coo71] STEPHEN A. COOK, *The complexity of theorem-proving procedures*, **STOC**, 1971, pp. 151–158.

[Deu85] DAVID DEUTSCH, *Quantum theory, the Church–Turing principle and the universal quantum computer*, **Proceedings of the Royal Society of London. Series A**, (1985).

[DJ92] DAVID DEUTSCH and RICHARD JOZSA, *Rapid solution of problems by quantum computation*, **Proceedings of the Royal Society of London. Series A**, vol. 493 (1992), no. 1907, pp. 553–558.

[DdW11] ANDREW DRUCKER and RONALD DE WOLF, *Quantum proofs for classical theorems*, **Theory of Computing**, vol. 2 (2011), pp. 1–54.

[EFMP11] M. D. EISAMAN, J. FAN, A. MIGDALL, and S. V. POLYAKOV, *Invited review article: Single-photon sources and detectors*, **Review of Scientific Instruments**, vol. 82 (2011), no. 7, p. 25.

[FU10] BILL FEFFERMAN and CHRIS UMANS, *Pseudorandom generators and the BQP vs. PH problem*, manuscript, 2010.

[Fey82] RICHARD FEYNMAN, *Simulating physics with computers*, **International Journal of Theoretical Physics**, vol. 21 (1982), no. 6–7, pp. 467–488.

[GJ79] M. R. GAREY and D. S. JOHNSON, *Computers and intractability : A guide to the theory of NP-completeness*, W. H. Freeman, New York, 1979.

[Gro96] L. K. GROVER, *A fast quantum mechanical algorithm for database search*, **Proceedings of 28th ACM STOC**, 1996, quant-ph/9605043, pp. 212–219.

[KL80] RICHARD M. KARP and RICHARD J. LIPTON, *Some connections between nonuniform and uniform complexity classes*, **STOC**, 1980, pp. 302–309.

[Lan61] ROLF LANDAUER, *Irreversibility and heat generation in the computing process*, **IBM Journal of Research and Development**, vol. 5 (1961), pp. 183–191.

[Lev73] LEONID LEVIN, *Universal search problems*, **Problems of Information Transmission**, vol. 9 (1973), no. 3, pp. 265–266.

[Moo65] GORDON MOORE, *Cramming more components onto integrated circuits*, **Electronics**, vol. 38 (1965), no. 8.

[NC00] MICHAEL A. NIELSEN and ISAAC L. CHUANG, **Quantum computation and quantum information**, Cambridge University Press, 2000.

[NW94] NOAM NISAN and AVI WIGDERSON, *Hardness vs randomness*, **Journal of Computer and System Sciences**, vol. 49 (1994), no. 2, pp. 149–167.

[Pos36] EMIL POST, *Finite combinatory processes-formulation 1*, **The Journal of Symbolic Logic**, vol. 1 (1936), no. 3, pp. 103–105.

[RSA78] R. L. RIVEST, A. SHAMIR, and L. ADLEMAN, *A method for obtaining digital signatures and public-key cryptosystems*, **Communications of the ACM**, vol. 21 (1978), no. 2, pp. 120–126.

[RGF+12] GUY ROPARS, GABRIEL GORRE, ALBERT LE FLOCH, JAY ENOCH, and VASUDEVAN LAKSHMINARAYANAN, *A depolarizer as a possible precise sunstone for viking navigation by polarized skylight*, **Proceedings of the Royal Society A**, vol. 468 (2012), no. 2139, pp. 671–684.

[Sch35] E. SCHRÖDINGER, *Die gegenwärtige Situation in der Quantenmechanik*, **Naturwissenschaften**, vol. 23 (1935), no. 48, pp. 807–812.

[Sho94] P. W. SHOR, *Algorithms for quantum computation: Discrete logarithms and factoring*, **Proceedings of the 35th annual symposium on the Foundations of Computer Science** (Los Alamitos, CA), IEEE, 1994, pp. 124–134.

[Sim97] DANIEL R. SIMON, *On the power of quantum computation*, **SIAM Journal on Computing**, vol. 26 (1997), no. 5, pp. 1474–1483.

[Sin00] SIMON SINGH, **The code book. The science of secrecy from ancient Egypt to quantum cryptography**, HarperCollins Publishers, 2000.

[Sip83] MICHAEL SIPSER, *A complexity theoretic approach to randomness*, **STOC**, 1983, pp. 330–335.

[Tur36] ALAN TURING, *On computable numbers, with an application to the Entscheidungsproblem*, **Proceedings of the London Mathematical Society**, vol. 2 (1936), no. 42, pp. 230–265, addendum 1937.

[vEB90] PETER VAN EMDE BOAS, *Machine models and simulation*, **Handbook of theoretical computer science, Volume A: Algorithms and complexity** (*A*), 1990, pp. 1–66.

[Zac86] STATHIS ZACHOS, *Probabilistic quantifiers, adversaries, and complexity classes: An overview*, **Structure in complexity theory conference**, 1986, pp. 383–400.

UNIVERSITY OF AMSTERDAM
and
  CENTRUM WISKUNDE & INFORMATICA
    SCIENCE PARK 123
    1098 XG AMSTERDAM
    THE NETHERLANDS
    E-mail: buhrman@cwi.nl

COMPUTABILITY THEORY, ALGORITHMIC RANDOMNESS AND TURING'S ANTICIPATION

ROD DOWNEY

Abstract. This article looks at the applications of Turing's legacy in computation, particularly to the theory of algorithmic randomness, where classical mathematical concepts such as measure could be made computational. It also traces Turing's *anticipation* of this theory in an early manuscript.

§1. Introduction. Beginning with the work of Church, Kleene, Post and particularly Turing, especially in the magic year of 1936, we know what computation means. Turing's theory has substantially developed under the names of *recursion theory* and *computability theory*. Turing's work can be seen as perhaps the high point in the confluence of ideas in 1936. This paper, and Turing's 1939 paper [141] (based on his PhD Thesis of the same name), laid solid foundations to the pure theory of computation. This article gives a brief history of some of the main lines of investigation in computability theory, a major part of Turing's legacy.

Computability theory and its tools for classifying computational tasks have seen applications in many areas such as analysis, algebra, logic, computer science and the like. Such applications will be discussed in articles in this volume. The theory even has applications into what is thought of as proof theory in what is called reverse mathematics. Reverse mathematics attempts to calibrate the logical strength of theorems of mathematics according to calibrations of comprehension axioms in second order mathematics. Generally speaking most separations, that is, proofs that a theorem is true in one system but not another, are performed in normal "ω" models rather than nonstandard ones. Hence, generally such proofs are results in computability theory which yield metamathematical proof theortical corollaries. Discussing reverse

---

Research supported by the Marsden Fund of New Zealand. Some of the work in this paper was done whilst the author was a visiting fellow at the Isaac Newton Institute, Cambridge, UK, as part of the Alan Turing "Semantics and Syntax" programme, in 2012. Some of this work was presented at CiE 2012 in Becher [7] and Downey [50]. Many thanks to Veronica Becher, Carl Jockusch, Paul Schupp, Ted Slaman and Richard Shore for numerous corrections.

Turing's Legacy: Developments from Turing's Ideas in Logic
Edited by Rod Downey
Lecture Notes in Logic, 42
© 2014, Association for Symbolic Logic

mathematics would take us a bit far afield, so we chose not to include this development in the present volume. In the present article, we we will look at the pure theory of computation.

As we later see, computability theory turned out to be the needed mathematical basis for the formalization of the old concept of *randomness* of individual objects. The theory of what we call today *algorithmic randomness* was anticipated by Turing in a manuscript that remained unpublished until its inclusion in the Collected Works [143]. The present article reviews the development of the theory of algorithmic randomness as part of Turing's legacy.

Mathematics has developed many tools to utilize randomness in the development of algorithms and in combinatorial (and other) techniques. For instance, these include Markov Chain Monte Carlo and the Metropolis algorithms, methods central to modern science, the probabilistic method is central to combinatorics. Quantum physics suggests to us that randomness is essential to our understanding of the Universe. Computer science uses randomness in cryptography, fast algorithms and proof techniques.

But the key question we need to ask is "What is randomness?". There are some in the physics community that suggest that the universe can generate "true randomness" which seems a philosophical notion, and this article is *not* concerned with this notion. Here we will be interested in what is called *algorithmic randomness*, which is not a philosophical notion, but a collection of precise mathematical notions.

The underlying idea in this area is that randomness should equate to some kind of inability to describe/predict/compress the random object using algorithmic means. We will use Turing's clarification of the notion of an algorithm to make this precise. For example, if we were presented with a very long string bit by bit, if it was random, then there would seem no way we should be able to predict, algorithmically, what the $n + 1$-st bit would be even knowing the first n bits.

The reader should note that this approach abandons the notion of "absolute randomness" since randomness depends on the algorithmic strength of the (algorithmic) predictor. The more powerful the algorithmic device, the fewer strings or reals will be random. The last decade has seen some quite dramatic advances in our understanding of algorithmic randomness. In particular, we have seen significant clarification as to the mathematical relationship between algorithmic computational power of infinite random sources and level algorithmic randomness. Much of this material has been reported in the short surveys Downey [40, 50], Nies [107] and long surveys [41, 46] and long monographs Downey and Hirschfeldt [45] and Nies [106]. Also the book edited by Hector Zenil [152] has a lot of discussion of randomness of varying levels of technicality, many aimed at a general audience.

To give a definition of algorithmic randomness and to understand questions like: "When is one real more random than another?"; and "What can be said about the algorithmic power of a random real?" we need a theory of computation. Fortunately this is readily at hand. We know what computation means. The theory has substantially developed, under the names of recursion theory and computability theory. As mentioned earlier, in this book, there are articles on the confluence of ideas in 1936, and the development of the theory at its roots. There are also articles on generalized computation complexity theory and applications of computability theory to algebra and model theory, complexity theory and also to analysis. However, there is none about the pure classical computability theory, underlying such applications and extensions. Thus this article will begin with a brief history of some of the main lines of investigation in this part of Turing's legacy.

Having done this, we will return to applying the theory to understanding algorithmic randomness.

To our knowledge, whilst he did have the notion of a pseudo-random number generator, Turing himself thought that randomness was a physical phenomenon, and certainly recognized the noncomputable nature of generating random strings. For example, from Turing [142], we have the following quote.

> " An interesting variant on the idea of a digital computer is a "digital computer with a random element." These have instructions involving the throwing of a die or some equivalent electronic process; one such instruction might for instance be, "Throw the die and put the-resulting number into store 1000." Sometimes such a machine is described as having free will (though I would not use this phrase myself)."

John von Neumann (e.g., [147]) also recognized the noncomputable nature of generating randomness.

> "Any one who considers arithmetical methods of producing random digits is, of course, in a state of sin."

Arguably this idea well predated any notion of computation, but the germ of this can be seen in the following quotation of Joseph Bertrand [11] in 1889.

> "How dare we speak of the laws of chance?
> Is not chance the antithesis of all law?"

There has been a developing body of work seeking to understand not just the theory of randomness but how it arises in mathematics; and in physics, biology and the like.

For example, we have also seen an initiative (whose roots go back to work of Demuth [38]) towards using these ideas in the understanding of almost everywhere behaviour and differentiation in analysis (such as Brattka, Miller,

Nies [19]). Also halting probabilities are natural and turn up in places apparently far removed from such considerations. For instance, as we later discuss, they turned up naturally in the study of subshifts of finite type (Hochman and Meyerovitch [70], Simpson [128, 130]), fractals (Braverman and Yampolsky [20, 21]). We also know that algorithmic randomness gives insight into Ergodic theory such as Avigad [5], Bienvenu et al. [12] and Franklin et al. [56].

§2. Classical computability theory. There are already long books devoted to classical computability theory such as Soare [135], Odifreddi [110, 111], Rogers [117], Griffor [64], and the subject is still being developed. In this section we aim at giving a "once over lightly" with an overview of what we believe are some highlights. As discussed by Sieg, Nerode and Soare in this volume, as well as extensively analysed in Davis [36] and Herken [69], we have seen how Turing's work has led to the birth of computation, and indeed, the digital computer. What about the pure theory of computation after Turing?

The work of Turing [141] led to the notion of relative computation. We imagine a machine M with an oracle (read only memory) A which can be consulted during the computation. This give rise to the the fundamental operator called the *jump* operator: A' is the halting problem with oracle A. Then $\emptyset'$ is the usual halting problem, and $(\emptyset')' = \emptyset^{(2)}$ would be the halting problem given an oracle for the halting problem.

The use of oracles also gives us a basic calibration of the complexity of sets (languages) $A \leq_T B$ means that (the characteristic function of) A can be computed from a machine with oracle access to B. This pre-ordering $\leq_T$ is called *Turing reducibility* and the equivalence classes are called (Turing) degrees.

The jump operator is monotone in the sense that if $X \leq_T Y$ then $X' \leq_T Y'$. Due to the work of Kleene and Post [76], as we discuss below, we know that it is not one to one on the degrees. For example, there are sets $X \not\equiv_T \emptyset$ with $X' \equiv_T \emptyset'$. We call such sets *low*, since we think of them as having low information content because the jump operator cannot distinguish them from having no oracle at all. The spine of the degrees is provided by the jump operator: start with $\emptyset$ and give it degree **0**. Form the halting problem $\emptyset'$ and its degree $\mathbf{0}'$. Then $\emptyset^{(2)}$ has degree $\mathbf{0}^{(2)}$. Iterate the construction and obtain any finite ordinal jump. Using what are called effective ordinal notations we can extend this to the transfinite: $\mathbf{0}^\omega$ is the effective join of $\mathbf{0}^{(n)}$ for all finite n and then ever upwards. Namely, $\mathbf{0}^{(\omega+1)}$ would be the degree of jump of some representative of $\mathbf{0}^\omega$. To work with larger ordinals, what is done is to represent ordinals for α via "notations" which are partial computable functions specifying sequences of smaller ordinals converging to α in the case that α is a limit, and the predecessor of α if α is a successor. In some sense this is the very least one would imagine needed for giving computable representations of ordinals. Remarkably, it is enough, in that for such representations, any two representations for the same ordinal allow us to define $\mathbf{0}^{(\alpha)}$ up to Turing degree, a result of Spector.

Returning to our story, it is certainly the case that Turing's original paper [140] is written very clearly. The same *cannot* be said about much of the early work in the theory of computation, particularly that of Kleene and Church. Most of it was couched in terms of lambda calculus or recursive functions, and it all seemed forbiddingly formal.

A great paper, following the early era of the 30s, was due to Emil Post [112], who returned to Turing's clear informal style. Whilst Turing [141] did define the notion of an oracle computation, it is only in Post's article that the notion of Turing reducibility was defined, and Post focused attention on recursively (= computably) enumerable sets. Post also demonstrated the connection between arithmetical definability and the hierarchies of computability theory, establishing that the n-th jump of the empty set was Σ_n^0 complete, etc. That is, he showed that if A is Σ_n^0 then A is many-one reducible to $\emptyset^{(n)}$, where $X \leq_m B$ means that there is a computable function f with $n \in X$ iff $f(n) \in B$. Many-one reducibility was one of the many refinements of Turing reducibility noted by Post.

Post also suggested the study of the ordering structures generated by Turing reducibilities and by many other refinements of these reducibility. Myhill [104] showed that if X and Y are two versions of the halting problem (for different universal machines) then $X \equiv_m Y$. Post also noted other reducibilities such as truth table reducibility and variants such as bounded truth table, weak truth table, etc. These reducibilities are commonly found in algebra. Truth table reducibility can me thought of as a reduction procedure which must be total for all oracles. It is extensively used in algorithmic randomness as it allows for translations of effective measures. The article of Homer and Selman in this volume discuss how miniaturizations of these ideas gave rise to computational complexity theory. The time bounded version of m-reducibility is used extensively in complexity theory where it is called Karp reducibility.

We concentrate now in describing the work done on Turing reducibilities. The work on other reducibilities is also vast.

2.1. The global degrees. Until mid 1950, it was consistent with all known facts that the ordering of the degrees was a linear ordering of length 2^ω with the countable predecessor property consisting of only iterated jumps of $\emptyset$. However, Kleene and Post [76] showed that this was not the case by exhibiting a pair of degrees $\mathbf{a}, \mathbf{b}$ ($\leq \mathbf{0}'$) which were incomparable. (i.e., $\mathbf{a} \not\leq \mathbf{b}$ and $\mathbf{a} \not\geq \mathbf{b}$, which is written as $\mathbf{a}|_T\mathbf{b}$.) The method of proof introduced by Kleene and Post is a kind of primitive Cohen forcing. Thus, the degrees are a nontrivial upper semi-lattice with join induced by $A \oplus B = \{2n \mid n \in A\} \cup \{2n+1 \mid n \in B\}$. Soon after, Spector [136] proved that there was a minimal degree $\mathbf{a}$; that is

$\mathbf{a} > \mathbf{0}$ and for all $\mathbf{c}$, it is not the case that $\mathbf{a} > \mathbf{c} > \mathbf{0}$. This paper implicitly uses another forcing technique which uses perfect closed sets as its conditions. In the same paper Spector proved an "exact pair" theorem showing that all countable ideals could be named by pairs of degrees as the elements below both, and the proof of this influential result introduces forcing with infinite conditions. This exact pair method allows us to show that the degrees are not a lattice.

Each of these two papers had very significant impact on the field. People showed that the degrees were very complicated indeed. The Kleene–Post method enabled the proof of the existence of low sets. This was extended by Friedberg [57] who showed that the range of the jump operator is as big as it can be: the Friedberg Jump Theorem says that if $\mathbf{a} \geq \emptyset'$ there is a degree $\mathbf{c}$ with $\mathbf{c} \cup \mathbf{0}' = \mathbf{c}' = \mathbf{a}$. If $\mathbf{c}' = \mathbf{a}$, we say that $\mathbf{a}$ inverts to $\mathbf{c}$. Friedberg observed a similar result for degrees $\mathbf{d} > \mathbf{0}^{(n)}$. The set C of degree $\mathbf{c}$ that the proof constructs is called 1-generic, meaning that it is Cohen generic for 1 quantifier arithmetic. The inversion to 1-generic degrees is not the only kind. Cooper [33] demonstrated that every degree above $\mathbf{0}'$ can be inverted to a minimal degree. This result uses a combination of the kind of coding used by Firedberg and Spector's methods. These Friedberg–Kleene–Post methods can also be pushed to the transfinite, as proven by Macintyre [93], so that given any $X > \emptyset^{(\alpha)}$, there is a set Y with $Y^{(\alpha)} \oplus \emptyset^{(\alpha)} \equiv_T Y^{(\alpha)} \equiv_T X$, for $\alpha < \omega_1^{CK}$ (the computable ordinals). Applications of such n-generic sets occur in many places in computability theory and its applications in, for instance, effective algebra, and randomness.

Spector's Theorem on minimal degrees was extended by many authors including Lachlan, Lerman, Lachlan–Lebeuf proving results on initial segments showing that these can be, respectively, all countable distributive lattices, countable lattices, and countable upper-semilattices, (see Lerman [87]) culminating in theorems like every partial ordering of size $\aleph_1$ with the countable predecessor property is an initial segment of the Turing degrees (Abraham–Shore [1]). Later it was shown that questions about further extensions often have to do with set theory (Groszek–Slaman [65]). There are still many questions open here. These results imply that the theory of the degrees is undecidable. There has been work towards understanding the quantifier level where undecidabilty occurs. The Kleene–Post theorem and Spector's exact pair theorem also had numerous extensions, heading towards definability results in the degrees, as well as combinations to extensions of embeddings, embeddings with jumps etc. Some noteworthy results here include Slaman–Woodin's proof of the definability from parameters of countable relations in the degrees, this leading to the (parameter-free) definability of the jump operator in the partial ordering of the degrees by Shore and Slaman [126] (also see Slaman [133]). Still open here is the longstanding question of Rogers: are the Turing degrees rigid?

Related here are results on relativization. Early on it was noted that most results relativized in the sense that if they were true then relativizing everything kept them true. For example, there are sets $A, B < \emptyset'$ with $A|_T B$. This result relativized in that the proof shows that for any oracle X, there are sets $X <_T A^X |_T B^X < X'$. One question was whether "everything relativizes" and, as a consequence, the cones of degrees above each degree would all be isomorphic, or perhaps elementary equivalent? The answer turned out to be no. Beginning with work of Feiner [52] who demonstrated that there were nonisomorphic cones if you had the jump operator, and culminating with work of Shore [123] who showed non-isomorphism in the language of partial ordering, and Shore [124] who demonstrated non-elementary equivalence in the same language.

2.2. Post's Problem and the priority method. Post observed that much of the work of undecidability proofs was in coding halting sets. He called sets A which were domains of partial computable functions *recursively enumerable* and now they are either known by this name or by the name *computably enumerable*, as suggested by Soare, since it captures the intentional meaning, and their degrees similarly. Post asked a very interesting question: does there exist a computable enumerable degree **a** with $\mathbf{0} < \mathbf{a} < \mathbf{0'}$?

This problem became known as Post's Problem, and its solution was highly influential. Post's problem was solved by two students, Friedberg [58] and Muchnik [102]. The method took the Kleene–Post method and added backtracking to give rise to a method known as the *priority method*.

Here is a brief description of the method applied in the setting of an old unpublished result of Tennenbaum. We construct a computable ordering of type $\omega + \omega^*$ with no infinite computable ascending or descending subsequences. We will build the ordering by adding two points at a time. We think of the points in the ω-part as blue and the ones in the ω^* part as red. Thus, if there were nothing happening, we would start with a blue and a red point $x_0 y_0$. At the next stage, we would add a red and a blue point to get $x_0 x_1 y_1 y_0$, etc. Now we must meet certain *requirements*, namely R_e that W_e, the e-th computably enumerable set, is not an infinite ascending sequence and B_e that W_e is not an infinite descending sequence. Lets consider R_e. This is saying that W_e if infinite is not all red. The way to force this to happen would be as follows. Suppose that at some stage we see some point x_n occur in W_e at stage s in its enumeration. Then if we changed the colours at this stage so that x_n was put into the blue section, we would be done since W_e would not be all red. That is, if we had at stage s, $x_0 \ldots x_m y_m \ldots y_0$, we could *recolour* so that at the next stage we would have $x_0 \ldots x_{n-1} x_{s+1} y_{s+1} x_n \ldots x_s y_s \ldots y_0$, that is *moving* the place we build the sequences to between x_{n-1} and x_n. That is, R_e seeks to make red things blue, and in the same spirit, B_e seeks to make blue things red. Furthermore, we need to make sure that from some point on all elements have as stable colour so that the order type is $\omega + \omega^*$. To to this we give each requirement some kind of *priority*. Say $R_0 < B_0 < R_1 < B_1 \ldots$. This means that R_0 has the highest priority and is allowed to make red elements blue, and if it does this, that action is not allowed to be undone. B_0 is allowed to make blue elements red, and this action cannot be undone by any other requirement *except* R_0. If it is undone by R_0 then the next element it makes red (which R_0 does not care about, as it has a satisfying element) will not be made blue by anyone. Finally, to make the order type $\omega + \omega^*$, we also ask that R_e and B_e only are allowed to move elements x_i, y_j for $i, j > e$.

The finite injury method is a mainstay of the area. It has applications in descriptive set theory, complexity theory and most other parts of computable mathematics. One longstanding question here is Sacks' questions about a *degree invariant* solution to Post's Problem. Is there a computably enumerable operator W such that for all X, $X <_T W^X <_T X'$, and for all $X \equiv_T Y$, $W^X \equiv_T W^Y$? Lachlan [81] showed that the answer is no if an index for the reductions witnessing $W^X \equiv_T W^Y$ can be read off from indices for the reductions witnessing $X \equiv_T Y$, and Downey and Shore [49] showed that the solution W, if there is one, needs to be reasonably constrained, low$_2$ or high. Martin has conjectured a very strong negative answer which says more or less that the only degree invariant operators on the degrees are jumps and their iterates. Slaman and Steel [131] have the strongest results here, showing, for instance, that there is no order preserving solution.

Powerful generalizations of the finite injury method came from constructions where each requirement could act infinitely often, but subsequent requirements could guess the activity and take it into account. This gave rise to *infinite injury* methods. There is no fixed method and these arguments have many classifications according to "how complex" they are. One method of classification was suggested by Leo Harrington. He said that priority arguments should be classified according to how many iterations of the jump are needed to produce an oracle which could compute how the requirements are satisfied in the construction. Finite injury arguments typically require one jump, and the easiest infinite injury arguments require 2 jumps. However, there are arguments requiring arbitrary numbers of jumps in both the pure theory and in applications such as computable model theory. A significant technical obstacle for such arguments is simply to find a way to coherently present the argument.

The early incarnations of the infinite injury method enabled the proof that the computably enumerable degrees are dense as a partial ordering (Sacks [118]), and that the diamond lattice is embeddable preserving 0 in the computably enumerable degrees. (Lachlan [80], Yates [151]). Sacks also used the method to prove the c.e. jump theorem, namely that if $X \geq \emptyset'$ is c.e. relative to $\emptyset'$ then there is a c.e. set Y with $Y' \equiv_T X$. In the c.e. case of jump theorems it is clearly necessary that "targets" be c.e. relative to $\emptyset'$. Again these results were pushed a long way. All (necessarily countable) distributive and some, but not all, finite non-distributive lattices are embeddable into the computably enumerable degrees (See, for instance, Lachlan [82], Lerman [88], Lempp–Lerman [86], and Lachlan–Soare [83]). Also, many lattices can be embedded densely, such as all distributive lattices (Slaman [132]), and some nondistributive lattices (Ambos-Spies, Hirschfeldt and Shore [2]) but not all embeddable lattices (Downey [39], Weinstein [150]).

We cannot expect that these embeddings can also jump invert, but as partial orderings, we can do embeddings with any reasonable expectation about the jumps consistent with the ordering relationships (Shore [125], building on earlier work of, for example, Robinson [116]). For example, if $\mathbf{a} < \mathbf{b}$ are computably enumerable degrees with jumps $\mathbf{c}$ and $\mathbf{d}$ respectively, and $\mathbf{e} < \mathbf{f}$ are degrees computably enumerable in $\mathbf{0}'$ with $\mathbf{c} \leq \mathbf{e} \leq \mathbf{f} \leq \mathbf{d}$, then there exist c.e. degrees $\mathbf{g}, \mathbf{h}$ with $\mathbf{a} < \mathbf{g}, \mathbf{h} < \mathbf{b}$, and (for example) $\mathbf{g}|\mathbf{h}$ and $\mathbf{g}' = \mathbf{e}$ and $\mathbf{h}' = \mathbf{f}$. Also many results were proven about the structure of the computably enumerable degrees, they are not a lattice, they have an undecidable first order theory, they have algebraic decompositions, etc. There are themes relating definability to enumerations. For example, any arithmetically definable class of computably enumerable degrees closed under double jump is definable in the c.e. degrees. (Nies, Shore and Slaman [108]) There were also a number of results relating the lattice of computably enumerable sets and the upper semi-lattice of computably enumerable degrees. For example, it was shown that maximal sets (that is co-atoms in the quotient structure of the computably enumerable sets modulo finite sets-a notion from Post's paper) always have high degrees (meaning $A' \equiv_T \emptyset''$) and include all high degrees; and form an orbit in the automorphism group of the lattice of computably enumerable sets. (Martin [95] and Soare [134], respectively.) Harrington and Soare [68] used the infinite injury method and great ingenuity to show that there is a definable property of the lattice of computably enumerable sets which solves Post's problem. Cholak and Harrington [29] proved a nice definability result for double jump classes in the c.e. degrees. Namely, suppose that $\mathcal{C} = \{\mathbf{a}\colon \mathbf{a}$ is the Turing degree of a Σ_3 set greater than $\mathbf{0}''\}$. Let $\mathcal{D} \subseteq \mathcal{C}$ such that $\mathcal{D}$ is upward closed. Then there is an non-elementary $(\mathcal{L}_{\omega_1,\omega})$ $\mathcal{L}(A)$ property $\varphi_\mathcal{D}(A)$ such that $D'' \in \mathcal{D}$ iff there is an A where $A \equiv D$ and $\varphi_\mathcal{D}(A)$. Double jumps are necessary since Rachel Epstein [51] recently showed that there is a c.e. degree $\mathbf{a}$ which is non-low and each of its members are automorphic to low sets. Cholak, Downey and Harrington [28] recently showed that determining orbits in the automorphism group of the lattice of computably enumerable sets is Σ_1^1 complete.

We refer the reader to Soare [135] for a somewhat dated but well-written account of results up to 1987.

More complex versions of the infinite injury method allowed for very complex results in involving n-th jumps, partial orderings and embeddings such as Lerman–Lempp [85], and things about arithmetical definability such as Harrington [67] (See Odifreddi [111] for this). These methods have been applied by Ash and Knight in effective algebra [3], and model theory (e.g., Marker [94]). The infinite injury method has also been applied in complexity theory such as Downey, Flum, Grohe and Weyer [42]. Modern computability theory could not exist without the infinite injury method.

2.3. Approximation techniques and Π_1^0 classes. A recurrent theme in computability theory is to use computable approximations to complex objects. This can have several forms.

For instance, the Limit Lemma of Shoenfield says that $A \leq_T \emptyset'$ iff there is a computable function $f(\cdot, \cdot)$ such that $\lim_s f(x, s)$ exists for all x and $x \in A$ iff $\lim_s f(x, s) = 1$. That is, A is computable from the halting problem iff A has a computable approximation which changes its mind only finitely often on each argument.

Another important example of approximation comes in the form of what are called the hyperimmune-free or computably dominated degrees. Such a degree **a** can be noncomputable but is defined to have the property that for all $f \leq_T \mathbf{a}$, there is a computable g with $f(x) \leq g(x)$ for all x. That is, we can compute a finite number of instances $\{0, \ldots, g(x)\}$ as the possible values of $f(x)$. The non-computably dominated degrees sort of resemble the ones below **0**′ to some extent, and the class has deep connections with algorithmic randomness. If A has non-computably dominated degree then there is a $f \leq_T A$, which "escapes" any computable function, meaning that if we run a construction with oracle f, then for any computable g, we know that there exist infinitely many n with $f(n) > g(n)$. Thus we run some kind of construction and construct g to measure when we need to perform some action. Then we will argue that $g(n)$ gives the relevant information that $f(n)$ encodes.

If something fails to be approximable then this fact of "escaping" can often be used in constructions. One illustration is the fact that high degrees compute functions that dominate every computable function. This allows us to show that, for instance, every high computably enumerable degree bounds a minimal pair of computably enumerable degrees. (Cooper [34].) Another such example concerns the non-low$_2$ (i.e., $\mathbf{a}'' > \mathbf{0}''$) degrees. Following work of Martin we know that A is non-low$_2$ iff A computes a function f which infinitely often escapes any $\emptyset'$ computable function. (That is, for all $g \leq_T \emptyset'$, $\exists^\infty n(f(n) > g(n))$.) This fact enables one to show, for instance, that any finite lattice can be embedded below such degrees preserving 0 and 1, and below and such degree we can find a 1-generic. With some more delicate techniques, such lon-low$_2$ methods can be adapted to the c.e. degrees, allowing lattice embeddings below such degrees, for instance (for example, Downey and Shore [48]). Work here is ongoing with new precise characterizations of what kinds of approximations allow us to embed, for example, the 5-element modular-nondiatributive lattice to be embedded below it, giving new definability results. (Downey and Greenberg [43])

Another recurrent approximation technique is the use of what are called Π_1^0 classes. (Computably bounded) Π_1^0 classes can be thought of as the collections of paths through an infinite computable binary tree. They occur often in effective mathematics. For example, if you give me a computable commutative ring then the ideals can be represented as a Π_1^0 class. The collection of complete extensions of Peano Arithmetic form a Π_1^0 class.

Many *basis* results can be proven for these useful classes. These assert that (perhaps under certain conditions) every nonempty Π_1^0 class has a member of a certain type. The classic result is the *Low Basis Theorem* of Jockusch and Soare [72] which asserts that every Π_1^0 class has a member of low degree (i.e., $A' \equiv_T \emptyset'$) and the Hyperimmune-free Basis Theorem which says it has one of computably dominated degree, and the Basis Theorem from the same paper that asserts that for every special Π_1^0 class $\mathcal{P}$ (i.e., with no computable members), and every $S \geq_T \emptyset'$, there is a a member $P \in \mathcal{P}$ with $P' \equiv_T S$.

The theory of Π_1^0 classes and algorithmic randomness interact very strongly. For example, the collection of Martin-Löf random reals (for a fixed constant c of randomness deficiency-as defined in the next section) forms a Π_1^0 class with no computable members, and which has positive measure. The basis theorem for special classes above therefore proves that there are random reals of low Turing degree and ones of every possible jump. Thus, tools from topology and from computability theory are simultaneously applicable. Π_1^0 classes and things like reverse mathematics are also intertwined, since Π_1^0 classes correspond to what is called *Weak König's Lemma*. For more we refer the reader to [24].

§3. Basics of algorithmic randomness.

3.1. Notation. We will refer to members of $\{0,1\}^* = 2^{<\omega}$ as *strings*, and infinite binary sequences (members of 2^ω, Cantor space) as *reals*. 2^ω is endowed with the tree topology, which has as basic clopen sets

$$[\sigma] := \{X \in 2^\omega : \sigma \prec X\},$$

where $\sigma \in 2^{<\omega}$. The *uniform* or *Lebesgue measure* on 2^ω is induced by giving each basic open set $[\sigma]$ measure $\mu([\sigma]) := 2^{-|\sigma|}$. This is simply the restatement that the uniform distribution has all strings of length n equally likely of probability 2^{-n}.

We identify an element X of 2^ω with the set $\{n: X(n) = 1\}$. The space 2^ω is measure-theoretically identical (via the usual mapping taking [0] to

$[0, \frac{1}{2})$ and $[1]$ to $[\frac{1}{2}, 1))$ with the real interval $[0, 1)$, although the two are not homeomorphic as topological spaces, so we can also think of elements of 2^ω as elements of $[0, 1]$. We will let $X \restriction n$ denote the first n bits of X.

3.2. von Mises. The theory of randomness of an individual sequence actually pre-dates the foundation of probability theory; and, arguably, one of the reasons for the latter was the unsatisfactory nature of the former until the 60s. The pioneer was Richard von Mises [146]. He said a random real should certainly obey the frequency laws like the law of large numbers for any reasonable sampling of the bits. Thus

$$\lim_{n\to\infty} \frac{|\{m \mid m < n \wedge X(m) = 1\}|}{n} = \frac{1}{2}.$$

This property is called *normality* and was studied by Borel and others. In fact, any random real clearly should be what is called *absolutely normal*, meaning it is normal to any basis (more on this later, when we return to Turing).

Inter alia, we mention that it is here that Turing later enters the picture. His interest was absolute normality and some of his ideas will anticipate those of the theory of algorithmic randomness as developed by Martin-Löf, Kolmogorov, Levin and others. We will return to this development in Section 6.

von Mises' idea was to consider any possible *selection* of a subsequence (i.e., of positions of the given real to sample) and ask that this selection be normal: Let $f: \omega \to \omega$ be an increasing injection, a selection function. Then a random X should satisfy the following.

$$\lim_{n\to\infty} \frac{|\{m \mid m \leq n \wedge X(f(m)) = 1\}|}{n} = \frac{1}{2}.$$

von Mises had no canonical choice for "acceptable selection rules". For example, if we take any real with infinitely many 1s, and make the selection the collection of places where the real is 1, then plainly the real fails to be random relative to that choice according to this criteria. Clearly that selection fails to realize the spirit of von Mises idea. What selection functions should be acceptable? Wald [148, 149] showed that for any *countable* collection of selection functions, there is a sequence that is random in the sense of von Mises. The problem is that von Mises work predated the work in the 30s of Church, Kleene, Post and Turing, culminating in the classic paper of Turing [140], clarifying the notion of computable function. Church [30] proposed restricting f to computable increasing functions. This incarnation of von Mises' idea gives rise to notions now called *computable stochasticity*, and, of we use partial computable selections, *partial computable stochasticity*.

Unfortunately, von Mises' approach, even with Church's reformulation using computability theory, was fatally injured (or at least seriously hurt) by the work of Ville [144]. In the following, $S(\alpha, n)$ is the number of 1s in the first n bits of α and similarly S_f for the selected places.

THEOREM 1 (Ville's Theorem [144]). *Let E be any countable collection of selection functions. Then there is a sequence $\alpha = \alpha_0 \alpha_1 \ldots$ such that the following hold.*

1. $\lim_n \frac{S(\alpha,n)}{n} = \frac{1}{2}$.
2. *For every $f \in E$ that selects infinitely many bits of α, we have* $\lim_n \frac{S_f(\alpha,n)}{n} = \frac{1}{2}$.
3. *For all n, we have* $\frac{S(\alpha,n)}{n} \leq \frac{1}{2}$.

The killer is item 3 which says that there are *never* situations with more 1s than 0s in the first n bits of α. Suppose you were betting on the outcomes of a sequence of coin tosses of a biased coin, where there are always fewer tails then heads. Certainly you could figure out a betting strategy to make a lot of money. This is the import of item 3.

Ville suggested adding a further statistical law, the law of the iterated logarithm, to von Mises' definition. However, we might well ask "How we can be sure that adding this law would be enough?". Why should we expect there not to be a further result like Ville's (which there is, see [45]) exhibiting a sequence that satisfies both the law of large numbers and the law of the iterated logarithm, yet clearly fails to have some other basic property that we would naturally associate with randomness?

We could add more and more statistical laws to our collection of desiderata for random sequences, but there is no reason to believe we would ever be done, and we certainly do not want a definition of randomness that changes with time, if we can avoid it.

3.3. Martin-Löf. One solution to this quandary was provided by the work of Per Martin-Löf [96], and as we later see somewhat anticipated by Turing. Martin-Löf's fundamental idea in [96] was to define an abstract notion of a performable statistical test for randomness, and require that a random sequence pass *all* such tests. He did so by effectivizing the notion of a set of measure 0. The way to think about Martin-Löf's definition below is that as we effectively shrink the measure of the open sets we regard as "tests", we are specifying reals satisfying them more and more.

In the below, a Σ_1^0 class is a computably enumerable collection $\{[\sigma] \mid \sigma \in W\}$ for some computably enumerable (c.e.) set W of strings. Alternatively think of this as a c.e. set of intervals in the interval $[0, 1]$.

DEFINITION 1 (Martin-Löf [96]).

1. *A* Martin-Löf test *is a sequence $\{U_n\}_{n \in \omega}$ of uniformly Σ_1^0 classes such that $\mu(U_n) \leq 2^{-n}$ for all n.*
2. *A class $C \subset 2^\omega$ is* Martin-Löf null *if there is a Martin-Löf test $\{U_n\}_{n \in \omega}$ such that $C \subseteq \bigcap_n U_n$.*
3. *A set $A \in 2^\omega$ is* Martin-Löf random *if $\{A\}$ is not Martin-Löf null.*

For example, think of the test that every second bit of the real is 1. It is okay for a random real to have this for a long time but at some stage it must abandon having every second bit 1. Thus we could specify this test by $U_1 = \{[01]\}$, $U_2 = \{[0001], [0101]\}$, etc. Even at this point we would like to make the reader aware of the calibrations of randomness possible. This test consists only of nested sequences of clopen sets. Thus any randomness notion defined by:

"X is random iff it passes all Martin-Löf tests but restricted to tests where each level is specified by a computable function giving a canonical index for a clopen set"

would be enough to pass this test and any "similar" tests. This "clopen" notion has a name and is called *Kurtz* or *weak* randomness. It is equivalent to saying X is in every Σ_1^0 class of measure 1.

3.4. Three approaches to algorithmic randomness. The modern viewpoint has three main paradigms for defining algorithmic randomness. Martin-Löf's above is called the *measure-theoretical paradigm*.

We briefly discuss the two other main paradigms in algorithmic randomness as they are crucial to our story. The first is the *computational paradigm*: random sequences are those whose initial segments are all hard to describe, or, equivalently, hard to compress.

We think of Turing machines U with input τ giving a string σ. We regard τ as a description of σ and the shortest such is regarded as the intrinsic information in σ. Kolmogorov [77] defined *plain U-Kolmogorov complexity* $C_U(\sigma)$ of σ as the *length* of the shortest τ with $U(\tau) = \sigma$. Turing machines can be enumerated $U_0, U_1, \ldots$ and hence we can remove the machine dependence by defining a new (universal) machine

$$U(0^e 1\tau) = U_e(\tau),$$

so that we can define for this machine M, $C(\sigma) = C_M(\sigma)$ and for all e, $C(\sigma) \leq C_{U_e}(\sigma) + e + 1$. We will use the notation $\leq^+$ to dispense with explicit mention of absolute additive constants in inequalities. For example, this inequaility would be written as $C(\sigma) \leq^+ C_{U_e}(\sigma)$.

A simple counting argument due to Kolmogorov [77] shows that as $C(\sigma) \leq^+ |\sigma|$ (using the identity machine), there must be strings of length n with $C(\sigma) \geq n$. We call such strings *C-random*. The intuition here is that the only way to describe σ would be to hardwire σ into the program. σ is *incompressible* and, in particular, has *no* regularities to allow for compression.

We would like to define a real, an infinite sequence, to be random by saying for all n, $C(\alpha \restriction n) \geq^+ n$. Unfortunately, there are no such random reals due to a phenomenon called complexity oscillations, which (in a quantitative way) say that in very long strings σ there must be segments with $C(\sigma \restriction n) < n$.

This oscillation really due to the fact that on input τ, we don't just get the *bits* of τ as information but the *length* of τ as well.

Specifically, imagine a sufficiently long string α. Now each initial segment σ of α has some shortest programme say σ^*. Now this program can be interpreted as a number $n^* = \sigma^*$. Consider τ the next segment of α after σ (i.e., $\sigma\tau \preceq \alpha$) *with τ having length n^**. Then the program that has input a string ρ and does the following. First it looks at its length and interprets this as a string. Taking that strings ν it computes $U(\nu)$ and if this halts outputs $U(\nu)\rho$.

Now assume we run this algorithm on τ. Then it computes $U(\sigma^*) = \sigma$ first, and then outputs $\sigma\tau$. This shows that $C(\sigma\tau) =^+ |\tau| =^+ C(\sigma)$. For long enough σ this is a compression. The key here is that we are using the *length* as well as the *bits* of τ. Thus we are losing the intentional meaning that the bits of τ are processed by U to produce σ. To get around this first Levin [89, 90] and later Chaitin [25] suggested using *prefix-free machines* to capture this intentional meaning that the *bits* of the input encode the information of the output.

One way is to use *prefix-free complexity* via machines whose domains are prefix-free sets of strings. That is, prefix free machines work like telephone numbers. If $U(\tau) \downarrow$ (i.e., halts) then for all $\hat{\tau} \neq \tau$ comparable with τ, $U(\hat{\tau}) \uparrow$.

Already we see a theme that there is not one but perhaps *many* notions of computational compressibility of relevance to understanding randomness. In the case of prefix-free complexity, in some sense we know we are on the correct track, due to the following theorem which can be interpreted as saying (for discrete spaces) that Occam's razor and Bayes' Theorem give the same result (in that the shortest description is essentially the probability that the string is output).

THEOREM 2 (Coding Theorem–Levin [89, 90], Chaitin [25]). *For all σ,*

$$K(\sigma) =^+ -\log(Q(\sigma))$$

where $Q(\sigma)$ is $\mu(\{\tau \mid U(\tau) = \sigma\})$, and of course, logs here are base 2.

Using this notion, and noticing that the universal machine above would be prefix-free if all the U_e were prefix free, we can define the prefix-free Kolmogorov complexity $K(\sigma)$.

DEFINITION 2 (Levin [90], Chaitin [25]). *A set A is 1-random if*

$$K(A \upharpoonright n) \geq^+ n.$$

Schnorr proved that we are on the right track here:

THEOREM 3 (Schnorr). *A real A is Martin-Löf random iff it is 1-random.*

It is not difficult to show that almost all reals are random, but Schnorr's Theorem give no explicit example. The oft-quoted example of a 1-random real is Chaitin's *halting probability* (for a universal prefix-free machine U):

$$\Omega = \sum_{\{\sigma \mid U(\sigma)\downarrow\}} 2^{-|\sigma|},$$

the measure of the domain of U (which is well-defined as the domain of U is a prefix free set of strings).

An easy proof of this fact is reminiscent of the fact that the halting problem is undecidable. We can use the Recursion Theorem to build part of the universal prefix-free machine U, via a prefix-free machine M with (known) coding constant e in U. Imagine we are monitoring $\Omega_s = \sum_{\{\sigma \mid U(\sigma)\downarrow\}}[s]$. Suppose that we see some $\sigma \preceq \Omega_s \upharpoonright s$ we see $K_s(\sigma) < |\sigma| - e - 1$. (That is, this segment does not look random.) This means that some ν of length $K_s(\sigma)$ has been enumerated into the domain of U describing σ. Then what we do is enumerate the same ν into the M_{s+1} describing σ also. Then, it follows that $\Omega_{s+1} \geq \Omega_s + 2^{-|\nu|}$, and in particular $\sigma \not\preceq \Omega_{s+1}$. The fact that Ω has a prefix-free domain means that M does too as we are simply recycling what U does.

The summary is "if the opponent says here's a short description of an initial segment of Ω_s, we act to show that it is not an initial segment after all."

It would seem that the definition of Ω is thoroughly machine dependent, but in the same spirit as Myhill's Theorem (showing that there is only one halting problem up to m-degree), we can define a reducibility on halting probabilities we call *Solovay reducibility*. $X \leq_S Y$ means there is a constant c and partial computable f such that for all rationals $q < Y$, $f(q) \downarrow < X$ and $c(Y - q) > (X - f(q))$. To wit, a good approximation for Y yields one for X. There is essentially one Ω: The approximation $\Omega = \lim_s \Omega_s$ is monotone from below, and Ω is what is called a *left c.e.-real*. Every left c.e. real is a halting probability in the same way that each c.e. set is a the domain of a Turing machine. Clearly $\leq_S$ is well-defined on left c.e. reals. The culmination of a series of papers is the Kučera–Slaman Theorem which states that there is really only one left-c.e. random real.

THEOREM 4 (Kučera–Slaman Theorem [79]). *A left c.e. real α is 1-random iff for all left c.e.-reals β, $\beta \leq_S \alpha$.*

3.5. Martingales. The final randomness paradigm is the one based on prediction. The *unpredictability paradigm* is that we should not be able to predict the next bit of a random sequence even if we know all preceding bits, in the same way that a coin toss is unpredictable even given the results of previous coin tosses.

DEFINITION 3 (Levy [92]). *A function* $d: 2^{<\omega} \to \mathbb{R}^{\geq 0}$ *is a* martingale[1] *if for all* σ,
$$d(\sigma) = \frac{d(\sigma 0) + d(\sigma 1)}{2}.$$
In addition d *is a* supermartingale *if for all* σ,
$$d(\sigma) \geq \frac{d(\sigma 0) + d(\sigma 1)}{2}.$$
A (super)martingale d *succeeds on a set* A *if* $\limsup_n d(A \upharpoonright n) = \infty$. *The collection of all sets on which* d *succeeds is called the* success set *of* d, *and is denoted by* $S[d]$.

The idea is that a martingale $d(\sigma)$ represents the capital that we have after betting on the bits of σ while following a particular betting strategy ($d(\lambda)$ being our starting capital). The *martingale condition* $d(\sigma) = \frac{d(\sigma 0) + d(\sigma 1)}{2}$ is a fairness condition, ensuring that the expected value of our capital after a bet is equal to our capital before the bet. Ville [144] proved that the success sets of (super)martingales correspond precisely to the sets of measure 0.

Now again we will need a notion of effective betting strategy. We will say that the martingale is computable if d is a computable function (with range $\mathbb{Q}$, without loss of generality), and we will say that d is c.e. iff d is given by an effective approximation $d(\sigma) = \lim_s d_s(\sigma)$ where $d_{s+1}(\sigma) \geq d_s(\sigma)$. This means that we are allowed to bet more as we become more confident of the fact that σ is the more likely outcome in the betting, as time goes on. The following result was anticipated in Ray Solomonoff's approach to randomness as discussed in e.g., Downey–Hirschfeldt [45].

THEOREM 5 (Schnorr [120, 121]). *A set is 1-random iff no c.e. (super)martingale succeeds on it.*

Schnorr argued that Theorem 5 showed that perhaps the notion of Martin-Löf randomness was not really capturing the notion of *effective randomness* as it was intrinsically *computably enumerable*. Schnorr argued that it seems strange that to define randomness we use c.e. martingales and not computable ones. Based on this possible defect, Schnorr defined two other notions of randomness, *computable randomness* (where the martingales are all computable) and Schnorr randomness (where we use the Martin-Löf definition but insist that $\mu(U_k) = 2^{-k}$ rather than $\leq 2^{-k}$ so, in particular, we know precisely the $[\sigma]$ in U_k uniformly in k and $[\sigma]$) meaning in each case that the randomness notion is a computable rather than a computably enumerable one. We know that Martin-Löf randomness implies computable randomness which implies Schnorr randomness, and none of these implications are reversible.

---
[1]A more complex notion of martingales (which are called martingale processes) is used in probability theory. We refer the reader to [45], where it is discussed how computable martingale processes can be used to characterize 1-random reals.

It seemed that Ville's Theorem was a fatal blow to von Mises' program. However, there seems to be a possible resurrection. Can we define 1-randomness using computable martingales somehow? The answer is "possibly" if we allow *non-monotonicity*. The idea is to use a computable *but non-monotonic* notion of randomness, where we have a betting strategy which bets on bits one at a time, but instead of being increasing, we can bet in some arbitrary order, and need not bet on all bits. The order is determined by what has happened so far. This gives a notion called *Kolmogorov–Loveland* (or non-monotonic) randomness and the following question has been open for quite a while.

Question 1 (Muchnik, Semenov, and Uspensky [103]).

Is every non-monotonically random sequence 1-random?

§4. **Developments.** The theory of algorithmic randomness has been widely developed. First many variations of the notions of a random real or string have been introduced. We have already seen three, Kurtz, Schnorr and computable randomness. Each of these notions has its own applications and gives its own insight into the level of randomness needed for measuring the randomness of some process.

4.1. Randomness is the same as differentiability. There has been quite a bit of recent work relating "almost everywhere" behaviour in analysis to levels of randomness. This is a program going back to the work of Oswald Demuth, a constructivist from Prague.

Here we will be working in computable analysis, a subject going back to Turing's original paper [140]. This area will be discussed in detail in the article by Avigad and Brattka in this volume. Briefly, if we are doing computable analysis, then we need some representation of the individual objects we will be analysing. For example a computable real is essentially a computably convergent Cauchy sequence. That is, we have $\{q_n \mid n \in \omega\}$ with limit α and for all j, we can effectively compute $g(j)$ such that $|\alpha - q_{g(j)}| < 2^{-j}$. In a function space, we would have a set of effectively described functions, such as polynomials with rational coefficients, effectively converging to the function in the sense of the appropriate norm. Then a typical definition of a computable function on such a space is that if I can approximate x in the input to within 2^{-j}, then I can compute a similar approximation to $f(x)$. When formalized this is implicit in Turing's orginal paper, and now usually called "type 2" computability. An alternative and weaker notion of computable function is that it takes computable reals to computable reals. This is referrred to as Markov computability. The reader is referred to the Avigad–Brattka paper for more details and the history of the development of the area.

Using such a setting, as an example of the relationship between randomness and differntiability, recall that the Denjoy upper and lower derivatives for a function f are defined as follows.

$$\overline{D}f(x) = \limsup_{h \to 0} \frac{f(x) - f(x+h)}{h} \text{ and } \underline{D}f(x) = \liminf_{h \to 0} \frac{f(x) - f(x+h)}{h}.$$

The Denjoy derivative exists iff both of the above quantities exist and are finite. The idea in this is that slopes like those in the definitions can be considered to be martingales. Using this for one direction, various notions of randomness can be characterized by: (i) varying the strength of the notion of computable real valued function (e.g., Markov computable, type 2 computable, etc.); and (ii) varying the theorem.

For an illustration, we have the following.

THEOREM 6 (Brattka, Miller and Nies [19]). *z is computably random iff every computable (in the type two sense) increasing function $f [0, 1] \to \mathbb{R}$ is Denjoy differentiable at z.*

There are similar results relating 1-randomness of z to the differentiability of functions of bounded variation at z. There is still a lot of activity here, and classes like Lipschitz functions and many other classical almost everywhere behaviour in analysis are found to correlate to various notions of randomness. The paper [19] is an excellent introduction to this material.

Other almost everywhere classical behaviour comes from ergodic theory. There is a great deal of current work exploring the relationship between ergodic theory and algorithmic randomness. The simplest example is an old theorem of Kučera which says that if $\mathcal{C}$ is a Π_1^0 class of measure 1, then for any 1-random X there must be a $Y \in \mathcal{C}$ with the "tail" of X in Y. By this we mean that there is some n which that for all $m > n$, $X(n) = Y(n)$. This is related to ergodic theory as can be seen by an analysis of the the Poincaré Ergodic Theorem. To wit, let (X, μ) be a probability space, and $T \colon X \to X$ measure preserving so that for measurable $A \subseteq X$, $\mu(T^{-1}A) = \mu(A)$. Such a map is called T-*invariant* if $T^{-1}A = A$ except on a measure 0 set. Finally the map is *ergodic* if every T-invariant subset is either null or co-null. The shift operator on Cantor space is the mapping $T(a_0 a_1 \dots) = a_1 a_2 \dots$ is an ergodic action with the Bernoulli product measure. The "tail" map above can be thought of as a statement of a statement about the shift operator.

A classic theorem of Poincaré is that if T is ergodic on (X, μ), then for all $E \subseteq X$ of positive measure and *almost all* $x \in X$, $T^n(x) \in E$ for infinitely many n. For a set E of measurable subsets of X, we call an x a *Poincaré point* if $T^n(x) \in Q$ for all $Q \in E$ of positive measure. Restating the theorem Kučera [78] we see the following: X is 1-random iff X is a Poincaré point for the shift operator with respect to the collection of effectively closed subsets of 2^ω.

Bienvenu et al. proved the following extension of this result.

THEOREM 7 (Bienvenu, et al. [12]). *Let T be computable ergodic on a computable probability space (X, μ). Then $x \in X$ is 1-random iff x is a Poincaré point for all effectively closed subsets of X.*

Again there is a lot of ongoing work here. For instance, one exciting development has seen the applications of algorithmic randomness to *symbolic dynamics* a sub-area of ergodic theory, with well-known applications in additive number theory and analysis. A d-dimensional *shift* of finite type is a collection of colourings of $\mathbb{Z}^d$ defined by local rules and a shift action (basically saying certain colourings are illegal). Its (Shannon) *entropy* is the asymptotic growth in the number of legal colourings. More formally, consider $G = (\mathbb{N}^d, +)$ or $(\mathbb{Z}^d, +)$, and A a finite set of symbols. We give A the discrete topology and A^G the product topology. The *shift action* of G on A^G is

$$(S^g x)(h) = x(h + g), \text{ for } g, h \in G \wedge x \in A^G.$$

A *subshift* is $X \subseteq A^G$ such that $x \in X$ implies $S^g x \in X$ (i.e., shift invariant). *Symbolic Dynamics* studies subshifts usually of "finite type." The following is a recent theorem showing that Ω occurs naturally in this setting.

THEOREM 8 (Hochman and Meyerovitch, [70]). *The values of entropies of subshifts of finite type over $\mathbb{Z}^d$ for $d \geq 2$ are exactly the complements of halting probabilities.*

In this area, Jan Reimann [114] gave a new and simpler proof of a classical theorem called Frostman's Lemma. An even more notable example is due to Simpson [128]. Simpson studies topological entropy for subshifts X and the relationship with Hausdorff dimension.

Here we pause to mention that, in the same way that we can suggest that an individual sequence can be thought to be random, the theory of effective Hausdorff dimension allows us to give an individual sequence effective *dimension*.

After effectivizing the the whole theory of Hausdorff using effective versions of "weighted" inner and outer "measures," it turns out that there are simple characterizations of these notions in terms of Kolmogorov complexity.

Mayordomo [97] proves that effective Hausdorff dimension of X is equal to $\liminf_{n \to \infty} \frac{K(X \upharpoonright n)}{n}$. Athreya, Hitchcock, Lutz, and Mayordomo [4] proved that the effective packing dimension is $\limsup_{n \to \infty} \frac{K(X \upharpoonright n)}{n}$ (C can replace K in both cases).

Again, there has been a long line of development seeking to understand algorithmic dimension. An easy way to make something of effective Hausdorff dimension $\frac{1}{2}$ is to take a 1-random real and "thin it out' by inserting a 0 in every second position. A longstanding question was whether in some sense this was necessary: could randomness be extracted from any a real of nonzero effective Hausdorff dimension? Miller [100] showed that the answer is no. A strong negative answer to this question could also be obtained by constructing a real of minimal Turing degree and of effective Hausdorff dimension 1, but this remains an open question. For packing dimension, either a Turing degree has only elements of effective packing dimension 0, or the sup of the packing dimensions of the members is 1 (Fortnow, Hitchcock, Aduri, Vinochandran, and Wang [53]). However, Conidis [32] showed that the degree did not need to have a real of effective packing dimension 1. In the case of effective Hausdorff dimension, Zimand [153] proved that dimension 1 can be extracted from two independent sources of nonzero Hausdorff entropy.

Looking at one use of these notions, we return to Simpson's work. If $X \subset A^G$ use the standard metric $\rho(x, y) = 2^{-|F_n|}$ where n is as large as possible with $x \restriction F_n = u \restriction F_n$ and $F_n = \{-n, \ldots, n\}^d$. In discussions with co-workers, Simpson proved that the classical dimension equals the entropy (generalizing a difficult result of Furstenburg 1967) using effective methods, which were much simpler.

THEOREM 9 (Simpson [128]). *If X is a subshift (closed and shift invariant), then the effective Hausdorff dimension of X is equal to the classical Hausdorff dimension of X is equal to the entropy, moreover there are calculable relationships between the effective and classical quantities. (See Simpson's home page for his recent talks and more precise details.)*

There are many other investigations looking into other Kolmogorov complexities, resource bounded versions such as polynomial time randomness, and the like, and randomness in other spaces than Cantor space. We will finish with a short section exploring the work of the last decade which seeks to understand how computability and randomness relate.

§5. **Computability and randomness.** Interactions of measure, randomness and computability go back to the early years of the study of degrees of unsolvability. The classical paper was de Leeuw et al. [37] where, amongst other things, it is proven that a set X is computably enumerable from a set of oracles of positive measure iff X is computably enumerable. As a consequence, we get a result later rediscovered by Sacks that if a real X is computable from a collection of sources of positive measure, then X must be computable. Nevertheless, another classical result is the following saying that 1-random sources can have computational power.

THEOREM 10 (Kučera [78], Gács [61]). *For every set X, there is a 1-random Y such that $X \leq_{wtt} Y$, where $\leq_{wtt}$ is Turing reducibility with use bounded by a computable function.*

Theorem 10 argues that 1-random reals are not random enough to correlate to the thesis that random reals should have no computational power. This intuition was clarified by Stephan who proved the following.[2]

---

[2]Interpreted by Hirschfeldt as saying that there are two methods of passing a stupidity test. One is the be the genuine article. The other is to be like Ω and be so smart that you know what a stupid person would say.

THEOREM 11 (Stephan [137]). *Suppose a random real is powerful enough to compute a $\{0, 1\}$-valued function f such that for all n, $f(n) \neq \varphi_n(n)$ (i.e., of PA degree). Then $\emptyset' \leq_T X$, so that it is a "false random."*

There is a lot of material on Chaitin's Omega suggesting that it is the "number of knowledge" and this has something to do with randomness. The result above more or less says that if you are a knowledgeable random then you are essentially code such an Omega. A remarkable theorem here is the following, demonstrating a deep relationship between PA degrees and random degrees (i.e., degrees containing randoms).

THEOREM 12 (Barmpalias, Lewis, and Ng [6]). *Every PA degree is the join of two 1-random degrees.*

We can strengthen the idea of randomness by giving the computational devices more compression power via Turing's notion of an oracle. Then if $\emptyset^{(n)}$ denotes the n-th iterate of the halting problem, we say that X is $n + 1$-random iff $K^{\emptyset^{(n)}}(X \upharpoonright n) \geq^+ n$ for all n. A pretty result proven by Miller and Yu [101] is that if $X \leq_T Y$ are both 1-random and X is n-random, so is Y.

We can similarly do this with other notions of randomness with a little care. For notions like Schnorr randomness we need stronger reducibilities reflecting the "totality" of the tests.

There are also many results concerning the relationships between the randomness notions and Turing (and other) degrees. For example, it can be shown that X is weakly 2-random (i.e., in every Σ_2^0 class of measure 1) iff X is 1-random and its degree forms a minimal pair with $\emptyset'$ (Downey, Nies, Weber, and Yu [47] plus Hirschfeldt and Miller (in [47]) for the hard direction). Hence no (weakly) 2-random real can bound a PA degree.

It is a surprising fact that for all n, n-randomness can be defined purely in terms of K with no oracle. This follows by the next result.

THEOREM 13 (Bienvenu, Muchnik, Shen, and Vereschagin [15]). $K^{\emptyset'}(\sigma) = \limsup_m K(\sigma \mid m) \pm O(1)$.

Hence A is 2-random iff for all n, $\limsup_m K(A \upharpoonright n \mid m) \geq^+ n$. For a small number of n, we know of "natural" definitions of n-randomness. For instance, we have seen that it is impossible for a real to have $C(X \upharpoonright n) \geq^+ n$ for *all n*, but Martin-Löf showed in his original paper [96] that there are reals X with $C(X \upharpoonright n) \geq^+ n$ for *infinitely many n*, and that these are all 1-random. Joe Miller [98] and later Nies, Stephan and Terwijn [109] showed that such randoms are precisely the 2-randoms, and later Miller [99] showed that the 2-randoms are exactly those that achieve maximal prefix-free complexity (which is $n + K(n)$) infinitely often. Also Becher and Gregorieff [10] have a kind of index set characterizations of higher notions of randomness. I know of no other natural definitions, such as for the 3-randoms. There has been a huge amount of work concerning the interplay between things like PA degrees and weakenings of the notion of fixed point free functions (that is, functions with $f(n) \neq \varphi_n(n)$ for all n). For example, you can show that this ability corresponds to tracing, and the speed of growth of the initial segment complexity of a real. As an illustration, A is *h-complex* if $C(A \upharpoonright n) \geq h(n)$ for all n. A is *autocomplex* if there is an A-computable order h such that A is h-complex, where an order is a nondecreasing unbounded function with $f(0) \geq 1$.

THEOREM 14 (Kjos-Hanssen, Merkle, and Stephan [73]). *A set is autocomplex iff it is of DNC degree.*

Another illustration of the interplay of notions of randomness and Turing degrees is the following theorem.

THEOREM 15 (Nies, Stephan, and Terwijn [109]). *If a nonhigh set (i.e., A' $\not\geq_T \emptyset^{(2)}$) is Schnorr random then it is 1-random.*

On the other hand, it is possible to show that within the high degrees the separations between computable, Schnorr, and Martin-Löf randomness all occuri ([109]). In the hyperimmune-free degrees, weak randomness coincides with all of these as well as weak 2-randomness. So the degree can have great effect on what a notion of randomness means.

One long sequence of results concerns lowness and randomness. For any reasonable property P we say that X is *low for P* if $P^X = P$. For example, being low for the Turing jump means that $X' \equiv_T \emptyset'$. A set A is low for 1-randomness iff A does not make any 1-randoms nonrandom. That is, if Y is 1-random then Y^A is $A - 1$-random. Normally we would expect an oracle A would enable us to compress some intital segment of Y for some Y allowing us to derandomize it. You can also have a notion of lowness for tests, meaning that every (effective nullset)^A can be covered by an effective nullset. In all cases the lowness notion for randomness and for tests have turned out to coincide with a single recent exception of "difference randomness" found by Diamondstone and Franklin (paper in preparation).

Now it is not altogether clear that noncomputable sets low for 1-randomness should exist. But they do and form a remarkable class called the K-trivials which had earlier and independently been defined purely in terms of their initial segment complexity. That is, the reals low for Martin-Löf randomness coincide with the class of reals A such that for all n, $K(A \upharpoonright n) \leq^+ K(n)$. (In fact Bienvenu and Downey [13] showed that it is enough to put a Solovay function[3] in place of $K(n)$.) Many properties of this class have been shown. The coincidence of these two concepts lowness and triviality is one of the jewels of the area. It was Andre Nies who proved the deep result that A is K-trivial iff A is low for Martin-Löf randomness iff A is useless as a compressor,

---

[3]That is a computable function f with $f(m) \geq K(m)$ for all m and $f(n) =^+ K(n)$ infinitely many n. See also Beinvenu and Merkle [14] and Hölzl, Kräling, and Merkle [71].

meaning that for all σ, $K^A(\sigma) =^+ K(\sigma)$. (Nies [105]). A good account of this material can be found in Nies [106, 107], but things are constantly changing, with perhaps seventeen characterizations of this class at present. We also refer to [45] for the situation up to mid-2010.

Other randomness notions give quite different lowness notions. For example, X is low for C, meaning $C^X =^+ C$, iff X is computable (essentially Chaitin [26]), and similarly Y is low for computable randomness iff Y is computable (Nies [105]). reals which are low for C nor any low for computable randomness. On the other hand, lowness for Schnorr and Kurtz randomness give interesting subclasses of the hyperimmune-free degrees characterized by notions of being computably dominated, and fixed point free functions in the case of Kurtz. (This is a detailed story with many references, beginning with the beautiful paper of Terwijn and Zambella [139], see Downey and Hirschfeldt [45].) Work here is still ongoing and many results have been proven, but the pattern remains very opaque. Even for a fixed real like Ω (i.e., when does Ω^X remain random?) results are quite interesting. In the case of Ω, X is low for Ω and X is computable from the halting problem, then X is K-trivial, but there are *random* reals low for Ω. In fact, X is 1-random and low for Ω, iff X is 2-random. (Result of Joe Miller, see [45].)

These classes again relate to various refinements of the jump and to "traceing" which means giving an effective collection of *possibilities* for (partial) functions computable from the degree at hand. Again this idea has taken on a life of its own, and such notions have been used to solve questions from classical computability theory. For instance, Downey and Greenberg [44] used "strong jump traceability" to solve a longstanding question of Jockusch and Shore on pseudo-jump operators and cone avoidance. Strongly jump traceable reals have their own techniques and theory and form a fascinating class, see e.g., [27].

We should also mention the the deep results of Reimann and Slaman who were looking at the question (first discussed by Levin):

"Given $X \neq_T \emptyset$, is there a measure relative to which X is random?"

Clearly we can trivially answer Levin's question: Every real is, we can concentrate a measure on a real. But clearly what is asked is for the situation where we are not allowed to do this concentration. If we allow atoms, then the answer is still that that every noncomputable real can be made random. On the other hand, if we ask that there are no atoms in the measure, the situation is very different. We get a nonempty class of *never continuously n-random* reals. For each n this class is countable, but the proof of this requires magical things like big fragments of Borel determinacy, *provably*. This metamathematical aspect of the answer seems strange in that the definitions of Martin-Löf randomness only needs a couple of quantifiers and hence we would expect a low level answer. But no. So algorithmic randomness not only interacts strongly with computability theory but also even with set theory. Reimann and Slaman's results use techniques involving models of ZFC and "master codes". The reader should look at Reimann and Slaman [115].

§6. Turing. What has this got to do with Turing? What was the anticipation we alluded to? Certainly, the very notion of algorithmic randomness needs the notion of algorithm and arguably there is this weak connection: Turing clarified the notion of algorithm. However, there is something rather more remarkable than that.

We return to the notion of (absolute) normality. Recall that X is Borel normal to base n if we represent X in base n, then for all $0 \leq i \leq n-1$,

$$\lim_s \frac{|\{X(k) = i \mid k \leq s\}|}{s} = \frac{1}{n}.$$

This notion was defined by Emil Borel in 1909. We have seen that variations of the notion of normality were the basis of von Mises approach to defining randomness.

Interestingly, normality *precisely defines* an algorithmic randomness notion. To give a machine (randomness) definition of normality, we change the computational device to that of a finite automaton. A real number is normal to a base b if, and only if, no finite state gambler can make infinite winnings when betting on its base b expansion, as we see more explicitly in Theorem 18 below. (See [122, 35, 18].)

Borel demonstrated that almost every real is absolutely normal, but asked the questions "How can we construct an *explicit* absolutely normal number?" and "Can a real be normal relative to one base and not another?"

Normality is a longstanding area of research in number theory. It is also yet another area of number theoretical research where the questions rapidly outrun our ability to prove theorems. For example, it is unknown if familiar reals like e and π and the like are normal to *any* base.

Returning to Borel's questions, how should we interpret "explicit construction of a normal number"? With the machinery of computability theory developed by Turing, Church, Kleene and others, we have at least one interpretation. From the material of the previous sections, it is obvious that Ω is normal. However, in some sense, this is cheating since it is not a computable, but a c.e. object so is hardly an *explicit construction*.

In an unpublished manuscript, Turing attacked the question of an explicit construction of an absolutely normal number by interpreting "explicit" to mean *computable*. His manuscript entitled *"A note on normal numbers"*, presumably written in 1938, presents the best answer to date to Borel's first question: an algorithm that produces absolutely normal numbers. This early proof of existence of computable normal numbers remained largely unknown because Turing's manuscript was only published in 1997 in his Collected Works, edited by J. L. Britton [143]. The editorial notes say that the proof given by Turing is inadequate and speculate that the theorem could be false. In [9] Becher, Figueira and Picci reconstructed and completed Turing's manuscript, trying to preserve his ideas as accurately as possible and correcting minor errors.

As Becher [7] remarks, the very first examples of normal numbers were independently given by Henri Lebesgue and Waclaw Sierpiński[4] in 1917 [84, 127]. They also lead to computable instances by giving a computable reformulation of the original constructions [8]. Together with Turing's algorithm these are the only known constructions of computable normal numbers. It is pretty clear that Turing was unaware of the limiting constructions given in [84, 127].

What does Turing's construction do? His paper says the following:

> Although it is known that almost all numbers are [absolutely] normal no example of [an absolutely] normal number has ever been given. I propose to show how [absolutely] normal numbers may be constructed and to prove that almost all numbers are [absolutely] normal constructively.

I won't reproduce Turing's construction, save to say that it makes an ingenious extension of the law of large numbers to blocks, and basically makes a low complexity Martin-Löf type test. The details can be found in Becher [7], for instance.

What Turing actually does is something very modern. He develops an effective version of measure theory (sound familiar?) and demonstrates that the reals which are *not* absolutely normal have *computable measure* 0. Therefore, there must be a computable real which is absolutely normal. Here is what Jack Lutz said of this in a lecture at the conference *Computability, Complexity and Randomness*, 2012 at Cambridge:

> Placing computability constraints on a nonconstructive theory like Lebesgue measure seems a priori to weaken the theory, but it may strengthen the theory for some purposes This vision is crucial for present-day investigations of
> - individual random sequences,
> - dimensions of individual sequences,
> - measure and category in complexity classes, etc.

§7. From a modern perspective. How have investigations into normality played out? Using polynomial martingales and hence a polynomial notion of randomness, we have the following.

---

[4]Both published their works in the same journal issue, but Lebesgue's dates back to 1909, immediately after Borel's question.

THEOREM 16 (Strauss [138]). *There are absolutely normal numbers computable in exponential time.*

Using a more delicate construction, Elvira Mayordomo brought the complexity of an *explicit* computable absolutely normal real down as follows.

THEOREM 17 (Mayordomo). *We can construct an absolutely normal number in time $O(n \log n)$.*

As mentioned earlier, this is all related to the theory of *finite state compressors* and the corresponding notion of dimension.

DEFINITION 4 (Schnorr and Stimm [122]).
1. *A finite state gambler is a quadruple $G = (Q, \delta, q_0, B)$ where (Q, δ, q_0) is a finite state automaton, and $B: Q \to \Delta_{\mathbb{Q}}(\Sigma)$ is a betting function, where $\Delta_{\mathbb{Q}}(\Sigma)$ is the collection of rational-valued probability measures on Σ.*
2. *A martingale is a function $d_G: \Sigma^* \to [0, \infty)$ with $d_G(\lambda) = 1$ (Here λ is the empty string), and again the fairness condition*:

$$d_G(wa) = |\Sigma| d_G(w) B(\delta(w))(a).$$

3. *for $s \in [0, \infty)$ the s-gale of G is $d_G^{(s)}(w) = 2^{(s-1)|w|} d_G(w)$.*

As usual, we say that d succeeds on X if $\limsup_{n \to \infty} d(X \restriction n) = \infty$ and that d succeeds strongly if $\liminf_{n \to \infty} d(X \restriction n) = \infty$. Then the *finite state dimension* of a real X is

$$\dim_F S(X) = \inf\{s \in [0, \infty) \mid \exists \text{ finite state } G \text{ s.t. } d_G^{(s)} \text{ succeeds on } X\}.$$

By a theorem of Dai, Lathrop, Lutz and Moyordomo [35], this quantity equals the infimum over all finite state compressors F of

$$\liminf_{n \to \infty} \frac{C_F(X \restriction n)}{n \log \Sigma},$$

aligning with the definition met before for effective Hausdorff dimension, with a similar formula holding for effective string dimension below. There is a similar definition for strong dimension and strong success.

The theorem is the following.

THEOREM 18 (Schnorr and Stimm [122]). *X is normal base b iff the base b finite state dimension of X is 1.*

There is a very active program concerned with the analysis of finite state dimensions. Many modern text compressors such as ZIP are examples of finite state compressors so this theory seems quite pertinent to applications. We refer the reader to Dai et al. [35] and to Lutz's home page for much more on this topic, and its relationship to things like DNA self-assembly.

REFERENCES

[1] U. ABRAHAM and R. SHORE, *Initial segments of the turing degrees of size* $\aleph_1$, **Israel Journal of Mathematics**, vol. 55 (1986), pp. 1–51.

[2] K. AMBOS-SPIES, D. HIRSCHFELDT, and R. SHORE, *Undecidability and 1-types in intervals of the c.e. degrees*, **Annals of Pure and Applied Logic**, vol. 106 (2000), pp. 1–48.

[3] CHRIS ASH and JULIA KNIGHT, **Computable structures and the hyperarithmetical hierarchy**, Elsevier, 2000.

[4] K. ATHREYA, J. HITCHCOCK, J. LUTZ, and E. MAYORDOMO, *Effective strong dimension in algorithmic information and computational complexity*, **SIAM Journal on Computing**, vol. 37 (2007), pp. 671–705.

[5] J. AVIGAD, *The metamathematics of ergodic theory*, **Annals of Pure and Applied Logic**, vol. 157 (2009), pp. 64–76.

[6] G. BARMPALIAS, A. LEWIS, and K. M. NG, *The importance of* Π_1^0 *classes in effective randomness*, **The Journal of Symbolic Logic**, vol. 75 (2010), no. 1, pp. 387–400.

[7] V. BECHER, *Turing's normal numbers: towards randomness*, **CiE 2012** (S. B. Cooper, A. Dawar, and B. Löwe, editors), Lecture Notes in Computer Science 7318, Springer, Heidelberg, 2012, pp. 35–45.

[8] V. BECHER and S. FIGUEIRA, *An example of a computable absolutely normal number*, **Theoretical Computer Science**, vol. 270 (2002), pp. 947–958.

[9] V. BECHER, S. FIGUEIRA, and R. PICCHI, *Turing's unpublished algorithm for normal numbers*, **Theoretical Computer Science**, vol. 377 (2007), pp. 126–138.

[10] V. BECHER and S. GRIGORIEFF, *From index sets to randomness in* $\emptyset^n$, *Random reals and possibly infinite computations*, **The Journal of Symbolic Logic**, vol. 74 (2009), no. 1, pp. 124–156.

[11] J. BERTRAND, **Calcul des probabilités**, Gauthier-Villars et fils, Paris, 1889.

[12] L. BIENVENU, A. DAY, M. HOYRUP, I. MEZHIROV, and A. SHEN, *Ergodic-type characterizations of algorithmic randomness*, **Information and Computation**, (to appear).

[13] L. BIENVENU and R. DOWNEY, *Kolmogorov complexity and Solovay functions*, **Symposium on Theoretical Aspects of Computer Science (STACS 2009)**, 2009, pp. 147–158.

[14] L. BIENVENU and W. MERKLE, *Reconciling data compression and Kolmogorov complexity*, **Proceedings of the 34th International Colloquium on Automata, Languages, and Programming (ICALP 2007)**, Lecture Notes in Computer Science 4596, Springer, 2007.

[15] L. BIENVENU, AN. A. MUCHNIK, A. SHEN, and N. VERESHCHAGIN, *Limit complexities revisited*, **Symposium on Theoretical Aspects of Computer Science (STACS 2008)**, 2008.

[16] ÉMIL BOREL, *Les probabilités dénombrables et leurs applications arithmétiques*, **Rendiconti del Circolo Matematico di Palermo**, vol. 27 (1909), pp. 247–271.

[17] ——, **Leçons sur la thèorie des fonctions**, 2nd ed., Gauthier Villars, 1914.

[18] C. BOURKE, J. HITCHCOCK, and N. VINODCHANDRAN, *Entropy rates and finite-state dimension*, **Theoretical Computer Science**, vol. 349 (2005), no. 3, pp. 392–406.

[19] V. BRATTKA, J. MILLER, and A. NIES, *Randomness and differentiability*, to appear.

[20] M. BRAVERMAN and M. YAMPOLSKY, *Non-computable Julia sets*, **Journal of the American Mathematical Society**, vol. 19 (2006), no. 3.

[21] ——, **Computability of Julia sets**, Springer-Verlag, 2008.

[22] YANN BUGEAUD, *Nombres de Liouville et nombres normaux*, **Comptes Rendus de l'Académie des Sciences de Paris**, vol. 335 (2002), pp. 117–120.

[23] ——, **Distribution modulo one and diophantine approximation**, Cambridge University Press, 2012.

[24] DOUGLAS CENZER and CARL JOCKUSCH, Π_1^0 *classes—structure and applications*, **Contemporary Mathematics**, vol. 257 (2000), pp. 39–59.

[25] G. CHAITIN, *A theory of program size formally identical to information theory*, **Journal of the ACM**, vol. 22 (1975), pp. 329–340.

[26] ———, *Information-theoretical characterizations of recursive infinite strings*, **Theoretical Computer Science**, vol. 2 (1976), pp. 45–48.

[27] P. CHOLAK, R. DOWNEY, and N. GREENBERG, *Strong-jump traceablilty. I. The computably enumerable case*, **Advances in Mathematics**, vol. 217 (2008), pp. 2045–2074.

[28] PETER CHOLAK, ROD DOWNEY, and LEO HARRINGTON, *On the orbits of computably enumerable sets*, **Journal of the American Mathematical Society**, vol. 21 (2008), no. 4, pp. 1105–1135.

[29] PETER CHOLAK and LEO HARRINGTON, *On the definability of the double jump in the computably enumerable sets*, **Journal of Mathematical Logic**, vol. 2 (2002), no. 2, pp. 261–296.

[30] ALONZO CHURCH, *On the concept of a random sequence*, **Bulletin of the American Mathematical Society**, vol. 46 (1940), pp. 130–135.

[31] R. CILIBRASI, P. M. B. VITANYI, , and R. DE WOLF, *Algorithmic clustering of music based on string compression*, **Computer Music Journal**, vol. 28 (2004), pp. 49–67.

[32] C. CONIDIS, *A real of strictly positive effective packing dimension that does not compute a real of effective packing dimension one*, **The Journal of Symbolic Logic**, vol. 77 (2012), no. 2, pp. 447–474.

[33] BARRY COOPER, *Minimal degrees and the jump operator*, **The Journal of Symbolic Logic**, vol. 38 (1973), pp. 249–271.

[34] ———, *Minimal pairs and high recursively enumerable degrees*, **The Journal of Symbolic Logic**, vol. 39 (1974), pp. 655–660.

[35] L. DAI, J. LUTZ, and E. MAYORDOMO, *Finite-state dimension*, **Theoretical Computer Science**, vol. 310 (2004), pp. 1–33.

[36] MARTIN DAVIS, **Computability and unsolvability**, Dover, 1985.

[37] K. DE LEEUW, E. F. MOORE, C. E. SHANNON, and N. SHAPIRO, *Computability by probabilistic machines*, **Automata studies** (C. E. Shannon and J. McCarthy, editors), Annals of Mathematics Studies 34, Princeton University Press, Princeton, NJ, 1956, pp. 183–212.

[38] O. DEMUTH, *The differentiability of constructive functions of weakly bounded variation on pseude-numbers*, **Commentationes Mathematicae Universitatis Carolinae**, vol. 16 (1975), pp. 583–599.

[39] R. DOWNEY, *Lattice nonembeddings and initial segments of the recursively enumerable degrees*, **Annals of Pure and Applied Logic**, vol. 49 (1990), pp. 97–119.

[40] ———, *Algorithmic randomness and computability*, **Proceedings of the 2006 International Congress of Mathematicians**, vol. 2, European Mathematical Society, 2006, pp. 1–26.

[41] ———, *Five lectures on algorithmic randomness*, **Computational prospects of infinity, Part I: Tutorials** (C. Chong, Q. Feng, T. A. Slaman, W. H. Woodin, and Y. Yang, editors), Lecture Notes Series, Institute for Mathematical Sciences, National University of Singapore, vol. 14, World Scientific, Singapore, 2008, pp. 3–82.

[42] R. DOWNEY, J. FLUM, M. GROHE, and M. WEYER, *Bounded fixed-parameter tractability and reducibility*, **Annals of Pure and Applied Logic**, vol. 148 (2007), pp. 1–19.

[43] R. DOWNEY and N. GREENBERG, *A transfinite hierarchy of computably enumerable degrees, unifying classes, and natural definability*, monograph, in preparation.

[44] ———, *Pseudo-jump operators and SJT Hard sets*, **Advances in Mathematics**, (to appear).

[45] R. DOWNEY and D. HIRSCHFELDT, **Algorithmic randomness and complexity**, Springer-Verlag, 2010.

[46] R. DOWNEY, D. HIRSCHFELDT, A. NIES, and S. TERWIJN, *Calibrating randomness*, **The Bulletin of Symbolic Logic**, vol. 12 (2006), pp. 411–491.

[47] R. DOWNEY, A. NIES, R. WEBER, and L. YU, *Lowness and Π_2^0 nullsets*, **The Journal of Symbolic Logic**, vol. 71 (2006), pp. 1044–1052.

[48] R. DOWNEY and R. SHORE, *Lattice embeddings below a non-low$_2$ recursively enumerable degree*, **Israel Journal of Mathematics**, vol. 94 (1996), pp. 221–246.

[49] ———, *There is no degree invariant half-jump*, **Proceedings of the American Mathematical Society**, vol. 125 (1997), pp. 3033–3037.

[50] ROD DOWNEY, *Randomness, computation and mathematics*, **CiE 2012** (S. B. Cooper, A. Dawar, and B. Löwe, editors), Lecture Notes in Computer Science 7318, Springer, Heidelberg, 2012.

[51] RACHEL EPSTEIN, *The nonlow computably enumerable degrees are not invariant in E*, **Transactions of the American Mathematical Society**, (to appear).

[52] LAWRENCE FEINER, *The strong homogeneity conjecture*, **The Journal of Symbolic Logic**, vol. 35 (1970), pp. 373–377.

[53] L. FORTNOW, J. HITCHCOCK, P. ADURI, V. VINOCHANDRAN, and F. WANG, *Extracting Kolmogorov complexity with applications to dimension zero-one laws*, **Proceedings of the 33rd International Colloquium on Automata, Languages, and Programming** (*ICALP 2006*), Lecture Notes in Computer Science 4051, Springer, 2006, pp. 335–345.

[54] W. FOUCHE, *The descriptive complexity of Brownian motion*, **Advances in Mathematics**, vol. 155 (2000), pp. 317–343.

[55] ——, *Dynamics of a generic Brownian motion: Recursive aspects*, **Theoretical Computer Science**, vol. 394 (2008), pp. 175–186.

[56] J. FRANKLIN, N. GREENBERG, J. MILLER, and KENG MENG NG, *Martin-Löf random points satisfy Birkhoff's ergodic theorem for effectively closed sets*, **Proceedings of the American Mathematical Society**, (to appear).

[57] R. FRIEDBERG, *A criterion for completeness of degrees of unsolvability*, **The Journal of Symbolic Logic**, vol. 22 (1957), pp. 159–160.

[58] ——, *Two recursively enumerable sets of incomparable degrees of unsolvability*, **Proceedings of the National Academy of Sciences of the United States of America**, vol. 43 (1957), pp. 236–238.

[59] H. FUCHS and C. SCHNORR, *Monte Carlo methods and patternless sequences*, **Operations Research Verfahren XXV, Symposium Heidelberg**, 1977, pp. 443–450.

[60] P. GÁCS, *On the relation between descriptional complexity and algorithmic probability*, **Theoretical Computer Science**, vol. 22 (1983), pp. 71–93.

[61] ——, *Every set is reducible to a random one*, **Information and Control**, vol. 70 (1986), pp. 186–192.

[62] P. GÁCS, M. HOYRUP, and C. ROJAS, *Randomness on computable probability spaces, a dynamical point of view*, **Theory of Computing Systems**, (to appear).

[63] S. GREGORIEFF and M. FERBUS, *Is randomness native to computer science? Ten years after*, In Zenil [152], pp. 243–263.

[64] E. GRIFFOR, **Handbook of computability theory**, Elsevier, 1999.

[65] MARCIA J. GROSZEK and THEODORE A. SLAMAN, *Independence results on the global structure of the Turing degrees*, **Transactions of the American Mathematical Society**, vol. 277 (1983), no. 2, pp. 579–588.

[66] G. H. HARDY and E. M. WRIGHT, **An introduction to the theory of numbers**, Oxford University Press, 1979, first edition in 1938.

[67] LEO HARRINGTON, *Maclachlin's conjecture*, handwritten notes 1970s.

[68] LEO HARRINGTON and ROBERT SOARE, *Post's Program and incomplete recursively enumerable sets*, **Proceedings of the National Academy of Sciences of the United States of America**, vol. 88 (1991), pp. 10242–10246.

[69] ROLF HERKEN, **The universal Turing Machine: A half-century survey**, Springer-Verlag, 1995.

[70] M. HOCHMAN and T. MEYEROVITCH, *A characterization of the entropies of multidimensional shifts of finite type*, **Annals of Mathematics**, vol. 171 (2010), no. 3, pp. 2011–2038.

[71] R. HÖLZL, T. KRÄLING, and W. MERKLE, *Time bounded Kolmogorov complexity and Solovay functions*, **Mathematical Foundations of Computer Science 2009**, Lecture Notes in Computer Science, vol. 5734, Springer, 2009, pp. 392–402.

[72] CARL JOCKUSCH and ROBERT SOARE, *Degrees of members of* Π_1^0 *classes*, **Pacific Journal of Mathematics**, vol. 40 (1972), pp. 605–616.

[73] B. KJOS-HANSSEN, W. MERKLE, and F. STEPHAN, *Kolmogorov complexity and the recursion theorem*, **Symposium on Theoretical Aspects of Computer Science (STACS 2006)**, Lecture Notes in Computer Science 3884, Springer, 2006, pp. 149–161.

[74] B. KJOS-HANSSEN and A. NERODE, *Effective dimension of points visited by Brownian motion*, **Theoretical Computer Science**, vol. 410 (2009), no. 4–5, pp. 347–354.

[75] B. KJOS-HANSSEN and T. SZABADOS, *Kolmogorov complexity and strong approximation of Brownian motion*, **Proceedings of the American Mathematical Society**, vol. 139 (2011), no. 9, pp. 3307–3316.

[76] STEPHEN KLEENE and EMIL POST, *The upper semi-lattice of degrees of recursive unsolvability*, **Annals of Mathematics**, vol. 59 (1954), no. 3, pp. 379–407.

[77] A. N. KOLMOGOROV, *Three approaches to the quantitative definition of information*, **Problems of Information Transmission**, vol. 1 (1965), pp. 1–7.

[78] A. KUČERA, *Measure,* Π_1^0 *classes, and complete extensions of* PA, **Recursion Theory Week: Proceedings, Oberwolfach, 1984**, Lecture Notes in Mathematics, vol. 1141, Springer, Berlin, 1985, pp. 245–259.

[79] A. KUČERA and T. SLAMAN, *Randomness and recursive enumerability*, **SIAM Journal on Computing**, vol. 31 (2001), pp. 199–211.

[80] A. LACHLAN, *Lower bounds for pairs of recursively enumerable degrees*, **Proceedings of the London Mathematical Society**, vol. 16 (1966), pp. 537–569.

[81] ———, *Uniform enumeration operators*, **The Journal of Symbolic Logic**, vol. 40 (1975), pp. 401–409.

[82] A. H. LACHLAN, *Embedding nondistributive lattices in the recursively enumerable degrees*, **Proceedings of the Conference in Mathematical Logic—London '70**, (*Bedford College, London, 1970*), Lecture Notes in Mathematics, vol. 255, Springer, Berlin, 1972, pp. 149–177.

[83] A. H. LACHLAN and R. SOARE, *Not every finite lattice is embeddable in the recursively enumerable degrees*, **Advances in Mathematics**, vol. 37 (1980), pp. 74–82.

[84] HENRI LEBESGUE, *Sur certaines démonstrations d'existence*, **Bulletin de la Société Mathématique de France**, vol. 45 (1917), pp. 132–144.

[85] STEFFEN LEMPP and MANUEL LERMAN, *The decidability of the existential theory of the poset of the recursively enumerable degrees with jump relations*, **Advances in Mathematics**, (1996).

[86] ———, *A finite lattice without critical triple that cannot be embedded into the enumerable Turing degrees*, **Annals of Pure and Applied Logic**, vol. 87 (1997), no. 2, pp. 167–185, Logic Colloquium '95 Haifa.

[87] MANUEL LERMAN, **Degrees of unsolvability: Local and global theory**, Springer-Verlag, 1983.

[88] ———, *The embedding problem for the recursively enumerable degrees*, **Recursion theory (Ithaca, NY, 1982)**, Proceedings of the Symposia in Pure Mathematics, vol. 42, American Mathematical Society, Providence, RI, 1985, pp. 13–20.

[89] L. LEVIN, **Some theorems on the algorithmic approach to probability theory and information theory**, Dissertation in Mathematics, Moscow University, 1971.

[90] ———, *Laws of information conservation (non-growth) and aspects of the foundation of probability theory*, **Problems of Information Transmission**, vol. 10 (1974), pp. 206–210.

[91] M. B. LEVIN, *On absolutely normal numbers*, **Moscow University Mathematics Bulletin**, vol. 34 (1979), pp. 32–39, English translation.

[92] P. LÉVY, *Théorie de l'addition des variables aléatoires*, Gauthier-Villars, 1937.

[93] ANGUS MACINTYRE, *Transfinite iterations of Friedberg's completeness criterion*, **The Journal of Symbolic Logic**, vol. 38 (1977), pp. 1–10.

[94] DAVID MARKER, *Degrees of models of true arithmetic*, **Proceedings Herbrand Symposio** (J. Stern, editor), North-Holland, Amsterdam, 1982, pp. 233–242.

[95] D. A. MARTIN, *Classes of recursively enumerable sets and degrees of unsolvability*, **Zeitschrift für Mathematische Logik und Grundlagen der Mathematik**, vol. 12 (1966), pp. 295–310.

[96] P. MARTIN-LÖF, *The definition of random sequences*, **Information and Control**, vol. 9 (1966), pp. 602–619.

[97] E. MAYORDOMO, *A Kolmogorov complexity characterization of constructive Hausdorff dimension*, **Information Processing Letters**, vol. 84 (2002), pp. 1–3.

[98] J. MILLER, *Kolmogorov random reals are 2-random*, **The Journal of Symbolic Logic**, vol. 69 (2004), no. 3, pp. 907–913.

[99] ———, *The K-degrees, low for K-degrees, and weakly low for K sets*, **Notre Dame Journal of Formal Logic**, vol. 50 (2010), no. 4, pp. 381–391.

[100] ———, *Extracting information is hard: a Turing degree of non-integral effective Hausdorff dimension*, **Advances in Mathematics**, vol. 226 (2011), no. 1, pp. 373–384.

[101] J. S. MILLER and L. YU, *On initial segment complexity and degrees of randomness*, **Transactions of the American Mathematical Society**, vol. 360 (2008), no. 6, pp. 3193–3210.

[102] A. MUCHNIK, *On the unsolvability of the problem of reducibility in the theory of algorithms*, **Doklady Akademii Nauk SSSR**, vol. 106 (1956), pp. 194–197.

[103] AN. A. MUCHNIK, A. SEMENOV, and V. USPENSKY, *Mathematical metaphysics of randomness*, **Theoretical Computer Science**, vol. 207 (1998), no. 2, pp. 263–317.

[104] J. MYHILL, *Creative sets*, **Zeitschrift für Mathematische Logik und Grundlagen der Mathematik**, vol. 1 (1955), pp. 97–108.

[105] A. NIES, *Lowness properties and randomness*, **Advances in Mathematics**, vol. 197 (2005), no. 1, pp. 274–305.

[106] ———, *Computability and randomness*, Oxford University Press, 2009.

[107] ———, *Interactions of computability and randomness*, **Proceedings of the International Congress of Mathematicians** (S. Ragunathan, editor), 2010, pp. 30–57.

[108] A. NIES, R. SHORE, and T. SLAMAN, *Interpretability and definability in the recursively enumerable degrees*, **Proceedings of the London Mathematical Society**, vol. 77 (1998), pp. 241–291.

[109] A. NIES, F. STEPHAN, and S. A. TERWIJN, *Randomness, relativization, and Turing degrees*, **The Journal of Symbolic Logic**, vol. 70 (2005), no. 2, pp. 515–535.

[110] P. ODIFREDDI, **Classical recursion theory**, vol. 1, North-Holland, 1989.

[111] ———, **Classical recursion theory**, vol. 2, North-Holland, 1999.

[112] EMIL POST, *Recursively enumerable sets of positive integers and their decision problems*, **Bulletin of the American Mathematical Society**, vol. 50 (1944), no. 5, pp. 284–316.

[113] M. POUR-EL and I. RICHARDS, **Computability in analysis and physics**, Springer-Verlag, 1989.

[114] J. REIMANN, *Effectively closed classes of measures and randomness*, **Annals of Pure and Applied Logic**, vol. 156 (2008), no. 1, pp. 170–182.

[115] J. REIMANN and T. SLAMAN, *Randomness for continuous measures*, (draft available from Reimann's web site), to appear.

[116] ROBERT ROBINSON, *Jump restricted interpolation in the recursively enumerable degrees*, **Annals of Mathematics**, vol. 93 (1971), no. 3, pp. 586–596.

[117] HARTLEY ROGERS, JR., **Theory of recursive functions and effective computabilty**, McGraw-Hill, 1967.

[118] GERALD SACKS, *The recursively enumerable degrees are dense*, **Annals of Mathematics**, vol. 80 (1964), pp. 300–312.

[119] W. M. SCHMIDT, *On normal numbers*, **Pacific Journal of Mathematics**, vol. 10 (1960), pp. 661–672.

[120] C. P. SCHNORR, *A unified approach to the definition of a random sequence*, **Mathematical Systems Theory**, vol. 5 (1971), pp. 246–258.

[121] ———, *Zufälligkeit und Wahrscheinlichkeit. Eine algorithmische Begründung der Wahrscheinlichkeitstheorie*, Lecture Notes in Mathematics, vol. 218, Springer-Verlag, Berlin-New York, 1971.

[122] C. P. SCHNORR and H. STIMM, *Endliche Automaten und Zufallsfolgen*, **Acta Informatica**, vol. 1 (1972), pp. 345–359.

[123] RICHARD SHORE, *The homogeneity conjecture*, **Proceedings of the National Academy of Sciences of the United States of America**, vol. 76 (1979), pp. 4218–4219.

[124] ———, *On homogeneity and definability in the first order theory of the Turing degrees*, **The Journal of Symbolic Logic**, vol. 47 (1982), pp. 8–16.

[125] ———, *A non-inversion theorem for the jump operator*, **Annals of Pure and Applied Logic**, vol. 40 (1988), pp. 277–303.

[126] RICHARD SHORE and THEODORE SLAMAN, *Defining the Turing jump*, **Mathematical Research Letters**, vol. 6 (1999), pp. 711–722.

[127] WACLAW SIERPIŃSKI, *Démonstration élémentaire du théorème de M. Borel sur les nombres absolument normaux et détermination effective d'un tel nombre*, **Bulletin de la Société Mathématique de France**, vol. 45 (1917), pp. 127–132.

[128] S. SIMPSON, *Symbolic dynamics: Entropy = dimension = complexity*, preprint (16 December 2011) submitted.

[129] ———, *Mass problems associated with effectively closed sets*, **Tohoku Mathematical Journal**, vol. 63 (2011), pp. 489–517.

[130] ———, *Medvedev degrees of 2-dimensional subshifts of finite type*, **Ergodic Theory and Dynamical Systems**, (to appear), published online 29 November 2012, http://dx.doi.org/10.1017/etds.2012.152.

[131] THEODORE SLAMAN and JOHN STEEL, *Definable functions on degrees*, **Cabal seminar 81–85**, Lecture Notes in Mathematics, vol. 1333, Springer, Berlin, 1988, pp. 37–55.

[132] THEODORE A. SLAMAN, *The density of infima in the recursively enumerable degrees*, **Annals of Pure and Applied Logic**, vol. 52 (1991), no. 1–2, pp. 155–179.

[133] ———, *Global properties of the Turing degrees and the Turing jump*, **Computational prospects of infinity. Part I. Tutorials**, Lecture Notes Series. Institute for Mathematical Sciences. National University of Singapore, vol. 14, World Science Publishing, Hackensack, NJ, 2008, pp. 83–101.

[134] ROBERT SOARE, *Automorphisms of the lattice of recursively enumerable sets I: maximal sets*, **Annals of Mathematics**, vol. 100 (1974), pp. 80–120.

[135] ———, *Recursively enumerable sets and degrees*, Springer-Verlag, 1987.

[136] CLIFFORD SPECTOR, *On the degrees of recursive unsolvability*, **Annals of Mathematics**, vol. 64 (1956), pp. 581–592.

[137] F. STEPHAN, *Martin-Löf random sets and PA-complete sets*, **Logic Colloquium '02**, Lecture Notes in Logic, vol. 27, Association for Symbolic Logic, 2006, pp. 342–348.

[138] MARTIN STRAUSS, *Normal numbers and sources for BPP*, **Theoretical Computer Science**, vol. 178 (1997), pp. 155–169.

[139] S. TERWIJN and D. ZAMBELLA, *Algorithmic randomness and lowness*, **The Journal of Symbolic Logic**, vol. 66 (2001), pp. 1199–1205.

[140] A. TURING, *On computable numbers with an application to the Entscheidungsproblem*, **Proceedings of the London Mathematical Society**, vol. 42 (1936), pp. 230–265, correction in **Proceedings of the London Mathematical Society**, vol. 43 (1937), pp. 544–546.

[141] ———, *Systems of logic based on ordinals*, **Proceedings of the London Mathematical Society. Second Series**, vol. 45 (1939), pp. 161–228.

[142] ———, *Computing machinery and intelligence*, **Mind**, vol. 59 (1950), pp. 433–460.

[143] ———, *A note on normal numbers*, **Collected works of A. M. Turing: Pure mathematics** (J. L. Britton, editor), North Holland, Amsterdam, 1992, with notes of the editor in pp. 263–265, pp. 117–119.

[144] J. VILLE, *Étude critique de la notion de collectif*, Gauthier-Villars, 1939.

[145] P. VITANYI, *Information distance in multiples*, **Institute of Electrical and Electronics Engineers. Transactions on Information Theory**, vol. 57 (2011), no. 4, pp. 2451–2456.

[146] R. VON MISES, *Grundlagen der Wahrscheinlichkeitsrechnung*, **Mathematische Zeitschrift**, vol. 5 (1919), pp. 52–99.

[147] J. VON NEUMANN, *Various techniques used in connection with random digits*, **Monte Carlo methods** (A. S. Householder, G. E. Forsythe, and H. H. Germond, editors), National Bureau of Standards Applied Mathematics Series, vol. 12, 1951, pp. 36–38.

[148] A. WALD, *Sur la notion de collectif dans le calcul des probabilités*, **Comptes Rendus des Séances de l'Académie des Sciences**, vol. 202 (1936), pp. 1080–1083.

[149] ———, *Die Weiderspruchsfreiheit des Kollektivbegriffes der Wahrscheinlichkeitsrechnung*, **Ergebnisse eines mathematischen Kolloquiums**, vol. 8 (1937), pp. 38–72.

[150] B. WEINSTEIN, *On embeddings of the 1–3–1 lattice into the recursively enumerable degrees*, Ph.D. thesis, University of California, Berkeley, 1988.

[151] C. YATES, *A minimal pair of recursively enumerable degrees*, **The Journal of Symbolic Logic**, vol. 31 (1966), no. 2, pp. 159–168.

[152] Hector Zenil (editor), **Randomness through computation: Some answers, more questions**, World Scientific, Singapore, 2011.

[153] M. ZIMAND, *Two sources are better than one for increasing the Kolmogorov complexity of infinite sequences*, **Computer science—theory and applications** (E. Hirsch, A. Razborov, A. Semenov, and A. Slissenko, editors), Lecture Notes in Computer Science 5010, Springer, Berlin, 2008, pp. 326–338.

SCHOOL OF MATHEMATICS, STATISTICS AND OPERATIONS RESEARCH
VICTORIA UNIVERSITY OF WELLINGTON
PO BOX 600
WELLINGTON, NEW ZEALAND
E-mail: rod.downey@vuw.ac.nz

COMPUTABLE MODEL THEORY

EKATERINA B. FOKINA, VALENTINA HARIZANOV, AND ALEXANDER MELNIKOV

CONTENTS

1. Introduction and preliminaries. 124
2. Degrees and jump degrees of structures and their isomorphism types. 129
3. Theories, types, models, and diagrams. 138
4. Small theories and their models. 144
5. Effective categoricity. 148
6. Automorphisms of effective structures. 156
7. Degree spectra of relations. 162
8. Families of relations on a structure. 167
9. Classes of structures and equivalence relations. 173

§1. Introduction and preliminaries. In the past few decades there has been increasing interest in computable model theory. Computable model theory uses the tools of computability theory to explore algorithmic content (effectiveness) of notions, theorems, and constructions in various areas of ordinary mathematics. In algebra this investigation based on intuitive notion of effectiveness dates back to van der Waerden who in his 1930 book *Modern Algebra* defined an *explicitly* given field as one the elements of which are uniquely represented by distinguishable symbols with which we can perform the field operations algorithmically. In his pioneering paper [329] on non-factorability of polynomials from 1930, van der Waerden essentially proved that an explicit field $(F, +, \cdot)$ does not necessarily have an algorithm for splitting polynomials in $F[x]$ into their irreducible factors.

Fokina was supported by the Austrian Science Fund (FWF) through the Elise Richter project V206-N13. Harizanov was partially supported by the NSF grant DMS-1202328 and by GWU Columbian College Facilitating Fund. Melnikov was partially supported by Tier 2 grant MOE2011-T2-1-071 (ARC 17/11, M45110030).

The authors are grateful to the anonymous referee, S. Goncharov, J. Knight, R. Miller, A. Montalbán, and A. Morozov for valuable suggestions.

Hilbert proposed in the early 1920s that the formalization of classical mathematical theories be based on consistent axiomatic systems, which are complete. Gödel's incompleteness theorem from 1931 showed that Hilbert's proposal was unattainable for a consistent system with an algorithmic set of axioms, capable of expressing arithmetic. Gödel's theorem is an astonishing early result of computable model theory. He showed that "there are in fact relatively simple problems in the theory of ordinary whole numbers which cannot be *decided* from the axioms."

In 1936, Turing invented the Turing machine, which marked the beginning of computability theory. The work of Church, Gödel, Kleene, Markov, Post, Turing and others in the next decade established the rigorous mathematical foundations for the computability theory. In the 1950s, a famous problem, involving the interplay of algebra and computability, the *word problem*, was resolved. It was shown independently by Novikov [279] and Boone [28] that there exists a finitely presented group G such that the word problem for G is *undecidable*. Adyan [1] further investigated the undecidability of various group-theoretic problems. In 1956, Fröhlich and Shepherdson [107] used the precise notion of a computable function to obtain a collection of results and examples about explicit rings and fields. For example, Fröhlich and Shepherdson proved that "there are two explicit fields that are isomorphic but not explicitly isomorphic." Several years later, Rabin [288] and Mal'cev [228, 229] studied more extensively computable groups and other *computable* (also called *recursive* or *constructive*) algebraic structures, including general structures. Another spectacular negative solution to a famous problem, which involves the interplay of number theory and computability, Hilbert's Tenth Problem, was completed by Matiyasevich [232] in 1970. Building on work of Davis, Putnam, and J. Robinson (see [233]), he established that there is no effective procedure to decide whether a given Diophantine equation has a solution in integers.

In the 1970s, Metakides and Nerode [239, 240] and other researchers in the United States (see [157, 26, 259, 289, 290, 296, 291, 245, 313, 225]) initiated a systematic study of computability in mathematical structures and constructions by using modern computability-theoretic tools, such as the priority method and various coding techniques. At the same time and independently, computable model theory was developed in the Siberian school of constructive mathematics (see [283, 280, 195, 120, 121, 122, 87, 123, 124, 284, 88] and also [92, 93]). While in classical mathematics we can replace some constructions by effective ones, for others such replacement is impossible in principle. For example, from the point of view of computable model theory, isomorphic structures may have very different properties.

Several different notions of effectiveness of structures have been investigated. The generalization and formalization of van der Waerden's intuitive notion of an explicitly given field led to the notion of a computable structure, which is one of the main notions in computable model theory. A structure is *computable* if its domain is computable and its relations and functions are uniformly computable. Further generalization led to a countable structure of a certain Turing degree $\mathbf{d}$. (Computable structures are of degree $\mathbf{0}$.) Henkin's construction of a model for a complete decidable theory is effective and produces a structure $\mathcal{A}$ with a computable domain such that the elementary diagram of $\mathcal{A}$ is decidable. Such a structure is called *decidable*. Thus, in the case of a computable structure, our starting point was semantic, while in the case of a decidable structure, the starting point was syntactic. It is easy to see that not every computable structure is decidable since for computable structures only the atomic (open) diagram has to be decidable. We can also assign Turing degrees or some other computability-theoretic degrees to isomorphisms, as well as to various relations on structures. We can also investigate structures, their theories, fragments of diagrams, relations, and isomorphisms within arithmetic and hyperarithmetic hierarchies.

Computability-theoretic notation in this paper is standard and as in [317]. We review some basic notions and notation. For $X \subseteq \omega$, let $\varphi_0^X, \varphi_1^X, \varphi_2^X, \ldots$ be a fixed effective enumeration of all unary X-computable functions. If X is computable, we omit the superscript X. For $e \in \omega$, let $W_e^X = \text{dom}(\varphi_e^X)$. Hence $W_0, W_1, W_2, \ldots$ is an effective enumeration of all computably enumerable (c.e.) sets. By $X \leq_T Y$ ($X \equiv_T Y$, respectively) we denote that X is Turing reducible to Y (X is Turing equivalent to Y, respectively). By $X <_T Y$ we denote that $X \leq_T Y$ but $Y \not\leq_T X$. We write $\mathbf{x} = \deg(X)$ for the Turing degree of X. Thus, $\mathbf{0} = \deg(\emptyset)$. Let $n \geq 1$. Then $\mathbf{x}^{(n)} = \deg(X^{(n)})$, where $X^{(n)}$ is the n-th Turing jump of X. A set is Σ_n^0 if it is c.e. relative to $\mathbf{0}^{(n-1)}$. A set is Π_n^0 if its negation is Σ_n^0, and a set is Δ_n^0 if it is both Σ_n^0 and Π_n^0. Let $\Delta_0^0 =_{\text{def}} \Delta_1^0$. A set X is *arithmetic* if $X \leq \emptyset^{(k)}$ for some $k \geq 0$. A set $X \leq_T \emptyset'$ and its Turing degree $\mathbf{x}$ are called *low* if $\mathbf{x}' \leq \mathbf{0}'$, and low_n if $\mathbf{x}^{(n)} \leq \mathbf{0}^{(n)}$. The *low basis theorem* of Jockusch and Soare [179], establishes that every infinite binary tree $\mathcal{T}$ has an infinite path f with $f' \leq_T \mathcal{T}'$. In particular, every infinite computable binary tree has a low path.

An ordinal is *computable* if it is finite or is the order type of a computable well order on ω. The computable ordinals form a countable initial segment of the ordinals. *Kleene's $\mathcal{O}$* is the set of notations for computable ordinals, with the corresponding partial order $<_\mathcal{O}$ (see [298, 301]). The ordinal 0 gets notation 1. If a is a notation for α, then 2^a is a notation for $\alpha + 1$. Then $a <_\mathcal{O} 2^a$, and also, if $b <_\mathcal{O} a$, then $b <_\mathcal{O} 2^a$. Suppose α is a limit ordinal. If φ_e is a total function, giving notations for an increasing sequence of ordinals with limit α, then $3 \cdot 5^e$ is a notation for α. For all n, we have $\varphi_e(n) <_\mathcal{O} 3 \cdot 5^e$, and if $b <_\mathcal{O} \varphi_e(n)$, then $b <_\mathcal{O} 3 \cdot 5^e$. Let $|a|$ denote the ordinal with notation a. If $a \in \mathcal{O}$, then the restriction of $<_\mathcal{O}$ to the set $\text{pred}(a) = \{b \in \mathcal{O}: b <_\mathcal{O} a\}$ is a well order of type $|a|$. For $a \in \mathcal{O}$, pred(a) is c.e., uniformly in a. The set $\mathcal{O}$ is Π_1^1-complete.

The least noncomputable ordinal is denoted by ω_1^{CK}, where CK stands for Church–Kleene. To extend the arithmetic hierarchy, we define the representative sets in the hyperarithmetic hierarchy, H_a for $a \in \mathcal{O}$. The definition is recursive, and is based on iterating Turing jump: $H_1 = \emptyset$, $H_{2^a} = (H_a)'$, and $H_{3 \cdot 5^e} = \{2^x \cdot 3^n : x \in H_{\varphi_e(n)}\}$. Let β be an infinite computable ordinal. Then a set is Σ_β^0 if it is c.e. relative to some H_a such that β is represented by notation a. A set is Π_β^0 if its negation is Σ_β^0, and a set is Δ_β^0 if it is both Σ_β^0 and Π_β^0. A set is *hyperarithmetic* if it is Δ_α^0 for some computable α. Hence, a set X is hyperarithmetic if $(\exists a \in \mathcal{O})[X \leq_T H_a]$. The hyperarithmetic sets coincide with Δ_1^1 sets.

Ershov classified Δ_2^0 sets as follows. Let α be a computable ordinal. A set $C \subseteq \omega$ is α-c.e. if there are: a computable function $f : \omega^2 \to \{0, 1\}$; and a computable function $o : \omega \times \omega \to \alpha + 1$ with the following properties:

$$(\forall x)[f(x, 0) = 0 \wedge \lim_{s \to \infty} f(x, s) = C(x)],$$
$$(\forall x)(\forall s)[o(x, 0) = \alpha \wedge o(x, s+1) \leq o(x, s)], \text{ and}$$
$$(\forall x)(\forall s)[f(x, s+1) \neq f(x, s) \Rightarrow o(x, s+1) < o(x, s)].$$

In particular, 1-c.e. sets are c.e. sets, and 2-c.e. sets are d.c.e. sets.

Several important notions of computability on effective structures have syntactic characterizations, which involve computable infinitary formulas introduced by Ash. Roughly speaking, these are infinitary formulas involving infinite conjunctions and disjunctions over c.e. sets. More precisely, let α be a computable ordinal. Ash defined computable Σ_α and Π_α formulas of $L_{\omega_1 \omega}$ recursively and simultaneously and together with their Gödel numbers. The computable Σ_0 and Π_0 formulas are the finitary quantifier-free formulas. The computable $\Sigma_{\alpha+1}$ formulas are of the form

$$\bigvee_{n \in W_e} \exists \overline{y_n} \psi_n(\overline{x}, \overline{y_n}),$$

where for $n \in W_e$, ψ_n is a Π_α formula indexed by its Gödel number, and $\exists \overline{y_n}$ is a finite block of existential quantifiers. That is, $\Sigma_{\alpha+1}$ formulas are c.e. disjunctions of $\exists \Pi_\alpha$ formulas. Similarly, $\Pi_{\alpha+1}$ formulas are c.e. conjunctions of $\forall \Sigma_\alpha$ formulas. It can be shown that a computable Σ_1 formula is of the form

$$\bigvee_{n \in \omega} \exists \overline{y_n} \theta_n(\overline{x}, \overline{y_n}),$$

where $(\theta_n(\overline{x}, \overline{y_n}))_{n \in \omega}$ is a computable sequence of quantifier-free formulas. If α is a limit ordinal, then Σ_α (Π_α, respectively) formulas are of the form $\bigvee_{n \in W_e} \psi_n$ ($\bigwedge_{n \in W_e} \psi_n$, respectively), such that there is a sequence $(\alpha_n)_{n \in W_e}$ of ordinals having limit α, given by the ordinal notation for α, and every ψ_n is a Σ_{α_n} (Π_{α_n}, respectively) formula. For a more precise definition of computable Σ_α and Π_α formulas see [17]. The important property of these formulas, due to Ash, is the following. For a structure $\mathcal{A}$, if $\theta(\overline{x})$ is a computable Σ_α formula, then the set $\{\overline{a}: \mathcal{A} \models \theta(\overline{a})\}$ is Σ_α^0 relative to $\mathcal{A}$. An analogous property holds for computable Π_α formulas. If $\mathcal{A}$ and $\mathcal{B}$ are hyperarithmetic structures satisfying the same computable infinitary sentences, then $\mathcal{A} \cong \mathcal{B}$ (see [140]).

The following is a *compactness theorem* due to Kreisel and Barwise.

THEOREM 1. *Let Γ be a Π_1^1 set of computable infinitary sentences. If every Δ_1^1 subset of Γ has a model, then Γ has a model.*

As a corollary we obtain that if Γ is a Π_1^1 set of computable infinitary sentences, and if every Δ_1^1 subset of Γ has a computable model, then Γ has a computable model (see [17]).

Complexity of a countable structure $\mathcal{A}$ can be measured by its Scott rank. There are several different definitions of Scott rank and we will use one in [17] (also see [41]). First we define a family of equivalence relations on finite tuples $\overline{a}$ and $\overline{b}$ of elements in $\mathcal{A}$, of the same length.

1. We say that $\overline{a} \equiv^0 \overline{b}$ if $\overline{a}$ and $\overline{b}$ satisfy the same quantifier-free formulas.
2. For $\alpha > 0$, we say that $\overline{a} \equiv^\alpha \overline{b}$ if for all $\beta < \alpha$, for every $\overline{c}$, there exists $\overline{d}$, and for every $\overline{d}$, there exists $\overline{c}$, such that $\overline{a}, \overline{c} \equiv^\beta \overline{b}, \overline{d}$.

The *Scott rank of a tuple* $\overline{a}$ in $\mathcal{A}$ is the least β such that for all $\overline{b}$, the relation $\overline{a} \equiv^\beta \overline{b}$ implies $(\mathcal{A}, \overline{a}) \cong (\mathcal{A}, \overline{b})$. The *Scott rank of $\mathcal{A}$*, $SR(\mathcal{A})$, is the least ordinal α greater than the ranks of all tuples in $\mathcal{A}$. For example, if $\mathcal{L}$ is a linear order of type ω, then $SR(\mathcal{L}) = 2$. For a hyperarithmetic structure, the Scott rank is at most $\omega_1^{CK} + 1$. It can be shown (see [17, 41]) that for a computable structure $\mathcal{A}$, we have:

(i) $SR(\mathcal{A}) < \omega_1^{CK}$ if there is a computable ordinal β such that the orbits of all tuples are defined by computable Π_β formulas;

(ii) $SR(\mathcal{A}) = \omega_1^{CK}$ if the orbits of all tuples are defined by computable infinitary formulas, but there is no bound on the complexity of these formulas; and

(iii) $SR(\mathcal{A}) = \omega_1^{CK} + 1$ if there is some tuple the orbit of which is not defined by any computable infinitary formula.

There are structures in natural classes, for example, abelian p-groups, where p is a prime number, with arbitrarily large computable ranks, and of rank $\omega_1^{CK} + 1$, but none of rank ω_1^{CK} (see [25]). Makkai [227] was the first to prove the existence of an arithmetic structure of Scott rank ω_1^{CK}, and in [210], J. Millar and Knight showed that such structure can be made computable. Through the recent work of Calvert, Knight, and J. Millar [42], Calvert, Goncharov, and Knight [37], and Freer [104], we started to better understand the structures of Scott rank ω_1^{CK}. Computable structures of Scott rank ω_1^{CK} were obtained in familiar classes such as trees, undirected graphs, fields of any fixed characteristic, and linear orders [42, 37]. Sacks asked whether for known examples of computable structures of Scott rank ω_1^{CK}, the computable infinitary theories are $\aleph_0$-categorical. In [36], Calvert, Goncharov, J. Millar, and Knight gave an affirmative answer for known examples. In [242], J. Millar and Sacks introduced an innovative technique that produced a countable structure $\mathcal{A}$ of Scott rank ω_1^{CK} such that $\omega_1^{\mathcal{A}} = \omega_1^{CK}$ and the $L_{\omega_1^{CK},\omega}$-theory of $\mathcal{A}$ is *not* $\aleph_0$-categorical. It is not known whether such a structure can be computable.

In this paper, we will not consider structures that are computable with bounds on the resources that algorithms can use, such as time and memory constraints. For a survey of polynomial time structures see the paper [48] by Cenzer and Remmel. Another approach that turned out to be very interesting, which is beyond the scope of this paper, is to consider functions representable by various types of finite automata. For instance, a function presented by a finite string automaton can be computed in linear time using a constant amount of memory. A seminal paper in this field is [202] by Khoussainov and Nerode. The most interesting property of automatic structures is that they have decidable model checking problems. We can use this property to prove the decidability of the first-order theories of many structures, e.g., Presburger arithmetic. There is also a class of tree automatic structures (see [300, 200]), which is richer than the class of automatic structures. Tree automatic structures have nice algorithmic properties, in particular, decidable model checking problem. Many interesting problems in this area remain open.

§2. Degrees and jump degrees of structures and their isomorphism types. We will assume that all structures are at most countable and their languages are computable. Clearly, finite structures are computable. Let **d** be a Turing degree. An infinite structure $\mathcal{M}$ is **d**-*computable* if its universe can be identified with the set of natural numbers ω in such a way that the relations and operations of $\mathcal{M}$ are uniformly **d**-computable. For example, we may consider structures computable in the halting problem, such as Σ_1^0 and Π_1^0 structures. See Higman [165], Feiner [94], Metakides and Nerode [239], Ershov and Goncharov [93], and Cenzer, Harizanov, and Remmel [45] for more on Σ_1^0 structures, and Remmel [289], Khoussainov, Slaman, and Semukhin [206], and Cenzer, Harizanov, and Remmel [45] for more on Π_1^0 structures.

If an algebraic structure is not computable, then it is natural to ask how close it is to a computable one. This property can be captured by the collection of all Turing degrees relative to which a given structure has a computable isomorphic copy. Thus, we have the following definition.

DEFINITION 1. The *degree spectrum* of a structure $\mathcal{A}$ is

$$\mathrm{DgSp}(\mathcal{A}) = \{\deg(D(\mathcal{B})) \colon \mathcal{B} \cong \mathcal{A}\},$$

where $D(\mathcal{B})$ is the the atomic diagram of $\mathcal{B}$.

Knight proved the following fundamental result about the degree spectrum of a structure.

THEOREM 2 ([209]). *The degree spectrum of any structure is either a singleton or is upward closed.*

A structure $\mathcal{A}$ is *automorphically trivial* if there is a finite subset C of its domain such that every permutation of the domain, which fixes C pointwise, is an automorphism of $\mathcal{A}$. Automorphically trivial structures include all finite structures, of course, and also some infinite structures, such as the complete graph on countably many vertices. If the structure is automorphically non-trivial, the degree spectrum is upward closed [209]. The degree spectrum of an automorphically trivial structure always contains exactly one Turing degree, and if the language is finite, that degree must be **0** (see [149]). Jockusch and Richter introduced the following notion.

DEFINITION 2 ([296]). If the degree spectrum of a structure $\mathcal{A}$ has a least element, then this element is called the *degree of the isomorphism type* of $\mathcal{A}$.

Richter [296, 297] initiated the systematic study of such degrees. Richter proved that if $\mathcal{A}$ is a structure without a computable copy, which satisfies the effective extendability condition, then the isomorphism type of $\mathcal{A}$ has no degree. A structure $\mathcal{A}$ satisfies the *effective extendability condition* if for every finite structure $\mathcal{M}$ isomorphic to a substructure of $\mathcal{A}$, and every embedding f of $\mathcal{M}$ into $\mathcal{A}$, there is an algorithm that determines whether a given finite structure $\mathcal{F}$ extending $\mathcal{M}$ can be embedded into $\mathcal{A}$ by an embedding extending f. Richter showed that every linear order, and every tree, as a partially ordered set, satisfy the effective extendability condition. More recently, A. Khisamiev [194] proved that every abelian p-group satisfies the effective extendability condition. Hence the isomorphism type of a countable linear order, a tree, or an abelian p-group, which is not isomorphic to a computable one, does not have a degree of its isomorphism type. Richter also showed that for any Turing degree **d**, there is a torsion abelian group the isomorphism type of which has the degree **d**, as well as that there is such a group the isomorphism type of which does not have a degree. Results of Richter motivated the study of jump degrees of structures. The following definition was also introduced by Jockusch and Richter.

DEFINITION 3 ([296]). Let $\mathcal{A}$ be a structure, and α a computable ordinal. We say that a Turing degree **d** is the αth *jump degree* of $\mathcal{A}$ if it is the least degree in

$$\{\mathbf{d}^{(\alpha)} : \mathbf{d} \in \mathrm{DgSp}(\mathcal{A})\}.$$

The degree **d** is said to be *proper* α^{th} *jump degree* of $\mathcal{A}$ if for every computable ordinal $\beta < \alpha$, the structure $\mathcal{A}$ has no β^{th} jump degree.

Given a class of structures, we may ask for which computable ordinals α there exist representatives of this class having (proper) α^{th} jump degrees.

The following theorem summarizes the results for linear orders due to Knight [209], Ash, Jockusch, and Knight [14], and Downey and Knight [77].

THEOREM 3 ([209, 14, 77]). *If a linear order has first jump degree, it must be* $\mathbf{0}'$. *In contrast, for each computable ordinal* $\alpha \geq 2$ *and every Turing degree* $\mathbf{d} \geq \mathbf{0}^{(\alpha)}$, *there exists a linear order having proper* α^{th} *jump degree* $\mathbf{d}$.

Ordinal jump degrees of Boolean algebras are well-understood as well, but the results differ from the ones for linear orders. Jockusch and Soare established the following result.

THEOREM 4 ([181]). *For* $n \in \omega$, *if a Boolean algebra has* n^{th} *jump degree, then it is* $\mathbf{0}^{(n)}$. *In contrast, for each* $\mathbf{d} \geq \mathbf{0}^{(\omega)}$, *there exists a Boolean algebra with proper* ω^{th} *jump degree* $\mathbf{d}$.

Oates investigated jump degrees of torsion abelian groups.

THEOREM 5 ([281]). *For every computable* α, *there is a torsion abelian group having proper* α^{th} *jump degree.*

The proof relies on algebraic properties of countable abelian p-groups, which are well-undestood.

The situation becomes more complex in the case of countable, torsion-free, abelian groups, where there is no suitable algebraic classification theory. Nevertheless, there has been a significant progress in this area. If $\mathcal{G} = (G, +)$ is a torsion-free abelian group, a set of nonzero elements $\{g_i : i \in I\} \subset G$ is *linearly independent* if $\alpha_1 g_{i_1} + \cdots + \alpha_k g_{i_k} = 0$ has no solution for $\{i_1, \ldots, i_k\} \subseteq I$, $\alpha_i \in \mathbb{Z}$ for each i, and $\alpha_i \neq 0$ for some i. A *basis* for $\mathcal{G}$ is a maximal linearly independent set, and the *rank* of $\mathcal{G}$ is the cardinality of a basis. Calvert, Harizanov, and Schlapentokh obtained the results about Turing degrees of isomorphism types for various familiar algebraic classes, including torsion-free abelian groups of finite rank.

THEOREM 6 ([39]). *There are algebraic fields and torsion-free abelian groups of any finite rank* > 1, *the isomorphism types of which have arbitrary Turing degrees. There are structures in each of these classes the isomorphism types of which do not have Turing degrees.*

For rank 1, torsion-free, abelian groups the result was previously obtained by Knight, Downey, and Jockusch (see [81]). Such groups are isomorphic to subgroups of $(\mathbb{Q}, +)$, and there is a known classification for these groups due to Baer.

Melnikov [238] showed that not every infinite-rank, torsion-free, abelian group has first jump degree. Results about the existence of proper jump degrees for torsion-free abelian groups were resolved by Downey and Jockusch for the first jump, and by Melnikov for the second and the third jump.

THEOREM 7 ([81, 238]). *For $n \in \{1, 2\}$ and every degree $\mathbf{d} \geq \mathbf{0}^{(n)}$, there is a torsion-free group having proper n^{th} jump degree $\mathbf{d}$. For every degree $\mathbf{d} > \mathbf{0}'''$, there is a torsion-free group having proper 3rd jump degree $\mathbf{d}$.*

The case of higher ordinals remained unresolved until the recent work of Andersen, Kach, Melnikov, and Solomon who obtained the following general result.

THEOREM 8 ([2]). *For every computable $\alpha > 3$, every $\mathbf{d} > \mathbf{0}^{(\alpha)}$ can be realized as a proper α^{th} jump degree of a torsion-free abelian group.*

It is not known whether the result can be strengthened to $\mathbf{d} = \mathbf{0}^{(\alpha)}$ for $\alpha > 2$. The groups from Theorem 7 are of the form $\bigoplus_{i \in \omega} \mathcal{H}_i$, where $\mathcal{H}_i \leq (\mathbb{Q}, +)$. Such groups, introduced by Baer in 1937, are called *completely decomposable* and have nice algebraic properties. In the case of only one summand, Coles, Downey, and Slaman [58] established the following theorem, as a consequence of their pure computability-theoretic result that for every set $C \subseteq \omega$, there is a Turing degree that is the least degree of the jumps of all sets X for which C is c.e. in X.

THEOREM 9 ([58]). *Every torsion-free abelian group of rank 1 has first jump degree.*

Theorem 9 can be extended to torsion-free abelian groups of any finite rank, as was observed in [39, 238]. It is not known which ordinals are realized as proper jump degrees of groups of the form $\bigoplus_{i \in \omega} \mathcal{H}_i$, where $\mathcal{H}_i \leq (\mathbb{Q}, +)$.

For certain classes of countable structures, we can use computable functors to translate results from one class of countable structures to another. A functor $\Phi \colon \mathcal{K} \to \mathcal{K}_1$ is *computable* if, given an enumeration of an open diagram of $\mathcal{A} \in \mathcal{K}$, we can enumerate the open diagram of $\Phi(\mathcal{A}) \in \mathcal{K}_1$, in a uniform fashion. Computable functors are also called effective transformations. Hirschfeldt, Khoussainov, Shore, and Slinko used injective effective transformations to transfer various computability-theoretic results from graphs to structures in other familiar algebraic classes.

THEOREM 10 ([174]). *For every automorphically nontrivial structure $\mathcal{A}$, there is a symmetric irreflexive graph, a partial order, a lattice, a ring, an integral domain of arbitrary characteristic, a commutative semigroup, or a 2-step nilpotent group the degree spectrum of which coincides with $\mathrm{DgSp}(\mathcal{A})$.*

As a consequence we obtain that these classes have structures with proper α^{th} jump degrees for all computable ordinals α. Frolov, Kalimullin, and R. Miller [109] investigated degree spectra of algebraic fields.

THEOREM 11 ([109]). *Every algebraic field has first jump degree.*

Not much is known about groups that are far from abelian. There are centerless groups that have arbitrary Turing degrees for their isomorphism classes, as well as no degrees [68]. Recently, Calvert, Harizanov, and Shlapentokh [38] started to investigate effective content of geometric objects, such as ringed spaces and schemes. In particular, they showed that ringed spaces corresponding to unions of varieties, ringed spaces corresponding to unions of subvarieties of certain fixed varieties, and schemes over a fixed field can have arbitrary Turing degrees for their isomorphism classes, as well as no degrees.

Lempp asked if there is a nontrivial sufficient condition on a structure, which will guarantee that its degree spectrum contains **0**. Slaman [315] and Wehner [332] independently obtained the following result, with different proofs.

THEOREM 12 ([315, 332]). *There exists a structure the degree spectrum of which is the set of all noncomputable Turing degrees.*

Wehner [332] constructed a family of sets that yields a structure with isomorphic copies in exactly the noncomputable Turing degrees. While Wehner's structure is elementarily equivalent to a computable structure, Slaman's is not. We will say that a structure such as one in Theorem 12 has *Slaman–Wehner degree spectrum*. More recently, Hirschfeldt [171] proved that there is a structure with Slaman–Wehner degree spectrum, which is a prime model of a complete decidable theory. This also gives another proof of Theorem 12. Hirschfeldt's structure is elementarily equivalent to a decidable structure. Hirschfeldt's degree spectrum result follows from his theorem in [171] that if $\mathcal{T}$ is a computable tree with no dead ends and with all infinite paths computable, and D is a noncomputable set, then there is a D-computable listing of the isolated paths in $\mathcal{T}$. Previously, Goncharov and Nurtazin [142] and T. Millar [245] established that this result does not hold if D is computable.

Downey asked if there exists a structure in a natural algebraic class of structures, such as a linear order or an abelian group, which has Slaman–Wehner spectrum. We can also ask which sets of degrees can be realized as degree spectra of structures. Since co-null collections of degrees are of a particular interest, we have the following definition due to Kalimullin.

DEFINITION 4 ([186]). An automorphically nontrivial structure $\mathcal{M}$ is called *almost computable* if the measure of $\text{DgSp}(\mathcal{M})$ is equal to 1 under the standard Lebesgue measure on the Cantor space.

For example, every structure with Slaman–Wehner spectrum is almost computable. More examples have been obtained recently. Kalimullin [187, 188, 189] investigated the relativization of Slaman–Wehner theorem to nonzero degrees. He showed that such a relativization holds for every *low* Turing degree, as well as every c.e. degree, but not for every Δ^0_3 Turing degree. Using the enumeration result of Wehner, also relativized, Goncharov, Harizanov, Knight, McCoy, R. Miller, and Solomon [116] showed that for every computable successor ordinal α, there is a structure with copies in just the degrees of sets X such that $\Delta^0_\alpha(X)$ is not Δ^0_α. As a consequence, they obtained the following result.

THEOREM 13 ([116]). *For each finite n, there is a structure with the degree spectrum consisting of exactly all non-low_n Turing degrees.*

Consequently, there are almost computable structures without arithmetic isomorphic copies. Csima and Kalimullin provided another interesting example of a possible degree spectrum.

THEOREM 14 ([63]). *The set of hyperimmune degrees is the degree spectrum of a structure.*

We could ask the following analogue of Lemmp's question for almost computable structures. If a structure is almost computable, must it contain a hyperarithmetic or a Π_1^1 degree? Greenberg, Montalbán, and Slaman [144] and independently Kalimullin and Nies (unpublished) obtained the following positive result.

THEOREM 15 ([144]). *If $\mathcal{M}$ is an almost computable structure, then there is some copy of $\mathcal{M}$ that is computable from Kleene's $\mathcal{O}$.*

This bound cannot be improved to be hyperarithmetic. Recently, Greenberg, Montalbán, and Slaman [143] constructed a linear order the degree spectrum of which is the set of all non-hyperarithmetic degrees. There are other examples of almost computable structures in various natural algebraic classes and we will discuss some of them.

Although the degree spectra of linear orders have been intensively studied, the following question remains open. Is there a linear order the degree spectrum of which is the set of all nonzero degrees? Jockusch and Soare [180] established that for every nonzero c.e. Turing degree $\mathbf{d}$, there is a linear order $\mathcal{L}$ of Turing degree $\mathbf{d}$ such that $\mathcal{L}$ does not have a computable copy. Downey, Seetapun, and Knight extended this result to an arbitrary nonzero Turing degree (see [81]). R. Miller [254] constructed a linear order with the spectrum containing all nonzero Δ_2^0 degrees but not $\mathbf{0}$. Recently, Frolov, Harizanov, Kalimullin, Kudinov, and R. Miller obtained the following examples.

THEOREM 16 ([108]). *Let $n \geq 2$. For every Turing degree $\mathbf{c}$, there is a linear order with spectrum $\{\mathbf{d} : \mathbf{d}^{(n)} > \mathbf{c}\}$. In particular, there is a linear order the spectrum of which contains exactly the non-low_n degrees.*

For a survey of related results on linear orders see [108].

Slaman–Wehner's degree spectrum is not possible when restricted to the class of countable Boolean algebras. Knight and Stob [212] obtained the following result about low_4 Boolean algebras, extending a result of Downey and Jockusch [75] for low Boolean algebras, and of Thurber [328] for low_2 Boolean algebras.

THEOREM 17 ([212]). *Every low_4 Boolean algebra has a computable isomorphic copy.*

One of the main open questions in this area is the following. Is every low_n, $n \geq 5$, Boolean algebra isomorphic to a computable one? The affirmative answer to this question is known as the *low$_n$ Boolean algebra conjecture*. There is some evidence that if every *low$_5$* Boolean algebra has a computable copy, then the proof of that statement should be different from the proof for *low$_4$* Boolean algebras. This follows from work of Harris and Montalbán in [161] where they showed that there are over 1000 invariants that have to be considered for the *low$_5$* case, as well as from work of Harris and Montalbán on the complexity of isomorphisms in [160].

Similarly to linear orders, the following question is open. Is there an abelian group having Slaman–Wehner degree spectrum? Recently, Khoussainov, Kalimullin, and Melnikov proved the following result about abelian p-groups.

THEOREM 18 ([185]). *There exists an abelian p-group, which has a **d**-computable copy relative to every noncomputable Δ_2^0 Turing degree **d**, but has no computable copy.*

In addition, Khoussainov, Kalimullin, and Melnikov [185] proved that there exists a noncomputable torsion abelian group the degree spectrum of which contains all hyperimmune degrees. They also showed that this result cannot be generalized to co-countable collections of degrees, when restricted to direct sums of cyclic groups. These results can be re-formulated in terms of effective monotonic approximations that we will later introduce. It is also known that there exists a torsion-free abelian group having exactly *nonlow* isomorphic copies [238]. Other structures studied in this context come from [174]. There are also some related results on equivalence structures (see [45, 185]).

In many cases, the existence of a computable copy of a structure is related to the ability to enumerate a certain invariant of the structure.

Examples.
(i) Given a set S, define the algebraic extension $\mathcal{F}_S$ of the prime field $\mathbb{Q}$ to be $\mathbb{Q}(\{\sqrt{p_x} : x \in S\})$. The field $\mathcal{F}_S$ has an X-computable copy if and only if S is c.e. in X.

(ii) Given a set S, define a subgroup $\mathcal{G}(S)$ of $(\mathbb{Q}, +)$ by having a generator $\frac{1}{p_x}$ for $\mathcal{G}(S)$ if and only if $x \in S$. Then $\mathcal{G}(S)$ has an X-computable copy if and only if S is c.e. in X.

It is well known that under an appropriate choice of S neither $\mathcal{F}_S$ nor $\mathcal{G}(S)$ has a Turing degree for its isomorphism type (see, for example, [39]). Nevertheless, in the examples above, we may define the *enumeration degree* of $\mathcal{F}_S$ or $\mathcal{G}(S)$ to be the degree of the set S under the enumeration reducibility $\leq_e$. There is also a direct way to define an enumeration degree spectrum of a structure, as A. Soskova and Soskov did in [322, 323]. More generally, we may view a degree spectrum as a mass problem. The following general definition is due to Medvedev.

DEFINITION 5 ([237]). A *mass problem* is a collection of total functions from ω to ω.

Stukachev defined various reducibilities between mass problems of structures, such as Muchnik reducibility. As usual, we identify the atomic diagram $\mathcal{D}(\mathcal{B})$ of a countable structure $\mathcal{B}$ with its characteristic function $\chi_{\mathcal{D}(\mathcal{B})} \in 2^\omega$, under Gödel coding of formulas.

DEFINITION 6 ([324]).
(i) The *mass problem* of a countable structure $\mathcal{A}$ is the set
$$\{\chi_{\mathcal{D}(\mathcal{B})} : \mathcal{B} \cong \mathcal{A}\}.$$
(ii) Given countable structures $\mathcal{A}$ and $\mathcal{B}$, we say that $\mathcal{A}$ is *Muchnik reducible* to $\mathcal{B}$, in symbols $\mathcal{A} \leq_w \mathcal{B}$, if $\mathrm{DgSp}(\mathcal{A}) \subseteq \mathrm{DgSp}(\mathcal{B})$.

Thus, $\mathcal{A}$ is Muchnik equivalent to $\mathcal{B}$, written as $\mathcal{A} =_w \mathcal{B}$, if $\mathcal{A} \leq_w \mathcal{B}$ and $\mathcal{B} \leq_w \mathcal{A}$. Selman's theorem [311] states that if a structure has an enumeration degree as defined above, then $\leq_w$ coincides with the enumeration reducibility $\leq_e$. Thus, the notion of enumeration degree is a special case of Definition 6. For other reducibilities on mass problems of structures see Stukachev [324, 325].

Whenever a reducibility is defined, we look for a suitable definition of the *jump*. Various authors recently and independently introduced the notion of the jump of an abstract structure: Baleva [22] and Soskov and A. Soskova [323] using Moschovakis extensions; Morozov [271] and Puzarenko [286] in the context of admissible sets; Montalbán [258] using predicates for computable infinitary Σ_1 formulas; Stukachev [326] using hereditarily finite extensions. It is remarkable that these different approaches turned out to be equivalent. We give the definition due to Montalbán.

DEFINITION 7 ([258]). Given a language L, let $\{\theta_i : i \in \omega\}$ be a computable enumeration of all computable infinitary Σ_1 formulas in L. Given a structure $\mathcal{A}$ for L, let $\mathcal{A}'$ be the structure obtained by adding to $\mathcal{A}$ infinitely many relations P_i, for $i \in \omega$, where
$$\mathcal{A} \models P_i(\overline{x}) \Leftrightarrow \theta_i(\overline{x}),$$
and the arity of P_i is the same as the length of $\overline{x}$ in $\theta_i(\overline{x})$.

Several results on degree spectra of structures can be re-formulated in terms of the jumps of structures. For instance, the result of Downey and Jockusch in [75] that every *low* Boolean algebra is isomorphic to a computable one follows from the following result. If $\mathcal{B}$ is a Boolean algebra, and $\mathbf{0}'$ computes a copy of $\mathcal{B}'$, then $\mathcal{B}$ has a computable copy. A better understanding of the jump operator on structures may help us establish or refute the low_n Boolean algebra conjecture.

A. Soskova and Soskov, and also Montalbán showed that the spectrum of a structure behaves well with respect to the jump operator of the structure. More precisely, they established the following *jump inversion theorem*.

THEOREM 19 ([323, 258]). *For every structure $\mathcal{A}$, we have*
$$\mathrm{DgSp}(\mathcal{A}') = \{\mathbf{d}' : \mathbf{d} \in \mathrm{DgSp}(\mathcal{A})\}.$$

Other authors also independently proved the jump inversion theorem. See Stukachev [326] for more on the jump inversion results. Recently, Puzarenko [287] and Montalbán [256] showed independently and simultaneously that the jump operator has a fixed point.

THEOREM 20 ([287, 256]). *There is a structure $\mathcal{A}$ such that $\mathcal{A} =_w \mathcal{A}'$.*

Montalbán proved this theorem under the assumption that "$0^{\#}$ exists", and Puzarenko obtained another proof that does not use this assumption.

Andrews and J. Miller [9] have recently defined the *spectrum of a theory T* to be the set of Turing degrees of models of T. The idea behind this notion is to better understand the relationship between the model-theoretic properties of a theory and the computability-theoretic complexity of its models. Theory spectra may coincide with degree spectra of structures, e.g., the cones above arbitrary Turing degrees are theory spectra, as well as the set of all noncomputable degrees. On the other hand, there are examples of theory spectra that are not degree spectra for any structure, and *vice versa*. We say that a real x is *Martin-Löf random* or *1-random* iff for every computable collection of c.e. open sets $\{U_n : n \in \omega\}$, with $\mu(U_n) \leq 2^{-n}$, $n \in \omega$, we have $x \notin \cap_{n \in \omega} U_n$, where μ is the standard Lebesgue measure on the Cantor space. A Turing degree is called *1-random* if it contains a set that is 1-random. For more on randomness see [278, 74].

THEOREM 21 ([9]). *The following sets of Turing degrees can be theory spectra:*
(a) *the degrees of complete extensions of Peano arithmetic,*
(b) *1-random degrees, and*
(c) *the union of the cones above two incomparable Turing degrees.*

However, as it follows from [322] and [9], these sets are not the degree spectra of any structures. On the other hand, by [144], there is a structure the degree spectrum of which consists of exactly the non-hyperarithmetic degrees.

THEOREM 22 ([9]). *The collection of non-hyperarithmetic degrees is not the spectrum of a theory.*

Further interesting examples can be found in [9], and for the case of atomic theories in [7].

The notion of the degree spectrum of a structure turned out to be useful in order to find a new approach to resolve one of the most famous conjectures in mathematical logic: Vaught's conjecture. Recall that *Vaught's conjecture* states that the number of countable models of a first-order theory is either countable or continuum. In [255], Montalbán analyzed computability-theoretic properties of a possible counterexample to Vaught's conjecture in terms of degree spectra of its models. The analysis is done under the assumption of projective determinacy (PD). The result of [255] is stated not only for finitary first-order theories, but for $L_{\omega_1\omega}$ sentences. When the continuum hypothesis (CH) does not hold, we say that an $L_{\omega_1\omega}$-theory T is a *counterexample* to Vaught's conjecture if it has uncountably many countable models but not continuum many. Montalbán [255] also gives another definition, which is equivalent to the given definition under $\neg$CH, and also makes sense when CH holds. He defines a class K of structures to satisfy the property *hyperarithmetic-is-computable on a cone* if there exists Y such that for all X with $X \geq_T Y$, every X-hyperarithmetic structure in K has an X-computable copy.

THEOREM 23 (ZFC + PD). ([255]) *Let T be an $L_{\omega_1\omega}$-sentence with uncountably many countable models. The following are equivalent*:

(i) *T is a counterexample to Vaught's conjecture*;

(ii) *The class of models of T satisfies the property* hyperarithmetic-is-computable on a cone; *and*

(iii) *There exists an oracle relative to which*

$$\{\mathrm{DgSp}(\mathcal{A}) \colon \mathcal{A} \models T\} = \{\{X \in 2^\omega \colon \omega_1^X \geq \alpha\} \colon \alpha \in \omega_1\}.$$

§3. Theories, types, models, and diagrams. We will assume that our theories are consistent, countable, and have infinite models. We will denote the *elementary* (*complete*) *diagram* of $\mathcal{A}$ by $D^c(\mathcal{A})$. It is easy to see that the theory of a structure $\mathcal{A}$ is computable in $D^c(\mathcal{A})$, and that $D^c(\mathcal{A})$ is computable in $(D(\mathcal{A}))^{(\omega)}$. The atomic diagram of a model of a theory may be of much lower Turing degree than the theory itself. Henkin's construction of models is effective and establishes that a decidable theory has a decidable model. The *low basis theorem* can be used to obtain for a theory S, a model $\mathcal{A}$ with

$$(D^c(\mathcal{A}))' \leq_T S'.$$

Harizanov, Knight, and Morozov [156] showed that for every automorphically nontrivial structure $\mathcal{A}$, and every set $X \geq_T D^c(\mathcal{A})$, there exists $\mathcal{B} \cong \mathcal{A}$ such that

$$D^c(\mathcal{B}) \equiv_T D(\mathcal{B}) \equiv_T X.$$

For every automorphically trivial structure $\mathcal{A}$, we have $D^c(\mathcal{A}) \equiv_T D(\mathcal{A})$.

A structure $\mathcal{A}$ is called *n-decidable* for $n \geq 1$ if the Σ_n-diagram of $\mathcal{A}$ is decidable. We will denote Σ_n-diagram $\mathcal{A}$ by $D_n(\mathcal{A})$. For sets X and Y, we say that Y is *c.e. in and above* (*c.e.a. in*) X if Y is c.e. relative to X, and $X \leq_T Y$. For any structure $\mathcal{A}$, $D_{n+1}(\mathcal{A})$ is c.e.a. in $D_n(\mathcal{A})$, uniformly in n, where $D_0(\mathcal{A}) = D(\mathcal{A})$. Chisholm and Moses [52] established that there is a linear order that is n-decidable for every $n \in \omega$, but has no decidable copy.

Goncharov [121] earlier obtained a similar result for Boolean algebras. There are familiar structures $\mathcal{A}$ such that for all $\mathcal{B} \cong \mathcal{A}$, we have $D^c(\mathcal{B}) \equiv_T D(\mathcal{B})$. In particular, this is true for algebraically closed fields, and for other structures for which we have effective elimination of quantifiers. In [156], Harizanov, Knight, and Morozov gave syntactic conditions on $\mathcal{A}$ under which for all $\mathcal{B} \cong \mathcal{A}$, we have $D^c(\mathcal{B}) \equiv_T D_n(\mathcal{B})$ for $n \in \omega$.

In the early 1960s, Vaught [331] developed the theory of prime, saturated, and homogeneous models using types. A countable structure $\mathcal{A}$ is *homogeneous* if for every two finite sequences $\overline{a}$ and $\overline{b}$ of the same length n, if $\overline{a}$ and $\overline{b}$ realize the same n-type in $\mathcal{A}$, then there is an automorphism of $\mathcal{A}$ taking $\overline{a}$ to $\overline{b}$. Every countable complete theory has a countable homogeneous model. Prime models and countable saturated models are examples of homogeneous models. The study of the computable content of these models was initiated in the 1970s. The set of all computable types of a complete decidable theory is a Π_2^0 set. Every principal type of such a theory is computable, and the set of all its principal types is Π_1^0.

A model $\mathcal{A}$ of a theory T is *prime* if for all models $\mathcal{B}$ of T, $\mathcal{A}$ elementarily embeds into $\mathcal{B}$. For example, the algebraic numbers form a prime model of the theory of algebraically closed fields of characteristic 0. All prime models of a given theory are isomorphic. It is well known that every complete atomic theory has a prime model. It is not difficult to show that if a complete decidable theory T has a decidable prime model, then the set of all principal types of T is uniformly computable. Goncharov and Nurtazin [142] and independently Harrington [157] established the converse.

THEOREM 24 ([142, 157]). *For a complete decidable theory T, the following are equivalent.*

1. *There is a uniform procedure that maps a formula consistent with T into a computable principal type of T, which contains this formula.*
2. *The theory T has a decidable prime model.*
3. *The theory T has a prime model and the set of all principal types of T is uniformly computable.*

For a set X and its Turing degree $\mathbf{x} = \deg(X)$, we say that a structure $\mathcal{A}$ is *decidable in X* or $\mathbf{x}$-*decidable* if $D^c(\mathcal{A}) \leq_T X$. Drobotun [87] and T. Millar [245] independently showed that a complete, atomic, decidable theory has a $\mathbf{0}'$-*decidable* prime model. More recently, Csima [64] strengthened this result by showing that every complete, atomic, decidable theory T has a prime model $\mathcal{A}$ such that $D^c(\mathcal{A})$ is *low*. Although Csima's result has the same flavor as the *low basis theorem*, it does not follow from it. Epstein extended Csima's result by establishing the following.

THEOREM 25 ([89]). *Let T be a complete, atomic, decidable theory with a prime model $\mathcal{A}$ such that $D^c(\mathcal{A})$ has a c.e. degree $\mathbf{c} > \mathbf{0}$. Then there is a prime model $\mathcal{B}$ of T such that $D^c(\mathcal{B})$ has a low c.e. degree $\mathbf{a}$, where $\mathbf{a} < \mathbf{c}$.*

On the other hand, there are theories with prime models the elementary diagrams of which have minimal degrees, but the theories have no decidable prime models.

Goncharov [128] proved that there is a complete, decidable, ω-stable theory in a finite language having no computable homogeneous model. A theory T is ω-stable if for every $\mathcal{M} \models T$ and every countable $X \subseteq M$, there are only countably many types of T over X. (Uncountably categorical theories, which will be investigated in the next section, are notable examples of ω-stable theories.) Goncharov's theory has infinitely many axioms. Peretyat'kin [285] constructed a complete, atomic, finitely axiomatizable (hence decidable) theory without a computable prime model. T. Millar [243] came up with a weaker notion of a decidable model, the notion of an *almost decidable model*, and showed that if a complete decidable theory has fewer than continuum many complete types, then the theory has an almost decidable prime model. Since not every decidable complete theory with only countably many complete types has a decidable model [128], T. Millar's result cannot be extended to decidable prime models. Hirschfeld obtained an interesting result about the degree spectrum of a prime model, already mentioned in the previous section.

THEOREM 26 ([171]). *There is a prime model of a complete decidable theory with Slaman–Wehner degree spectrum.*

We can also consider theories of algebraic structures from natural classes, such as groups or linear orders. Even if their theories are not necessarily decidable, they can have computable models. N. Khisamiev obtained the following negative result.

THEOREM 27 ([197]). *There is a complete theory of abelian groups with both a computable model and a prime model, but no computable prime model.*

Interestingly, the proof of this result has influenced other investigations in computable model theory, outside group theory. Khisamiev's proof uses the concept of a limitwise monotonic function, which he introduced in [196] to study which abelian p-groups have computable isomorphic copies.

DEFINITION 8 ([196]). A total function $F\colon \omega \to \omega$ is *limitwise monotonic* if there is a computable function $f\colon \omega^2 \to \omega$ such that for all $i, s \in \omega$, we have $f(i, s) \leq f(i, s + 1)$, the limit $\lim_{s \to \infty} f(i, s)$ exists, and $F(i) = \lim_{s \to \infty} f(i, s)$.

See [185] for more on limitwise monotonic functions. Using limitwise monotonic functions, Hirschfeldt obtained a negative solution to a long-standing problem posed by Rosenstein [299].

THEOREM 28 ([169]). *There is a complete theory of linear orders having a computable model and a prime model, but no computable prime model.*

A set X and its Turing degree are called *prime bounding* if every complete, atomic, decidable theory has a prime model $\mathcal{A}$ such that $D^c(\mathcal{A}) \leq_T X$. Thus,

$\emptyset'$ is prime bounding. Csima, Hirschfeldt, Knight, and Soare obtained the following equivalence.

THEOREM 29 ([66]). *Let $X \leq_T \emptyset'$. Then X is prime bounding if and only if X is not* low$_2$.

This theorem gives an interesting characterization of *low$_2$* sets in terms of prime models of certain theories, thus providing a link between computable model theory and degree theory. To prove that a *low$_2$* set X is not prime bounding, we use a $\emptyset'$-computable listing of the array of sets $\{Y : Y \leq_T X\}$ to find a complete, atomic, decidable theory T, which diagonalizes against all potential prime models of T the elementary diagrams of which are computable in X. To prove that any set X that is not *low$_2$* is indeed prime bounding, we fix a function $f \leq_T X$ that dominates every total $\emptyset'$-computable function. Given a complete, atomic, decidable theory T, we use f to build a prime model of T. In addition to the two properties in Theorem 29, Csima, Hirschfeldt, Knight, and Soare [66] consider a number of other properties equivalent to these two, some of which are related to limitwise monotonic functions.

Recall that a countable *saturated* model is a model realizing every type of its language augmented by any finite tuple of constants for its elements. The earliest effective notion related to saturated models was the notion of a recursively saturated model introduced and first studied by Barwise and Schlipf in [26]. A *recursively saturated* model is a model (of a computable language) realizing every *computable* set of formulas consistent with its theory, in the language expanded by any finite set of constants. Note that every saturated model is recursively saturated. It is well known that a complete theory has a countable saturated model if and only if the theory has only countably many n-types for every $n \geq 1$. On the other hand, *every* complete theory in a computable language with infinite models has a countable recursively saturated model. In fact, in the case of a computable language, early proofs of several classical results in model theory can be simplified using recursively saturated models (see [49]). The simplification is done by replacing "large" models by recursively saturated models in the proofs [26]. The "large" models exist only under certain set-theoretic restrictions [49]. Being a computable language is often not a severe restriction since many important languages are computable or even finite. These remarkable results provide an application of computability theory to classical model theory. However, a recursively saturated model does not have to be decidable or even computable, so we will turn our attention to decidable saturated models.

Decidable saturated models of complete decidable theories are fairly well-understood. There is a complete description of decidable saturated models in terms of types, due to Morley [259] and T. Millar [245] independently.

THEOREM 30 ([259, 245]). *Let T be a complete decidable theory. The set of all types of T is uniformly computable if and only if T has a decidable saturated model.*

Thus, a complete theory with a decidable saturated model also has a decidable prime model. Morozov obtained a general positive result for Boolean algebras.

THEOREM 31 ([261]). *Every countable saturated Boolean algebra has a decidable isomorphic copy.*

If the types are not uniformly computable, then the existence of a decidable saturated model is not guaranteed, as shown independently by Goncharov and Nurtazin [142], Morley [259] and T. Millar [245], who constructed counterexamples.

THEOREM 32 ([142, 259, 245]). *There is a complete decidable theory with all types computable, which does not have a decidable saturated model.*

Any saturated model of a complete decidable theory with all types computable has a $0'$-decidable isomorphic copy [142, 259, 245]. This result leads to the investigation of the effective content of saturated models using degree-theoretic concepts and machinery. The following definition was introduced by Harris and is similar to the one for prime models. A Turing degree **d** is *saturated bounding* if every complete decidable theory with types all computable has a **d**-decidable saturated model. Macintyre and Marker [226] showed that the degrees of complete extensions of Peano arithmetic are saturated bounding. There is a recent negative result due to Harris.

THEOREM 33 ([159]). *For every $n \in \omega$, no low$_n$ c.e. degree is saturated bounding.*

For a structure $\mathcal{A}$, the *type spectrum* of $\mathcal{A}$ is the set of all types realized in $\mathcal{A}$. Since a countable homogeneous structure is uniquely determined, up to isomorphism, by the set of types it realizes, Morley posed the following natural question for a complete decidable theory T. If the type spectrum of a countable homogeneous model $\mathcal{A}$ of T consists only of computable types and is computable, does $\mathcal{A}$ have a decidable isomorphic copy? Independently, Goncharov [124], Peretyat'kin [284], and T. Millar [246] answered Morley's question negatively.

THEOREM 34 ([124, 284, 246]). *There exists a complete decidable theory T having a homogeneous model $\mathcal{M}$ without a decidable copy, such that the type spectrum of $\mathcal{M}$ consists only of computable types and is computable.*

In fact, Goncharov [124] and Peretyat'kin [284] provided a criterion for a homogeneous model to be decidable. Their criterion can be stated in terms of the effective extension property. A computable set of computable types of a theory has the *effective extension property* if there is a partial computable function f which, given a type Γ_n of arity k and a formula θ_i of arity $k+1$

(identified with their indices), outputs the index for a type containing Γ_n and θ_i, if there exists such a type.

It is well known that every countable model has a countable homogeneous elementary extension. Ershov conjectured that every decidable model can be elementary embedded into a decidable homogeneous elementary extension. Peretyat'kin refuted Ershov's conjecture in a strong way.

THEOREM 35 ([283]). *There exists a decidable model, which does not have a* computable *homogeneous elementary extension.*

Goncharov and Drobotun [133] constructed a computable linear order that does not have a computable homogeneous elementary extension.

Regarding more recent investigation of degree-theoretic content of homogeneous models, similarly to prime bounding and saturated bounding degrees, we have the following definition. A Turing degree **d** is *homogeneous bounding* if every complete decidable theory has a **d**-decidable homogeneous model. Csima, Harizanov, Hirschfeldt, and Soare obtained the following result about homogeneous bounding degrees.

THEOREM 36 ([61]). *There is a complete decidable theory T such that every countable homogeneous model of T has the degree of a complete extension of Peano arithmetic.*

This theorem implies that every homogeneous bounding degree is the degree of a complete extension of Peano arithmetic, but it is in fact stronger, since we build a *single* theory T such that the use of the degrees of complete extensions of Peano arithmetic is necessary to compute even the atomic diagram of a homogeneous model of T. Together with the converse of Theorem 36 due to Macintyre and Marker [226], we have the following consequence.

COROLLARY 1. *A Turing degree **d** is homogeneous bounding if and only **d** is the degree of a complete extension of Peano arithmetic.*

Lange introduced the following definition of a **0**-homogeneous bounding degree.

DEFINITION 9 ([219]).

1. A countable structure $\mathcal{A}$ has a **d**-*basis* if the types realized in $\mathcal{A}$ are all computable and the Turing degree **d** can list Δ_0^0-indices for all types realized in $\mathcal{A}$.
2. A Turing degree **c** is **0**-*basis homogeneous bounding* if for every automorphically nontrivial homogeneous model $\mathcal{A}$ with a **0**-basis, there exists $\mathcal{B}$ such that $\mathcal{B} \cong \mathcal{A}$ and $\mathcal{B}$ is **c**-decidable.

Now we can restate Theorem 34 as follows: There exists a homogeneous model $\mathcal{A}$ having a **0**-basis but no decidable isomorphic copy.

THEOREM 37 ([219]). *Let T be a complete decidable theory and let $\mathcal{A}$ be a homogeneous model of T with a $\mathbf{0}'$-basis. Then $\mathcal{A}$ has an isomorphic copy decidable in a* low *degree.*

This theorem implies Csima's result that every complete, atomic, decidable theory T has a prime model decidable in a *low* degree (see [64]).

THEOREM 38 ([219]). *Let T be a complete decidable theory with all types computable. Let $\mathcal{A}$ be a homogeneous model of T with a $\mathbf{0}$-basis. Then $\mathcal{A}$ has an isomorphic copy $\mathcal{B}$ decidable in any nonzero degree.*

Lange also gave a characterization of $\mathbf{0}$-basis homogeneous bounding degrees.

THEOREM 39 ([219, 220]). *A degree $\mathbf{d} \leq \mathbf{0}'$ is $\mathbf{0}$-basis homogeneous bounding if and only if $\mathbf{d}$ is* nonlow$_2$.

§4. **Small theories and their models.** We now consider the question of the existence of effective (computable, decidable, etc.) models for *small theories*, that is, theories with at most countably many countable models.

DEFINITION 10. Let κ be a cardinal. A theory is called *κ-categorical* if it has exactly one model of cardinality κ, up to isomorphism.

The following result is well known as *Morley's categoricity theorem* (see [49]).

THEOREM 40 (Morley). *If a theory T is κ-categorical for some uncountable cardinal κ, then T is λ-categorical for all uncountable λ.*

Hence, theories categorical in an uncountable cardinal are also called *uncountably categorical*. The theories that are $\aleph_0$-categorical are also called *countably categorical*. A theory that is both countably and uncountably categorical is simply called *totally categorical*. For the case of an uncountably categorical but not countably categorical theory, Baldwin and Lachlan [21] established that its countable models can be listed in a chain of proper elementary embeddings:

$$\mathcal{A}_0 \preceq \mathcal{A}_1 \preceq \mathcal{A}_2 \preceq \cdots \preceq \mathcal{A}_\omega,$$

where $\mathcal{A}_0$ is a prime model, and $\mathcal{A}_\omega$ is a saturated model of the theory. Thus, an uncountably categorical theory has either only one countable model or countably many countable models, up to isomorphism.

DEFINITION 11. A theory is called *Ehrenfeucht* if it has finitely many but more than one countable models, up to isomorphism.

By Vaught's theorem, if a theory has two nonisomorphic models, then it has at least three nonisomorphic models. An example of a theory with exactly three countable models was given by Ehrenfeucht. His result can be easily generalized to obtain a theory with exactly n countable models, for any finite $n \geq 3$.

An important question in computable model theory is when a small theory has a computable model. For the case of countably categorical theories, Lerman and Schmerl [225] gave sufficient conditions, which were later extended by Knight as follows.

THEOREM 41 ([208]). *Let T be a countably categorical theory. If $T \cap \Sigma_{n+2}$ is Σ^0_{n+1} uniformly in n, then T has a computable model.*

The natural question posed by Knight is whether there exist countably categorical theories of high complexity, which satisfy the conditions of the previous theorem. First examples were given by Goncharov and Khoussainov in [139], and then generalized by Fokina as follows.

THEOREM 42 ([95]). *There exists a countably categorical theory of arbitrary arithmetic complexity, which has a computable model.*

The proof is based on the method of Marker's extensions from [139]. (This method was later applied to investigate various other properties of computable structures, such as in [100, 103].)

The case of a countably categorical theory with a nonarithmetic complexity was resolved by Khoussainov and Montalbán [201]. The unique model of their theory, up to isomorphism, is a modification of the random graph.

THEOREM 43 ([201]). *There exists a countably categorical theory S with a computable model such that $S \equiv_T \mathbf{0}^{(\omega)}$.*

Another proof of Theorem 43 can be found in [6].

Recall that a consistent decidable theory always has a decidable model. For small theories we can say more. Obviously, if a theory is countably categorical and decidable, then its only (up to isomorphism) countable model always has a decidable copy. For the case of uncountably categorical but not countably categorical theories, Harrington [157] and N. Khisamiev [195] showed that such a theory T is decidable if and only if all countable models of T have decidable isomorphic copies. If T is uncountably categorical but not decidable, then it is possible that some of its models can be isomorphic to computable models, while the others cannot be isomorphic to computable ones.

The following definition of a spectrum of computable models was introduced by Khoussainov, Nies, and Shore.

DEFINITION 12 ([203]). Let T be an uncountably categorical theory with Baldwin–Lachlan elementary chain of countable models:
$$\mathcal{A}_0 \preceq \mathcal{A}_1 \preceq \mathcal{A}_2 \preceq \cdots \preceq \mathcal{A}_\omega.$$

The *spectrum of computable models* of the theory T is the set:
$$\mathrm{SCM}(T) = \{i \leq \omega : \mathcal{A}_i \text{ has a computable isomorphic copy}\}.$$

A number of researchers investigated which sets can be realized as spectra of computable models of uncountably categorical theories. The first example of a nontrivial spectrum of computable models for uncountably categorical theories was given by Goncharov in [123], where he produced a theory with only the prime model $\mathcal{A}_0$ being isomorphic to a computable one. Goncharov's example was followed by a series of results about various spectra by Kudaibergenov [214], Khoussainov, Nies, and Shore [203], Nies [277], Herwig, Lempp, and Ziegler [164], Hirschfeldt, Khoussainov, and Semukhin [172], and Andrews [5, 4]. All these spectra of computable models are finite or co-finite. On the other hand, the upper bound Nies gave in [277] is $\Sigma^0_{\omega+3}$. The above mentioned uncountably categorical theories are $\mathbf{0}''$-decidable; in particular, all their countable models are isomorphic to $\mathbf{0}''$-decidable ones. Two natural questions arise:

1. What could be the complexity of an uncountably categorical theory with a computable model?
2. Is there a bound on the complexity of all countable models, up to isomorphism, of an uncountably categorical theory with a computable model?

Concerning the first question, the examples of arbitrary arithmetic complexity were given in [95, 139]. Again, the authors used Marker's extensions to build the structures. Andrews [6] resolved the nonarithmetic case by adapting famous Hrushovski's examples from [177] to computable model-theoretic setting.

THEOREM 44 ([6]). *There exist uncountably categorical theories of arbitrary arithmetic complexity, as well as of nonarithmetic complexity, which have computable models.*

Andrews used the same method to obtain the spectra of computable models in [5, 4]. The original Hrushovski's construction [177] is a powerful model-theoretic tool for building strongly minimal theories. Its modification by Andrews allows us to carry out the construction effectively, and with much greater control, thus providing a remarkable application of model-theoretic methods to solve computability-theoretic problems.

The second question was raised in the mid 1990s by Lempp. He asked whether it was possible to construct an uncountably categorical theory T with a computable prime model such that none of the countable nonprime models is even arithmetic. The answer to this question is negative for a subclass of uncountably categorical theories (see [118]). As usual, *acl* stands for the algebraic closure operation.

DEFINITION 13. (i) A complete theory T is *strongly minimal* if any definable subset of any model $\mathcal{M}$ of T is finite or co-finite. A structure $\mathcal{M}$ is *strongly minimal* if it has a strongly minimal theory.

(ii) A strongly minimal model $\mathcal{M}$ is *trivial* if for all subsets $A \subseteq M$,

$$acl(A) = \bigcup_{a \in A} acl(\{a\}).$$

Goncharov, Harizanov, Lempp, Laskowski, and McCoy established the following result for trivial, strongly minimal models.

THEOREM 45 ([118]). *Let $\mathcal{M}$ be a computable, trivial, strongly minimal model. Then $Th(\mathcal{M})$ forms a $\mathbf{0}''$-computable set of sentences, and thus all countable models of $Th(\mathcal{M})$ are isomorphic to $\mathbf{0}''$-decidable ones.*

In particular, all countable models of $Th(\mathcal{M})$ are isomorphic to $\mathbf{0}''$-computable models. The proof of Theorem 45 shows an interesting interplay between algorithmic and model-theoretic properties of structures. Namely, the authors proved that for any trivial, strongly minimal theory T in language L, the elementary diagram of any model $\mathcal{M}$ of T is a *model complete* L-theory. This implies that T is $\forall\exists$-axiomatizable, which in turn implies $\mathbf{0}''$-decidability. Furthermore, it was established in [118] that for any strongly minimal, trivial, not totally categorical theory T, the spectrum of computable models is Σ_5^0.

As Khoussainov, Laskowski, Lempp, and Solomon showed in [199], the result in Theorem 45 is best possible in the following sense.

THEOREM 46 ([199]). *There exists a trivial, strongly minimal (and hence uncountably categorical) theory, which has a computable prime model and each of the other countable models computes $\mathbf{0}''$.*

In [73], Dolich, Laskowski, and Raichev generalized the results of [118] to any uncountably categorical, trivial theory of Morley rank 1. A new, more constructive proof of the same results can be found in [222].

In the case of Ehrenfeucht theories, the question which models can be computable or decidable also has a long history. In the mid 70s, Nerode asked whether all models of a decidable Ehrenfeucht theory must be decidable, by analogy with the results in [157, 195]. Morley [259] gave an example of a theory with six models, of which only the prime model was decidable. A good overview of further related results can be found in [113].

Sudoplatov [327] gave a model-theoretic characterization of Ehrenfeucht models, that is, models of Ehrenfeucht theories. In particular, he introduced the notion of a *limit model*, and a special kind of a pre-order on the set of almost prime models. Recall that a model is *almost prime* if it becomes prime after an enrichment by finitely many constants. Analogously to the case of uncountably categorical theories, Gavryushkin introduced in [114] a notion of the spectrum of computable models for Ehrenfeucht theories. He characterized these spectra in Sudoplatov's terms of pre-orders on almost prime models and the number of limit models over almost prime models. Moreover, Gavryushkin constructed examples of computable Ehrenfeucht models of arbitrarily high arithmetic and nonarithmetic complexity.

THEOREM 47 ([114]). *For every $n \geq 3$, there exists an Ehrenfeucht theory T of arbitrary arithmetic complexity such that it has n countable models, up to isomorphism, and it has a computable model among them. There also exists such a theory, which is Turing equivalent to the true first-order arithmetic.*

For further examples of Ehrenfeucht theories with various spectra of computable models see [113].

§5. **Effective categoricity.** We are interested in the complexity of isomorphisms between a computable structure and its computable and noncomputable copies. The main notion in this area of investigation is that of computable categoricity. A computable structure $\mathcal{M}$ is *computably categorical* if for every computable structure $\mathcal{A}$ isomorphic to $\mathcal{M}$, there exists a computable isomorphism from $\mathcal{M}$ onto $\mathcal{A}$. This concept has been part of computable model theory since 1956 when Fröhlich and Shepherdson [107] produced examples of computable fields, extensions of the rationals, of both finite and infinite transcendence degrees, which were not computably categorical. These examples refute the natural conjecture that a computable field is computably categorical exactly when it has finite transcendence degree over its prime subfield (which is either $\mathbb{Q}$ or the p-element $\mathbb{F}_p$, depending on characteristic). Later, Ershov [90] showed that an algebraically closed field is computably categorical if and only if it has finite transcendence degree over its prime subfield. This also follows from work of Nurtazin [280] and can be found in Metakides and Nerode [240]. In [229], Mal'cev considered the notion of a recursively (computably) stable structure. A computable structure $\mathcal{M}$ is *computably stable* if every isomorphism from $\mathcal{M}$ to another computable structure is computable. In the same paper Mal'cev investigated the notion of *autostability* of structures, which is equivalent to that of computably categoricity. Since then computable categoricity has been studied extensively. It has been extended to arbitrary levels of hyperarithmetic hierarchy, and more precisely to Turing degrees $\mathbf{d}$. Computable categoricity of a computable structure $\mathcal{M}$ can also be relativized to all (including noncomputable) structures $\mathcal{A}$ isomorphic to $\mathcal{M}$ (see [17]).

DEFINITION 14. A computable structure $\mathcal{M}$ is $\mathbf{d}$-*computably categorical* if for every computable structure $\mathcal{A}$ isomorphic to $\mathcal{M}$, there exists a $\mathbf{d}$-computable isomorphism from $\mathcal{M}$ onto $\mathcal{A}$.

In the case when $\mathbf{d} = \mathbf{0}^{(n-1)}$, $n \geq 1$, we also say that $\mathcal{M}$ is Δ_n^0-*categorical*. Thus, computably categorical is the same as $\mathbf{0}$-computably categorical or Δ_1^0-categorical. We can similarly define Δ_α^0-categorical structures for any computable ordinal α.

Computably categorical structures tend to be quite rare. For a structure in a typical algebraic class, being computably categorical is usually equivalent to having a finite basis or a finite generating set (for example, in the case of a vector space), or to being highly homogeneous (for example, in the case of a random graph). For instance, Goncharov and Dzgoev [134], and Remmel [293] independently proved that a computable linear order is computably categorical if and only if it has only finitely many successor pairs (also called adjacencies). They also established that a computable Boolean algebra is computably categorical if and only if it has finitely many atoms (see also LaRoche [221]). As usual, by $\mathbb{Z}(p^n)$ we denote the cyclic group of order p^n, and by $\mathbb{Z}(p^\infty)$ the quasicyclic (Prüfer) abelian p-group. The length of an abelian p-group G, $\lambda(G)$, is the least ordinal α such that $p^{\alpha+1}G = p^\alpha G$. The divisible part of G is $\text{Div}(G) = p^{\lambda(G)}G$ and is a direct summand of G. The group G is said to be reduced if $\text{Div}(G) = \{0\}$. Goncharov [125] and Smith [316] independently characterized computably categorical abelian p-groups as those that can be written in one of the following forms: $(\mathbb{Z}(p^\infty))^l \oplus F$ for $l \in \omega \cup \{\infty\}$ and F is a finite group, or $(\mathbb{Z}(p^\infty))^n \oplus H \oplus (\mathbb{Z}(p^k))^\infty$, where $n, k \in \omega$ and H is a finite group. Goncharov, Lempp, and Solomon [119] proved that a computable, ordered, abelian group is computably categorical if and only if it has finite rank. Similarly, they showed that a computable, ordered, Archimedean group is computably categorical if and only if it has finite rank. Lempp, McCoy, R. Miller, and Solomon [223] characterized computably categorical trees of finite height. R. Miller [250] previously established that no computable tree of infinite height is computably categorical.

An *equivalence structure* is a structure with a single equivalence relation. Calvert, Cenzer, Harizanov, and Morozov [32] established that a computable equivalence structure $\mathcal{A}$ is computably categorical if and only if either $\mathcal{A}$ has finitely many finite equivalence classes, or $\mathcal{A}$ has finitely many infinite classes, upper bound on the size of finite classes, and exactly one finite k with infinitely many classes of size k. An *injection structure* $\mathcal{A} = (A, f)$ consists of a nonempty set A and a 1-1 function $f \colon A \to A$. Given $a \in A$, the *orbit* $O_f(a)$ of a under f is $\{b \in A \colon (\exists n \in \mathbb{N})[f^n(a) = b \vee f^n(b) = a]\}$. An injection structure (A, f) may have two types of infinite orbits: Z-orbits, which are isomorphic to $(\mathbb{Z}, S)$, and ω-orbits, which are isomorphic to (ω, S). Cenzer, Harizanov, and Remmel [46] characterized computably categorical injection structures as those that have finitely many infinite orbits.

R. Miller and Schoutens [252] solved a long-standing problem by constructing a computable field that has *infinite* transcendence degree over the rationals, yet is computably categorical. Their idea uses a computable set of rational polynomials (more specifically, the Fermat polynomials) to "tag" elements of a transcendence basis. Hence their field has an infinite intrinsically computable transcendence basis (that is, computable in every isomorphic computable copy of the field), with each single element effectively distinguishable from the others.

Very little is known about Δ_n^0-categoricity, for $n \geq 2$, of structures from natural classes of algebraic structures. Obtaining their classification is usually a difficult task. The reason is either the absence of invariants (such as for linear orders, abelian and nilpotent groups), or the lack of a suitable computability-theoretic notion which would capture the property of being Δ_n^0-categorical (see discussion of Δ_2^0-categoricity for equivalence structures below). There is a complete description of higher levels categoricity (in fact, stability) for well-orders due to Ash [11]. Harris [158] has recently announced a description of Δ_n^0-categorical Boolean algebras, for any $n < \omega$. McCoy [236] characterized, under certain restrictions, Δ_2^0-categorical linear orders and Boolean algebras. Barker [24] proved that for every computable ordinal α, there are $\Delta_{2\alpha+2}^0$-categorical but not $\Delta_{2\alpha+1}^0$-categorical abelian p-groups. Lempp, McCoy, R. Miller, and Solomon [223] proved that for every $n \geq 1$, there is a computable tree of finite height, which is Δ_{n+1}^0-categorical but not Δ_n^0-categorical.

The following problems remain open. Describe Δ_2^0-categorical linear orders. Describe Δ_2^0-categorical equivalence relations. Describe Δ_2^0-categorical abelian p-groups. Resolving these problems may require new algebraic invariants or new computability-theoretic notions.

In the next theorem we present several recent results on the upper bounds for categoricity. Recall that a set X is *semi-low* if $\{e : W_e \cap X \neq \emptyset\}$ is Δ_2^0.

THEOREM 48. (i) (*follows from* [44, 234]) *Every computable, free, non-abelian group is* Δ_4^0-*categorical, and the result cannot be improved to* Δ_3^0.

(ii) ([80]) *Every computable, free, abelian group is* Δ_2^0-*categorical, and the result cannot be improved to computable categoricity.*

(iii) ([80]) *Every computable abelian group of the form* $\bigoplus_{i \in \omega} H_i$, *where* $H_i \leq (\mathbb{Q}, +)$ *for* $i \in \omega$, *is* Δ_3^0-*categorical. A computable group of this form is* Δ_2^0-*categorical if and only if it is isomorphic to a free module over a localization of* $\mathbb{Z}$ *by a set of primes with a* semi-low *complement.*

(iv) ([32]) *Every computable equivalence relation is* Δ_3^0-*categorical, and the result cannot be improved to* Δ_2^0.

We may compare these results with those stated in Theorems 90 and 91. More generally, the study of higher categoricity is often equivalent to the study of algebraic properties of a family of relations specific for a given class (such as independence relations, back-and-forth relations, etc.). The result in Theorem 48 (iii) has been recently extended to arbitrary direct sums of rational subgroups [79], for which the sharp upper bound is Δ_5^0.

We can relativize the notion of Δ_α^0-categoricity by studying the complexity of isomorphisms from a computable structure to any countable isomorphic structure.

DEFINITION 15. A computable structure $\mathcal{M}$ is *relatively Δ_α^0-categorical* if for every $\mathcal{A}$ isomorphic to $\mathcal{M}$, there is an isomorphism from $\mathcal{M}$ to $\mathcal{A}$, which is Δ_α^0 relative to the atomic diagram of $\mathcal{A}$.

Clearly, a relatively Δ_α^0-categorical structure is Δ_α^0-categorical. For linear orders [134, 293], Boolean algebras [134, 293], trees of finite height [223], abelian p-groups [125, 316, 33], equivalence structures [32], and injection structures [46], computable categoricity implies relative computable categoricity. R. Miller and Shlapentokh [253] proved that a computable algebraic field F with a splitting algorithm is computably categorical iff it is decidable which pairs of elements of F belong to the same orbit under automorphisms. They also showed that this criterion is equivalent to relative computable categoricity of F.

A remarkable feature of relative Δ_α^0-categoricity is that it admits a syntactic characterization. This characterization involves the existence of certain effective Scott families. Scott families come from *Scott isomorphism theorem*, which says that for a countable structure $\mathcal{A}$, there is an $L_{\omega_1 \omega}$-sentence the countable models of which are exactly the isomorphic copies of $\mathcal{A}$. For proof of Scott isomorphism theorem see [17]. A *Scott family* for a structure $\mathcal{A}$ is a countable family Φ of $L_{\omega_1 \omega}$-formulas with finitely many fixed parameters from A such that:

(i) Each finite tuple in $\mathcal{A}$ satisfies some $\psi \in \Phi$; and
(ii) If $\overline{a}, \overline{b}$ are tuples in $\mathcal{A}$, of the same length, satisfying the same formula in Φ, then there is an automorphism of $\mathcal{A}$, which maps $\overline{a}$ to $\overline{b}$.

If we strengthen condition (ii) to require that the formulas in Φ define each tuple in $\mathcal{A}$, then Φ is called a *defining family* for $\mathcal{A}$. A *formally Σ_α^0 Scott family* is a Σ_α^0 Scott family consisting of computable Σ_α formulas. In particular, it follows that a formally c.e. Scott family is a c.e. Scott family consisting of finitary existential formulas. The following equivalence was established by Goncharov [120] for $\alpha = 1$, and by Ash, Knight, Manasse, and Slaman [10] and independently by Chisholm [50] for any computable ordinal α.

THEOREM 49 ([10, 50]). *The following are equivalent for a computable structure* $\mathcal{A}$.

1. *The structure* $\mathcal{A}$ *is relatively* Δ_α^0-*categorical.*
2. *The structure* $\mathcal{A}$ *has a formally* Σ_α^0 *Scott family* Φ *with finitely many fixed parameters.*
3. *The structure* $\mathcal{A}$ *has a c.e. Scott family consisting of computable* Σ_α *formulas with finitely many fixed parameters.*

Infinitary language is essential for Scott families. Cholak, Shore, and Solomon [55] proved the existence of a computably stable rigid graph that does not have a Scott family of finitary formulas.

In [236], McCoy characterized relatively Δ_2^0-categorical linear orders and Boolean algebras. In [235], McCoy gave a complete description of relatively Δ_3^0-categorical Boolean algebras, and proved that there are $2^{\aleph_0}$ relatively Δ_3^0-categorical linear orders. More recently, Calvert, Cenzer, Harizanov, and Morozov investigated relative Δ_2^0-categoricity for equivalence structures [32] and abelian p-groups [33], and Cenzer, Harizanov, and Remmel [46] investigated relative Δ_2^0-categoricity for injection structures. In the following theorem we state some of these characterizations of relative Δ_2^0-categoricity. As usual, by ω^* we denote the reverse order of ω, and by η the order type of rationals. For a group G, the *period* of G is $max\{order(g) \colon g \in G\}$ if this quantity is finite, and ∞ otherwise.

THEOREM 50. (i) ([236]) *A computable linear order is relatively Δ_2^0-categorical if and only if it is a sum of finitely many intervals, each of type m, ω, ω^*, $\mathbb{Z}$, or $n \cdot \eta$, so that each interval of type $n \cdot \eta$ has a supremum and infimum.*

(ii) ([236]) *A computable Boolean algebra is relatively Δ_2^0-categorical if and only if it can be expressed as a finite direct sum $c_1 \vee \cdots \vee c_n$, where each c_i is either atomless, an atom, or a 1-atom.*

(iii) ([32]) *A computable equivalence structure is relatively Δ_2^0-categorical if and only if it either has finitely many infinite equivalence classes, or there is an upper bound on the size of its finite equivalence classes.*

(iv) ([46]) *A computable injection structure is relatively Δ_2^0-categorical if and only if it has finitely many orbits of type ω, or finitely many orbits of type Z.*

(v) ([33]) *A computable abelian p-group G is relatively Δ_2^0-categorical if and only if G is reduced and $\lambda(G) \leq \omega$, or G is isomorphic to $\bigoplus_\alpha \mathbb{Z}(p^\infty) \oplus H$, where $\alpha \leq \omega$ and H has finite period.*

Every Δ_2^0-categorical injection structure is relatively Δ_2^0-categorical (see [46]). Every computable injection structure is relatively Δ_3^0-categorical. Every computable equivalence structure is relatively Δ_3^0-categorical. There is no such bound for a computable abelian p-group G. For example, it follows from the index set results in [40] that if $\lambda(G) = \omega \cdot n$ and $m \leq 2n - 1$, or if $\lambda(G) > \omega \cdot n$ and $m \leq 2n - 2$, then G is not Δ_m^0-categorical.

Goncharov [122] was the first to show that computable categoricity of a computable structure does not imply its relative computable categoricity. The main idea of his proof was to code a special kind of family of sets into a computable structure. Such families were constructed independently by Badaev [20] and Selivanov [310]. The result of Goncharov was lifted to higher levels in the hyperarithmetic hierarchy by Goncharov, Harizanov, Knight, McCoy, R. Miller, and Solomon for successor ordinals [116], and by Chisholm, Fokina, Goncharov, Harizanov, Knight, and Quinn for limit ordinals [51].

THEOREM 51 ([116, 51]). *For every computable ordinal α, there is a Δ_α^0-categorical but not relatively Δ_α^0-categorical structure.*

It is not known whether every (computable) Δ_1^1-categorical structure must be relatively Δ_1^1-categorical (see [137]). Kach and Turetsky [183] showed that there exists a Δ_2^0-categorical equivalence structure, which is not relatively Δ_2^0-categorical. Hirschfeldt, Kramer, R. Miller, and Shlapentokh [168] characterized relative computable categoricity for computable algebraic fields and used their characterization to construct a field with the following property.

THEOREM 52 ([168]). *There is a computably categorical algebraic field, which is not relatively computably categorical.*

The notions of computable categoricity and relative computable categoricity coincide if we add more effectiveness requirements on the structure. Goncharov [120] proved that in the case of 2-decidable structures, computable categoricity and relative computable categoricity coincide. Kudinov showed that the assumption of 2-decidability cannot be weakened, by giving in [216] an example of 1-decidable and computably categorical structure, which is not relatively computably categorical. Ash [12] established that for every computable ordinal α, under certain decidability conditions on $\mathcal{A}$, if $\mathcal{A}$ is Δ^0_α-categorical, then $\mathcal{A}$ is relatively Δ^0_α-categorical.

T. Millar [244] proved that if a structure $\mathcal{A}$ is 1-decidable, then any expansion of $\mathcal{A}$ by finitely many constants remains computably categorical. Cholak, Goncharov, Khoussainov, and Shore showed that the assumption of 1-decidability is important.

THEOREM 53 ([54]). *There is a computable structure, which is computably categorical, but ceases to be after naming any element of the structure.*

Clearly, the structure in this theorem is not relatively computably categorical. Khoussainov and Shore [205] proved that there is a computably categorical structure $\mathcal{A}$ without a formally c.e. Scott family such that the expansion of $\mathcal{A}$ by any finite number of constants is computably categorical.

Downey, Kach, Lempp, and Turetsky have recently obtained the following result.

THEOREM 54 ([83]). *Any* 1-*decidable computably categorical structure is relatively Δ^0_2-categorical.*

Based on this theorem, we could conjecture that every computable structure that is computably categorical should be relatively Δ^0_3-categorical. However, this is not the case, as recently announced by Downey, Kach, Lempp, Lewis, Montalbán, and Turetsky.

THEOREM 55 ([76]). *For every computable ordinal α, there is a computably categorical structure that is not relatively Δ^0_α-categorical.*

Thus, a natural question arises whether there is a computably categorical structure that is not relatively hyperarithmetically categorical. In [76], the uniformity of the constructed structures together with an overspill argument allowed the authors to establish that the problem of computable categoricity is Π^1_1-complete, which was a long-standing open question.

DEFINITION 16. The **d**-*computable dimension* of a computable structure $\mathcal{M}$ is the number of computable isomorphic copies of $\mathcal{M}$, up to **d**-computable isomorphism.

Hence, a computably categorical structure has computable dimension 1. Many natural structures have computable dimension 1 or ω. For example, it was shown in [240] that it is impossible for a computable algebraic field to have finite computable dimension greater than 1. Goncharov was the first to produce examples of computable structures of finite computable dimension greater than 1.

THEOREM 56 ([127, 129]). *For every finite $n \geq 2$, there is a computable structure of computable dimension n.*

After Goncharov's examples, structures of finite computable dimension $n \geq 2$ were found in several familiar classes, such as 2-step nilpotent groups [141] and other classes [174].

For a computable structure $\mathcal{A}$, some Turing degree, which is not necessarily $\mathbf{0}^{(n)}$, may compute an isomorphism between any two computable copies of the structure. The following notion of the categoricity spectrum, introduced by Fokina, Kalimullin, and R. Miller, aims to capture the set of all Turing degrees capable of computing isomorphisms between arbitrary computable copies of $\mathcal{A}$.

DEFINITION 17 ([103]). Let $\mathcal{A}$ be a computable structure.
(i) The *categoricity spectrum* of $\mathcal{A}$ is

$$\mathrm{CatSpec}(\mathcal{A}) = \{\mathbf{x} : \mathcal{A} \text{ is } \mathbf{x}\text{-computably categorical}\}.$$

(ii) A Turing degree $\mathbf{d}$ is the *degree of categoricity* of $\mathcal{A}$, if it exists, if $\mathbf{d}$ is the least degree in $\mathrm{CatSpec}(\mathcal{A})$.

(iii) A Turing degree $\mathbf{d}$ is *categorically definable* if it is the degree of categoricity of some computable structure.

This terminology intends to parallel the notions of the degree spectrum of a structure $\mathcal{A}$, and the degree of the isomorphism class of $\mathcal{A}$. Since there are only countably many computable structures, most Turing degrees are not categorically definable. Fokina, Kalimullin, and R. Miller investigated which Turing degrees are categorically definable. Their main result in [103] gives a partial answer for the case of arithmetic degrees, and was later extended by Csima, Franklin, and Shore to hyperarithmetic degrees.

THEOREM 57 ([65]). (i) *For every computable ordinal α, $\mathbf{0}^{(\alpha)}$ is the degree of categoricity of a computable structure.*

(ii) *For a computable successor ordinal α, every degree $\mathbf{d}$ that is c.e.a. in $\mathbf{0}^{(\alpha)}$ is a degree of categoricity.*

Negative results were also obtained in [103, 65]. Namely, if $\mathbf{d}$ is a non-hyperarithmetic degree, then $\mathbf{d}$ cannot be the degree of categoricity of a computable structure. Furthermore, Anderson and Csima showed that not all hyperarithmetic degrees are degrees of categoricity.

THEOREM 58 ([3]). (i) *There exists a Σ_2^0 degree that is not categorically definable.*

(ii) *Every degree of a set that is 2-generic relative to some perfect tree is not a degree of categoricity.*

(iii) *Every noncomputable hyperimmune-free degree is not a degree of categoricity.*

Thus, it is natural to ask whether all Δ_2^0 degrees are categorically definable.

Not every computable structure has a degree of categoricity. The first negative example was built by R. Miller.

THEOREM 59 ([251]). *There exists a computable field with a splitting algorithm, which is not computably categorical, and such that its categoricity spectrum must contain degrees $\mathbf{d}_0$ and $\mathbf{d}_1$ with $\mathbf{d}_0 \wedge \mathbf{d}_1 = \mathbf{0}$.*

Subsequently, R. Miller built another computable field the categoricity spectrum of which has no least degree and does not contain $\mathbf{0}'$. R. Miller used the algebraicity of the field to present the isomorphisms between it and a computable isomorphic copy as infinite paths through a finite-branching computable tree. If the field has a splitting algorithm, then the branching of this tree is computable, and we can apply the *low basis theorem*. If the field does not have a splitting algorithm, then we relativize to the degree of the branching and apply the relativized *low basis theorem*.

Further interesting examples of structures without the degree of categoricity were built by Fokina, Frolov, and Kalimullin [98]. The main property of their structures is that they are *rigid*, that is, they have no nontrivial automorphisms, which was not the case for the examples in [251]. If a rigid structure $\mathcal{M}$ is $\mathbf{d}$-categorical, then it is also $\mathbf{d}$-*stable*, i.e., every isomorphism from $\mathcal{M}$ onto a computable copy is $\mathbf{d}$-computable. (The converse is not true, for example, a computable copy of a two-dimensional vector space over $\mathbb{Q}$ is computably stable but not rigid.) Constructions from [98] give for every nonzero c.e. degree $\mathbf{d}$, a rigid $\mathbf{d}$-computably categorical structure with no degree of categoricity. The authors construct similar rigid structures for all degrees $\mathbf{d}$ that are c.e.a. in $\mathbf{0}^{(n)}$, for any $n \in \omega$. When we pass to d.c.e. structures, we lose the property of rigidity. It is natural to ask whether there is a computable structure the categoricity spectrum of which is the set of all noncomputable Turing degrees. It is also interesting to find out whether the union of two cones of Turing degrees can be a categoricity spectrum.

In recent papers [132, 131, 130], Goncharov investigated categoricity restricted to decidable structures.

DEFINITION 18. A decidable structure $\mathcal{A}$ is called *decidably categorical* if every two decidable copies of $\mathcal{A}$ are computably isomorphic.

Nurtazin gave the following characterization of decidably categorical structures. Recall that for a complete theory T, a formula $\theta(\overline{x})$ is called *complete* if for every formula $\psi(\overline{x})$, either $T \vdash \theta(\overline{x}) \Rightarrow \psi(\overline{x})$ or $T \vdash \theta(\overline{x}) \Rightarrow \neg\psi(\overline{x})$.

THEOREM 60 ([280]). *Let $\mathcal{A}$ be a decidable structure. Then $\mathcal{A}$ is decidably categorical if and only if there is a finite tuple $\bar{c}$ of elements in A such that $(\mathcal{A}, \bar{c})$ is a prime model of the theory $\mathrm{Th}(\mathcal{A}, \bar{c})$ and the set of complete formulas of this theory is computable.*

Moreover, Nurtazin proved that if there is no such $\bar{c}$, then there are infinitely many decidable copies of $\mathcal{A}$, no two of which are computably isomorphic.

Similarly to the case of computable categoricity, we define *decidable categoricity spectrum* of $\mathcal{A}$ to be the collection of degrees that can compute at least one isomorphism between *decidable* copies of $\mathcal{M}$. In [132], Goncharov studied decidable categoricity of almost prime models. It is not difficult to see that the collection of atomic formulas in a decidable almost prime model $\mathcal{M}$ is co-c.e. Therefore, a c.e. degree is always contained in the decidable categoricity spectrum of $\mathcal{M}$. Goncharov established the following result.

THEOREM 61 ([132]). *Every c.e. degree $\mathbf{d}$ is the degree of decidable categoricity of some decidable almost prime model.*

Goncharov also investigated decidable categoricity of Ehrenfeucht models.

THEOREM 62 ([131]). *There exists a decidable Ehrenfeucht theory T such that T has a decidable prime model that is decidably categorical, and T has a decidable almost prime model that is not decidably categorical.*

Effective categoricity of computable structures has also been recently investigated within Ershov's difference hierarchy: for graphs by Khoussainov, Stephan, and Yang [207], and for the equivalence structures by Cenzer, LaForte, and Remmel [47].

§6. Automorphisms of effective structures. In algebra, automorphism groups of structures often reflect the algebraic properties of structures (for example, as in Galois theory). In computable model theory, the study of *effective* automorphisms help us better understand computability-theoretic properties of countable structures. The set of all automorphisms of a computable structure forms a group under composition, and we may ask questions about the isomorphism types of this group and its natural subgroups. Thus, the theory of automorphisms of effective structures provides another link between computable algebra and classical group theory. We may also study the Turing degrees of members of the automorphism group. This line of investigation is related to the study of effective categoricity of structures. Finally, we may restrict ourselves to computable structures from familiar classes (such as Boolean algebras, linear orders, etc.) and study groups of effective automorphisms for these structures. As usual, we assume that all infinite computable structures have ω as their domains. The next definition captures one of the main notions of this investigation.

DEFINITION 19. For an infinite computable structure $\mathcal{M}$ (with domain ω) and a Turing degree $\mathbf{d}$, we define $\mathrm{Aut}_{\mathbf{d}}(\mathcal{M})$ to be the set of all permutations of ω, which are computable in $\mathbf{d}$ and induce automorphisms of $\mathcal{M}$.

We write $\mathrm{Aut}_c(\mathcal{M})$ for $\mathrm{Aut}_\mathbf{0}(\mathcal{M})$ (the subscript c stands for *computable*). For every Turing degree $\mathbf{d}$, the set $\mathrm{Aut}_{\mathbf{d}}(M)$ forms a group under composition. In contrast, the set $\mathrm{Aut}_p(\omega)$ of all primitive recursive permutations of ω is not a group under composition, as shown by Kuznetsov [218]. One of the central objectives here is to study classical and effective properties of the group $\mathrm{Aut}_{\mathbf{d}}(\mathcal{M})$ for various $\mathcal{M}$ and $\mathbf{d}$. We can start with a structure in the empty language, that is, ω with equality, and consider its automorphism group $\mathrm{Aut}_{\mathbf{d}}(\omega)$ as a structure. Recall that the degree of the isomorphism type of a structure, if it exists, is the least Turing degree in its Turing degree spectrum. Morozov established the following result.

THEOREM 63 ([264]). *For every Turing degree* $\mathbf{d}$, *the degree of the isomorphism type of the group* $\mathrm{Aut}_{\mathbf{d}}(\omega)$ *is* $\mathbf{d}''$.

Morozov showed that the embedding $\mathcal{F} \colon \mathbf{d} \to \mathrm{Aut}_{\mathbf{d}}(\omega)$ can be used to substitute Turing reducibility with the group-theoretic embedding.

THEOREM 64 ([268]). *For every pair* $\mathbf{c}, \mathbf{d}$ *of Turing degrees, we have*

$$(\mathrm{Aut}_{\mathbf{d}}(\omega) \leq \mathrm{Aut}_{\mathbf{c}}(\omega)) \Leftrightarrow (\mathbf{d} \leq \mathbf{c}),$$

where $\leq$ *stands for the usual group-theoretic embedding.*

It follows from this theorem that $\mathbf{c} = \mathbf{d}$ if and only if $\mathrm{Aut}_{\mathbf{d}}(\omega) \cong \mathrm{Aut}_{\mathbf{c}}(\omega)$. In contrast, there exists a Turing degree $\mathbf{a}$ such that $\mathrm{Aut}_{\mathbf{a}}(\omega)$ and $\mathrm{Aut}_{\mathbf{b}}(\omega)$ are elementary equivalent for all $\mathbf{b} \geq \mathbf{a}$ (see [269]). Intuitively, the last statement says that this first-order theory cannot recognize the difference between very "large" Turing degrees. Kent investigated group-theoretic properties of $\mathrm{Aut}_{\mathbf{d}}(\omega)$.

THEOREM 65 ([192]). *For every Turing degree* $\mathbf{d}$, *the unique normal series for* $\mathrm{Aut}_{\mathbf{d}}(\omega)$ *has the form*

$$\{1\} \triangleleft E \triangleleft F \triangleleft \mathrm{Aut}_{\mathbf{d}}(\omega),$$

where F is the subgroup of permutations that change only finitely many numbers, E is the subgroup of even permutations of F, and 1 is the identity permutation.

Notice that a finitely generated subgroup of $\mathrm{Aut}_c(\omega)$ has to be a Π_1^0 group. Higman asked if every Π_1^0 finitely generated group can be isomorphically embedded into $\mathrm{Aut}_c(\omega)$. The following result of Morozov answers Higman's question negatively.

THEOREM 66 ([270]). *There exists a 2-generated* Π_1^0 *group G such that* $G \nleq \mathrm{Aut}_c(\omega)$.

Morozov syntactically characterized subgroups of $\mathrm{Aut}_c(\omega)$, which are isomorphic to the whole $\mathrm{Aut}_c(\omega)$.

THEOREM 67 ([265]). *There exists a first-order sentence in the language of groups such that for every $G \leqq \mathrm{Aut}_c(\omega)$,*

$$(G \models \phi) \Leftrightarrow (G \cong \mathrm{Aut}_c(\omega)).$$

More specifically, Morozov [265] proved that the class of all groups of the form $\mathrm{Aut}_c(\mathcal{M})$, where $\mathcal{M}$ is a computable structure, is definable in the monadic second-order language within $\mathrm{Aut}_c(\omega)$. He also showed that the theories of the following three classes of groups are all distinct and differ from the theory of all groups: (i) groups that can be embedded into $\mathrm{Aut}_c(\omega)$, (ii) groups that are $\mathrm{Aut}_c(\mathcal{M})$ for computable $\mathcal{M}$, and (iii) computable groups. The first class cannot be axiomatized by a hyperarithmetic set of axioms, the other two cannot be axiomatized by any arithmetic set of axioms. Furthermore, Morozov [265] proved that there exists a single sentence, consistent with the theory of groups, which is not true in any group $\mathrm{Aut}_c(\mathcal{M})$ where $\mathcal{M}$ is a computable structure.

Now, for various computable structures $\mathcal{M}$, we compare $\mathrm{Aut}_\mathbf{d}(\mathcal{M})$ and $\mathrm{Aut}(\mathcal{M})$. For $\mathbf{d} = \mathbf{0}$, Dzgoev [88], and independently Manaster and Remmel [231] established the following result.

THEOREM 68 ([88, 231]). *There exists a computable structure $\mathcal{M}$ such that $\mathrm{Aut}(\mathcal{M})$ has 2^ω elements, while $\mathrm{Aut}_c(\mathcal{M})$ has only one element.*

The previous theorem can be strengthened in several ways. Kudaibergenov [215] showed that we can make such $\mathcal{M}$ decidable and homogeneous. Morozov [267] proved that there exists a computable structure $\mathcal{M}$ with $\mathrm{card}(\mathrm{Aut}(\mathcal{M})) = 2^\omega$ such that every hyperarithmetic structure isomorphic to $\mathcal{M}$ has no nontrivial hyperarithmetic automorphisms. For a criterion for the existence of two isomorphic but not hyperarithmetically isomorphic tuples in a hyperarithmetic structure, and examples of well-known structures with this phenomenon see [117].

For a computable structure $\mathcal{M}$, the group $\mathrm{Aut}_c(\mathcal{M})$ does not have to be isomorphic to a computable one. Morozov [263] gave the following characterization of $\mathrm{Aut}_c(\mathcal{M})$ having a computable copy.

THEOREM 69 ([263]). *For a computable structure $\mathcal{M}$, the group $\mathrm{Aut}_c(\mathcal{M})$ is isomorphic to a computable one if and only if there exists a finite tuple $\overline{p}$ such that $\mathrm{Aut}(\mathcal{M}, \overline{p}) = \{1\}$, and the set $\{(\overline{m}, \overline{n}) : \overline{m} \cong_c \overline{n}\}$ is c.e., where*

$$\overline{m} \cong_c \overline{n} \Leftrightarrow (\exists f \in \mathrm{Aut}_c(\mathcal{M}))[f : \overline{m} \to \overline{n}].$$

This theorem has some interesting corollaries.

COROLLARY 2 ([263]). *A finitely generated group G is isomorphic to $\mathrm{Aut}_c(\mathcal{M})$ for some computable structure $\mathcal{M}$ if and only if G has a decidable word problem.*

For groups that are not finitely generated the situation is rather complex. Even if a group is abelian, not much can be said. It is not difficult to show that

$\bigoplus_{p \in S} \mathbb{Z}_p$, where S is a set of primes, is isomorphic to $\mathrm{Aut}_c(\mathcal{M})$ for some computable structure $\mathcal{M}$ if and only if S is Σ_3^0 (see Morozov and Buzykaeva [272]). The general case of arbitrary abelian groups is unresolved. Theorem 69 also implies that for every infinite computable Boolean algebra $\mathcal{B}$, the group $\mathrm{Aut}_c(\mathcal{B})$ is not computable, and the same is true for every decidable infinite model of an $\aleph_0$-categorical theory with a computable set of atomic formulas.

We can show that the group $\mathrm{Aut}_c(\mathcal{M})$ for a computable structure $\mathcal{M}$ is $\mathbf{0}''$-computable (folklore). This upper bound is sharp, as shown in the following theorem due to Morozov.

THEOREM 70 ([260]). *For every Turing degree* $\mathbf{d} \leq \mathbf{0}''$, *there exists a computable structure* $\mathcal{M}$ *such that* $\deg(D(\mathrm{Aut}_c(\mathcal{M}))) = \mathbf{d}$.

We may ask whether for various computable $\mathcal{M}$, the group $\mathrm{Aut}_c(\mathcal{M})$ has a degree of its isomorphism type. As we have seen earlier, this was the case when $\mathcal{M}$ is ω with equality. Nonetheless, Morozov [260] constructed a computable structure $\mathcal{M}$ such that $\mathrm{Aut}_c(\mathcal{M})$ has no degree of its isomorphism type. We may also ask which Turing degrees contain only groups isomorphic to $\mathrm{Aut}_c(\mathcal{M})$ for some computable $\mathcal{M}$. Morozov [263, 260] proved that this collection of degrees is the singleton $\{\mathbf{0}\}$.

Recently Harizanov, Morozov, and R. Miller [150] introduced another notion in the study of $\mathrm{Aut}(\mathcal{M})$.

DEFINITION 20 ([150]). The *automorphism (Turing) degree spectrum* of a computable structure $\mathcal{M}$, in symbols $\mathrm{AutSp}(\mathcal{M})$, is the set

$$\{\deg(f) \colon f \in \mathrm{Aut}(\mathcal{M}) - \{1_\mathcal{M}\}\},$$

where $1_\mathcal{M}$ is the identity automorphism of $\mathcal{M}$.

Harizanov, Morozov, and R. Miller [150] showed that various collections of Turing degrees, including many upper cones, can be realized as automorphism degree spectra. Let $\mathcal{M}$ be a computable structure. If $\mathrm{AutSp}(\mathcal{M})$ is the upper cone of degrees $\geq \mathbf{d}$, then $\mathbf{d}$ is hyperarithmetic. Harizanov, Morozov, and R. Miller [150] showed that for any computable ordinal α, and any Turing degree $\mathbf{d}$ with $\mathbf{0}^{(\alpha)} \leq \mathbf{d} \leq \mathbf{0}^{(\alpha+1)}$, the upper cone of degrees $\geq \mathbf{d}$ forms an automorphism spectrum. They also showed that there exists a computable structure $\mathcal{A}$ the automorphism spectrum of which is the union of the upper cones above each degree of an infinite antichain of Σ_n^0 degrees for $n \geq 1$. The spectrum $\mathrm{AutSp}(\mathcal{M})$ is at most countable if and only if it contains only hyperarithmetic degrees. Since for every $f, g \in \mathrm{Aut}(\mathcal{M})$ the composition fg is also an automorphism, the automorphism degree spectrum cannot contain exactly two incomparable degrees, as Harizanov, Morozov, and R. Miller showed.

THEOREM 71 ([150]).
1. Let $\mathbf{d}_0$ and $\mathbf{d}_1$ be incomparable Turing degrees. Then no computable structure $\mathcal{M}$ has $\operatorname{AutSp}(\mathcal{M}) = \{\mathbf{d}_0, \mathbf{d}_1\}$ or $\operatorname{AutSp}(\mathcal{M}) = \{\mathbf{0}, \mathbf{d}_0, \mathbf{d}_1\}$.
2. There exist pairwise incomparable Δ_2^0 Turing degrees $\mathbf{d}_0$, $\mathbf{d}_1$, $\mathbf{d}_2$, and computable structures $\mathcal{A}$ and $\mathcal{B}$ such that $\operatorname{AutSp}(\mathcal{A}) = \{\mathbf{d}_0, \mathbf{d}_1, \mathbf{d}_2\}$ and $\operatorname{AutSp}(\mathcal{B}) = \{\mathbf{0}, \mathbf{d}_0, \mathbf{d}_1, \mathbf{d}_2\}$.

It was shown in [150] that there exists a computable structure $\mathcal{A}$ such that for every c.e. degree $\mathbf{d}$, some computable copy of $\mathcal{A}$ has automorphism degree spectrum $\{\mathbf{d}\}$. If $\mathbf{0}^{(\alpha)} \leq \mathbf{d} \leq \mathbf{0}^{(\alpha+1)}$ for some computable ordinal α, then there exists a computable structure with the automorphism degree spectrum $\{\mathbf{d}\}$. A total function $f \colon \omega \to \omega$ is said to be a Π_1^0-*function singleton* if there exists a computable tree $\mathcal{T} \subseteq \omega^{<\omega}$ through which f is a unique infinite path. It was proved in [150] that a Turing degree $\mathbf{d}$ contains a Π_1^0-function singleton if and only if $\{\mathbf{d}\}$ is the automorphism spectrum of some computable structure.

For a computable structure $\mathcal{M}$ from some well-known algebraic class of structures, the typical question we might ask is: Given $\operatorname{Aut}(\mathcal{M})$, what can we say about the isomorphism type of $\mathcal{M}$? Obtaining a satisfactory answer to this question is usually a difficult task. The effective analogue of the question—when $\mathcal{M}$ is computable and $\operatorname{Aut}(\mathcal{M})$ is replaced by $\operatorname{Aut}_c(\mathcal{M})$—is not any easier. In the case of computable Boolean algebras, Morozov [262] obtained a positive partial result. By $\mathcal{B} \cong_c \mathcal{A}$ we denote that $\mathcal{B}$ and $\mathcal{A}$ are computably isomorphic.

THEOREM 72 ([262]). *Let $\mathcal{A}$ be an atomic decidable Boolean algebra. For every computable Boolean algebra $\mathcal{B}$, we have*

$$(\operatorname{Aut}_c(\mathcal{B}) \cong \operatorname{Aut}_c(\mathcal{A})) \Rightarrow (\mathcal{B} \cong_c \mathcal{A}).$$

In contrast, Remmel [294] showed that for every computable Boolean algebra $\mathcal{B}$, there exists $\mathcal{C} \cong \mathcal{B}$ such that every $f \in \operatorname{Aut}_c(\mathcal{C})$ moves only finitely many atoms of $\mathcal{C}$. It is also proven in [262] that there exist two decidable Boolean algebras, $\mathcal{B}_0$ and $\mathcal{B}_1$, such that $\mathcal{B}_0 \not\cong \mathcal{B}_1$ and $\operatorname{Aut}_c(\mathcal{B}_0) \cong \operatorname{Aut}_c(\mathcal{B}_1)$. Morozov [262] also showed that there exists a computable Boolean algebra $\mathcal{B}$, and a Boolean algebra $\mathcal{C}$ having no computable copy, such that $\operatorname{Aut}(\mathcal{B}) \cong \operatorname{Aut}(\mathcal{C})$.

In [57], Chubb, Harizanov, Morozov, Pingrey, and Ufferman investigated the relationship between algebraic structures and their inverse semigroups of partial automorphisms. An *inverse semigroup* is a semigroup where for each element f there is a unique g so that $gfg = g$ and $fgf = f$. For a structure $\mathcal{M}$, the authors considered the semigroup $I_{fin}(\mathcal{M})$ of all finite automorphisms, and, in the case of a computable structure $\mathcal{M}$, the semigroup of all partial computable automorphisms, $I_{pc}(\mathcal{M})$. As usual, $\equiv$ stands for elementary equivalence of structures. In [57], it was shown that structures from certain classes can be recovered, up to isomorphism or elementary equivalence, from these semigroups. For example, for all nontrivial countable equivalence structures $\mathcal{A}_0$ and $\mathcal{A}_1$, we have:

(i) $(I_{fin}(\mathcal{A}_0) \cong I_{fin}(\mathcal{A}_1)) \Leftrightarrow (\mathcal{A}_0 \cong \mathcal{A}_1)$; and

(ii) $(I_{fin}(\mathcal{A}_0) \equiv I_{fin}(\mathcal{A}_1)) \Leftrightarrow (\mathcal{A}_0 \equiv \mathcal{A}_1)$.

We call an equivalence structure (A, E) *nontrivial* if E differs from the diagonal relation $\{(a, a): a \in A\}$ and from the set $A \times A$. It was shown in [57] that for a nontrivial computable equivalence structure $\mathcal{E}_0$, there is a first-order sentence σ in the language of inverse semigroups such that for any nontrivial computable equivalence structure $\mathcal{E}_1$, we have

$$(I_{pc}(\mathcal{E}_1) \models \sigma) \Rightarrow (\mathcal{E}_1 \cong_c \mathcal{E}_0).$$

The authors of [57] also considered partial orders, relatively complemented distributive lattices, and Boolean algebras. It would be interesting to investigate for other natural algebraic structures how structures themselves can be recovered, up to isomorphism or elementary equivalence, from various inverse semigroups of their partial automorphisms.

There are also interesting results about computable automorphisms of computable linear orders. Schwartz obtained the following characterization of computable linear orders containing dense intervals.

THEOREM 73 ([302]). *A computable linear order $\mathcal{A}$ contains a dense interval if and only if* $\mathrm{card}(\mathrm{Aut}_c(\mathcal{L})) > 1$ *for every computable $\mathcal{L}$ such that $\mathcal{L} \cong \mathcal{A}$.*

In order to state the next result by Morozov and Truss [273], we will first introduce some notation. For a computable structure $\mathcal{M}$ and a *Turing ideal* I, let $\mathrm{Aut}_I(\mathcal{M})$ be the collection of all automorphisms of $\mathcal{M}$ computable from members of I. Let $\mathcal{Q} = (\mathbb{Q}, \leq)$.

THEOREM 74 ([273]). *For Turing ideals I and J we have*:

$$(\mathrm{Aut}_I(\mathcal{Q}) \leq \mathrm{Aut}_J(\mathcal{Q})) \Leftrightarrow (I \subseteq J), \text{ and}$$

$$(\mathrm{Aut}_I(\mathcal{Q}) \cong \mathrm{Aut}_J(\mathcal{Q})) \Leftrightarrow (I = J).$$

The proof uses techniques from the theory of ordered abelian groups (see [115]). It is interesting to compare Theorem 74 with Theorem 64. The next result of Morozov and Truss can be compared with Theorem 67.

THEOREM 75 ([274]). *There is a first-order sentence τ such that, up to isomorphism, the group $\mathrm{Aut}_c(\mathcal{Q})$ is the only model of τ among all subgroups of $\mathrm{Aut}_c(\omega)$.*

Lempp, McCoy, Morozov, and Solomon studied the algebraic properties of $\mathrm{Aut}_c(\mathcal{Q})$ and compared them with those of $\mathrm{Aut}(\mathcal{Q})$. They obtained the following result distinguishing $\mathrm{Aut}_c(\mathcal{Q})$ from $\mathrm{Aut}(\mathcal{Q})$.

THEOREM 76 ([224]). *The following three properties, known to be true for $\mathrm{Aut}(\mathcal{Q})$, fail for $\mathrm{Aut}_c(\mathcal{Q})$*:

(a) *the group is divisible*;

(b) *every element is a commutator of itself with some other element; and*

(c) *two elements are conjugate if and only if they have isomorphic orbital structures.*

Not much is known about effective automorphisms of computable modules, including vector spaces and abelian groups. Many algebraic difficulties arise in the study of their automorphism groups. The following result about modules, due to Morozov, is similar to Theorem 73.

THEOREM 77 ([266]). *For every computable division ring $\mathcal{R}$, there exists a computable copy of the module $\mathcal{M} = \bigoplus_{i \in \omega} \mathcal{R}$ such that $\mathrm{Aut}_c(\mathcal{M})$ contains only multiplications by scalars from $\mathcal{R}$.*

Further related results can be found in [217].

§7. Degree spectra of relations. One of the important questions in computable model theory is how a specific property of a computable structure may change if the structure is isomorphically transformed so that it remains computable. A computable property of a computable structure $\mathcal{A}$, which Ash and Nerode [19] considered, is given by an additional computable relation R on the domain of $\mathcal{A}$. (That is, R is not named in the language of $\mathcal{A}$.) Ash and Nerode investigated syntactic conditions on $\mathcal{A}$ and R under which for every isomorphism f from $\mathcal{A}$ onto a computable structure $\mathcal{B}$, $f(R)$ is c.e. Such relations are called *intrinsically c.e.* on $\mathcal{A}$. In general, we have the following definition. Let $\mathcal{P}$ be a certain complexity class.

DEFINITION 21 ([19]). An additional relation R on the domain of a computable structure $\mathcal{A}$ is called *intrinsically $\mathcal{P}$* on $\mathcal{A}$ if the image of R under every isomorphism from $\mathcal{A}$ to a computable structure belongs to $\mathcal{P}$.

For example, the successor relation, and being an even number are not intrinsically computable relations on $(\omega, <)$. Clearly, if $\mathcal{A}$ is a computably stable structure, then every computable relation on its domain is intrinsically computable.

If R is definable in $\mathcal{A}$ by a computable Σ_1 formula with finitely many parameters, then R is intrinsically c.e. Ash and Nerode [19] proved that, under a certain extra decidability condition on $\mathcal{A}$ and R, the relation R is intrinsically c.e. on $\mathcal{A}$ iff R is definable by a computable Σ_1 formula with finitely many parameters. The Ash–Nerode condition for an m-ary relation R says that there is an algorithm, which determines for every existential formula $\psi(x_0, \ldots, x_{m-1}, \overline{y})$ and every $\overline{c} \in A^{lh(\overline{y})}$, whether the following implication holds for every $\overline{a} \in A^m$:

$$(\mathcal{A} \vDash \psi(\overline{a}, \overline{c})) \Rightarrow R(\overline{a}).$$

Barker [23] extended this result by showing that for every computable ordinal α, under certain additional decidability conditions on $\mathcal{A}$, the relation R is intrinsically Σ_α^0 on $\mathcal{A}$ iff R is definable by a computable Σ_α formula with finitely many parameters. For the relative notions, the effectiveness conditions are not needed. Let $\mathcal{P}$ be a certain complexity class, which can be relativized, such as the class of all Σ_α^0 sets.

DEFINITION 22. An additional relation R on the domain of a computable structure $\mathcal{A}$ is called *relatively intrinsically* $\mathcal{P}$ on $\mathcal{A}$ if the image of R under every isomorphism from $\mathcal{A}$ to any structure $\mathcal{B}$ is $\mathcal{P}$ relative to the atomic diagram of $\mathcal{B}$.

The following equivalence is due to Ash, Knight, Manasse, and Slaman [10], and independently Chisholm [50].

THEOREM 78 ([10, 50]). *Let $\mathcal{A}$ be a computable structure. A relation R on $\mathcal{A}$ is relatively intrinsically Σ_α^0 iff R is definable by a computable Σ_α formula with finitely many parameters.*

Goncharov [122] and Manasse [230] gave examples of intrinsically c.e. relations on computable structures, which are not relatively intrinsically c.e. This result was lifted to higher levels in the hyperarithmetic hierarchy by Goncharov, Harizanov, Knight, McCoy, R. Miller, and Solomon for successor ordinals [116], and by Chisholm, Fokina, Goncharov, Harizanov, Knight, and Quinn for limit ordinals [51].

THEOREM 79 ([116, 51]). *For every computable ordinal α, there is a computable structure $\mathcal{A}$ with an intrinsically Σ_α^0 relation R such that R is not definable by a computable Σ_α formula with finitely many parameters.*

In addition to considering the complexity of relations on computable structures within hyperarithmetic hierarchy, we can also consider their degrees, such as Turing degrees or strong degrees. Harizanov introduced the following notion.

DEFINITION 23 ([151]). The *Turing degree spectrum* of R on $\mathcal{A}$, in symbols $\mathrm{DgSp}_\mathcal{A}(R)$, is the set of all Turing degrees of the images of R under all isomorphisms from $\mathcal{A}$ onto computable structures.

If for some isomorphism f from $\mathcal{A}$ to a computable structure, we have $X = f(R)$ and $\mathbf{x} = \deg(X)$, then we say that $\mathbf{x}$ is realized in $\mathrm{DgSp}_\mathcal{A}(R)$ *via* X, or *via* f. Uncountable degree spectra of relations were studied by Harizanov [153, 155], and Ash, Cholak, and Knight [13]. In particular, they showed independently that if every Turing degree $\leq \mathbf{0}''$ can be realized in $\mathrm{DgSp}_\mathcal{A}(R)$ *via* an isomorphism of the same Turing degree as its image of R, then $\mathrm{DgSp}_\mathcal{A}(R)$ contains every Turing degree. In [53], the authors investigated the spectra of relations on computable structures under strong reducibilities such as *weak truth-table* (wtt) reducibility and *truth-table* (tt) reducibility.

In [152], Harizanov studied when every c.e. degree can be obtained in $\mathrm{DgSp}_\mathcal{A}(R)$ *via* an isomorphism of the same degree as its image of R. Ash, Cholak, and Knight [13] lifted her result to arbitrary α-c.e. degrees, where α

is a computable ordinal, in Ershov's difference hierarchy. For example, the degree spectrum of the successor relation on $(\omega, \leq)$ contains all c.e. degrees, and the same holds for the set of all even numbers. The degree spectrum of the set of algebraic elements in an algebraically closed field of infinite transcendence degree contains all c.e. Turing degrees.

One of the general results by Harizanov about $\mathrm{DgSp}_{\mathcal{A}}(R)$ containing all c.e. degrees is the following theorem, which requires extra effectiveness condition—it is enough that the existential diagram of $(\mathcal{A}, R)$ is computable.

THEOREM 80 ([152]). *Let $\mathcal{A}$ be a computable structure, and let R be a relation that is intrinsically c.e. on $\mathcal{A}$, while $\neg R$ is not. Then, under a certain extra decidability condition, for any c.e. degree $\mathbf{d}$, we have $\mathbf{d} \in \mathrm{DgSp}_{\mathcal{A}}(R)$.*

Ash and Knight [16] generalized the previous theorem. Their generalization involves degrees that are coarser than Turing degrees. In the following definition we will use the symbol Δ^0_α to denote a complete Δ^0_α set.

DEFINITION 24 ([16]). (i) $A \leq_{\Delta^0_\alpha} B$ iff $A \leq_T B \oplus \Delta^0_\alpha$.
(ii) $A \equiv_{\Delta^0_\alpha} B$ iff $(A \leq_{\Delta^0_\alpha} B$ and $B \leq_{\Delta^0_\alpha} A)$.
(iii) The equivalence classes under $\equiv_{\Delta^0_\alpha}$ are called α-degrees.

Note that $\leq_{\Delta^0_1}$ is the same as $\leq_T$.

THEOREM 81 ([16]). *Let $\mathcal{A}$ be a computable structure, and let R be a relation that is not intrinsically Δ^0_α on $\mathcal{A}$. Then, under certain extra effectiveness conditions, for any Σ^0_α set C, there is an isomorphism f from $\mathcal{A}$ onto a computable copy with $f(R) \equiv_{\Delta^0_\alpha} C$.*

Ash and Knight also showed that it is not possible to substitute Turing degrees for α-degrees. In [15], they produced examples of structures $\mathcal{A}$ and relations R, satisfying a great deal of effectiveness, in which certain Σ^0_α Turing degrees, in particular, minimal degrees, are impossible for the image of R. Hirschfeldt and White [175] constructed a family of relations on computable structures, the degrees of which coincide with the levels of the hyperarithmetic hierarchy. Their examples are built up from back-and-forth trees, which explicitly code the alternations of quantifiers.

Using Goncharov's result from the theory of numberings [126], we can show that there is a computable non-intrinsically c.e. relation R on a computable structure $\mathcal{A}$ such that $\mathrm{DgSp}_{\mathcal{A}}(R) = \{\mathbf{0}, \mathbf{d}\}$, where $\mathbf{d} \leq \mathbf{0}''$ but $\mathbf{d} \not\leq \mathbf{0}'$ (see [152]). Harizanov [154] showed that there is a two-element degree spectrum $\mathrm{DgSp}_{\mathcal{A}}(R) = \{\mathbf{0}, \mathbf{d}\}$, such that $\mathbf{0} < \mathbf{d} \leq \mathbf{0}'$ where $\mathbf{d}$ cannot be realized *via* a c.e. set. Goncharov and Khoussainov [138], and Khoussainov and Shore [205] proved that there is a two-element degree spectrum $\mathrm{DgSp}_{\mathcal{A}}(R) = \{\mathbf{0}, \mathbf{c}\}$ such that $\mathbf{c}$ is a nonzero degree realized *via* a c.e. set. Khoussainov and Shore broadly generalized this result.

THEOREM 82 ([205]). *Let $(P, \preceq)$ be a computable partially ordered set. There are a computable structure $\mathcal{A}$ and a computable unary relation R on its domain such that $(\mathrm{DgSp}_{\mathcal{A}}(R), \leq) \cong (P, \preceq)$ and every degree in $\mathrm{DgSp}_{\mathcal{A}}(R)$ is realized via a c.e. set.*

For some familiar relations on computable structures, their Turing degree spectra exhibit the dichotomy: either singletons or infinite. Harizanov [152] established that if for a non-intrinsically c.e. relation R on $\mathcal{A}$, the Ash–Nerode decidability condition holds, then $\mathrm{DgSp}_{\mathcal{A}}(R)$ must be infinite. Hirschfeldt [170] gave a sufficient condition for a relation to have infinite degree spectrum. Applying this condition to linear orders and using the proof of a result of Moses [275], Hirschfeldt established that a computable relation on a computable linear order is either intrinsically computable or has an infinite Turing degree spectrum. Downey, Goncharov, and Hirschfeldt proved the same dichotomy for relations on Boolean algebras.

THEOREM 83 ([82]). *A computable relation on a computable Boolean algebra is either intrinsically computable or has infinite Turing degree spectrum.*

A similar question can be asked for computable relations on other classes of structures such as computable abelian groups. Another interesting question from [82] is whether the degree spectrum of an intrinsically Δ_2^0 relation on a computable linear order is always a singleton or infinite.

Degree spectra have also been investigated for specific important relations on natural classes of structures. One such relation is the successor relation S on a computable linear order $\mathcal{L}$. There are two known examples of singleton degree spectra of the successor relation. If $\mathcal{L}$ has only finitely many successor pairs, then the order is computably categorical, hence the successor relation is intrinsically computable. Downey and Moses [85] constructed a linear order $\mathcal{L}$ having an intrinsically complete successor relation, that is, $\mathrm{DgSp}_{\mathcal{L}}(S) = \{\mathbf{0'}\}$. It was a long-standing open question to investigate upward closure in c.e. degrees of the degree spectrum of the successor relation in computable linear orders. Harizanov, Chubb, and Frolov [56] showed that if $\mathcal{A}$ is a computable linear order with domain A where for all $x \in A$ there is a successor pair (a, b) in $\mathcal{A}$ with $x < a$, then the degree spectrum of the successor relation of $\mathcal{A}$ is closed upward in the c.e. Turing degrees. As a consequence, they established that for every c.e. Turing degree $\mathbf{b}$, the upper cone of c.e. Turing degrees determined by $\mathbf{b}$ is the degree spectrum of the successor relation of some computable linear order. Downey, Lempp, and Wu [78] established the positive result in full generality by developing a new method of constructing Δ_3^0 isomorphisms. Their proof uses a result from [56].

THEOREM 84 ([78]). *If a computable linear order has infinitely many successor pairs, then the degree spectrum of the successor relation is closed upward in the c.e. Turing degrees.*

In [320], Soskov established that a Δ_1^1 relation on computable $\mathcal{A}$, which is invariant under automorphisms of $\mathcal{A}$, is definable in $\mathcal{A}$ by a computable infinitary formula with no parameters. This led to the following characterization of intrinsically Δ_1^1 relations.

THEOREM 85 ([320]). *For a computable structure $\mathcal{A}$, and a relation R on $\mathcal{A}$, the following are equivalent*:
  (i) *R is intrinsically Δ_1^1 on $\mathcal{A}$*;
  (ii) *R is relatively intrinsically Δ_1^1 on $\mathcal{A}$*; *and*
  (iii) *R is definable in $\mathcal{A}$ by a computable infinitary formula with finitely many parameters.*

In the following theorem characterizing intrinsically Π_1^1 relations, Soskov [321] established the equivalence (ii) ⇔ (iii), while (i) ⇔ (ii) was established in [137].

THEOREM 86 ([321, 137]). *For a computable structure $\mathcal{A}$ and relation R on $\mathcal{A}$, the following are equivalent*:
  (i) *R is intrinsically Π_1^1 on $\mathcal{A}$*;
  (ii) *R is relatively intrinsically Π_1^1 on $\mathcal{A}$*; *and*
  (iii) *R is definable in $\mathcal{A}$ by a Π_1^1 disjunction of computable infinitary formulas with finitely many parameters.*

Goncharov, Harizanov, Knight, and Shore [137] considered a general family of examples of intrinsically Π_1^1 relations arising in computable structures of Scott rank $\omega_1^{CK} + 1$. A *Harrison order* is a computable linear order of type $\omega_1^{CK}(1 + \eta)$. Harrison [162] showed that such an order exists. The initial segment of this order of type ω_1^{CK} is intrinsically Π_1^1 since it is defined by the disjunction of computable infinitary formulas saying that the interval to the left of x has order type α, for computable ordinals α. A *Harrison Boolean algebra* is a computable Boolean algebra of type $I(\omega_1^{CK}(1 + \eta))$, where for an order $\mathcal{L}$, the interval algebra $I(\mathcal{L})$ is the algebra generated, under finite union, by the intervals $[a, b)$, $(-\infty, b)$, $[a, \infty)$, with endpoints in $\mathcal{L}$. The set of superatomic elements of this Boolean algebra is intrinsically Π_1^1. A *Harrison group* is a countable abelian p-group G such that $\lambda(G) = \omega_1^{CK}$, every element in its Ulm sequence $(u_\alpha(G))_{\alpha < \omega_1^{CK}}$ is ∞, and the divisible part has infinite dimension. Recall that the Ulm subgroups G_α are defined by $G_\alpha = p^{\omega\alpha}G$, and $u_\alpha(G) =_{\text{def}} \dim_{\mathbb{Z}_p} P_\alpha(G)/P_{\alpha+1}(G)$, where $P_\alpha(G) = G_\alpha \cap \{x \in G : px = 0\}$. The set of elements of a Harrison group, which have computable ordinal heights, is intrinsically Π_1^1. It is the complement of the divisible part. By a *path* through Kleene's $\mathcal{O}$ we mean a subset of $\mathcal{O}$ that is linearly ordered under $<_\mathcal{O}$ and includes a notation for every computable ordinal.

THEOREM 87 ([137]). *The following sets are equal*:
  1. *the set of Turing degrees of Π_1^1 paths through $\mathcal{O}$*;

2. the set of Turing degrees of left-most paths of computable trees $\mathcal{T} \subseteq \omega^{<\omega}$ such that $\mathcal{T}$ has a path, but no hyperarithmetic path;
3. the set of Turing degrees of maximal well-ordered initial segments of Harrison orders;
4. the set of Turing degrees of superatomic parts of Harrison Boolean algebras; and
5. the set of Turing degrees of divisible parts of Harrison groups.

For certain types of structures, there is a close connection between the notions of degree spectra of structures and of relations. Harizanov and R. Miller [149] defined a computable structure $\mathcal{U}$ to be *spectrally universal* for a theory T if for every automorphically nontrivial countable model $\mathcal{A}$ of T, there is an embedding $f\colon \mathcal{A} \to \mathcal{U}$ such that $\mathcal{A}$ as a structure, has the same degree spectrum as $f(\mathcal{A})$, as a relation on the domain of $\mathcal{U}$. Spectrally universal structures investigated in [149] are the countable dense linear order and the random graph. Both are Fraïssé limits. This led Csima, Harizanov, R. Miller, and Montalbán to develop the theory of computable Fraïssé limits in [62]. They gave a sufficient condition for certain Fraïssé limits to be spectrally universal, which they used to show that the countable atomless Boolean algebra is spectrally universal.

For syntactic characterizations of relations having Post-type properties on structures, or their degree-theoretic complexity see [166, 167, 18, 145, 135, 146, 136].

§8. **Families of relations on a structure.** Many important algebraic properties can be investigated by considering natural families of relations on a structure. For example, for a vector space V we can consider the family of its bases:

$$\mathcal{B}(V) = \{X \subseteq V \colon X \text{ is a basis of } V\}.$$

For an orderable field F we can consider the set of all linear orders on its domain, which are invariant under the field operations:

$$O(F) = \{R \subseteq F \times F \colon R \text{ is an order on } F\}.$$

Such a family of relations does not necessarily have a computable member even when the structure is computable. Mal'cev [229] showed that there exists a computable vector space without a computable basis. Metakides and Nerode [240] and Ershov [90] showed that there exists a computable orderable field that cannot be computably ordered. We could ask for a sufficient condition on a family of relations on a computable structure to have a computable member. More generally, we may ask what the collection of Turing degrees of its members is.

DEFINITION 25 ([70]). Given a family of relations $\mathcal{R}$ on a computable structure $\mathcal{M}$, define

$$\mathrm{DgSp}(\mathcal{R}) = \{\deg(R) \colon R \in \mathcal{R}\}.$$

In the next definition we will be computing all relations simultaneously (uniformly).

DEFINITION 26. Let $\mathcal{A}$ be a computable structure, and let $\mathcal{R} = (R_i)_{i \in I}$ be a family of relations on $\mathcal{A}$, where $l(i)$ is the arity of R_i. Define

$$\mathrm{DgSp}(\mathcal{R}; \mathcal{A}) = \deg\{\overline{a} \subseteq A^{l(i)} \colon \mathcal{A} \models R_i(\overline{a}), i \in I\}.$$

In many interesting examples, the index set I and the arities of relations are computable. The previous two definitions are dependent on a given presentation of a structure. We could let the definitions range over all computable copies of $\mathcal{A}$. However, this approach is not common.

Let us consider the problem of computing a generating set (or a basis) of a given computable structure. The definition of a basis depends on the class of structures. The study of the problem of computing a basis in several classical algebraic examples provides a natural link between Definition 26 and Definition 25. More specifically, to build a basis stage-by-stage (Definition 25), one usually needs a corresponding notion of independence (Definition 26). Consider the following example.

Example. Let V be a countable vector space of infinite dimension. Define the following sets of relations on V.
1. For every $i \in \omega$, and any $x_0, \ldots, x_i \in V$, we set $P_i(x_0, \ldots, x_i) = 1$ if and only if $x_0, \ldots, x_i$ are linearly independent.
2. Let $\mathbb{B}$ be the collection of maximal linearly independent sets (bases) in V.

If $\mathcal{P} = (P_i)_{i \in \omega}$ is uniformly computable, then we say that V has an algorithm for linear independence.

THEOREM 88 (folklore; see [229, 239]). *Every computable vector space over a computable field has a $\mathbf{0}'$-computable basis, and this bound is sharp.*

Let us now consider another natural example from algebra.

Example. Let F be a countable algebraically closed field of infinite transcendence degree. Define the following sets of relations on F.
1. For every $i \in \omega$, and any $a_0, \ldots, a_i \in F$, we set $R_i(a_1, \ldots, a_i) = 1$ if and only if, $a_1, \ldots, a_i$ are algebraically independent.
2. Let $\mathbb{A}$ be the collection of maximal algebraically independent subsets of F.

If $\mathcal{R} = (R_i)_{i \in \omega}$ is uniformly computable in F, then we say that F has an algorithm for algebraic independence.

THEOREM 89 (folklore; see [107, 240, 288]). *The algebraic closure of $\mathbb{Q}(x_i : i \in \omega)$ has a $\mathbf{0}'$-maximal algebraically independent set, and this bound is sharp.*

It is clear that independence can be formalized using families of relations as in Definition 26, and the collection of bases should be studied according to Definition 25. It is important to observe that in the context of vector spaces and algebraically closed fields, the existence of a generating set is equivalent to the problem of computable categoricity relative to an oracle. The same can be said about many other natural examples.

A number of researchers investigated complexity of independent sets and other subsets and subspaces of c.e. vector spaces and c.e. algebraically closed fields (see, for example, [239, 184, 290, 313, 241, 71]). In many of their results the operations (vector addition and scalar multiplication, or field operations, respectively) play no direct role. For instance, in the proofs of Theorems 88 and 89 only the phenomenon of independence occurs. In fact, Metakides and Nerode [241] initiated the study of the effective content of abstract independence relations (Steinitz closure systems). For an extended survey of the results about computable Steinitz closure systems, see the paper [86] by Downey and Remmel.

We will now discuss recent results about bases of various structures. Downey and Melnikov [80] studied free modules over localizations of integers.

THEOREM 90 ([80]). *Let $S \subseteq \omega$ be a c.e. set of primes.*

(i) *Every computable free module $\mathcal{F}(S)$ over the localization of $\mathbb{Z}$ by S has a Σ_3^0 (actually, Π_2^0 in S) set of generators.*

(ii) *Every computable copy of $\mathcal{F}(S)$ has a Σ_2^0 set of generators if and only if the complement of S is semi-low.*

The theorem can be equivalently re-formulated in terms of computable categoricity relative to an oracle. The corresponding analogue of linear independence for free modules of this kind is *S-independence*, which is a generalization of the classical notion of *p*-independence (see [80]). As a consequence of Theorem 90 with $S = \emptyset$, it follows that every free abelian group has a Π_1^0 generating set.

Algebraic structure becomes more complex in the case of free nonabelian groups. Relatively recently, Sela in a series of papers [303, 304, 305, 306, 307, 308, 309] gave a positive solution to the problem of elementary equivalence of free groups of different finite ranks greater than 1, posed by Tarski in the 1940s. (See also Kharlampovich and Myasnikov [193].) Inspired by this result, Carson, Harizanov, Knight, Lange, McCoy, Morozov, Safranski, Quinn, and Wallbaum [44], and McCoy and Wallbaum [234] investigated free groups in the context of computable model theory. Let F_∞ be the free group of rank $\aleph_0$.

THEOREM 91 ([44, 234]). *Every computable copy of F_∞ has a Π_2^0 basis, and the result cannot be improved to Σ_2^0.*

The proof of the theorem uses deep results in algebra. The corresponding notion of independence is what is called *primitiveness* in every finitely generated subgroup (see [44, 234]).

In general, not every family of unary relations (Definition 25) has a hyperarithmetic "notion of independence" (Definition 26). For example, consider the collection of paths on $\mathcal{T} \subset \omega^{<\omega}$, where $\mathcal{T}$ codes a Σ_1^1-complete set. In contrast, we have seen that natural structures well-understood in algebra tend to have arithmetic bases. Thus, we can ask whether there is a natural structure (such as a ring, a module, or a group) for which finding a generating set is not (hyper)arithmetic. A possible candidate is the pure transcendental ring over the rationals, $\mathbb{Q}[x_i : i \in \omega]$. Does every computable copy of $\mathbb{Q}[x_i : i \in \omega]$ have a (hyper)arithmetic basis? Describing automorphism orbits of generators in $\mathbb{Q}[x_i : i \in \omega]$ is a long-standing open problem in algebra. There has been some progress in this direction; see the recent paper by Shestakov and Umirbaev [312].

We will now discuss some old and recent results on orders on orderable groups and fields. Recall that a left order on a group $\mathcal{G} = (G, \cdot)$ is a linear order $<$ of its elements, which is left-invariant under the group operation:

$$(\forall x, y)(\forall z)[x < y \Rightarrow z \cdot x < z \cdot y].$$

A right order is defined similarly. A bi-order (or simply order) is invariant under both left and right multiplication. The definition of an order for a field is similar. Clearly, every left order on an abelian group is a bi-order. Every left order $<_l$ on $\mathcal{G}$ induces a right order $<_r$ on $\mathcal{G}$ as follows:

$$a <_r b \Leftrightarrow b^{-1} <_l a^{-1}.$$

It is well known that an abelian group is orderable if and only if it is torsion-free. A field is orderable exactly when it is formally real (see [110]). As for fields, in the case of computable orderable groups, the effective analogue of the classical result fails. Downey and Kurtz [84] showed that there exists a computable group isomorphic to $\mathbb{Z}^\omega = \bigoplus_{i \in \omega} \mathbb{Z}$, which does not have a computable order. On the other hand, Dobritsa [72] previously showed that every computable, torsion-free, abelian group is isomorphic to a computable group with a computable order.

For a group $\mathcal{G}$, by $\mathrm{LO}(\mathcal{G})$ we denote the set of all left orders on $\mathcal{G}$, and by $\mathrm{BiO}(\mathcal{G})$ the set of all bi-orders on $\mathcal{G}$. There is a natural topology on these sets (when nonempty), making the topological spaces compact, even when $\mathcal{G}$ is a semigroup instead of a group, or just a structure with a single binary operation (see [67]). In some cases this space is homeomorphic to the Cantor set. Sikora [314] established that the space $\mathrm{BiO}(\mathbb{Z}^n)$ for $n > 1$ is homeomorphic to the Cantor set. Dabkowska [69] established that the space $\mathrm{BiO}(\mathbb{Z}^\omega)$ is homeomorphic to the Cantor set. (Her result can also be obtained from [84].) Solomon [319] obtained the following results about Turing degrees of orders on abelian groups.

THEOREM 92 ([319]).

1. *A computable, torsion-free, abelian group of finite rank greater than 1 has an order in every Turing degree.*
2. *A computable, torsion-free, abelian group of infinite rank has an order in every Turing degree* $\mathbf{d} \geq \mathbf{0}'$.
3. *Let $n > 1$. A computable, torsion-free, properly n-step nilpotent group has an order in every Turing degree* $\mathbf{d} \geq \mathbf{0}^{(n)}$.

The positive cone of an order $<$ on a group $\mathcal{G}$ is $P = \{a \in G : e \leq a\}$, where $e \in G$ is the identity element. The negative cone is $P^{-1} = \{a \in G : a \leq e\}$. Clearly, $a \leq b$ iff $a^{-1}b \in P$. Hence, we can effectively pass from binary relations (orders) to unary relations (positive cones) and *vice versa*. We can easily verify that if $P \subseteq G$ is a *subsemigroup* of $\mathcal{G}$ (i.e., $PP \subseteq P$), which satisfies $P \cap P^{-1} = \{e\}$, then P defines a left order on $\mathcal{G}$ if and only if P is *total* (i.e., $P \cup P^{-1} = G$). Moreover, P defines a bi-order on $\mathcal{G}$ if, in addition, P is a *normal* subsemigroup (i.e., $g^{-1}Pg \subseteq P$ for every $g \in G$). Denote by $\mathbb{C}(\mathcal{G})$ the set of all positive cones of orders on $\mathcal{G}$. Clearly, $\mathrm{DgSp}(\mathrm{BiO}(\mathcal{G})) = \{\deg(C) : C \in \mathbb{C}(\mathcal{G})\}$.

Solomon [318] established that for every orderable computable group $\mathcal{G}$, there is a computable binary tree $\mathcal{T}$ and a Turing degree preserving bijection from $\mathbb{C}(\mathcal{G})$ to the set of all infinite paths of $\mathcal{T}$. Hence $\mathbb{C}(\mathcal{G})$ corresponds to a Π_1^0 class, and, by the *low basis theorem*, $\mathrm{BiO}(\mathcal{G})$ contains an order of *low* Turing degree. Previously, Metakides and Nerode [240] established the same results for computable orderable fields. Moreover, they showed that the sets of orders of computable orderable fields are in exact correspondence to the collections of Π_1^0 subsets of 2^ω.

THEOREM 93 ([240]). *For every nonempty Π_1^0 class $\mathbb{P}$, there is a computable orderable field $\mathcal{F}$ and a Turing degree preserving bijection $f : \mathbb{P} \to \mathbb{C}(\mathcal{F})$.*

The proof is based on a result by Craven [60] that for every Boolean topological space $\mathcal{T}$, there is a formally real field $\mathcal{F}$ such that $\mathbb{C}(\mathcal{F})$ is homeomorphic to $\mathcal{T}$. Many corollaries about degree spectra of orders on fields follow from Theorem 93. It is not hard to see that the situation is different for torsion-free abelian groups. Solomon [318], using a result by Jockusch and Soare [179], showed that there is a Π_1^0 class $\mathbb{P}$ such that for any computable, torsion free, abelian group $\mathcal{G}$, we have $\{\deg(f) : f \in \mathbb{P}\} \neq \mathrm{DgSp}(\mathrm{BiO}(\mathcal{G}))$.

More recently, Dabkowska, Dabkowski, Harizanov, and Togha [70] studied topological and computability-theoretic properties of left orders and bi-orders on (not necessarily abelian) groups. They obtained general sufficient conditions for the degree spectra of orders on groups to contain upper cones of

Turing degrees. As a corollary they established the following result about the free groups F_n of rank n.

THEOREM 94 ([70]). *Every computable copy of F_n, where $n > 1$, has an order in every Turing degree.*

Sikora [314] conjectured that $\mathrm{BiO}(F_n)$ for $n > 1$ is homeomorphic to the Cantor set. The conjecture still remains open. It was shown in [148] that there is a computable copy of F_∞ with no computable left order, and hence the space $\mathrm{BiO}(F_\infty)$ (as well as $\mathrm{LO}(F_\infty)$) is homeomorphic to the Cantor set.

Kach, Lange, and Solomon [182] constructed computable, torsion-free, abelian groups such that the degree spectra of their orders are not upward closed. The groups are isomorphic to effectively completely decomposable groups. N. Khisamiev and Krykpaeva [198] defined a computable, infinite-rank, torsion-free, abelian group $\mathcal{H}$ to be *effectively completely decomposable* if there is a uniformly computable sequence of rank one groups $\mathcal{H}_i$, $i \in \omega$, such that $\mathcal{H}$ is equal to $\oplus_{i \in \omega} \mathcal{H}_i$.

THEOREM 95 ([182]). *Let $\mathcal{H}$ be a computable and effectively completely decomposable group. Then there is a computable copy $\mathcal{G}$ of $\mathcal{H}$ such that $\mathrm{DgSp}(\mathrm{BiO}(\mathcal{G}))$ contains $\mathbf{0}$, but is not upward closed.*

More precisely, Kach, Lange, and Solomon showed that there is a non-computable, c.e. set C such that $\mathcal{G}$ has exactly two computable orders, and every C-computable order on $\mathcal{G}$ is computable. On the other hand, since $\mathcal{H}$ is effectively completely decomposable, it has a computable basis formed by choosing a nonzero element h_i from every $\mathcal{H}_i$. Hence $\mathrm{DgSp}(\mathrm{BiO}(\mathcal{H}))$ contains every Turing degree, and $\mathcal{G}$ is not effectively completely decomposable. Kach, Lange, and Solomon [182] conjectured that the conclusion of Theorem 95 holds for all computable, infinite-rank, torsion-free, abelian groups $\mathcal{H}$.

Natural relations in partial orders are their chains and antichains. Complexity of infinite chains and antichains in computable partial orders was studied by Herrmann [163] and Harizanov, Jockusch, and Knight [147]. It follows from an effective version of Ramsey's theorem for pairs, due to Jockusch [178], that a computable partial order of ω has either an infinite Δ_2^0 chain, or an infinite Δ_2^0 antichain, or else both an infinite Π_2^0 chain and an infinite Π_2^0 antichain. On the other hand, Herrmann [163] showed that there is a computable partial order of ω with no infinite Σ_2^0 chain or antichain. Harizanov, Jockusch, and Knight [147] showed that there is a computable partial order with an infinite chain but none that is Σ_1^1 or Π_1^1, and they obtained the analogous result for antichains. They also showed that there is a computably axiomatizable theory T of partial orders such that T has a computable model with arbitrarily long finite chains but no computable model with an infinite chain. They also established the corresponding result for antichains.

§9. **Classes of structures and equivalence relations.** Our goal is to measure the complexity of classes of computable structures and equivalence relations on these classes. More precisely, we want to know how complex the answers to the following types of questions are. Does a computable structure belong to a particular class of structures with fixed algebraic, model-theoretic, or algorithmic properties (e.g., a class of groups, uncountably categorical structures, decidable structures, etc.)? Are two structures from such a class isomorphic, computably isomorphic, bi-embeddable, etc.? We are looking for a criterion that will allow us to say whether such questions have "nice" answers.

There are many papers investigating the complexity of classes of countable structures. There is earlier work in descriptive set theory [248, 249] investigating subsets of the Polish space of structures with universe ω for a given countable relational language. Concerning the possible complexity (in the noneffective Borel hierarchy) of the set of copies of a given structure, D. Miller [249] showed that if this set is $\Delta^0_{\alpha+1}$, then it is $d\text{-}\Sigma^0_\alpha$. In [248], A. Miller showed that this set cannot be properly Σ^0_2. There are also examples illustrating other possibilities.

The main issue here is to find an optimal definition of the class of structures under investigation. This often requires the use of various internal properties of the structures in the class. After a reasonable definition is found, it is necessary to prove its sharpness. Usually, this is done by proving completeness in some complexity class.

In the case of equivalence relations, the study of Borel reducibility has developed into a rich area of descriptive set theory. The notion of Borel reducibility allows us to compare the complexity of equivalence relations on Polish spaces (see [190, 111]). In particular, natural equivalence relations on classes of countable structures, such as isomorphism and bi-embeddability, have been widely studied; for example, see [105, 176, 106]. An effective version of this study was introduced by Calvert, Cummins, Knight, and S. Miller (Quinn) [34], and Knight, S. Miller (Quinn), and Vanden Boom [211]. The main idea is that the complexity of the isomorphism relation on various classes of countable structures can be measured using the effective transformations. The introduced c-embeddings and tc-embeddings are based on uniform enumeration reducibility and uniform Turing reducibility, respectively. The main advantage of this approach is that it allows distinctions among classes with countably many isomorphism types.

In computable model theory, we may state our goal as follows. Let K be a class of structures. We denote by K^c the set of computable structures in K. A *computable characterization* of K should separate computable structures in K from all other structures (those not in K, or noncomputable ones). A *computable classification* for K up to an equivalence relation E (isomorphism, computable isomorphism, etc.) should determine each computable element, up to the equivalence E, in terms of relatively simple invariants. In [140], Goncharov and Knight presented three possible approaches to the study of computable characterizations of classes of structures.

Within the framework of the first approach, we say that K has a *computable characterization* if K^c is the set of computable models of a computable infinitary sentence.

PROPOSITION 1. (i) *The class of linear orders can be characterized by a single first-order sentence.*

(ii) *The class of abelian p-groups is characterized by a single computable Π_2 sentence.*

(iii) *The classes of well orders and reduced abelian p-groups cannot be characterized by single computable infinitary sentences.*

Furthermore, we say that there is a *computable classification* for K if there is a computable bound on the ranks of elements of K^c. By a *computable rank* $R^c(\mathcal{A})$ of a structure $\mathcal{A}$ we mean the least ordinal α such that for all tuples $\overline{a}$ and $\overline{b}$ in $\mathcal{A}$, of the same length, if for all $\beta < \alpha$, all computable Π_β formulas that true of $\overline{a}$ are also true of $\overline{b}$, then there is an automorphism of $\mathcal{A}$ taking $\overline{a}$ to $\overline{b}$. For example, the computable rank of a vector space over $\mathbb{Q}$ is 1. There is no computable bound on computable ranks of linear orders and abelian p-groups. The computable rank is not the same as the Scott rank. However, for a hyperarithmetic structure, its computable rank is a computable ordinal just in case its Scott rank is computable (see [140]). If $\mathcal{A}$ is hyperarithmetic, then $R^c(\mathcal{A}) \leq \omega_1^{CK}$.

The second approach involves the notion of an index set. A *computable index* for a structure $\mathcal{A}$ is a number e such that $D(\mathcal{A}) = W_e$, where $D(\mathcal{A})$ is the atomic diagram of $\mathcal{A}$. We denote the structure with index e by $\mathcal{A}_e$. For a class K of structures, the *index set* $I(K)$ is the set of computable indices of members of K^c:

$$I(K) = \{e \colon W_e = D(\mathcal{A}) \wedge \mathcal{A} \in K\}.$$

For an equivalence relation E on a class K, we define

$$I(E, K) = \{(m, n) \colon m, n \in I(K) \wedge \mathcal{A}_m E \mathcal{A}_n\}.$$

Within this approach, we say that K has a *computable characterization* if $I(K)$ is hyperarithmetic. The class K has a *computable classification* up to E if $I(E, K)$ is hyperarithmetic.

The first and the second approach are known to be equivalent [140]. In fact, we do not know a better way to estimate the complexity of an index set than by giving a description by a computable infinitary formula.

PROPOSITION 2 ([140]). (i) *For the following classes K, the index set $I(K)$ is Π_2^0:*

(a) *linear orders,*

(b) *Boolean algebras,*
(c) *abelian p-groups, and*
(d) *vector spaces over* $\mathbb{Q}$.

(ii) (*Kleene, Spector.*) *For the following classes* K, *the index set* $I(K)$ *is not hyperarithmetic*:

(a) *well-orders,*
(b) *superatomic Boolean algebras, and*
(c) *reduced abelian p-groups.*

In the next theorem, the calculations of the complexity of index sets for classes of structures with interesting model-theoretic properties are due to White [333], Calvert, Fokina, Goncharov, Knight, Kudinov, Morozov, and Puzarenko [35], Fokina [99], and Pavlovskii [282]. In (v), $\Sigma_3^0 - \Sigma_3^0$ denotes the difference of two Σ_3^0 sets.

THEOREM 96.

(i) ([333, 282]) *The index set of computable prime models is an m-complete* $\Pi_{\omega+2}^0$ *set.*
(ii) ([333]) *The index set of computable homogeneous models is an m-complete* $\Pi_{\omega+2}^0$ *set.*
(iii) ([282]) *The index set of structures with uncountably categorical theories is a* Δ_ω^0-*hard* $\Sigma_{\omega+1}^0$ *set.*
(iv) ([282]) *The index set of structures with countably categorical theories is a* Δ_ω^0-*hard* $\Pi_{\omega+2}^0$ *set.*
(v) ([99]) *The index set of structures with decidable countably categorical theories is an m-complete* $\Sigma_3^0 - \Sigma_3^0$ *set.*
(vi) ([35]) (a) *The index set of computable structures with noncomputable Scott ranks is m-complete* Σ_1^1.
(b) *The index set of structures with the Scott rank* ω_1^{CK} *is m-complete* Π_2^0 *relative to Kleene's* $\mathcal{O}$.
(c) *The index set of structures with the Scott rank* $\omega_1^{CK}+1$ *is m-complete* Σ_2^0 *relative to Kleene's* $\mathcal{O}$.

The index sets for structures with specific algorithmic properties were also studied by White [333], Fokina [100], and Downey, Kach, Lempp, and Turetsky [83].

THEOREM 97. (i) ([100]) *The index set of decidable structures is* Σ_3^0-*complete.*
(ii) ([333]) *The index set of hyperarithmetically categorical structures is* Π_1^1-*complete.*
(iii) ([83]) *The index set of relatively computably categorical structures is* Σ_3^0-*complete.*

The following result of Downey, Kach, Lempp, Lewis, Montalbán, and Turetsky resolves an important old problem.

THEOREM 98 ([76]). *The index set of computably categorical structures is Π_1^1-complete.*

The structures constructed to establish this result are computable trees of a special kind. It would be worthwhile to calculate the complexity of the index sets of other classes of computable structures having interesting algebraic, model-theoretic, or algorithmic properties.

The third approach of Goncharov and Knight [140] to computable characterization of classes of structures involves the notion of *enumeration*. A class of structures has a good characterization if all its structures are represented in the list, up to isomorphism or some other equivalence relation. A good classification of the class would mean listing each equivalence class only once.

DEFINITION 27. (i) An *enumeration* of K^c/E is a sequence $(\mathcal{M}_n)_{n\in\omega}$ representing all E-equivalence classes in K^c.
(ii) A *Friedberg enumeration* of K^c/E is an enumeration in which every E-equivalence class is represented only once.
(iii) An enumeration is Δ_α^0-*computable* if there is a Δ_α^0-computable sequence of computable indices for the structures.

We say that K has a *computable characterization* if there is a hyperarithmetic enumeration of $K^c/\cong$. We say that K has a *computable classification* up to E if there is a hyperarithmetic Friedberg enumeration of K^c/E. It is known that this approach is not equivalent to the previous two approaches, but is only implied by them. Recall that a Harrison order is a computable linear order of type $\omega_1^{CK}(1+\eta)$.

PROPOSITION 3 ([140]). *Consider the class K consisting of copies of the Harrison order and of the linear orders of rank at most ω. Then $K^c/\cong$ has a hyperarithmetic Friedberg enumeration, but the index set $I(K)$ is not hyperarithmetic.*

We will now focus on the classification problems up to important equivalence relations. The most interesting cases are isomorphism, bi-embeddability, and isomorphism of bounded algorithmic complexity. Possible ways to compare the complexity of various equivalence relations are:

1. comparison among sets; and
2. comparison among equivalence relations.

The former case was discussed above. It corresponds to the second approach from [140]. Within this approach, we usually prove m-completeness among sets in some complexity class. There has been quite a lot of work on the isomorphism problem for various classes of computable structures by Goncharov and Knight [140], Calvert [29, 30, 31], and Calvert, Harizanov, Knight, and S. Miller (Quinn) [40].

THEOREM 99. (i) ([30]) *The isomorphism problem for computable vector spaces over $\mathbb{Q}$ is m-complete among Π_3^0 sets.*

(ii) ([30]) *The isomorphism problem for torsion-free abelian groups of finite characteristic is m-complete among Σ_3^0 sets.*

(iii) ([140]) (a) *The isomorphism problem for abelian p-groups is m-complete among Σ_1^1 sets.*

(b) *The isomorphism problem for trees is m-complete among Σ_1^1 sets.*

Recently, Carson, Fokina, Harizanov, Knight, Safranski, Quinn, and Wallbaum initiated the study of the *computable embedding problem*. In [43], they investigated the relation between the isomorphism problem and the embedding problem for some well-known classes of structures. The isomorphism problem and the embedding problem were compared as sets, that is, using the standard m-reducibility. While for some classes of structures the two problems have the same complexity, for other classes the isomorphism problem is more complicated than the embedding problem, or *vice versa*.

Further comparison of complexity of equivalence relations was done using the 2-dimensional versions of reducibilities. This approach can be seen as an analogue of investigation done in descriptive set theory. Recall that in descriptive set theory, two equivalence relations, E and F, on Borel classes of structures, K and L, respectively, can be compared using Borel reducibility. In the computable case, instead of arbitrary invariant Borel classes of countable structures, we consider classes of computable structures with hyperarithmetic index sets. In other words, we consider classes consisting of computable models of computable infinitary sentences. As mentioned above, this corresponds to a "nice" characterization of a class.

A straightforward analogue of the Borel reducibility is the hyperarithmetic reducibility.

DEFINITION 28. For equivalence relations E_1, E_2 on (hyperarithmetic subsets of) ω, we say that E_1 is *h-reducible to* E_2, in symbols $E_1 \leq_h E_2$, if there is a hyperarithmetic function f such that for all x, y,

$$x \ E_1 \ y \Leftrightarrow f(x) \ E_2 \ f(y).$$

A stronger reducibility would be a 2-dimensional version of the m-reducibility. This reducibility is traditionally used in the general study of equivalence relations on ω. It was introduced by Ershov in [91] where he studied properties of numberings. Later it was used, for example, in [27, 112, 59, 8] and denoted simply by $\leq$. As sometimes we need to emphasize the difference between m-reducibility and h-reducibility, we will denote the reducibility *via* a computable function by $\leq_m$, specifying when necessary that we consider the 2-dimensional version of m-reducibility among relations. When the results hold for both h-reducibility and m-reducibility we will use the symbol $\leq$.

DEFINITION 29. Let E_1, E_2 be equivalence relations on hyperarithmetic subsets $X, Y \subseteq \omega$, respectively. The relation E_1 is *m-reducible to* E_2, in symbols

$E_1 \leq_m E_2$, iff there exists a partial computable function f with $X \subseteq \text{dom}(f)$ and $X \subseteq f(Y)$ such that for all $x, y \in X$,

$$x \, E_1 \, y \Leftrightarrow f(x) \, E_2 \, f(y).$$

Each notion of reducibility generates the corresponding notion of completeness.

DEFINITION 30. A relation E on a hyperarithmetic subset of ω is an *h-complete* Σ_1^1 equivalence relation, or *m-complete* Σ_1^1 equivalence relation, if E is Σ_1^1 and every Σ_1^1 equivalence relation E_1 on a hyperarithmetic subset of ω is h-reducible to E, or m-reducible to E, respectively.

We use the previous definitions to compare equivalence relations on classes of computable structures. Recall that each such relation E on a class K has the index set $I(E, K)$. We make no distinction between E and $I(E, K)$ in the following sense. If E_1 is an arbitrary equivalence relation on ω, then we say that E_1 h-reduces to E, or m-reduces to E, iff there exists a hyperarithmetic, or computable, respectively, sequence of computable structures $\{\mathcal{A}_x\}_{x \in \omega}$ from K such that for all x, y, we have $x \, E_1 \, y$ iff $\mathcal{A}_x \, E \, \mathcal{A}_y$. (This is equivalent to $E_1 \leq_h I(E, K)$ or $E_1 \leq_m I(E, K)$ in the sense of Definitions 28 and 29.) From now on we will write $\leq$ to denote either of $\leq_h$, $\leq_m$. We will use the terms "reduces," "complete," etc. for the corresponding notion of reducibility.

The following result is due to Fokina and S. Friedman.

PROPOSITION 4 ([101]). *There is a class K of structures with hyperarithmetic index set such that the bi-embeddability relation on K^c is complete among Σ_1^1 equivalence relations.*

This result corresponds to the analogous result in descriptive set theory due to S. Friedman and Motto Ros [106]. However, the theory of Σ_1^1 equivalence relations on ω under $\leq$-reducibility behaves very differently from the theory of Borel equivalence relations on Polish spaces. In particular, Fokina, S. Friedman, Harizanov, Knight, McCoy, and Montalbán [96] established the following completeness result.

THEOREM 100 ([96]). *The isomorphism of computable graphs is complete with respect to the chosen effective reducibility in the context of all Σ_1^1 equivalence relations on ω.*

This is false in the context of countable structures and Borel reducibility since Kechris and Louveau [191] showed that there are examples of Borel equivalence relations that are not Borel-reducible to isomorphism of graphs. Moreover, the authors of [96] proved that the isomorphism relation on computable torsion abelian groups is complete among Σ_1^1 equivalence relations on ω, while in the classical case it is known to be incomplete among isomorphism relations on classes of countable structures, as established by H. Friedman and Stanley [105]. In [96], the authors also established that the isomorphism relation on computable, torsion-free, abelian groups is complete among Σ_1^1 equivalence relations on ω, while in the case of countable structures it is not known to be complete for isomorphism relations.

Regarding bounding the complexity of the isomorphism relation, Fokina, S. Friedman, and Nies obtained the following result.

THEOREM 101 ([97]). *The computable isomorphism relation on computable structures from classes including predecessor trees, Boolean algebras, and metric spaces is a complete Σ_3^0 equivalence relation under the computable reducibility.*

To prove their result, the authors first showed that *one-one equivalence* relation of c.e. sets, as an equivalence relation on indices, is Σ_3^0 complete, and then reduced this equivalence relation to the computable isomorphism on predecessor trees. Using the technique developed by Hirschfeldt and White in [175] and Csima, Franklin, and Shore in [65], the result of Theorem 101 can be lifted to hyperarithmetic levels.

It follows from [106] by S. Friedman and Motto Ros that the following result holds for the bi-embeddability relation on computable structures.

THEOREM 102 ([106]). *For every Σ_1^1 equivalence relation E on ω, there exists a hyperarithmetic class K of structures, which is closed under isomorphism, and such that E is h-equivalent to the bi-embeddability relation on computable structures from K.*

In fact, the reduction functions have complexity at most $\mathbf{0}'$. In [102], Fokina and S. Friedman showed that the general structure of Σ_1^1 equivalence relations on hyperarithmetic subsets of ω is rich. Theorem 102 states that the structure of bi-embeddability relations on hyperarithmetic classes of computable structures is as complex as the whole structure of Σ_1^1 equivalence relations under h-reducibility. It would be interesting to answer the following question and possibly get a refinement of Theorem 102. If E is a Σ_1^1 equivalence relation on ω, does there exist a hyperarithmetic class K of structures, which is closed under isomorphism, and such that E is equivalent to the bi-embeddability relation on computable structures from K *via* computable functions?

It is not known whether there exists a hyperarithmetic class of computable structures with Σ_1^1, but not Δ_1^1 isomorphism relation, which is not complete among all isomorphism relations on hyperarithmetic classes of computable structures. An affirmative answer to the following question may help solve this problem. Does there exist a hyperarithmetic class K of computable structures, which contains a unique structure of noncomputable Scott rank (up to isomorphism)? If such a class exists, then the isomorphism relation on the class of computable graphs cannot be reduced to the isomorphism relation on K. Indeed, there exist nonisomorphic graphs of high (that is, ω_1^{CK} or $\omega_1^{CK} + 1$) Scott rank. They must be mapped to nonisomorphic structures in K. However, no computable structure of high Scott rank can be mapped to a computable structure of computable Scott rank under a hyperarithmetic reducibility. This question is closely connected with many important open questions in computable model theory concerning computable structures of high Scott rank, such as the question of strong computable approximation (see [140, 35]). It is known that, up to bi-embeddability, this is true in the following sense. In the class of computable linear orders, the equivalence class of linear orders bi-embeddable with the rationals is Σ_1^1-complete, but every computable scattered linear order (that is, one not bi-embeddable with the rationals) has a hyperarithmetic equivalence class. For more information on the bi-embeddability relation in the class of countable linear orders see the paper [257] by Montalbán.

REFERENCES

[1] S. I. ADYAN, *Algorithmic unsolvability of problems of recognition of certain properties of groups*, **Doklady Akademii Nauk SSSR**, vol. 103 (1955), pp. 533–535, (Russian).

[2] B. M. ANDERSEN, A. M. KACH, A. G. MELNIKOV, and D. R. SOLOMON, *Jump degrees of torsion-free abelian groups*, **The Journal of Symbolic Logic**, (to appear).

[3] B. A. ANDERSON and B. F. CSIMA, *Degrees that are not degrees of categoricity*, preprint.

[4] U. ANDREWS, *New spectra of strongly minimal theories in finite languages*, **Annals of Pure and Applied Logic**, vol. 162 (2011), pp. 367–372.

[5] ———, *A new spectrum of recursive models using an amalgamation construction*, **The Journal of Symbolic Logic**, vol. 76 (2011), pp. 883–896.

[6] ———, *The degrees of categorical theories with recursive models*, **Proceedings of the American Mathematical Society**, vol. 131 (2013), pp. 2501–2514.

[7] U. ANDREWS and J. F. KNIGHT, *Spectra of atomic theories*, **The Journal of Symbolic Logic**, vol. 78 (2013), pp. 1189–1198.

[8] U. ANDREWS, S. LEMPP, J. S. MILLER, K. M. NG, L. S. MAURO, and A. SORBI, *Universal computably enumerable equivalence relations*, preprint.

[9] U. ANDREWS and J. S. MILLER, *Spectra of theories and structures*, **Proceedings of the American Mathematical Society**, (to appear).

[10] C. ASH, J. KNIGHT, M. MANASSE, and T. SLAMAN, *Generic copies of countable structures*, **Annals of Pure and Applied Logic**, vol. 42 (1989), pp. 195–205.

[11] C. J. ASH, *Recursive labeling systems and stability of recursive structures in hyperarithmetical degrees*, **Transactions of the American Mathematical Society**, vol. 298 (1986), pp. 497–514.

[12] ———, *Categoricity in hyperarithmetical degrees*, **Annals of Pure and Applied Logic**, vol. 34 (1987), pp. 1–14.

[13] C. J. ASH, P. CHOLAK, and J. F. KNIGHT, *Permitting, forcing, and copying of a given recursive relation*, **Annals of Pure and Applied Logic**, vol. 86 (1997), pp. 219–236.

[14] C. J. ASH, C. G. JOCKUSCH, JR., and J. F. KNIGHT, *Jumps of orderings*, **Transactions of the American Mathematical Society**, vol. 319 (1990), pp. 573–599.

[15] C. J. ASH and J. F. KNIGHT, *Possible degrees in recursive copies*, **Annals of Pure and Applied Logic**, vol. 75 (1995), pp. 215–221.

[16] ———, *Possible degrees in recursive copies II*, **Annals of Pure and Applied Logic**, vol. 87 (1997), pp. 151–165.

[17] ———, *Computable structures and the hyperarithmetical hierarchy*, Elsevier, Amsterdam, 2000.

[18] C. J. ASH, J. F. KNIGHT, and J. B. REMMEL, *Quasi-simple relations in copies of a given recursive structure*, **Annals of Pure and Applied Logic**, vol. 86 (1997), pp. 203–218.

[19] C. J. ASH and A. NERODE, *Intrinsically recursive relations*, **Aspects of effective algebra** (J. N. Crossley, editor), U. D. A. Book Company, Steel's Creek, Australia, 1981, pp. 26–41.

[20] S. A. BADAEV, *Computable enumerations of families of general recursive functions*, **Algebra and Logic**, vol. 16 (1977), pp. 129–148, (Russian); (1978) pp. 83–98 (English translation).

[21] J. BALDWIN and A. LACHLAN, *On strongly minimal sets*, **The Journal of Symbolic Logic**, vol. 36 (1971), pp. 79–96.

[22] V. BALEVA, *The jump operation for structure degrees*, **Archive for Mathematical Logic**, vol. 45 (2006), pp. 249–265.

[23] E. BARKER, *Intrinsically Σ_α^0 relations*, **Annals of Pure and Applied Logic**, vol. 39 (1988), pp. 105–130.

[24] ———, *Back and forth relations for reduced abelian p-groups*, **Annals of Pure and Applied Logic**, vol. 75 (1995), pp. 223–249.

[25] J. BARWISE, *Infinitary logic and admissible sets*, **The Journal of Symbolic Logic**, vol. 34 (1969), pp. 226–252.

[26] J. BARWISE and J. SCHLIPF, *On recursively saturated models of arithmetic*, **Model theory and algebra**, Lecture Notes in Mathematics 498, Springer, Berlin, 1975, pp. 42–55.

[27] C. BERNARDI and A. SORBI, *Classifying positive equivalence relations*, **The Journal of Symbolic Logic**, vol. 48 (1983), pp. 529–538.

[28] W. W. BOONE, *The word problem*, **Proceedings of the National Academy of Sciences**, vol. 44 (1958), pp. 1061–1065.

[29] W. CALVERT, *The isomorphism problem for classes of computable fields*, **Archive for Mathematical Logic**, vol. 43 (2004), pp. 327–336.

[30] ———, *Algebraic structure and computable structure*, PhD dissertation, University of Notre Dame, 2005.

[31] ———, *The isomorphism problem for computable abelian p-groups of bounded length*, **The Journal of Symbolic Logic**, vol. 70 (2005), pp. 331–345.

[32] W. CALVERT, D. CENZER, V. HARIZANOV, and A. MOROZOV, *Effective categoricity of equivalence structures*, **Annals of Pure and Applied Logic**, vol. 141 (2006), pp. 61–78.

[33] ———, *Effective categoricity of Abelian p-groups*, **Annals of Pure and Applied Logic**, vol. 159 (2009), pp. 187–197.

[34] W. CALVERT, D. CUMMINS, J. F. KNIGHT, and S. MILLER, *Comparing classes of finite structures*, **Algebra and Logic**, vol. 43 (2004), pp. 374–392.

[35] W. CALVERT, E. FOKINA, S. GONCHAROV, J. KNIGHT, O. KUDINOV, A. MOROZOV, and V. PUZARENKO, *Index sets for classes of high rank structures*, **The Journal of Symbolic Logic**, vol. 72 (2007), pp. 1418–1432.

[36] W. CALVERT, S. GONCHAROV, J. MILLAR, and J. KNIGHT, *Categoricity of computable infinitary theories*, **Archive for Mathematical Logic**, vol. 48 (2009), pp. 25–38.

[37] W. CALVERT, S. S. GONCHAROV, and J. F. KNIGHT, *Computable structures of Scott rank ω_1^{CK} in familiar classes*, **Contemporary mathematics** (S. Gao, S. Jackson, and Y. Zhang, editors), Advances in Logic, vol. 425, American Mathematical Society, Providence, RI, 2007, pp. 49–66.

[38] W. CALVERT, V. HARIZANOV, and A. SHLAPENTOKH, *Turing degrees of the isomorphism types of geometric objects*, **Computability**, (to appear).

[39] ———, *Turing degrees of isomorphism types of algebraic objects*, **Journal of the London Mathematical Society**, vol. 75 (2007), pp. 273–286.

[40] W. CALVERT, V. S. HARIZANOV, J. F. KNIGHT, and S. MILLER, *Index sets of computable structures*, **Algebra and Logic**, vol. 45 (2006), pp. 306–325.

[41] W. CALVERT and J. F. KNIGHT, *Classification from a computable point of view*, **The Bulletin of Symbolic Logic**, vol. 12 (2006), pp. 191–218.

[42] W. CALVERT, J. F. KNIGHT, and J. MILLAR, *Computable trees of Scott rank ω_1^{CK}, and computable approximability*, **The Journal of Symbolic Logic**, vol. 71 (2006), pp. 283–298.

[43] J. CARSON, E. FOKINA, V. HARIZANOV, J. KNIGHT, S. QUINN, C. SAFRANSKI, and J. WALLBAUM, *The computable embedding problem*, **Algebra and Logic**, vol. 50 (2011), pp. 707–732.

[44] J. CARSON, V. HARIZANOV, J. KNIGHT, K. LANGE, C. MCCOY, A. MOROZOV, S. QUINN, C. SAFRANSKI, and J. WALLBAUM, *Describing free groups*, **Transactions of the American Mathematical Society**, vol. 364 (2012), pp. 5715–5728.

[45] D. CENZER, V. HARIZANOV, and J. REMMEL, Σ_1^0 *and* Π_1^0 *equivalence structures*, **Annals of Pure and Applied Logic**, vol. 162 (2011), pp. 490–503.

[46] ———, *Effective categoricity of injection structures*, **Algebra and Logic**, (to appear).

[47] D. CENZER, G. LAFORTE, and J. REMMEL, *Equivalence structures and isomorphisms in the difference hierarchy*, **The Journal of Symbolic Logic**, vol. 74 (2009), pp. 535–556.

[48] D. CENZER and J. B. REMMEL, *Complexity-theoretic model theory and algebra*, **Handbook of recursive mathematics, volume 1** (Yu. L. Ershov, S. S. Goncharov, A. Nerode, and J. B. Remmel, editors), Studies in Logic and the Foundations of Mathematics 139, North-Holland, Amsterdam, 1998, pp. 381–513.

[49] C. C. CHANG and H. J. KEISLER, *Model theory*, North-Holland, Amsterdam, 1973.

[50] J. CHISHOLM, *Effective model theory vs. recursive model theory*, **The Journal of Symbolic Logic**, vol. 55 (1990), pp. 1168–1191.

[51] J. CHISHOLM, E. FOKINA, S. GONCHAROV, V. HARIZANOV, J. KNIGHT, and S. QUINN, *Intrinsic bounds on complexity and definability at limit levels*, **The Journal of Symbolic Logic**, vol. 74 (2009), pp. 1047–1060.

[52] J. CHISHOLM and M. MOSES, *An undecidable linear order that is n-decidable for all n*, **Notre Dame Journal of Formal Logic**, vol. 39 (1998), pp. 519–526.

[53] J. A. CHISHOLM, J. CHUBB, V. S. HARIZANOV, D. R. HIRSCHFELDT, C. G. JOCKUSCH, JR., T. H. MCNICHOLL, and S. PINGREY, Π_1^0 *classes and strong degree spectra of relations*, **The Journal of Symbolic Logic**, vol. 72 (2007), pp. 1003–1018.

[54] P. CHOLAK, S. GONCHAROV, B. KHOUSSAINOV, and R. A. SHORE, *Computably categorical structures and expansions by constants*, **The Journal of Symbolic Logic**, vol. 64 (1999), pp. 13–37.

[55] P. CHOLAK, R. A. SHORE, and R. SOLOMON, *A computably stable structure with no Scott family of finitary formulas*, **Archive for Mathematical Logic**, vol. 45 (2006), pp. 519–538.

[56] J. CHUBB, A. FROLOV, and V. HARIZANOV, *Degree spectra of the successor relation on computable linear orderings*, **Archive for Mathematical Logic**, vol. 48 (2009), pp. 7–13.

[57] J. CHUBB, V. HARIZANOV, A. MOROZOV, S. PINGREY, and E. UFFERMAN, *Partial automorphism semigroups*, **Annals of Pure and Applied Logic**, vol. 156 (2008), pp. 245–258.

[58] R. J. COLES, R. G. DOWNEY, and T. A. SLAMAN, *Every set has a least jump enumeration*, **Journal of the London Mathematical Society**, vol. 62 (2000), pp. 641–649.

[59] S. COSKEY, J. HAMKINS, and R. MILLER, *The hierarchy of equivalence relations on the natural numbers under computable reducibility*, **Computability**, vol. 1 (2012), pp. 15–38.

[60] T. C. CRAVEN, *The Boolean space of orderings of a field*, **Transactions of the American Mathematical Society**, vol. 209 (1975), pp. 225–235.

[61] B. CSIMA, V. HARIZANOV, D. HIRSCHFELDT, and R. SOARE, *Bounding homogeneous models*, **The Journal of Symbolic Logic**, vol. 72 (2007), pp. 305–323.

[62] B. CSIMA, V. HARIZANOV, R. MILLER, and A. MONTALBÁ, *Computability of Fraïssé limits*, **The Journal of Symbolic Logic**, vol. 76 (2011), pp. 66–93.

[63] B. CSIMA and I. SH. KALIMULLIN, *Degree spectra and immunity properties*, **Mathematical Logic Quarterly**, vol. 56 (2010), pp. 67–77.

[64] B. F. CSIMA, *Degree spectra of prime models*, **The Journal of Symbolic Logic**, vol. 69 (2004), pp. 430–412.

[65] B. F. CSIMA, J. N. Y. FRANKLIN, and R. A. SHORE, *Degrees of categoricity and the hyperarithmetic hierarchy*, **Notre Dame Journal of Formal Logic**, vol. 54 (2013), pp. 215–231.

[66] B. F. CSIMA, D. R. HIRSCHFELDT, J. F. KNIGHT, and R. I. SOARE, *Bounding prime models*, **The Journal of Symbolic Logic**, vol. 69 (2004), pp. 1117–1142.

[67] M. DABKOWSKA, M. DABKOWSKI, V. HARIZANOV, J. PRZYTYCKI, and M. VEVE, *Compactness of the space of left orders*, **Journal of Knot Theory and Its Ramifications**, vol. 16 (2007), pp. 257–366.

[68] M. DABKOWSKA, M. DABKOWSKI, V. HARIZANOV, and A. SIKORA, *Turing degrees of nonabelian groups*, **Proceedings of the American Mathematical Society**, vol. 135 (2007), pp. 3383–3391.

[69] M. A. DABKOWSKA, *Turing degree spectra of groups and their spaces of orders*, PhD dissertation, George Washington University, 2006.

[70] M. A. DABKOWSKA, M. K. DABKOWSKI, V. S. HARIZANOV, and A. A. TOGHA, *Spaces of orders and their Turing degree spectra*, **Annals of Pure and Applied Logic**, vol. 161 (2010), pp. 1134–1143.

[71] R. DIMITROV, V. HARIZANOV, and A. S. MOROZOV, *Dependence relations in computably rigid computable vector spaces*, **Annals of Pure and Applied Logic**, vol. 132 (2005), pp. 97–108.

[72] V. P. DOBRITSA, *Some constructivizations of abelian groups*, **Siberian Mathematical Journal**, vol. 24 (1983), pp. 167–173, (English translation).

[73] A. DOLICH, C. LASKOWSKI, and A. RAICHEV, *Model completeness for trivial, uncountably categorical theories of Morley rank one*, **Archive for Mathematical Logic**, vol. 45 (2006), pp. 931–945.

[74] R. DOWNEY and D. HIRSCHFELDT, *Algorithmic randomness and complexity*, Springer, 2010.

[75] R. DOWNEY and C. G. JOCKUSCH, JR., *Every low Boolean algebra is isomorphic to a recursive one*, **Proceedings of the American Mathematical Society**, vol. 122 (1994), pp. 871–880.

[76] R. DOWNEY, A. KACH, S. LEMPP, A. LEWIS, A. MONTALBÁN, and D. TURETSKY, *The complexity of computable categoricity*, preprint.

[77] R. DOWNEY and J. KNIGHT, *Orderings with αth jump degree $\mathbf{0}^{(\alpha)}$*, **Proceedings of the American Mathematical Society**, vol. 114 (1992), pp. 545–552.

[78] R. DOWNEY, S. LEMPP, and G. WU, *On the complexity of the successivity relation in computable linear orderings*, **Journal of Mathematical Logic**, vol. 10 (2010), pp. 83–99.

[79] R. DOWNEY and A. G. MELNIKOV, *Computable completely decomposable groups*, preprint.

[80] ———, *Effectively categorical abelian groups*, **Journal of Algebra**, (to appear).

[81] R. G. DOWNEY, *On presentations of algebraic structures*, **Complexity, logic and recursion theory** (A. Sorbi, editor), Lecture Notes in Pure and Applied Mathematics 187, Marcel Dekker, New York, 1997, pp. 157–205.

[82] R. G. DOWNEY, S. S. GONCHAROV, and D. R. HIRSCHFELDT, *Degree spectra of relations on Boolean algebras*, **Algebra and Logic**, vol. 42 (2003), pp. 105–111.

[83] R. G. DOWNEY, A. M. KACH, S. LEMPP, and D. D. TURETSKY, *Computable categoricity versus relative computable categoricity*, **Fundamenta Mathematicae**, vol. 221 (2013), pp. 129–159.

[84] R. G. DOWNEY and S. A. KURTZ, *Recursion theory and ordered groups*, **Annals of Pure and Applied Logic**, vol. 32 (1986), pp. 137–151.

[85] R. G. DOWNEY and M. F. MOSES, *Recursive linear orders with incomplete successivities*, **Transactions of the American Mathematical Society**, vol. 326 (1991), pp. 653–668.

[86] R. G. DOWNEY and J. B. REMMEL, *Computable algebras and closure systems: coding properties*, **Handbook of recursive mathematics, volume 2** (Yu. L. Ershov, S. S. Goncharov, A. Nerode, and J. B. Remmel, editors), Studies in Logic and the Foundations of Mathematics 139, North-Holland, Amsterdam, 1998, pp. 997–1039.

[87] B. N. DROBOTUN, *Enumerations of simple models*, **Siberian Mathematical Journal**, vol. 18 (1977), pp. 707–716, (English translation).

[88] V. D. DZGOEV, *Recursive automorphisms of constructive models*, **Proceedings of the 15th All-Union Algebraic Conference (Novosibirsk, 1979), Part 2**, (Russian), p. 52.

[89] R. EPSTEIN, *Computably enumerable degrees of prime models*, **The Journal of Symbolic Logic**, vol. 73 (2008), pp. 1373–1388.

[90] YU. L. ERSHOV, *Theorie der Numierungen III*, **Zeitschrift für Mathematische Logik und Grundlagen der Mathematik**, vol. 23 (1977), pp. 289–371.

[91] ———, *Theory of numberings*, Nauka, Moscow, 1977, (Russian).

[92] ———, *Decidability problems and constructive models*, Nauka, Moscow, 1980, (Russian).

[93] YU. L. ERSHOV and S. S. GONCHAROV, *Constructive models*, Siberian School of Algebra and Logic, Kluwer Academic/Plenum Publishers, 2000, (English translation).

[94] L. FEINER, *Hierarchies of Boolean algebras*, **The Journal of Symbolic Logic**, vol. 35 (1970), pp. 365–374.

[95] E. FOKINA, *On complexity of categorical theories with computable models*, **Vestnik NGU**, vol. 5 (2005), pp. 78–86, (Russian).

[96] E. FOKINA, S. FRIEDMAN, V. HARIZANOV, J. KNIGHT, C. MCCOY, and A. MONTALBÁN, *Isomorphism relations on computable structures*, **The Journal of Symbolic Logic**, vol. 77 (2012), pp. 122–132.

[97] E. FOKINA, S. D. FRIEDMAN, and A. NIES, *Equivalence relations that are Σ^0_3 complete for computable reducibility*, **Logic, Language, Information and Computation, 19th International Workshop, WoLLIC 2012** (C.-H. L. Ong and R. J. G. B. de Queiroz, editors), Springer, Berlin, 2012, pp. 26–33.

[98] E. FOKINA, A. FROLOV, and I. KALIMULLIN, *Spectra of categoricity for rigid structures*, **Notre Dame Journal of Formal Logic**, (to appear).

[99] E. B. FOKINA, *Index sets of decidable models*, **Siberian Mathematical Journal**, vol. 48 (2007), pp. 939–948, (English translation).

[100] ———, *Index sets for some classes of structures*, **Annals of Pure and Applied Logic**, vol. 157 (2009), pp. 139–147.

[101] E. B. FOKINA and S. D. FRIEDMAN, *Equivalence relations on classes of computable structures*, **Proceedings of Computability in Europe 2009**, Lecture Notes in Computer Science 5635, Springer, Heidelberg, 2009, pp. 198–207.

[102] ———, *On Σ^1_1 equivalence relations over the natural numbers*, **Mathematical Logic Quarterly**, vol. 58 (2012), pp. 113–124.

[103] E. B. FOKINA, I. KALIMULLIN, and R. MILLER, *Degrees of categoricity of computable structures*, **Archive for Mathematical Logic**, vol. 49 (2010), pp. 51–67.

[104] C. FREER, *Models with high Scott rank*, PhD dissertation, Harvard University, 2008.

[105] H. FRIEDMAN and L. STANLEY, *A Borel reducibility theory for classes of countable structures*, **The Journal of Symbolic Logic**, vol. 54 (1989), pp. 894–914.

[106] S. D. FRIEDMAN and L. MOTTO ROS, *Analytic equivalence relations and bi-embeddability*, **The Journal of Symbolic Logic**, vol. 76 (2011), pp. 243–266.

[107] A. FRÖHLICH and J. SHEPHERDSON, *Effective procedures in field theory*, **Philosophical Transactions of the Royal Society. Series A**, vol. 248 (1956), pp. 407–432.

[108] A. FROLOV, V. HARIZANOV, I. KALIMULLIN, O. KUDINOV, and R. MILLER, *Degree spectra of high$_n$ and nonlow$_n$ degrees*, **Journal of Logic and Computation**, vol. 22 (2012), pp. 755–777.

[109] A. FROLOV, I. KALIMULLIN, and R. MILLER, *Spectra of algebraic fields and subfields*, **Mathematical theory and computational practice, Computability in Europe** (K. Ambos-Spies, B. Löwe, and W. Merkle, editors), Lecture Notes in Computer Science 5635, Springer, Berlin, 2009, pp. 232–241.

[110] L. FUCHS, *Partially ordered algebraic systems*, Pergamon Press, Oxford, 1963.

[111] S. GAO, *Invariant descriptive set theory*, Pure and Applied Mathematics, CRC Press/Chapman & Hall, 2009.

[112] S. GAO and P. GERDES, *Computably enumerable equivalence relations*, **Studia Logica**, vol. 67 (2001), pp. 27–59.

[113] A. GAVRYUSHKIN, *Computable models of Ehrenfeucht theories*, **Centre de Recerca Matemática (Barcelona)**, vol. 11 (2012), pp. 67–77.

[114] A. N. GAVRYUSHKIN, *Spectra of computable models for Ehrenfeucht theories*, **Algebra and Logic**, vol. 46 (2007), pp. 149–157, (English translation).

[115] A. M. W. GLASS, **Ordered permutation groups**, London Mathematical Society Lecture Note Series, vol. 55, Cambridge University Press, 1981.

[116] S. GONCHAROV, V. HARIZANOV, J. KNIGHT, C. MCCOY, R. MILLER, and R. SOLOMON, *Enumerations in computable structure theory*, **Annals of Pure and Applied Logic**, vol. 136 (2005), pp. 219–246.

[117] S. GONCHAROV, V. HARIZANOV, J. KNIGHT, A. MOROZOV, and A. ROMINA, *On automorphic tuples of elements in computable models*, **Siberian Mathematical Journal**, vol. 46 (2005), pp. 405–412, (English translation).

[118] S. GONCHAROV, V. HARIZANOV, C. LASKOWSKI, S. LEMPP, and C. MCCOY, *Trivial, strongly minimal theories are model complete after naming constants*, **Proceedings of the American Mathematical Society**, vol. 131 (2003), pp. 3901–3912.

[119] S. GONCHAROV, S. LEMPP, and R. SOLOMON, *The computable dimension of ordered abelian groups*, **Advances in Mathematics**, vol. 175 (2003), pp. 102–143.

[120] S. S. GONCHAROV, *Selfstability and computable families of constructivizations*, **Algebra and Logic**, vol. 14 (1975), pp. 647–680, (Russian).

[121] ———, *Restricted theories of constructive Boolean algebras*, **Siberian Mathematical Journal**, vol. 17 (1976), pp. 601–611, (English translation).

[122] ———, *The quantity of nonautoequivalent constructivizations*, **Algebra and Logic**, vol. 16 (1977), pp. 169–185, (English translation).

[123] ———, *Constructive models of and $\aleph_1$-categorical theories*, **Matematicheskie Zametki**, vol. 23 (1978), pp. 885–888, (Russian). (English translation in **Mathematical Notes**, vol. 23 (1978), no. 5–6, pp. 486–488).

[124] ———, *Strong constructivizability of homogeneous models*, **Algebra and Logic**, vol. 17 (1978), pp. 247–263, (English translation).

[125] ———, *Autostability of models and abelian groups*, **Algebra and Logic**, vol. 19 (1980), pp. 13–27, (English translation).

[126] ———, *Computable single valued numerations*, **Algebra and Logic**, vol. 19 (1980), pp. 325–356, (English translation).

[127] ———, *Problem of number of nonautoequivalent constructivizations*, **Algebra and Logic**, vol. 19 (1980), pp. 401–414, (English translation).

[128] ———, *A totally transcendental decidable theory without constructivizable homogeneous models*, **Algebra and Logic**, vol. 19 (1980), pp. 85–93, (English translation).

[129] ———, *Autostable models and algorithmic dimensions*, **Handbook of recursive mathematics, volume 1** (Yu. L. Ershov, S. S. Goncharov, A. Nerode, and J. B. Remmel, editors), North-Holland, Amsterdam, 1998, pp. 261–287.

[130] ———, *Autostability of prime models with respect to strong constructivizations*, **Algebra and Logic**, vol. 48 (2009), pp. 410–417, (English translation).

[131] ———, *On the autostability of almost prime models with respect to strong constructivizations*, **Russian Mathematical Surveys**, vol. 65 (2010), pp. 901–935, (English translation).

[132] ———, *Degrees of autostability relative to strong constructivizations*, **Proceedings of the Steklov Institute of Mathematics**, vol. 274 (2011), pp. 105–115, (English translation).

[133] S. S. GONCHAROV and B. N. DROBOTUN, *Numerations of saturated and homogeneous models*, **Siberian Mathematical Journal**, vol. 21 (1980), pp. 164–176, (English translation).

[134] S. S. GONCHAROV and V. D. DZGOEV, *Autostability of models*, **Algebra and Logic**, vol. 19 (1980), pp. 28–37, (English translation).

[135] S. S. GONCHAROV, V. S. HARIZANOV, J. F. KNIGHT, and C. F. D. MCCOY, *Simple and immune relations on countable structures*, **Archive for Mathematical Logic**, vol. 42 (2003), pp. 279–291.

[136] ——, *Relatively hyperimmune relations on structures*, **Algebra and Logic**, vol. 43 (2004), pp. 94–101, (English translation).

[137] S. S. GONCHAROV, V. S. HARIZANOV, J. F. KNIGHT, and R. A. SHORE, Π_1^1 *relations and paths through* $\mathcal{O}$, **The Journal of Symbolic Logic**, vol. 69 (2004), pp. 585–611.

[138] S. S. GONCHAROV and B. KHOUSSAINOV, *On the spectrum of degrees of decidable relations*, **Doklady Mathematics**, vol. 55 (1997), pp. 55–57, (English translation).

[139] ——, *Complexity of theories of computable categorical models*, **Algebra and Logic**, vol. 43 (2004), pp. 365–373, (English translation).

[140] S. S. GONCHAROV and J. F. KNIGHT, *Computable structure and non-structure theorems*, **Algebra and Logic**, vol. 41 (2002), pp. 351–373.

[141] S. S. GONCHAROV, A. V. MOLOKOV, and N. S. ROMANOVSKII, *Nilpotent groups of finite algorithmic dimension*, **Siberian Mathematical Journal**, vol. 30 (1989), pp. 63–68.

[142] S. S. GONCHAROV and A. T. NURTAZIN, *Constructive models of complete decidable theories*, **Algebra and Logic**, vol. 12 (1973), pp. 125–142 (Russian); (1974) pp. 67–77, (English translation).

[143] N. GREENBERG, A. MONTALBÁN, and T. A. SLAMAN, *Relative to any non-hyperarithmetic set*, **Journal of Mathematical Logic**, vol. 13 (2013), pp. 1–26.

[144] ——, *The Slaman–Wehner theorem in higher recursion theory*, **Proceedings of the American Mathematical Society**, vol. 139 (2011), pp. 1865–1869.

[145] V. HARIZANOV, *Effectively nowhere simple relations on computable structures*, **Recursion theory and complexity** (M. M. Arslanov and S. Lempp, editors), Walter de Gruyter, Berlin, 1999, pp. 59–70.

[146] ——, *Turing degrees of hypersimple relations on computable structures*, **Annals of Pure and Applied Logic**, vol. 121 (2003), pp. 209–226.

[147] V. HARIZANOV, C. JOCKUSCH, JR., and J. KNIGHT, *Chains and antichains in computable partial orderings*, **Archive for Mathematical Logic**, vol. 48 (2009), pp. 39–53.

[148] V. HARIZANOV, J. KNIGHT, C. MCCOY, V. PUZARENKO, R. SOLOMON, and J. WALLBAUM, *Orders on* F_∞, preprint.

[149] V. HARIZANOV and R. MILLER, *Spectra of structures and relations*, **The Journal of Symbolic Logic**, vol. 72 (2007), pp. 324–348.

[150] V. HARIZANOV, R. MILLER, and A. S. MOROZOV, *Simple structures with complex symmetry*, **Algebra and Logic**, vol. 49 (2010), pp. 98–134, (English translation).

[151] V. S. HARIZANOV, *Degree spectrum of a recursive relation on a recursive structure*, PhD dissertation, University of Wisconsin, Madison, 1987.

[152] ——, *Some effects of Ash–Nerode and other decidability conditions on degree spectra*, **Annals of Pure and Applied Logic**, vol. 55 (1991), pp. 51–65.

[153] ——, *Uncountable degree spectra*, **Annals of Pure and Applied Logic**, vol. 54 (1991), pp. 255–263.

[154] ——, *The possible Turing degree of the nonzero member in a two-element degree spectrum*, **Annals of Pure and Applied Logic**, vol. 60 (1993), pp. 1–30.

[155] ——, *Turing degrees of certain isomorphic images of computable relations*, **Annals of Pure and Applied Logic**, vol. 93 (1998), pp. 103–113.

[156] V. S. HARIZANOV, J. F. KNIGHT, and A. S. MOROZOV, *Sequences of n-diagrams*, **The Journal of Symbolic Logic**, vol. 67 (2002), pp. 1227–1247.

[157] L. HARRINGTON, *Recursively presentable prime models*, **The Journal of Symbolic Logic**, vol. 39 (1974), pp. 305–309.

[158] K. HARRIS, *Categoricity in Boolean algebras*, preprint.

[159] ——, *On bounding saturated models*, preprint.

[160] K. HARRIS and A. MONTALBÁN, *Boolean algebra approximations*, preprint.

[161] ——, *On the n-back-and-forth types of Boolean algebras*, **Transactions of the American Mathematical Society**, vol. 364 (2012), pp. 827–866.

[162] J. HARRISON, *Recursive pseudo-well-orderings*, **Transactions of the American Mathematical Society**, vol. 131 (1968), pp. 526–543.

[163] E. HERRMANN, *Infinite chains and antichains in computable partial orderings*, **The Journal of Symbolic Logic**, vol. 66 (2001), pp. 923–934.

[164] B. HERWIG, S. LEMPP, and M. ZIEGLER, *Constructive models of uncountably categorical theories*, **Proceedings of the American Mathematical Society**, vol. 127 (1999), pp. 3711–3719.

[165] G. HIGMAN, *Subgroups of finitely presented groups*, **Proceedings of the Royal Society of London**, vol. 262 (1961), pp. 455–475.

[166] G. HIRD, *Recursive properties of intervals of recursive linear orders*, **Logical methods** (J. N. Crossley, J. B. Remmel, R. A. Shore, and M. E. Sweedler, editors), Birkhäuser, 1993, pp. 422–437.

[167] G. R. HIRD, *Recursive properties of relations on models*, **Annals of Pure and Applied Logic**, vol. 63 (1993), pp. 241–269.

[168] D. HIRSCHFELDT, K. KRAMER, R. MILLER, and A. SHLAPENTOKH, *Categoricity properties for computable algebraic fields*, **Transactions of the American Mathematical Society**, (to appear).

[169] D. R. HIRSCHFELDT, *Prime models of theories of computable linear orderings*, **Proceedings of the American Mathematical Society**, vol. 129 (2001), pp. 3079–3083.

[170] ———, *Degree spectra of relations on computable structures in the presence of Δ_2^0 isomorphisms*, **The Journal of Symbolic Logic**, vol. 67 (2002), pp. 697–720.

[171] ———, *Computable trees, prime models and relative decidability*, **Proceedings of the American Mathematical Society**, vol. 134 (2006), pp. 1495–1498.

[172] D. R. HIRSCHFELDT, B. KHOUSSAINOV, and P. SEMUKHIN, *An uncountably categorical theory whose only computably presentable model is saturated*, **Notre Dame Journal of Formal Logic**, vol. 47 (2006), pp. 63–71.

[173] D. R. HIRSCHFELDT, B. KHOUSSAINOV, and R. A. SHORE, *A computably categorical structure whose expansion by a constant has infinite computable dimension*, **The Journal of Symbolic Logic**, vol. 68 (2003), pp. 1199–1241.

[174] D. R. HIRSCHFELDT, B. KHOUSSAINOV, R. A. SHORE, and A. M. SLINKO, *Degree spectra and computable dimensions in algebraic structures*, **Annals of Pure and Applied Logic**, vol. 115 (2002), pp. 71–113.

[175] D. R. HIRSCHFELDT and W. M. WHITE, *Realizing levels of the hyperarithmetic hierarchy as degree spectra of relations on computable structures*, **Notre Dame Journal of Formal Logic**, vol. 43 (2002), pp. 51–64.

[176] G. HJORTH, *The isomorphism relation on countable torsion-free Abelian groups*, **Fundamenta Mathematicae**, vol. 175 (2002), pp. 241–257.

[177] E. HRUSHOVSKI, *A new strongly minimal set*, **Annals of Pure and Applied Logic**, vol. 62 (1993), pp. 147–166, **Stability in model theory, III (Trento, 1991)**.

[178] C. G. JOCKUSCH, JR., *Ramsey's theorem and recursion theory*, **The Journal of Symbolic Logic**, vol. 37 (1972), pp. 268–280.

[179] C. G. JOCKUSCH, JR. and R. I. SOARE, Π_1^0 *classes and degrees of theories*, **Transactions of the American Mathematical Society**, vol. 173 (1972), pp. 33–56.

[180] ———, *Degrees of orderings not isomorphic to recursive linear orderings*, **Annals of Pure and Applied Logic**, vol. 52 (1991), pp. 39–64.

[181] ———, *Boolean algebras, Stone spaces, and the iterated Turing jump*, **The Journal of Symbolic Logic**, vol. 59 (1994), pp. 1121–1138.

[182] A. M. KACH, K. LANGE, and R. SOLOMON, *Degrees of orders on torsion-free abelian groups*, **Annals of Pure and Applied Logic**, (to appear).

[183] A. M. KACH and D. TURETSKY, Δ_2^0-*categoricity of equivalence structures*, **New Zealand Journal of Mathemtics**, vol. 39 (2009), pp. 143–149.

[184] I. KALANTARI and A. RETZLAFF, *Maximal vector spaces under automorphisms of the lattice of recursively enumerable vector spaces*, **The Journal of Symbolic Logic**, vol. 42 (1977), pp. 481–491.

[185] I. KALIMULLIN, B. KHOUSSAINOV, and A. MELNIKOV, *Limitwise monotonic sequences and degree spectra of structures*, **Proceedings of the American Mathematical Society**, (to appear).

[186] I. SH. KALIMULLIN, *Some notes on degree spectra of structures*, **Computation and logic in the real world, Computability in Europe** (S. B. Cooper, B. Löwe, and A. Sorbi, editors), Lecture Notes in Computer Science 4497, Springer, Berlin, 2007, pp. 389–397.

[187] ———, *Spectra of degrees of some algebraic structures*, **Algebra and Logic**, vol. 46 (2007), pp. 399–408, (English translation).

[188] ———, *Almost computably enumerable families of sets*, **Sbornik. Mathematics**, vol. 199 (2008), pp. 1451–1458, (English translation).

[189] ———, *Restrictions on the spectra of degrees of algebraic structures*, **Siberian Mathematical Journal**, vol. 49 (2008), pp. 1034–1043, (English translation).

[190] V. KANOVEI, **Borel equivalence relations. structure and classification**, University Lecture Series 44, American Mathematical Society, Providence, RI, 2008.

[191] A. KECHRIS and A. LOUVEAU, *The classification of hypersmooth Borel equivalence relations*, **Journal of the American Mathematical Society**, vol. 10 (1997), pp. 215–242.

[192] C. F. KENT, *Constructive analogues of the group of permutations of the natural numbers*, **Transactions of the American Mathematical Society**, vol. 104 (1962), pp. 347–362.

[193] O. KHARLAMPOVICH and A. MYASNIKOV, *Elementary theory of free non-abelian groups*, **Journal of Algebra**, vol. 302 (2006), pp. 451–552.

[194] A. N. KHISAMIEV, *On the upper semilattice L_E*, **Siberian Mathematical Journal**, vol. 45 (2004), pp. 173–187, (English translation).

[195] N. G. KHISAMIEV, *Strongly constructive models of a decidable theory*, **Izvestiya Akademii Nauk Kazakhskoj SSR, Seriya Fiziko-Matematicheskaya**, vol. 1 (1974), pp. 83–84, (Russian).

[196] ———, *A constructibility criterion for the direct product of cyclic p-groups*, **Izvestiya Akademii Nauk Kazakhskoj SSR, Seriya Fiziko-Matematicheskaya**, vol. 51 (1981), pp. 51–55, (Russian).

[197] ———, *Theory of abelian groups with constructive models*, **Siberian Mathematical Journal**, vol. 27 (1986), pp. 572–585, (English translation).

[198] N. G. KHISAMIEV and A. A. KRYKPAEVA, *Effectively totally decomposable abelian groups*, **Siberian Mathematical Journal**, vol. 38 (1997), pp. 1227–1229, (English translation).

[199] B. KHOUSSAINOV, C. LASKOWSKI, S. LEMPP, and R. SOLOMON, *On the computability-theoretic complexity of trivial, strongly minimal models*, **Proceedings of the American Mathematical Society**, vol. 135 (2007), pp. 3711–3721.

[200] B. KHOUSSAINOV and M. MINNES, *Three lectures on automatic structures*, **Logic Colloquium '07** (F. Delon, U. Kohlenbach, P. Maddy, and F. Stephan, editors), Lecture Notes in Logic 35, Cambridge University Press, 2010, pp. 132–176.

[201] B. KHOUSSAINOV and A. MONTALBÁN, *A computable $\aleph_0$-categorical structure whose theory computes true arithmetic*, **The Journal of Symbolic Logic**, vol. 75 (2010), pp. 728–740.

[202] B. KHOUSSAINOV and A. NERODE, *Automatic presentations of structures*, **Logic and Computational Complexity: International Workshop, LCC '94, Indianapolis** (D. Leivant, editor), Lecture Notes in Computer Science 960, Springer, Berlin, 1995, pp. 367–395.

[203] B. KHOUSSAINOV, A. NIES, and R. SHORE, *On recursive models of theories*, **Notre Dame Journal of Formal Logic**, vol. 38 (1997), pp. 165–178.

[204] B. KHOUSSAINOV, P. SEMUKHIN, and F. STEPHAN, *Applications of Kolmogorov complexity to computable model theory*, **The Journal of Symbolic Logic**, vol. 72 (2007), pp. 1041–1054.

[205] B. KHOUSSAINOV and R. A. SHORE, *Computable isomorphisms, degree spectra of relations and Scott families*, **Annals of Pure and Applied Logic**, vol. 93 (1998), pp. 153–193.

[206] B. KHOUSSAINOV, T. SLAMAN, and P. SEMUKHIN, *Π^0_1-presentations of algebras*, **Archive for Mathematical Logic**, vol. 45 (2006), pp. 769–781.

[207] B. KHOUSSAINOV, F. STEPHAN, and Y. YANG, *Computable categoricity and the Ershov hierarchy*, **Annals of Pure and Applied Logic**, vol. 156 (2008), pp. 86–95.

[208] J. KNIGHT, *Nonarithmetical $\aleph_0$-categorical theories with recursive models*, **The Journal of Symbolic Logic**, vol. 59 (1994), pp. 106–112.

[209] J. F. KNIGHT, *Degrees coded in jumps of orderings*, **The Journal of Symbolic Logic**, vol. 51 (1986), pp. 1034–1042.

[210] J. F. KNIGHT and J. MILLAR, *Computable structures of Scott rank ω_1^{CK}*, **Journal of Mathematical Logic**, vol. 10 (2010), pp. 31–43.

[211] J. F. KNIGHT, S. MILLER, and M. VANDEN BOOM, *Turing computable embeddings*, **The Journal of Symbolic Logic**, vol. 73 (2007), pp. 901–918.

[212] J. F. KNIGHT and M. STOB, *Computable Boolean algebras*, **The Journal of Symbolic Logic**, vol. 65 (2000), pp. 1605–1623.

[213] G. KREISEL, *Note on arithmetic models for consistent formulae of the predicate calculus*, **Fundamenta Mathematicae**, vol. 37 (1950), pp. 265–285.

[214] K. ZH. KUDAIBERGENOV, *Constructivizable models of undecidable theories*, **Siberian Mathematical Journal**, vol. 21 (1980), pp. 155–158, (Russian).

[215] ———, *Effectively homogenous models*, **Siberian Mathematical Journal**, vol. 27 (1986), pp. 180–182, (Russian).

[216] O. KUDINOV, *An autostable 1-decidable model without a computable Scott family of ∃-formulas*, **Algebra and Logic**, vol. 35 (1996), pp. 458–467.

[217] V. A. KUZICHEVA, *Inverse isomorphisms of rings of recursive endomorphisms*, **Moscow University Mathematics Bulletin**, vol. 41 (1986), pp. 82–84, (English translation).

[218] A. V. KUZNETSOV, *On primitive recursive functions of large oscillation*, **Doklady Akademii Nauk SSSR**, vol. 71 (1950), pp. 233–236, (Russian).

[219] K. LANGE, *The degree spectra of homogeneous models*, **The Journal of Symbolic Logic**, vol. 73 (2008), pp. 1009–1028.

[220] ———, *A characterization of the **0**-basis homogeneous bounding degrees*, **The Journal of Symbolic Logic**, vol. 75 (2010), pp. 971–995.

[221] P. LAROCHE, *Recursively presented Boolean algebras*, **Notices of the American Mathematical Society**, vol. 24 (1977), pp. A552–A553.

[222] C. LASKOWSKI, *Characterizing model completeness among mutually algebraic structures*, **The Journal of Symbolic Logic**, vol. 78 (2013), pp. 185–194.

[223] S. LEMPP, C. MCCOY, R. MILLER, and R. SOLOMON, *Computable categoricity of trees of finite height*, **The Journal of Symbolic Logic**, vol. 70 (2005), pp. 151–215.

[224] S. LEMPP, C. F. D. MCCOY, A. S. MOROZOV, and R. SOLOMON, *Group theoretic properties of the group of computable automorphisms of a countable dense linear order*, **Order**, vol. 19 (2002), pp. 343–364.

[225] M. LERMAN and J. SCHMERL, *Theories with recursive models*, **The Journal of Symbolic Logic**, vol. 44 (1979), pp. 59–76.

[226] A. J. MACINTYRE and D. MARKER, *Degrees of recursively saturated models*, **Transactions of the American Mathematical Society**, vol. 282 (1984), pp. 539–554.

[227] M. MAKKAI, *An example concerning Scott heights*, **The Journal of Symbolic Logic**, vol. 46 (1981), pp. 301–318.

[228] A. I. MAL'CEV, *Constructive algebras I*, **Russian Mathematical Surveys**, vol. 16 (1961), pp. 77–129, (English translation).

[229] ———, *On recursive Abelian groups*, **Doklady Akademii Nauk SSSR**, vol. 3 (1962), pp. 1431–1434, (English translation).

[230] M. MANASSE, *Techniques and counterexamples in almost categorical recursive model theory*, PhD dissertation, University of Wisconsin, Madison, 1982.

[231] A. B. MANASTER and J. B. REMMEL, *Some recursion theoretic aspects of dense two-dimensional partial orderings*, **Aspects of effective algebra** (J. N. Crossley, editor), U. D. A. Book Co., Steel's Creek, Australia, 1981, pp. 161–188.

[232] Yu. V. MATIYASEVICH, *The diophantineness of enumerable sets*, **Doklady Akademii Nauk SSSR**, vol. 191 (1970), pp. 279–282, (Russian).

[233] ———, *Hilbert's tenth problem*, The MIT Press, Cambridge, Massachusetts, 1993, (English translation).

[234] C. MCCOY and J. WALLBAUM, *Describing free groups, part II*: Π_4^0-*hardness and no* Σ_2^0 *basis*, **Transactions of the American Mathematical Society**, vol. 364 (2012), pp. 5729–5734.

[235] C. F. D. MCCOY, *On* Δ_3^0-*categoricity for linear orders and Boolean algebras*, **Algebra and Logic**, vol. 41 (2002), pp. 295–305, (English translation).

[236] ———, Δ_2^0-*categoricity in Boolean algebras and linear orderings*, **Annals of Pure and Applied Logic**, vol. 119 (2003), pp. 85–120.

[237] Yu. T. MEDVEDEV, *Degrees of difficulty of the mass problem*, **Doklady Akademii Nauk SSSR**, vol. 104 (1955), pp. 501–504, (Russian).

[238] A. G. MELNIKOV, *Enumerations and completely decomposable torsion-free abelian groups*, **Theory of Computing Systems**, vol. 45 (2009), pp. 897–916.

[239] G. METAKIDES and A. NERODE, *Recursively enumerable vector spaces*, **Annals of Pure and Applied Logic**, vol. 11 (1977), pp. 147–171.

[240] ———, *Effective content of field theory*, **Annals of Mathematical Logic**, vol. 17 (1979), pp. 289–320.

[241] ———, *Recursion theory on fields and abstract dependence*, **Journal of Algebra**, vol. 65 (1980), pp. 36–59.

[242] J. MILLAR and G. E. SACKS, *Atomic models higher up*, **Annals of Pure and Applied Logic**, vol. 155 (2008), pp. 225–241.

[243] T. MILLAR, *Prime models and almost decidability*, **The Journal of Symbolic Logic**, vol. 51 (1986), pp. 412–420.

[244] ———, *Recursive categoricity and persistence*, **The Journal of Symbolic Logic**, vol. 51 (1986), pp. 430–434.

[245] T. S. MILLAR, *Foundations of recursive model theory*, **Annals of Mathematical Logic**, vol. 13 (1978), pp. 45–72.

[246] ———, *Homogeneous models and decidability*, **Pacific Journal of Mathematics**, vol. 91 (1980), pp. 407–418.

[247] ———, *Type structure complexity and decidability*, **Transactions of the American Mathematical Society**, vol. 271 (1982), pp. 73–81.

[248] A. W. MILLER, *On the Borel classification of the isomorphism class of a countable model*, **Notre Dame Journal of Formal Logic**, vol. 24 (1983), pp. 22–34.

[249] D. E. MILLER, *The invariant* Π_α^0 *separation principle*, **Transactions of the American Mathematical Society**, vol. 242 (1978), pp. 185–204.

[250] R. MILLER, *The computable dimension of trees of infinite height*, **The Journal of Symbolic Logic**, vol. 70 (2005), pp. 111–141.

[251] ———, **d**-*computable categoricity for algebraic fields*, **The Journal of Symbolic Logic**, vol. 74 (2009), pp. 1325–1351.

[252] R. MILLER and H. SCHOUTENS, *Computably categorical fields via Fermat's Last Theorem*, **Computability**, vol. 2 (2013), pp. 51–65.

[253] R. MILLER and A. SHLAPENTOKH, *Computable categoricity for algebraic fields with splitting algorithms*, **Transactions of the American Mathematical Society**, (to appear).

[254] R. G. MILLER, *The* Δ_2^0-*spectrum of a linear order*, **The Journal of Symbolic Logic**, vol. 66 (2001), pp. 470–486.

[255] A. MONTALBÁN, *A computability theoretic equivalent to Vaught's conjecture*, preprint.

[256] ———, *A fixed point for the jump operator on structures*, preprint.

[257] ———, *On the equimorphism types of linear orderings*, **The Bulletin of Symbolic Logic**, vol. 13 (2007), pp. 71–99.

[258] ——, *Notes on the jump of a structure*, **Mathematical theory and computational practice, CiE 2009** (K. Ambos-Spies, B. Löwe, and W. Merkle, editors), Lecture Notes in Computer Science, vol. 5635, Springer, 2009, pp. 372–378.

[259] M. MORLEY, *Decidable models*, **Israel Journal of Mathematics**, vol. 25 (1976), pp. 233–240.

[260] A. S. MOROZOV, **On degrees of the recursive automorphism groups**, Algebra, Logic, and Applications, in memoriam of A. I. Kokorin (Irkutsk University, 1994), pp. 79–85 (Russian).

[261] ——, *Strong constructivizability of countable saturated Boolean algebras*, **Algebra and Logic**, vol. 21 (1982), pp. 130–137, (English translation).

[262] ——, *Groups of recursive automorphisms of constructive Boolean algebras*, **Algebra and Logic**, vol. 22 (1983), pp. 95–112, (English translation).

[263] ——, *Computable groups of automorphisms of models*, **Algebra and Logic**, vol. 25 (1986), pp. 261–266, (English translation).

[264] ——, *Permutations and implicit definability*, **Algebra and Logic**, vol. 27 (1988), pp. 12–24, (English translation).

[265] ——, *On theories of classes of groups of recursive permutations*, **Trudy Instituta Matematiki**, vol. 12 (1989), pp. 91–104, (Russian). (English translation in **Siberian Advances in Mathematics**, vol. 1 (1991), pp. 138–153.).

[266] ——, *Rigid constructive modules*, **Algebra and Logic**, vol. 28 (1989), pp. 379–387, (English translation).

[267] ——, *Functional trees and automorphisms of models*, **Algebra and Logic**, vol. 32 (1993), pp. 28–38, (English translation).

[268] ——, *Turing reducibility as algebraic embeddability*, **Siberian Mathematical Journal**, vol. 38 (1997), pp. 312–313, (English translation).

[269] ——, *Groups of computable automorphisms*, **Handbook of recursive mathematics, volume 1** (Yu. L. Ershov, S. S. Goncharov, A. Nerode, and J. B. Remmel, editors), Studies in Logic and the Foundations of Mathematics 139, North-Holland, Amsterdam, 1998, pp. 311–345.

[270] ——, *Once again on the Higman question*, **Algebra and Logic**, vol. 39 (2000), pp. 78–83, (English translation).

[271] ——, *On the relation of Σ-reducibility between admissible sets*, **Siberian Mathematical Journal**, vol. 45 (2004), pp. 634–652, (English translation).

[272] A. S. MOROZOV and A. N. BUZYKAEVA, *On a hierarchy of groups of computable automorphisms*, **Siberian Mathematical Journal**, vol. 43 (2002), pp. 124–127, (English translation).

[273] A. S. MOROZOV and J. K. TRUSS, *On computable automorphisms of the rational numbers*, **The Journal of Symbolic Logic**, vol. 66 (2001), pp. 1458–1470.

[274] ——, *On the categoricity of the group of all computable automorphisms of the rational numbers*, **Algebra and Logic**, vol. 46 (2007), pp. 354–361, (English translation).

[275] M. MOSES, *Relations intrinsically recursive in linear orders*, **Zeitschrift für Mathematische Logik und Grundlagen der Mathematik**, vol. 32 (1986), pp. 467–472.

[276] A. MOSTOWSKI, *A formula with no recursively enumerable model*, **Fundamenta Mathematicae**, vol. 42 (1955), pp. 125–140.

[277] A. NIES, *A new spectrum of recursive models*, **Notre Dame Journal of Formal Logic**, vol. 40 (1999), pp. 307–314.

[278] ——, **Computability and randomness**, Oxford University Press, 2009.

[279] P. S. NOVIKOV, *On the algorithmic unsolvability of the word problem in group theory*, **Proceedings of the Steklov Institute of Mathematics**, vol. 44 (1955), pp. 1–143, (Russian).

[280] A. T. NURTAZIN, *Strong and weak constructivizations and computable families*, **Algebra and Logic**, vol. 13 (1974), pp. 177–184, (English translation).

[281] S. OATES, **Jump degrees of groups**, PhD dissertation, University of Notre Dame, 1989.

[282] E. N. PAVLOVSKII, *An estimate for the algorithmic complexity of classes of computable models*, **Siberian Mathematical Journal**, vol. 49 (2008), pp. 512–523, (English translation).

[283] M. G. PERETYAT'KIN, *Strongly constructive models and enumerations of the Boolean algebra of recursive sets*, **Algebra and Logic**, vol. 10 (1971), pp. 535–557, (Russian); (1973) pp. 332–345, (English translation).

[284] ———, *Criterion for strong constructivizability of a homogeneous model*, **Algebra and Logic**, vol. 17 (1978), pp. 290–301, (English translation).

[285] ———, *Turing machine computations in finitely axiomatizable theories*, **Algebra and Logic**, vol. 21 (1982), pp. 272–295, (English translation).

[286] V. G. PUZARENKO, *On a certain reducibility on admissible sets*, **Siberian Mathematical Journal**, vol. 50 (2009), pp. 330–340, (English translation).

[287] ———, *Fixed points for the jump operator*, **Algebra and Logic**, vol. 50 (2011), pp. 418–438, (English translation).

[288] M. O. RABIN, *Computable algebra, general theory and theory of computable fields*, **Transactions of the American Mathematical Society**, vol. 95 (1960), pp. 341–360.

[289] J. B. REMMEL, *Combinatorial functors on co-r.e. structures*, **Annals of Mathematical Logic**, vol. 10 (1976), pp. 261–287.

[290] ———, *Maximal and cohesive vector spaces*, **The Journal of Symbolic Logic**, vol. 42 (1977), pp. 400–418.

[291] ———, *Recursively enumerable Boolean algebras*, **Annals of Mathematical Logic**, vol. 15 (1978), pp. 75–107.

[292] ———, *Recursive isomorphism types of recursive Boolean algebras*, **The Journal of Symbolic Logic**, vol. 46 (1981), pp. 572–594.

[293] ———, *Recursively categorical linear orderings*, **Proceedings of the American Mathematical Society**, vol. 83 (1981), pp. 387–391.

[294] ———, *Recursively rigid Boolean algebras*, **Annals of Pure Applied Logic**, vol. 36 (1987), pp. 39–52.

[295] J.-P. RESSAYRE, *Boolean models and infinitary first order languages*, **Annals of Mathematical Logic**, vol. 6 (1973), pp. 41–92.

[296] L. J. RICHTER, **Degrees of unsolvability of models**, PhD dissertation, University of Illinois at Urbana-Champaign, 1977.

[297] ———, *Degrees of structures*, **The Journal of Symbolic Logic**, vol. 46 (1981), pp. 723–731.

[298] H. ROGERS, **Theory of recursive functions and effective computability**, McGraw-Hill, New York, 1967, 2nd edition: MIT Press, Cambridge, MA, 1987.

[299] J. ROSENSTEIN, **Linear orderings**, Academic Press, New York, 1982.

[300] S. RUBIN, *Automata presenting structures: a survey of the finite string case*, **The Bulletin of Symbolic Logic**, vol. 14 (2008), pp. 169–209.

[301] G. E. SACKS, **Higher recursion theory**, Springer, Berlin, 1990.

[302] S. SCHWARZ, *Recursive automorphisms of recursive linear orderings*, **Annals of Pure and Applied Logic**, vol. 26 (1984), pp. 69–73.

[303] Z. SELA, *Diophantine geometry over groups I: Makanin–Razborov diagrams*, **Publications Mathématiques. Institute de Hautes Études Scientifiques**, vol. 93 (2001), pp. 31–105.

[304] ———, *Diophantine geometry over groups II: Completions, closures, and formal solutions*, **Israel Journal of Mathematics**, vol. 134 (2003), pp. 173–254.

[305] ———, *Diophantine geometry over groups IV: An iterative procedure for validation of a sentence*, **Israel Journal of Mathematics**, vol. 143 (2004), pp. 1–130.

[306] ———, *Diophantine geometry over groups III: Rigid and solid solutions*, **Israel Journal of Mathematics**, vol. 147 (2005), pp. 1–73.

[307] ———, *Diophantine geometry over groups V_1: Quantifier elimination I*, **Israel Journal of Mathematics**, vol. 150 (2005), pp. 1–197.

[308] ———, *Diophantine geometry over groups V_2: Quantifier elimination II*, **Geometric and Functional Analysis**, vol. 16 (2006), pp. 537–706.

[309] ——, *Diophantine geometry over groups VI: The elementary theory of a free group*, **Geometric and Functional Analysis**, vol. 16 (2006), pp. 707–730.

[310] V. L. SELIVANOV, *Enumerations of families of general recursive functions*, **Algebra and Logic**, vol. 15 (1976), pp. 128–141, (English translation).

[311] A. SELMAN, *Arithmetical reducibilities I*, **Zeitschrift für Mathematische Logik und Grundlagen der Mathematik**, vol. 17 (1971), pp. 335–370.

[312] I. P. SHESTAKOV and U. U. UMIRBAEV, *The tame and the wild automorphisms of polynomial rings in three variables*, **Journal of the American Mathematical Society**, vol. 17 (2004), pp. 197–227.

[313] R. A. SHORE, *Controlling the dependence degree of a recursively enumerable vector space*, **The Journal of Symbolic Logic**, vol. 43 (1978), pp. 13–22.

[314] A. S. SIKORA, *Topology on the spaces of orderings of groups*, **Bulletin of the London Mathematical Society**, vol. 36 (2004), pp. 519–526.

[315] T. SLAMAN, *Relative to any nonrecursive set*, **Proceedings of the American Mathematical Society**, vol. 126 (1998), pp. 2117–2122.

[316] R. L. SMITH, *Two theorems on autostability in p-groups*, **Logic Year 1979–80, University of Connecticut, Storrs**, Lecture Notes in Mathematics 859, Springer, Berlin, 1981, pp. 302–311.

[317] R. I. SOARE, **Recursively enumerable sets and degrees**, Springer, Berlin, 1987.

[318] D. R. SOLOMON, **Reverse mathematics and ordered groups**, PhD dissertation, Cornell University, 1998.

[319] R. SOLOMON, Π_1^0 *classes and orderable groups*, **Annals of Pure and Applied Logic**, vol. 115 (2002), pp. 279–302.

[320] I. N. SOSKOV, *Intrinsically hyperarithmetical sets*, **Mathematical Logic Quarterly**, vol. 42 (1996), pp. 469–480.

[321] ——, *Intrinsically Π_1^1 relations*, **Mathematical Logic Quarterly**, vol. 42 (1996), pp. 109–126.

[322] ——, *Degree spectra and co-spectra of structures*, **Annuaire de l'Université de Sofia "St. Kliment Ohridski," Faculté de Mathématiques et Informatique**, vol. 96 (2004), pp. 45–68.

[323] A. A. SOSKOVA and I. N. SOSKOV, *A jump inversion theorem for the degree spectra*, **Journal of Logic and Computation**, vol. 19 (2009), pp. 199–215.

[324] A. I. STUKACHEV, *Degrees of presentability of structures. I*, **Algebra and Logic**, vol. 46 (2007), pp. 419–432, (English translation).

[325] ——, *Degrees of presentability of structures. II*, **Algebra and Logic**, vol. 47 (2008), pp. 65–74, (English translation).

[326] ——, *A jump inversion theorem for the semilattices of Σ-degrees*, **Sibirskie Èlektronnye Matematicheskie Izvestiya**, vol. 6 (2009), pp. 182–190, (Russian). (English translation in **Siberian Advances in Mathematics**, vol. 20 (2010), pp. 68–74.).

[327] S. V. SUDOPLATOV, *Complete theories with finitely many countable models. I*, **Algebra and Logic**, vol. 43 (3004), pp. 62–69, (English translation).

[328] J. J. THURBER, *Every low$_2$ Boolean algebra has a recursive copy*, **Proceedings of the American Mathematical Society**, vol. 123 (1995), pp. 3859–3866.

[329] B. L. VAN DER WAERDEN, *Eine Bemerkung über die Unzerlegbarkeit von Polynomen*, **Mathematische Annalen**, vol. 102 (1930), pp. 738–739.

[330] R. L. VAUGHT, *Sentences true in all constructive models*, **The Journal of Symbolic Logic**, vol. 25 (1960), pp. 39–58.

[331] ——, *Denumerable models of complete theories*, **Proceedings of Symposium on Foundations of Mathematics: Infinitistic Methods**, Pergamon Press, London, 1961, pp. 301–321.

[332] S. WEHNER, *Enumerations, countable structures and Turing degrees*, **Proceedings of the American Mathematical Society**, vol. 126 (1998), pp. 2131–2139.

[333] W. WHITE, **Characterization for computable structures**, PhD dissertation, Cornell University, 2000.

KURT GÖDEL RESEARCH CENTER
  UNIVERSITY OF VIENNA
  VIENNA, AUSTRIA
  E-mail: efokina@logic.univie.ac.at

DEPARTMENT OF MATHEMATICS
  GEORGE WASHINGTON UNIVERSITY
  WASHINGTON, DC 20052, USA
  E-mail: harizanv@gwu.edu

SCHOOL OF MATHEMATICS, STATISTICS AND OPERATIONS RESEARCH
  VICTORIA UNIVERSITY OF WELLINGTON
  WELLINGTON, NEW ZEALAND
  E-mail: alexander.melnikov@vuw.ac.nz

TOWARDS COMMON-SENSE REASONING VIA CONDITIONAL SIMULATION: LEGACIES OF TURING IN ARTIFICIAL INTELLIGENCE

CAMERON E. FREER, DANIEL M. ROY, AND JOSHUA B. TENENBAUM

Abstract. The problem of replicating the flexibility of human common-sense reasoning has captured the imagination of computer scientists since the early days of Alan Turing's foundational work on computation and the philosophy of artificial intelligence. In the intervening years, the idea of cognition as computation has emerged as a fundamental tenet of Artificial Intelligence (AI) and cognitive science. But what kind of computation is cognition?

We describe a computational formalism centered around a probabilistic Turing machine called QUERY, which captures the operation of probabilistic conditioning via *conditional simulation*. Through several examples and analyses, we demonstrate how the QUERY abstraction can be used to cast common-sense reasoning as probabilistic inference in a statistical model of our observations and the uncertain structure of the world that generated that experience. This formulation is a recent synthesis of several research programs in AI and cognitive science, but it also represents a surprising convergence of several of Turing's pioneering insights in AI, the foundations of computation, and statistics.

§1. Introduction.

In his landmark paper *Computing Machinery and Intelligence* [Tur50], Alan Turing predicted that by the end of the twentieth century, "general educated opinion will have altered so much that one will be able to speak of machines thinking without expecting to be contradicted." Even if Turing has not yet been proven right, the idea of *cognition as computation* has emerged as a fundamental tenet of Artificial Intelligence (AI) and cognitive science. But what kind of computation—what kind of computer program—is cognition?

AI researchers have made impressive progress since the birth of the field over 60 years ago. Yet despite this progress, no existing AI system can reproduce any nontrivial fraction of the inferences made regularly by children. Turing himself appreciated that matching the capability of children, e.g., in language, presented a key challenge for AI:

> We hope that machines will eventually compete with men in all purely intellectual fields. But which are the best ones to start with? Even this is a difficult decision. Many people think that a very abstract activity, like the playing of chess, would be best. It can also be maintained that it is best to provide the machine with the

best sense organs money can buy, and then teach it to understand and speak English. This process could follow the normal teaching of a child. Things would be pointed out and named, etc. Again I do not know what the right answer is, but I think both approaches should be tried. [Tur50, p. 460]

Indeed, many of the problems once considered to be grand AI challenges have fallen prey to essentially brute-force algorithms backed by enormous amounts of computation, often robbing us of the insight we hoped to gain by studying these challenges in the first place. Turing's presentation of his "imitation game" (what we now call "the Turing test"), and the problem of common-sense reasoning implicit in it, demonstrates that he understood the difficulty inherent in the open-ended, if commonplace, tasks involved in conversation. Over a half century later, the Turing test remains resistant to attack.

The analogy between minds and computers has spurred incredible scientific progress in both directions, but there are still fundamental disagreements about the nature of the computation performed by our minds, and how best to narrow the divide between the capability and flexibility of human and artificial intelligence. The goal of this article is to describe a computational formalism that has proved useful for building simplified models of common-sense reasoning. The centerpiece of the formalism is a universal probabilistic Turing machine called QUERY that performs *conditional simulation*, and thereby captures the operation of conditioning probability distributions that are themselves represented by probabilistic Turing machines. We will use QUERY to model the inductive leaps that typify common-sense reasoning. The distributions on which QUERY will operate are models of latent unobserved processes in the world and the sensory experience and observations they generate. Through a running example of medical diagnosis, we aim to illustrate the flexibility and potential of this approach.

The QUERY abstraction is a component of several research programs in AI and cognitive science developed jointly with a number of collaborators. This chapter represents our own view on a subset of these threads and their relationship with Turing's legacy. Our presentation here draws heavily on both the work of Vikash Mansinghka on "natively probabilistic computing" [Man09, MJT08, Man11, MR13] and the "probabilistic language of thought" hypothesis proposed and developed by Noah Goodman [KGT08, GTFG08, GG12, GT12]. Their ideas form core aspects of the picture we present. The *Church* probabilistic programming language (introduced in [GMRBT08] by Goodman, Mansinghka, Roy, Bonawitz, and Tenenbaum) and various Church-based cognitive science tutorials (in particular, [GTO11], developed by Goodman, Tenenbaum, and O'Donnell) have also had a strong influence on the presentation.

This approach also draws from work in cognitive science on "theory-based Bayesian models" of inductive learning and reasoning [TGK06] due to Tenenbaum and various collaborators [GKT08, KT08, TKGG11]. Finally, some of the theoretical aspects that we present are based on results in computable probability theory by Ackerman, Freer, and Roy [Roy11, AFR11].

While the particular formulation of these ideas is recent, they have antecedents in much earlier work on the foundations of computation and computable analysis, common-sense reasoning in AI, and Bayesian modeling and statistics. In all of these areas, Turing had pioneering insights.

1.1. A convergence of Turing's ideas. In addition to Turing's well-known contributions to the philosophy of AI, many other aspects of his work—across statistics, the foundations of computation, and even morphogenesis—have converged in the modern study of AI. In this section, we highlight a few key ideas that will frequently surface during our account of common-sense reasoning via conditional simulation.

An obvious starting point is Turing's own proposal for a research program to pass his eponymous test. From a modern perspective, Turing's focus on learning (and in particular, induction) was especially prescient. For Turing, the idea of programming an intelligent machine entirely by hand was clearly infeasible, and so he reasoned that it would be necessary to construct a machine with the ability to adapt its own behavior in light of experience—i.e., with the ability to learn:

> Instead of trying to produce a programme to simulate the adult mind, why not rather try to produce one that simulates the child's? If this were then subjected to an appropriate course of education one would obtain the adult brain. [Tur50, p. 456]

Turing's notion of learning was inductive as well as deductive, in contrast to much of the work that followed in the first decade of AI. In particular, he was quick to explain that such a machine would have its flaws (in reasoning, quite apart from calculational errors):

> [A machine] might have some method for drawing conclusions by scientific induction. We must expect such a method to lead occasionally to erroneous results. [Tur50, p. 449]

Turing also appreciated that a machine would not only have to learn facts, but would also need to learn *how to learn*:

> Important amongst such imperatives will be ones which regulate the order in which the rules of the logical system concerned are to be applied. For at each stage when one is using a logical system, there is a very large number of alternative steps, any of which one is permitted to apply [...] These choices make the difference between a brilliant and a footling reasoner, not the difference between a

sound and a fallacious one. [...] [Some such imperatives] may be 'given by authority', but others may be produced by the machine itself, *e.g.*, by scientific induction. [Tur50, p. 458]

In addition to making these more abstract points, Turing presented a number of concrete proposals for how a machine might be programmed to learn. His ideas capture the essence of supervised, unsupervised, and reinforcement learning, each major areas in modern AI.[1] In Sections 5 and 7 we will return to Turing's writings on these matters.

One major area of Turing's contributions, while often overlooked, is statistics. In fact, Turing, along with I. J. Good, made key advances in statistics in the course of breaking the Enigma during World War II. Turing and Good developed new techniques for incorporating evidence and new approximations for estimating parameters in hierarchical models [Goo79, Goo00] (see also [Zab95, §5] and [Zab12]), which were among the most important applications of Bayesian statistics at the time [Zab12, §3.2]. Given Turing's interest in learning machines and his deep understanding of statistical methods, it would have been intriguing to see a proposal to combine the two areas. Yet if he did consider these connections, it seems he never published such work. On the other hand, much of modern AI rests upon a statistical foundation, including Bayesian methods. This perspective permeates the approach we will describe, wherein learning is achieved via Bayesian inference, and in Sections 5 and 6 we will re-examine some of Turing's wartime statistical work in the context of hierarchical models.

A core latent hypothesis underlying Turing's diverse body of work was that processes in nature, including our minds, could be understood through mechanical—in fact, *computational*—descriptions. One of Turing's crowning achievements was his introduction of the *a*-machine, which we now call the Turing machine. The Turing machine characterized the limitations and possibilities of computation by providing a mechanical description of a human computer. Turing's work on morphogenesis [Tur52] and AI each sought mechanical explanations in still further domains. Indeed, in all of these areas, Turing was acting as a natural scientist [Hod97], building models of natural phenomena using the language of computational processes.

In our account of common-sense reasoning as conditional simulation, we will use probabilistic Turing machines to represent mechanical descriptions of the world, much like those Turing sought. In each case, the stochastic machine represents one's uncertainty about the generative process giving rise to

---

[1] Turing also developed some of the early ideas regarding *neural networks*; see the discussions in [Tur48] about "unorganized machines" and their education and organization. This work, too, has grown into a large field of modern research, though we will not explore neural nets in the present article. For more details, and in particular the connection to work of McCulloch and Pitts [MP43], see Copeland and Proudfoot [CP96] and Teuscher [Teu02].

some pattern in the natural world. This description then enables probabilistic inference (via QUERY) about these patterns, allowing us to make decisions and manage our uncertainty in light of new evidence. Over the course of the article we will see a number of stochastic generative processes of increasing sophistication, culminating in models of decision making that rely crucially on recursion. Through its emphasis on inductive learning, Bayesian statistical techniques, universal computers, and mechanical models of nature, this approach to common-sense reasoning represents a convergence of many of Turing's ideas.

1.2. Common-sense reasoning via QUERY. For the remainder of the paper, our focal point will be the probabilistic Turing machine QUERY, which implements a generic form of probabilistic conditioning. QUERY allows one to make predictions using complex probabilistic models that are themselves specified using probabilistic Turing machines. By using QUERY appropriately, one can describe various forms of learning, inference, and decision-making. These arise via Bayesian inference, and common-sense behavior can be seen to follow *implicitly* from past experience and models of causal structure and goals, rather than explicitly via rules or purely deductive reasoning. Using the extended example of medical diagnosis, we aim to demonstrate that QUERY is a surprisingly powerful abstraction for expressing common-sense reasoning tasks that have, until recently, largely defied formalization.

As with Turing's investigations in AI, the approach we describe has been motivated by reflections on the details of human cognition, as well as on the nature of computation. In particular, much of the AI framework that we describe has been inspired by research in cognitive science attempting to model human inference and learning. Indeed, hypotheses expressed in this framework have been compared with the judgements and behaviors of human children and adults in many psychology experiments. Bayesian generative models, of the sort we describe here, have been shown to predict human behavior on a wide range of cognitive tasks, often with high quantitative accuracy. For examples of such models and the corresponding experiments, see the review article [TKGG11]. We will return to some of these more complex models in Section 8. We now proceed to define QUERY and illustrate its use via increasingly complex problems and the questions these raise.

§2. Probabilistic reasoning and QUERY. The specification of a probabilistic model can implicitly define a space of complex and useful behavior. In this section we informally describe the universal probabilistic Turing machine QUERY, and then use QUERY to explore a medical diagnosis example that captures many aspects of common-sense reasoning, but in a simple domain. Using QUERY, we highlight the role of conditional independence and conditional probability in building compact yet highly flexible systems.

2.1. An informal introduction to QUERY. The QUERY formalism was originally developed in the context of the Church probabilistic programming language [GMRBT08], and has been further explored by Mansinghka [Man11] and Mansinghka and Roy [MR13].

At the heart of the QUERY abstraction is a probabilistic variation of Turing's own mechanization [Tur36] of the capabilities of human "computers", the Turing machine. A Turing machine is a finite automaton with read, write, and seek access to a finite collection of infinite binary tapes, which it may use throughout the course of its execution. Its input is loaded onto one or more of its tapes prior to execution, and the output is the content of (one or more of) its tapes after the machine enters its halting state. A *probabilistic* Turing machine (PTM) is simply a Turing machine with an additional read-only tape comprised of a sequence of independent random bits, which the finite automaton may read and use as a source of randomness.

Turing machines (and their probabilistic generalizations) capture a robust notion of deterministic (and probabilistic) computation: Our use of the Turing machine abstraction relies on the remarkable existence of *universal* Turing machines, which can simulate all other Turing machines. More precisely, there is a PTM UNIVERSAL and an encoding $\{e_s : s \in \{0,1\}^*\}$ of all PTMs, where $\{0,1\}^*$ denotes the set of finite binary strings, such that, on inputs s and x, UNIVERSAL halts and outputs the string t if and only if (the Turing machine encoded by) e_s halts and outputs t on input x. Informally, the input s to UNIVERSAL is analogous to a program written in a programming language, and so we will speak of (encodings of) Turing machines and programs interchangeably.

QUERY is a PTM that takes two inputs, called the *prior program* P and *conditioning predicate* C, both of which are themselves (encodings of) PTMs that take no input (besides the random bit tape), with the further restriction that the predicate C return only a 1 or 0 as output. The semantics of QUERY are straightforward: first generate a sample from P; if C is satisfied, then output the sample; otherwise, try again. More precisely:

1. Simulate the predicate C on a random bit tape R (i.e., using the existence of a universal Turing machine, determine the output of the PTM C, if R were its random bit tape);
2. If (the simulation of) C produces 1 (i.e., if C *accepts*), then simulate and return the output produced by P, using the *same* random bit tape R; and
3. Otherwise (if C *rejects* R, returning 0), return to step 1, using an independent sequence R' of random bits.

It is important to stress that P and C share a random bit tape on each iteration, and so the predicate C may, in effect, act as though it has access to any intermediate value computed by the prior program P when deciding whether to accept or reject a random bit tape. More generally, any value computed by P can be recomputed by C and vice versa. We will use this fact to simplify the description of predicates, informally referring to values computed by P in the course of defining a predicate C.

As a first step towards understanding QUERY, note that if ⊤ is a PTM that always accepts (i.e., always outputs 1), then QUERY(P, ⊤) produces the same distribution on outputs as executing P itself, as the semantics imply that QUERY would halt on the first iteration.

Predicates that are not identically 1 lead to more interesting behavior. Consider the following simple example based on a remark by Turing [Tur50, p. 459]: Let N_{180} be a PTM that returns (a binary encoding of) an integer N drawn uniformly at random in the range 1 to 180, and let $DIV_{2,3,5}$ be a PTM that accepts (outputs 1) if N is divisible by 2, 3, and 5; and rejects (outputs 0) otherwise. Consider a typical output of

$$\text{QUERY}(N_{180}, DIV_{2,3,5}).$$

Given the semantics of QUERY, we know that the output will fall in the set

$$\{30, 60, 90, 120, 150, 180\} \tag{1}$$

and moreover, because each of these possible values of N was *a priori* equally likely to arise from executing N_{180} alone, this remains true *a posteriori*. You may recognize this as the conditional distribution of a uniform distribution conditioned to lie in the set (1). Indeed, QUERY performs the operation of conditioning a distribution.

The behavior of QUERY can be described more formally with notions from probability theory. In particular, from this point on, we will think of the output of a PTM (say, P) as a random variable (denoted by φ_P) defined on an underlying probability space with probability measure $\mathbb{P}$. (We will define this probability space formally in Section 3.1, but informally it represents the random bit tape.) When it is clear from context, we will also regard any named intermediate value (like N) as a random variable on this same space. Although Turing machines manipulate binary representations, we will often gloss over the details of how elements of other countable sets (like the integers, naturals, rationals, etc.) are represented in binary, but only when there is no risk of serious misunderstanding.

In the context of QUERY(P, C), the output distribution of P, which can be written $\mathbb{P}(\varphi_P \in \cdot)$, is called the *prior* distribution. Recall that, for all measurable sets (or simply *events*) A and C,

$$\mathbb{P}(A \mid C) := \frac{\mathbb{P}(A \cap C)}{\mathbb{P}(C)}, \tag{2}$$

is the conditional probability of the event A given the event C, provided that $\mathbb{P}(C) > 0$. Then the distribution of the output of QUERY(P, C), called the *posterior* distribution of φ_P, is the conditional distribution of φ_P given the event $\varphi_C = 1$, written

$$\mathbb{P}(\varphi_P \in \cdot \mid \varphi_C = 1).$$

Then returning to our example above, the prior distribution, $\mathbb{P}(N \in \cdot)$, is the uniform distribution on the set $\{1, \ldots, 180\}$, and the posterior distribution,

$$\mathbb{P}(N \in \cdot \mid N \text{ divisible by 2, 3, and 5}),$$

is the uniform distribution on the set given in (1), as can be verified via equation (2).

Those familiar with statistical algorithms will recognize the mechanism of QUERY to be exactly that of a so-called "rejection sampler". Although the definition of QUERY describes an explicit algorithm, we do not actually intend QUERY to be executed in practice, but rather intend for it to define and represent complex distributions. (Indeed, the description can be used by algorithms that work by very different methods than rejection sampling, and can aid in the communication of ideas between researchers.)

The actual implementation of QUERY in more efficient ways than via a rejection sampler is an active area of research, especially via techniques involving Markov chain Monte Carlo (MCMC); see, e.g., [GMRBT08, WSG11, WGSS11, SG12]. Turing himself recognized the potential usefulness of randomness in computation, suggesting:

> It is probably wise to include a random element in a learning machine. A random element is rather useful when we are searching for a solution of some problem. [Tur50, p. 458]

Indeed, some aspects of these algorithms are strikingly reminiscent of Turing's description of a random system of rewards and punishments in guiding the organization of a machine:

> The character may be subject to some random variation. Pleasure interference has a tendency to fix the character, i.e., towards preventing it changing, whereas pain stimuli tend to disrupt the character, causing features which had become fixed to change, or to become again subject to random variation. [Tur48, §10]

However, in this paper, we will not go into further details of implementation, nor the host of interesting computational questions this endeavor raises.

Given the subtleties of conditional probability, it will often be helpful to keep in mind the behavior of a rejection-sampler when considering examples of QUERY. (See [SG92] for more examples of this approach.) Note that, in our example, every simulation of N_{180} generates a number "accepted by" $DIV_{2,3,5}$ with probability $\frac{1}{30}$, and so, on average, we would expect the loop within QUERY to repeat approximately 30 times before halting. However, there is no finite bound on how long the computation could run. On the other hand,

(a) Disease marginals		
n	Disease	p_n
1	Arthritis	0.06
2	Asthma	0.04
3	Diabetes	0.11
4	Epilepsy	0.002
5	Giardiasis	0.006
6	Influenza	0.08
7	Measles	0.001
8	Meningitis	0.003
9	MRSA	0.001
10	Salmonella	0.002
11	Tuberculosis	0.003

(b) Unexplained symptoms		
m	Symptom	ℓ_m
1	Fever	0.06
2	Cough	0.04
3	Hard breathing	0.001
4	Insulin resistant	0.15
5	Seizures	0.002
6	Aches	0.2
7	Sore neck	0.006

(c) Disease-symptom rates

$c_{n,m}$	1	2	3	4	5	6	7
1	.1	.2	.1	.2	.2	.5	.5
2	.1	.4	.8	.3	.1	.0	.1
3	.1	.2	.1	.9	.2	.3	.5
4	.4	.1	.0	.2	.9	.0	.0
5	.6	.3	.2	.1	.2	.8	.5
6	.4	.2	.0	.2	.0	.7	.4
7	.5	.2	.1	.2	.1	.6	.5
8	.8	.3	.0	.3	.1	.8	.9
9	.3	.2	.1	.2	.0	.3	.5
10	.4	.1	.0	.2	.1	.1	.2
11	.3	.2	.1	.2	.2	.3	.5

TABLE 1. Medical diagnosis parameters. (These values are fabricated.) (a) p_n is the marginal probability that a patient has a disease n. (b) ℓ_m is the probability that a patient presents symptom m, assuming they have no disease. (c) $c_{n,m}$ is the probability that disease n causes symptom m to present, assuming the patient has disease n.

one can show that QUERY(N_{180}, $DIV_{2,3,5}$) eventually halts with probability one (equivalently, it halts *almost surely*, sometimes abbreviated "a.s.").

Despite the apparent simplicity of the QUERY construct, we will see that it captures the essential structure of a range of common-sense inferences. We now demonstrate the power of the QUERY formalism by exploring its behavior in a medical diagnosis example.

2.2. Diseases and their symptoms. Consider the following prior program, DS, which represents a simplified model of the pattern of *Diseases and Symptoms* we might find in a typical patient chosen at random from the population. At a high level, the model posits that the patient may be suffering from some, possibly empty, set of diseases, and that these diseases can cause symptoms. The prior program DS proceeds as follows: For each disease n listed in Table 1a, sample an independent binary random variable D_n with mean p_n, which we will interpret as indicating whether or not a patient has disease n depending on whether $D_n = 1$ or $D_n = 0$, respectively. For each symptom m listed in Table 1b, sample an independent binary random variable L_m with mean ℓ_m and for each pair (n, m) of a disease and symptom, sample an independent binary random variable $C_{n,m}$ with mean $c_{n,m}$, as listed in Table 1c. (Note that the numbers in all three tables have been fabricated.) Then, for each symptom m, define

$$S_m = \max\{L_m, D_1 \cdot C_{1,m}, \ldots, D_{11} \cdot C_{11,m}\},$$

so that $S_m \in \{0, 1\}$. We will interpret S_m as indicating that a patient has symptom m; the definition of S_m implies that this holds when any of the variables on the right hand side take the value 1. (In other words, the max operator is playing the role of a logical OR operation.) Every term of the form $D_n \cdot C_{n,m}$ is interpreted as indicating whether (or not) the patient has disease n and disease n has caused symptom m. The term L_m captures the possibility that the symptom may present itself despite the patient having none of the listed diseases. Finally, define the output of DS to be the vector $(D_1, \ldots, D_{11}, S_1, \ldots, S_7)$.

If we execute DS, or equivalently QUERY(DS, ⊤), then we might see outputs like those in the following array:

	Diseases											Symptoms						
	1	2	3	4	5	6	7	8	9	10	11	1	2	3	4	5	6	7
1	0	0	0	0	0	0	0	0	0	0	0	0	0	0	0	0	0	0
2	0	0	0	0	0	0	0	0	0	0	0	0	0	0	0	0	0	0
3	0	0	1	0	0	0	0	0	0	0	0	0	0	0	1	0	0	0
4	0	0	1	0	0	1	0	0	0	0	0	1	0	0	1	0	0	0
5	0	0	0	0	0	0	0	0	0	0	0	0	0	0	0	0	1	0
6	0	0	0	0	0	0	0	0	0	0	0	0	0	0	0	0	0	0
7	0	0	1	0	0	0	0	0	0	0	0	0	0	0	1	0	1	0
8	0	0	0	0	0	0	0	0	0	0	0	0	0	0	0	0	0	0

We will interpret the rows as representing eight patients chosen independently at random, the first two free from disease and not presenting any symptoms; the third suffering from diabetes and presenting insulin resistance; the fourth suffering from diabetes and influenza, and presenting a fever and insulin resistance; the fifth suffering from unexplained aches; the sixth free from disease and symptoms; the seventh suffering from diabetes, and presenting insulin resistance and aches; and the eighth also disease and symptom free.

This model is a toy version of the real diagnostic model QMR-DT [SMH+91]. QMR-DT is probabilistic model with essentially the structure of DS, built from data in the Quick Medical Reference (QMR) knowledge base of hundreds of diseases and thousands of findings (such as symptoms or test results). A key aspect of this model is the disjunctive relationship between the diseases and the symptoms, known as a "noisy-OR", which remains a popular modeling idiom. In fact, the structure of this model, and in particular the idea of layers of disjunctive causes, goes back even further to the "causal calculus" developed by Good [Goo61], which was based in part on his wartime work with Turing on the weight of evidence, as discussed by Pearl [Pea04, §70.2].

Of course, as a model of the latent processes explaining natural patterns of diseases and symptoms in a random patient, DS still leaves much to be desired. For example, the model assumes that the presence or absence of any two diseases is independent, although, as we will see later on in our analysis, diseases are (as expected) typically not independent conditioned on symptoms. On the other hand, an actual disease might cause another disease, or might cause a symptom that itself causes another disease, possibilities that this model does not capture. Like QMR-DT, the model DS avoids simplifications made by many earlier expert systems and probabilistic models to not allow for the simultaneous occurrence of multiple diseases [SMH+91]. These caveats notwithstanding, a close inspection of this simplified model will demonstrate a surprising range of common-sense reasoning phenomena.

Consider a predicate OS, for *Observed Symptoms*, that accepts if and only if $S_1 = 1$ and $S_7 = 1$, i.e., if and only if the patient presents the symptoms of a fever and a sore neck. What outputs should we expect from QUERY(DS, OS)? Informally, if we let μ denote the distribution over the combined outputs of DS and OS on a shared random bit tape, and let $A = \{(x, c): c = 1\}$ denote the set of those pairs that OS accepts, then QUERY(DS, OS) generates samples from the conditioned distribution $\mu(\cdot \mid A)$. Therefore, to see what the condition $S_1 = S_7 = 1$ implies about the plausible execution of DS, we must consider the conditional distributions of the diseases given the symptoms. The following conditional probability calculations may be very familiar to some readers, but will be less so to others, and so we present them here to give a more complete picture of the behavior of QUERY.

2.2.1. *Conditional execution.* Consider a $\{0, 1\}$-assignment d_n for each disease n, and write $D = d$ to denote the event that $D_n = d_n$ for every such n. Assume for the moment that $D = d$. Then what is the probability that OS accepts? The probability we are seeking is the conditional probability

$$\mathbb{P}(S_1 = S_7 = 1 \mid D = d) = \mathbb{P}(S_1 = 1 \mid D = d) \cdot \mathbb{P}(S_7 = 1 \mid D = d), \quad (3)$$

where the equality follows from the observation that once the D_n variables are fixed, the variables S_1 and S_7 are independent. Note that $S_m = 1$ if and only if $L_m = 1$ or $C_{n,m} = 1$ for some n such that $d_n = 1$. (Equivalently, $S_m = 0$ if and only if $L_m = 0$ and $C_{n,m} = 0$ for all n such that $d_n = 1$.) By the independence of each of these variables, it follows that

$$\mathbb{P}(S_m = 1 \mid D = d) = 1 - (1 - \ell_m) \prod_{n:\, d_n=1} (1 - c_{n,m}). \quad (4)$$

Let d' be an alternative $\{0, 1\}$-assignment. We can now characterize the *a posteriori* odds

$$\frac{\mathbb{P}(D = d \mid S_1 = S_7 = 1)}{\mathbb{P}(D = d' \mid S_1 = S_7 = 1)}$$

of the assignment d versus the assignment d'. By Bayes' rule, this can be rewritten as

$$\frac{\mathbb{P}(S_1 = S_7 = 1 \mid D = d) \cdot \mathbb{P}(D = d)}{\mathbb{P}(S_1 = S_7 = 1 \mid D = d') \cdot \mathbb{P}(D = d')}, \quad (5)$$

where $\mathbb{P}(D = d) = \prod_{n=1}^{11} \mathbb{P}(D_n = d_n)$ by independence. Using (3), (4) and (5), one may calculate that

$$\frac{\mathbb{P}(\text{Patient only has influenza} \mid S_1 = S_7 = 1)}{\mathbb{P}(\text{Patient has no listed disease} \mid S_1 = S_7 = 1)} \approx 42,$$

i.e., it is forty-two times more likely that an execution of DS satisfies the predicate OS via an execution that posits the patient only has the flu than an execution which posits that the patient has no disease at all. On the other hand,

$$\frac{\mathbb{P}(\text{Patient only has meningitis} \mid S_1 = S_7 = 1)}{\mathbb{P}(\text{Patient has no listed disease} \mid S_1 = S_7 = 1)} \approx 6,$$

and so

$$\frac{\mathbb{P}(\text{Patient only has influenza} \mid S_1 = S_7 = 1)}{\mathbb{P}(\text{Patient only has meningitis} \mid S_1 = S_7 = 1)} \approx 7,$$

and hence we would expect, over many executions of QUERY(DS, OS), to see roughly seven times as many explanations positing only influenza than positing only meningitis.

Further investigation reveals some subtle aspects of the model. For example, consider the fact that

$$\frac{\mathbb{P}(\text{Patient only has meningitis and influenza} \mid S_1 = S_7 = 1)}{\mathbb{P}(\text{Patient has meningitis, maybe influenza, but nothing else} \mid S_1 = S_7 = 1)}$$
$$= 0.09 \approx \mathbb{P}(\text{Patient has influenza}), \quad (6)$$

which demonstrates that, once we have observed some symptoms, diseases are no longer independent. Moreover, this shows that once the symptoms have been "explained" by meningitis, there is little pressure to posit further causes, and so the posterior probability of influenza is nearly the prior probability of influenza. This phenomenon is well-known and is called *explaining away*; it is also known to be linked to the computational hardness of computing probabilities (and generating samples as QUERY does) in models of this variety. For more details, see [Pea88, §2.2.4].

2.2.2. *Predicates give rise to diagnostic rules.* These various conditional probability calculations, and their ensuing explanations, all follow from an analysis of the DS model conditioned on one particular (and rather simple) predicate OS. Already, this gives rise to a picture of how QUERY(DS, OS) implicitly captures an elaborate system of rules for what to believe following the observation of a fever and sore neck in a patient, assuming the background knowledge captured in the DS program and its parameters. In a similar way, every diagnostic scenario (encodable as a predicate) gives rise to its own complex set of inferences, each expressible using QUERY and the model DS.

As another example, if we look (or test) for the remaining symptoms and find them to all be absent, our new beliefs are captured by QUERY(DS, OS*) where the predicate OS* accepts if and only if

$$(S_1 = S_7 = 1) \wedge (S_2 = \cdots = S_6 = 0).$$

We need not limit ourselves to reasoning about diseases given symptoms. Imagine that we perform a diagnostic test that rules out meningitis. We could represent our new knowledge using a predicate capturing the condition

$$(D_8 = 0) \wedge (S_1 = S_7 = 1) \wedge (S_2 = \cdots = S_6 = 0).$$

Of course this approach would not take into consideration our uncertainty regarding the accuracy or mechanism of the diagnostic test itself, and so, ideally, we might expand the DS model to account for how the outcomes of diagnostic tests are affected by the presence of other diseases or symptoms. In Section 6, we will discuss how such an extended model might be learned from data, rather than constructed by hand.

We can also reason in the other direction, about symptoms given diseases. For example, public health officials might wish to know about how frequently those with influenza present no symptoms. This is captured by the conditional probability

$$\mathbb{P}(S_1 = \cdots = S_7 = 0 \mid D_6 = 1),$$

and, via QUERY, by the predicate for the condition $D_6 = 1$. Unlike the earlier examples where we reasoned backwards from effects (symptoms) to their likely causes (diseases), here we are reasoning in the same forward direction as the model DS is expressed.

The possibilities are effectively inexhaustible, including more complex states of knowledge such as, *there are at least two symptoms present*, or *the patient does not have both salmonella and tuberculosis*. In Section 4 we will consider the vast number of predicates and the resulting inferences supported by QUERY and DS, and contrast this with the compact size of DS and the table of parameters.

In this section, we have illustrated the basic behavior of QUERY, and have begun to explore how it can be used to decide what to believe in a given scenario. These examples also demonstrate that rules governing behavior need not be explicitly described as rules, but can arise implicitly via other mechanisms, like QUERY, paired with an appropriate prior and predicate. In this example, the diagnostic rules were determined by the definition of DS and the table of its parameters. In Section 5, we will examine how such a table of probabilities itself might be learned. In fact, even if the parameters are learned from data, the structure of DS itself still posits a strong structural relationship among the diseases and symptoms. In Section 6 we will explore how this structure could be learned. Finally, many common-sense reasoning tasks involve making a *decision*, and not just determining what to believe. In Section 7, we will describe how to use QUERY to make decisions under uncertainty.

Before turning our attention to these more complex uses of QUERY, we pause to consider a number of interesting theoretical questions: What kind of probability distributions can be represented by PTMs that generate samples? What kind of conditional distributions can be represented by QUERY? Or represented by PTMs in general? In the next section we will see how Turing's work formed the foundation of the study of these questions many decades later.

§3. **Computable probability theory.** We now examine the QUERY formalism in more detail, by introducing aspects of the framework of computable probability theory, which provides rigorous notions of computability for probability distributions, as well as the tools necessary to identify probabilistic operations that can and cannot be performed by algorithms. After giving a formal description of probabilistic Turing machines and QUERY, we relate them to the concept of a computable measure on a countable space. We then explore the representation of points (and random points) in uncountable spaces, and examine how to use QUERY to define models over uncountable spaces like the reals. Such models are commonplace in statistical practice, and thus might be expected to be useful for building a statistical mind. In fact, no generic and computable QUERY formalism exists for conditioning on observations taking values in uncountable spaces, but there are certain circumstances in which we can perform probabilistic inference in uncountable spaces.

Note that although the approach we describe uses a universal Turing machine (QUERY), which can take an arbitrary pair of programs as its prior and predicate, we do not make use of a so-called *universal* prior program (itself necessarily noncomputable). For a survey of approaches to inductive reasoning involving a universal prior, such as Solomonoff induction [Sol64], and computable approximations thereof, see Rathmanner and Hutter [RH11].

Before we discuss the capabilities and limitations of QUERY, we give a formal definition of QUERY in terms of probabilistic Turing machines and conditional distributions.

3.1. A formal definition of QUERY**.** Randomness has long been used in mathematical algorithms, and its formal role in computations dates to shortly after the introduction of Turing machines. In his paper [Tur50] introducing the Turing test, Turing informally discussed introducing a "random element", and in a radio discussion c. 1951 (later published as [Tur96]), he considered placing a random string of 0s and 1s on an additional input bit tape of a Turing machine. In 1956, de Leeuw, Moore, Shannon, and Shapiro [dMSS56] proposed probabilistic Turing machines (PTMs) more formally, making use of Turing's formalism [Tur39] for oracle Turing machines: a PTM is an oracle Turing machine whose oracle tape comprises independent random bits. From this perspective, the output of a PTM is itself a random variable and so we may speak of *the distribution of (the output of) a* PTM. For the PTM QUERY, which simulates other PTMs passed as inputs, we can express its distribution in terms of the distributions of PTM inputs. In the remainder of this section, we describe this formal framework and then use it to explore the class of distributions that may be represented by PTMs.

Fix a canonical enumeration of (oracle) Turing machines and the corresponding partial computable (oracle) functions $\{\varphi_e\}_{e \in \mathbb{N}}$, each considered as a partial function

$$\{0,1\}^\infty \times \{0,1\}^* \to \{0,1\}^*,$$

where $\{0,1\}^\infty$ denotes the set of countably infinite binary strings and, as before, $\{0,1\}^*$ denotes the set of finite binary strings. One may think of each such partial function as a mapping from an oracle tape and input tape to an output tape. We will write $\varphi_e(x,s) \downarrow$ when φ_e is defined on oracle tape x and input string s, and $\varphi_e(x,s) \uparrow$ otherwise. We will write $\varphi_e(x)$ when the input string is empty or when there is no input tape. As a model for the random bit tape, we define an independent and identically distributed (i.i.d.) sequence $R = (R_i : i \in \mathbb{N})$ of binary random variables, each taking the value 0 and 1 with equal probability, i.e, each R_i is an independent Bernoulli(1/2) random variable. We will write $\mathbb{P}$ to denote the distribution of the random bit tape R. More formally, R will be considered to be the identity function on the Borel probability space $(\{0,1\}^\infty, \mathbb{P})$, where $\mathbb{P}$ is the countable product of Bernoulli(1/2) measures.

Let s be a finite string, let $e \in \mathbb{N}$, and suppose that

$$\mathbb{P}\{r \in \{0,1\}^\infty : \varphi_e(r,s)\downarrow\} = 1.$$

Informally, we will say that the probabilistic Turing machine (indexed by) e halts almost surely on input s. In this case, we define the *output distribution of the eth (oracle) Turing machine on input string s* to be the distribution of the random variable

$$\varphi_e(R,s);$$

we may directly express this distribution as

$$\mathbb{P} \circ \varphi_e(\cdot, s)^{-1}.$$

Using these ideas we can now formalize QUERY. In this formalization, both the prior and predicate programs P and C passed as input to QUERY are finite binary strings interpreted as indices for a probabilistic Turing machine with no input tape. Suppose that P and C halt almost surely. In this case, the output distribution of QUERY(P, C) can be characterized as follows: Let

$R = (R_i : i \in \mathbb{N})$ denote the random bit tape, let $\pi \colon \mathbb{N} \times \mathbb{N} \to \mathbb{N}$ be a standard pairing function (i.e., a computable bijection), and, for each $n, i \in \mathbb{N}$, let $R_i^{(n)} := R_{\pi(n,i)}$ so that $\{R^{(n)} : n \in \mathbb{N}\}$ are independent random bit tapes, each with distribution $\mathbb{P}$. Define the nth *sample from the prior* to be the random variable

$$X_n := \varphi_\mathsf{P}(R^{(n)}),$$

and let

$$N := \inf \{n \in \mathbb{N} \colon \varphi_\mathsf{C}(R^{(n)}) = 1\}$$

be the first iteration n such that the predicate C evaluates to 1 (i.e., accepts). The *output distribution of* QUERY(P, C) is then the distribution of the random variable

$$X_N,$$

whenever $N < \infty$ holds with probability one, and is undefined otherwise. Note that $N < \infty$ a.s. if and only if C accepts with non-zero probability. As above, we can give a more direct characterization: Let

$$\mathcal{A} := \{R \in \{0,1\}^\infty \colon \varphi_\mathsf{C}(R) = 1\}$$

be the set of random bit tapes R such that the predicate C accepts by outputting 1. The condition "$N < \infty$ with probability one" is then equivalent to the statement that $\mathbb{P}(\mathcal{A}) > 0$. In that case, we may express the output distribution of QUERY(P, C) as

$$\mathbb{P}_\mathcal{A} \circ \varphi_\mathsf{P}^{-1}$$

where $\mathbb{P}_\mathcal{A}(\cdot) := \mathbb{P}(\cdot \mid \mathcal{A})$ is the distribution of the random bit tape conditioned on C accepting (i.e., conditioned on the event $\mathcal{A}$).

3.2. Computable measures and probability theory. Which probability distributions are the output distributions of *some* PTM? In order to investigate this question, consider what we might learn from simulating a given PTM P (on a particular input) that halts almost surely. More precisely, for a finite bit string $r \in \{0,1\}^*$ with length $|r|$, consider simulating P, replacing its random bit tape with the finite string r: If, in the course of the simulation, the program attempts to read beyond the end of the finite string r, we terminate the simulation prematurely. On the other hand, if the program halts and outputs a string t then we may conclude that all simulations of P will return the same value when the random bit tape begins with r. As the set of random bit tapes beginning with r has $\mathbb{P}$-probability $2^{-|r|}$, we may conclude that the distribution of P assigns at least this much probability to the string t.

It should be clear that, using the above idea, we may enumerate the (prefix-free) set of strings $\{r_n\}$, and matching outputs $\{t_n\}$, such that P outputs t_n

when its random bit tape begins with r_n. It follows that, for all strings t and $m \in \mathbb{N}$,

$$\sum_{\{n \leq m:\, t_n = t\}} 2^{-|r_n|}$$

is a lower bound on the probability that the distribution of P assigns to t, and

$$1 - \sum_{\{n \leq m:\, t_n \neq t\}} 2^{-|r_n|}$$

is an upper bound. Moreover, it is straightforward to show that as $m \to \infty$, these converge monotonically from above and below to the probability that P assigns to the string t.

This sort of effective information about a real number precisely characterizes the *computable real numbers*, first described by Turing in his paper [Tur36] introducing Turing machines. For more details, see the survey by Avigad and Brattka connecting computable analysis to work of Turing, elsewhere in this volume [AB13].

DEFINITION 3.1 (computable real number). A real number $r \in \mathbb{R}$ is said to be *computable* when its left and right cuts of rationals $\{q \in \mathbb{Q}: q < r\}$, $\{q \in \mathbb{Q}: r < q\}$ are computable (under the canonical computable encoding of rationals). Equivalently, a real is computable when there is a computable sequence of rationals $\{q_n\}_{n \in \mathbb{N}}$ that *rapidly converges to r*, in the sense that $|q_n - r| < 2^{-n}$ for each n.

We now know that the probability of each output string t from a PTM is a computable real (in fact, *uniformly in t*, i.e., this probability can be computed for each t by a single program that accepts t as input). Conversely, for every computable real $\alpha \in [0, 1]$ and string t, there is a PTM that outputs t with probability α. In particular, let $R = (R_1, R_2, \ldots)$ be our random bit tape, let $\alpha_1, \alpha_2, \ldots$ be a uniformly computable sequence of rationals that rapidly converges to α, and consider the following simple program: On step n, compute the rational $A_n := \sum_{i=1}^{n} R_i \cdot 2^{-i}$. If $A_n < \alpha_n - 2^{-n}$, then halt and output t; If $A_n > \alpha_n + 2^{-n}$, then halt and output $t0$. Otherwise, proceed to step $n + 1$. Note that $A_\infty := \lim A_n$ is uniformly distributed in the unit interval, and so $A_\infty < \alpha$ with probability α. Because $\lim \alpha_n \to \alpha$, the program eventually halts for all but one (or two, in the case that α is a dyadic rational) random bit tapes. In particular, if the random bit tape is the binary expansion of α, or equivalently, if $A_\infty = \alpha$, then the program does not halt, but this is a $\mathbb{P}$-measure zero event.

Recall that we assumed, in defining the output distribution of a PTM, that the program halted almost surely. The above construction illustrates why the stricter requirement that PTMs halt always (and not just almost surely) could be very limiting. In fact, one can show that there is no PTM that halts always and whose output distribution assigns, e.g., probability 1/3 to 1 and 2/3 to 0. Indeed, the same is true for all non-dyadic probability values (for details see [AFR11, Prop. 9]).

We can use this construction to sample from any distribution v on $\{0, 1\}^*$ for which we can compute the probability of a string t in a uniform way. In particular, fix an enumeration of all strings $\{t_n\}$ and, for each $n \in \mathbb{N}$, define the distribution v_n on $\{t_n, t_{n+1}, \dots\}$ by $v_n = v/(1 - v\{t_1, \dots, t_{n-1}\})$. If v is computable in the sense that for any t, we may compute real $v\{t\}$ uniformly in t, then v_n is clearly computable in the same sense, uniformly in n. We may then proceed in order, deciding whether to output t_n (with probability $v_n\{t_n\}$) or to recurse and consider t_{n+1}. It is straightforward to verify that the above procedure outputs a string t with probability $v\{t\}$, as desired.

These observations motivate the following definition of a computable probability measure, which is a special case of notions from computable analysis developed later; for details of the history see [Wei99, §1].

DEFINITION 3.2 (computable probability measure). A probability measure on $\{0, 1\}^*$ is said to be *computable* when the measure of each string is a computable real, uniformly in the string.

The above argument demonstrates that the samplable probability measures —those distributions on $\{0, 1\}^*$ that arise from sampling procedures performed by probabilistic Turing machines that halt a.s.—coincide with computable probability measures.

While in this paper we will not consider the efficiency of these procedures, it is worth noting that while the class of distributions that can be sampled by Turing machines coincides with the class of computable probability measures on $\{0, 1\}^*$, the analogous statements for polynomial-time Turing machines fail. In particular, there are distributions from which one can efficiently sample, but for which output probabilities are not efficiently computable (unless $\mathbf{P} = \mathbf{PP}$), for suitable formalizations of these concepts [Yam99].

3.3. Computable probability measures on uncountable spaces. So far we have considered distributions on the space of finite binary strings. Under a suitable encoding, PTMs can be seen to represent distributions on general countable spaces. On the other hand, many phenomena are naturally modeled in terms of continuous quantities like real numbers. In this section we will look at the problem of representing distributions on uncountable spaces, and then consider the problem of extending QUERY in a similar direction.

To begin, we will describe distributions on the space of *infinite* binary strings, $\{0, 1\}^\infty$. Perhaps the most natural proposal for representing such distributions is to again consider PTMs whose output can be interpreted as representing a random point in $\{0, 1\}^\infty$. As we will see, such distributions will have an equivalent characterization in terms of uniform computability of the measure of a certain class of sets.

Fix a computable bijection between $\mathbb{N}$ and finite binary strings, and for $n \in \mathbb{N}$, write $\bar{n}$ for the image of n under this map. Let e be the index of some PTM, and suppose that $\varphi_e(R, \bar{n}) \in \{0, 1\}^n$ and $\varphi_e(R, \bar{n}) \sqsubseteq \varphi_e(R, \overline{n+1})$ almost surely for all $n \in \mathbb{N}$, where $r \sqsubseteq s$ for two binary strings r and s when r is a prefix of s. Then the *random point in* $\{0, 1\}^\infty$ *given by* e is defined to be

$$\lim_{n \to \infty} (\varphi_e(R, \bar{n}), 0, 0, \dots). \tag{7}$$

Intuitively, we have represented the (random) infinite object by a program (relative to a fixed random bit tape) that can provide a convergent sequence of finite approximations.

It is obvious that the distribution of $\varphi_e(R, \bar{n})$ is computable, uniformly in n. As a consequence, for every basic clopen set $A = \{s : r \sqsubseteq s\}$, we may compute the probability that the limiting object defined by (7) falls into A, and thus we may compute arbitrarily good lower bounds for the measure of unions of computable sequences of basic clopen sets, i.e., c.e. open sets.

This notion of computability of a measure is precisely that developed in computable analysis, and in particular, via the Type-Two Theory of Effectivity (TTE); for details see Edalat [Eda96], Weihrauch [Wei99], Schröder [Sch07], and Gács [Gac05]. This formalism rests on Turing's oracle machines [Tur39]; for more details, again see the survey by Avigad and Brattka elsewhere in this volume [AB13]. The representation of a measure by the values assigned to basic clopen sets can be interpreted in several ways, each of which allows us to place measures on spaces other than just the set of infinite strings. From a topological point of view, the above representation involves the choice of a particular basis for the topology, with an appropriate enumeration, making $\{0, 1\}^\infty$ into a *computable topological space*; for details, see [Wei00, Def. 3.2.1] and [GSW07, Def. 3.1].

Another approach is to place a metric on $\{0, 1\}^\infty$ that induces the same topology, and that is computable on a dense set of points, making it into a *computable metric space*; see [Hem02] and [Wei93] on approaches in TTE, [Bla97] and [EH98] in effective domain theory, and [Wei00, Ch. 8.1] and [Gac05, §B.3] for more details. For example, one could have defined the distance between two strings in $\{0, 1\}^\infty$ to be 2^{-n}, where n is the location of the first bit on which they differ; instead choosing $1/n$ would have given a different metric space but would induce the same topology, and hence the same notion of computable measure. Here we use the following definition of a computable metric space, taken from [GHR10, Def. 2.3.1].

DEFINITION 3.3 (computable metric space). A *computable metric space* is a triple $(S, \delta, \mathcal{D})$ for which δ is a metric on the set S satisfying:
1. (S, δ) is a complete separable metric space;
2. $\mathcal{D} = \{s(1), s(2), \dots\}$ is an enumeration of a dense subset of S; and,
3. the real numbers $\delta(s(i), s(j))$ are computable, uniformly in i and j.

We say that an S-valued random variable X (defined on the same space as R) is an (*almost-everywhere*) *computable S-valued random variable* or *random point in S* when there is a PTM e such that $\delta(X_n, X) < 2^{-n}$ almost surely for all $n \in \mathbb{N}$, where $X_n := s(\varphi_e(R, \bar{n}))$. We can think of the random sequence $\{X_n\}$ as a *representation* of the random point X. A *computable probability measure* on S is precisely the distribution of such a random variable.

For example, the real numbers form a computable metric space $(\mathbb{R}, d, \mathbb{Q})$, where d is the Euclidean metric, and $\mathbb{Q}$ has the standard enumeration. One can show that computable probability measures on $\mathbb{R}$ are then those for which the measure of an arbitrary finite union of rational open intervals admits arbitrarily good lower bounds, uniformly in (an encoding of) the sequence of intervals. Alternatively, one can show that the space of probability measures on $\mathbb{R}$ is a computable metric space under the Prokhorov metric, with respect to (a standard enumeration of) a dense set of atomic measures with finite support in the rationals. The notions of computability one gets in these settings align with classical notions. For example, the set of naturals and the set of finite binary strings are indeed both computable metric spaces, and the computable measures in this perspective are precisely as described above.

Similarly to the countable case, we can use QUERY to sample points in *uncountable spaces* conditioned on a predicate. Namely, suppose the prior program P represents a random point in an uncountable space with distribution ν. For any string s, write P(s) for P with the input fixed to s, and let C be a predicate that accepts with non-zero probability. Then the PTM that, on input $\bar{n}$, outputs the result of simulating QUERY(P($\bar{n}$), C) is a representation of ν conditioned on the predicate accepting. When convenient and clear from context, we will denote this derived PTM by simply writing QUERY(P, C).

3.4. Conditioning on the value of continuous random variables. The above use of QUERY allows us to condition a model of a computable real-valued random variable X on a predicate C. However, the restriction on predicates (to accept with non-zero probability) and the definition of QUERY itself do not, in general, allow us to condition on X itself taking a specific value. Unfortunately, the problem is not superficial, as we will now relate.

Assume, for simplicity, that X is also continuous (i.e., $\mathbb{P}\{X = x\} = 0$ for all reals x). Let x be a computable real, and for every computable real $\varepsilon > 0$, consider the (partial computable) predicate C_ε that accepts when $|X - x| < \varepsilon$, rejects when $|X - x| > \varepsilon$, and is undefined otherwise. (We say that such a predicate *almost* decides the event $\{|X - x| < \varepsilon\}$ as it decides the set outside a measure zero set.) We can think of QUERY(P, C_ε) as a "positive-measure approximation" to conditioning on $X = x$. Indeed, if P is a prior program that samples a computable random variable Y and $B_{x,\varepsilon}$ denotes the closed ε-ball around x, then this QUERY corresponds to the conditioned distribution $\mathbb{P}(Y \mid X \in B_{x,\varepsilon})$, and so provided $\mathbb{P}\{X \in B_{x,\varepsilon}\} > 0$, this is well-defined and evidently computable. But what is its relationship to the original problem?

While one might be inclined to think that QUERY(P, $C_{\varepsilon=0}$) represents our original goal of conditioning on $X = x$, the continuity of the random variable X implies that $\mathbb{P}\{X \in B_{x,0}\} = \mathbb{P}\{X = x\} = 0$ and so C_0 rejects with probability one. It follows that QUERY(P, $C_{\varepsilon=0}$) does not halt on any input, and thus does not represent a distribution.

The underlying problem is that, in general, conditioning on a null set is mathematically undefined. The standard measure-theoretic solution is to consider the so-called "regular conditional distribution" given by conditioning on the σ-algebra generated by X—but even this approach would in general fail to solve our problem because the resulting disintegration is only defined up to a null set, and so is undefined at points (including x). (For more details, see [AFR11, §III] and [Tju80, Ch. 9].)

There have been various attempts at more constructive approaches, e.g., Tjur [Tju74, Tju75, Tju80], Pfanzagl [Pfa79], and Rao [Rao88, Rao05]. One approach worth highlighting is due to Tjur [Tju75]. There he considers additional hypotheses that are equivalent to the existence of a continuous *disintegration*, which must then be unique at all points. (We will implicitly use this notion henceforth.) Given the connection between computability and continuity, a natural question to ask is whether we might be able to extend QUERY along the lines.

Despite various constructive efforts, no general method had been found for computing conditional distributions. In fact, conditional distributions are not in general computable, as shown by Ackerman, Freer, and Roy [AFR11, Thm. 29], and it is for this reason we have defined QUERY in terms of conditioning on the event $C = 1$, which, provided that C accepts with non-zero probability as we have required, is a positive-measure event. The proof of the noncomputability of conditional probability [AFR11, §VI] involves an encoding of the halting problem into a pair (X, Y) of computable (even, absolutely continuous) random variables in [0, 1] such that no "version" of the conditional distribution $\mathbb{P}(Y \mid X = x)$ is a computable function of x.

What, then, is the relationship between conditioning on $X = x$ and the approximations C_ε defined above? In sufficiently nice settings, the distribution represented by QUERY(P, C_ε) converges to the desired distribution as $\varepsilon \to 0$. But as a corollary of the aforementioned noncomputability result, one sees that it is noncomputable in general to determine a value of ε from a desired level of accuracy to the desired distribution, for if there were such a general and computable relationship, one could use it to compute conditional distributions, a contradiction. Hence although such a sequence of approximations might converge in the limit, one cannot in general compute how close it is to convergence.

On the other hand, the presence of noise in measurements can lead to computability. As an example, consider the problem of representing the distribution of Y conditioned on $X + \xi = x$, where Y, X, and x are as above, and ξ is independent of X and Y and uniformly distributed on the interval $[-\varepsilon, \varepsilon]$. While conditioning on continuous random variables is not computable in general, here it is possible. In particular, note that $\mathbb{P}(Y \mid X + \xi = x) = \mathbb{P}(Y \mid X \in B_{x,\varepsilon})$ and so QUERY(P, C_ε) represents the desired distribution.

This example can be generalized considerably beyond uniform noise (see [AFR11, Cor. 36]). Many models considered in practice posit the existence of independent noise in the quantities being measured, and so the QUERY formalism can be used to capture probabilistic reasoning in these settings as well. However, in general we should not expect to be able to reliably approximate noiseless measurements with noisy measurements, lest we contradict the noncomputability of conditioning. Finally, it is important to note that the computability that arises in the case of certain types of independent noise is a special case of the computability that arises from the existence and computability of certain conditional probability densities [AFR11, §VII]. This final case covers most models that arise in statistical practice, especially those that are finite-dimensional.

In conclusion, while we cannot hope to condition on arbitrary computable random variables, QUERY covers nearly all of the situations that arise in practice, and suffices for our purposes. Having laid the theoretical foundation for QUERY and described its connection with conditioning, we now return to the medical diagnosis example and more elaborate uses of QUERY, with a goal of understanding additional features of the formalism.

§4. Conditional independence and compact representations. In this section, we return to the medical diagnosis example, and explain the way in which conditional independence leads to compact representations, and conversely, the fact that efficient probabilistic programs, like DS, exhibit many conditional independencies. We will do so through connections with the Bayesian network formalism, whose introduction by Pearl [Pea88] was a major advancement in AI.

4.1. The combinatorics of QUERY. Humans engaging in common-sense reasoning often seem to possess an unbounded range of responses or behaviors; this is perhaps unsurprising given the enormous variety of possible situations that can arise, even in simple domains.

Indeed, the small handful of potential diseases and symptoms that our medical diagnosis model posits already gives rise to a combinatorial explosion of potential scenarios with which a doctor could be faced: among 11 potential diseases and 7 potential symptoms there are

$$3^{11} \cdot 3^7 = 387\,420\,489$$

partial assignments to a subset of variables.

Building a table (i.e., function) associating every possible diagnostic scenario with a response would be an extremely difficult task, and probably nearly impossible if one did not take advantage of some structure in the domain to devise a more compact representation of the table than a structureless, huge list. In fact, much of AI can be interpreted as proposals for specific structural assumptions that lead to more compact representations, and the QUERY framework can be viewed from this perspective as well; the prior program DS implicitly defines a full table of responses, and the predicate can be understood as a way to index into this vast table.

This leads us to three questions: Is the table of diagnostic responses induced by DS any good? How is it possible that so many responses can be encoded so compactly? And what properties of a model follow from the existence of an efficient prior program, as in the case of our medical diagnosis example and the prior program DS? In the remainder of the section we will address the latter two questions, returning to the former in Section 5 and Section 6.

4.2. Conditional independence. Like DS, every probability model of 18 binary variables implicitly defines a gargantuan set of conditional probabilities. However, unlike DS, most such models have no compact representation. To see this, note that a probability distribution over k outcomes is, in general, specified by $k - 1$ probabilities, and so in principle, in order to specify a distribution on $\{0, 1\}^{18}$, one must specify

$$2^{18} - 1 = 262\,143$$

probabilities. Even if we discretize the probabilities to some fixed accuracy, a simple counting argument shows that most such distributions have no short description.

In contrast, Table 1 contains only

$$11 + 7 + 11 \cdot 7 = 95$$

probabilities, which, via the small collection of probabilistic computations performed by DS and described informally in the text, parameterize a distribution over 2^{18} possible outcomes. What properties of a model can lead to a compact representation?

The answer to this question is *conditional independence*. Recall that a collection of random variables $\{X_i : i \in I\}$ is *independent* when, for all finite subsets $J \subseteq I$ and measurable sets A_i where $i \in J$, we have

$$\mathbb{P}(\bigwedge_{i \in J} X_i \in A_i) = \prod_{i \in J} \mathbb{P}(X_i \in A_i). \tag{8}$$

If X and Y were binary random variables, then specifying their distribution would require 3 probabilities in general, but only 2 if they were independent. While those savings are small, consider instead n binary random variables X_j, $j = 1, \ldots, n$, and note that, while a generic distribution over these random variables would require the specification of $2^n - 1$ probabilities, only n probabilities are needed in the case of full independence.

Most interesting probabilistic models with compact representations will not exhibit enough independence between their constituent random variables to explain their own compactness in terms of the factorization in (8). Instead, the slightly weaker (but arguably more fundamental) notion of conditional independence is needed. Rather than present the definition of conditional independence in its full generality, we will consider a special case, restricting our attention to conditional independence with respect to a discrete random variable N taking values in some countable or finite set $\mathcal{N}$. (For the general case, see Kallenberg [Kal02, Ch. 6].) We say that a collection of random variables $\{X_i : i \in I\}$ is *conditionally independent* given N when, for all $n \in \mathcal{N}$, finite subsets $J \subseteq I$ and measurable sets A_i, for $i \in J$, we have

$$\mathbb{P}\Big(\bigwedge_{i \in J} X_i \in A_i \mid N = n\Big) = \prod_{i \in J} \mathbb{P}(X_i \in A_i \mid N = n).$$

To illustrate the potential savings that can arise from conditional independence, consider n binary random variables that are conditionally independent given a discrete random variable taking k values. In general, the joint distribution over these $n + 1$ variables is specified by $k \cdot 2^n - 1$ probabilities, but, in light of the conditional independence, we need specify only $k(n + 1) - 1$ probabilities.

4.3. Conditional independencies in DS. In Section 4.2, we saw that conditional independence gives rise to compact representations. As we will see, the variables in DS exhibit many conditional independencies.

To begin to understand the compactness of DS, note that the 95 variables

$$\{D_1, \ldots, D_{11};\ L_1, \ldots, L_7;\ C_{1,1}, C_{1,2}, C_{2,1}, C_{2,2}, \ldots, C_{11,7}\}$$

are independent, and thus their joint distribution is determined by specifying only 95 probabilities (in particular, those in Table 1). Each symptom S_m is then derived as a deterministic function of a 23-variable subset

$$\{D_1, \ldots, D_{11};\ L_m;\ C_{1,m}, \ldots, C_{11,m}\},$$

which implies that the symptoms are conditionally independent given the diseases. However, these facts alone do not fully explain the compactness of DS. In particular, there are

$$2^{2^{23}} > 10^{10^6}$$

binary functions of 23 binary inputs, and so by a counting argument, most have no short description. On the other hand, the max operation that defines S_m does have a compact *and efficient* implementation. In Section 4.5 we will see that this implies that we can introduce additional random variables representing intermediate quantities produced in the process of computing each symptom S_m from its corresponding collection of 23-variable "parent" variables, and that these random variables exhibit many more conditional independencies than exist between S_m and its parents. From this perspective, the compactness of DS is tantamount to there being only a small number of such variables that need to be introduced. In order to simplify our explanation of this connection, we pause to introduce the idea of representing conditional independencies using graphs.

4.4. Representations of conditional independence. A useful way to represent conditional independence among a collection of random variables is in terms of a directed acyclic graph, where the vertices stand for random variables, and the collection of edges indicates the presence of certain conditional independencies. An example of such a graph, known as a directed graphical model or Bayesian network, is given in Figure 1. (For more details on Bayesian networks, see the survey by Pearl [Pea04]. It is interesting to note that Pearl cites Good's "causal calculus" [Goo61]—which we have already encountered in connection with our medical diagnosis example, and which was based in part on Good's wartime work with Turing on the weight of evidence—as a historical antecedent to Bayesian networks [Pea04, §70.2].)

Directed graphical models often capture the "generative" structure of a collection of random variables: informally, by the direction of arrows, the diagram captures, for each random variable, which other random variables were directly implicated in the computation that led to it being sampled. In order to understand exactly which conditional independencies are formally encoded in such a graph, we must introduce the notion of d-separation.

We determine whether a pair (x, y) of vertices are d-separated by a subset of vertices $\mathcal{E}$ as follows: First, mark each vertex in $\mathcal{E}$ with a ×, which we will indicate by the symbol $\otimes$. If a vertex with (any type of) mark has an unmarked parent, mark the parent with a +, which we will indicate by the symbol $\oplus$. Repeat until a fixed point is reached. Let $\odot$ indicate unmarked vertices. Then x and y are d-separated if, for all (undirected) paths from x to y through the graph, one of the following patterns appears.

$$\to \otimes \to$$
$$\leftarrow \otimes \leftarrow$$
$$\leftarrow \otimes \to$$
$$\to \odot \leftarrow$$

More generally, if $\mathcal{X}$ and $\mathcal{E}$ are disjoint sets of vertices, then the graph encodes the conditional independence of the vertices $\mathcal{X}$ given $\mathcal{E}$ if every pair of vertices in $\mathcal{X}$ is d-separated given $\mathcal{E}$. If we fix a collection V of random variables, then we say that a directed acyclic graph G over V is a *Bayesian network* (equivalently, a directed graphical model) when the random variables in V indeed posses all of the conditional independencies implied

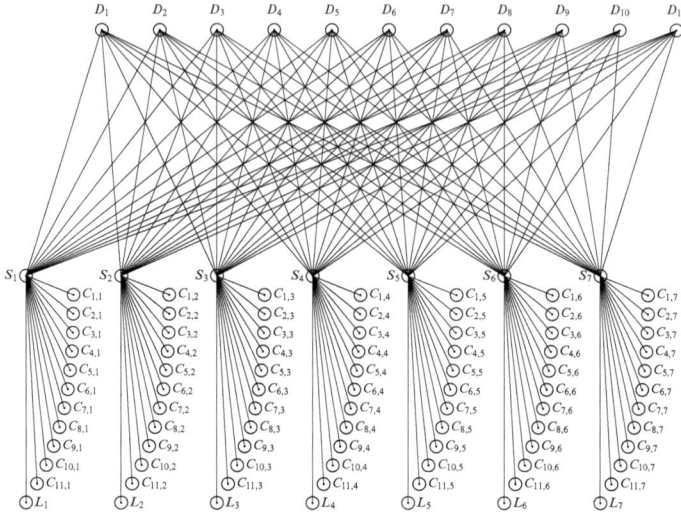

FIGURE 1. Directed graphical model representations of the conditional independence underlying the medical diagnosis example. (Note that the directionality of the arrows has not been rendered as they all simply point towards the symptoms S_m.)

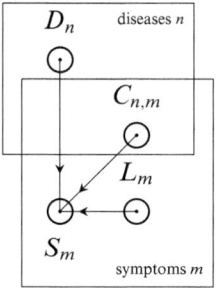

FIGURE 2. The repetitive structure Figure 1 can be partially captured by so-called "plate notation", which can be interpreted as a primitive for-loop construct. Practitioners have adopted a number of strategies like plate notation for capturing complicated structures.

by the graph by d-separation. Note that a directed graph G says nothing about which conditional independencies do *not* exist among its vertex set.

Using the notion of d-separation, we can determine from the Bayesian network in Figure 1 that the diseases $\{D_1, \ldots, D_{11}\}$ are independent (i.e., conditionally independent given $\mathcal{E} = \emptyset$). We may also conclude that the symptoms $\{S_1, \ldots, S_7\}$ are conditionally independent given the diseases $\{D_1, \ldots, D_{11}\}$.

In addition to encoding a set of conditional independence statements that hold among its vertex set, directed graphical models demonstrate that the joint distribution over its vertex set admits a concise factorization: For a collection of binary random variables $X_1, \ldots, X_k$, write $p(X_1, \ldots, X_k): \{0, 1\}^k \to [0, 1]$ for the probability mass function (p.m.f.) taking an assignment $x_1, \ldots, x_k$ to its probability $\mathbb{P}(X_1 = x_1, \ldots, X_k = x_k)$, and write

$$p(X_1, \ldots, X_k \mid Y_1, \ldots, Y_m): \{0, 1\}^{k+m} \to [0, 1]$$

for the *conditional* p.m.f. corresponding to the conditional distribution

$$\mathbb{P}(X_1, \ldots, X_k \mid Y_1, \ldots, Y_m).$$

It is a basic fact from probability that

$$p(X_1, \ldots, X_k) = p(X_1) \cdot p(X_2 \mid X_1) \cdots p(X_k \mid X_1, \ldots, X_{k-1}) \quad (9)$$

$$= \prod_{i=1}^{k} p(X_i \mid X_j, j < i).$$

Such a factorization provides no advantage when seeking a compact representation, as a conditional p.m.f. of the form $p(X_1, \ldots, X_k \mid Y_1, \ldots, X_m)$ is determined by $2^m \cdot (2^k - 1)$ probabilities. On the other hand, if we have a directed graphical model over the same variables, then we may have a much more concise factorization. In particular, let G be a directed graphical model over $\{X_1, \ldots, X_k\}$, and write $\text{Pa}(X_j)$ for the set of vertices X_i such that $(X_i, X_j) \in G$, i.e., $\text{Pa}(X_j)$ are the parent vertices of X_j. Then the joint p.m.f. may be expressed as

$$p(X_1, \ldots, X_k) = \prod_{i=1}^{k} p(X_i \mid \text{Pa}(X_i)). \quad (10)$$

Whereas the factorization given by (9) requires the full set of

$$\sum_{i=1}^{k} 2^{i-1} = 2^k - 1$$

probabilities to determine, this factorization requires $\sum_{i=1}^{k} 2^{|\text{Pa}(X_i)|}$ probabilities, which in general can be exponentially smaller in k.

4.5. Efficient representations and conditional independence. As we saw at the beginning of this section, models with only a moderate number of variables can have enormous descriptions. Having introduced the directed graphical model formalism, we can use DS as an example to explain why, roughly speaking, the output distributions of efficient probabilistic programs exhibit many conditional independencies.

What does the efficiency of DS imply about the structure of its output distribution? We may represent DS as a small boolean circuit whose inputs are random bits and whose 18 output lines represent the diseases and symptom indicators. Specifically, assuming the parameters in Table 1 were dyadics, there would exist a circuit composed of constant-fan-in elements implementing DS whose size grows linearly in the number of diseases and in the number of symptoms.

If we view the input lines as random variables, then the output lines of the logic gates are also random variables, and so we may ask: what conditional independencies hold among the circuit elements? It is straightforward to show that the circuit diagram, viewed as a directed acyclic graph, is a directed graphical model capturing conditional independencies among the inputs, outputs, and internal gates of the circuit implementing DS. For every gate, the conditional probability mass function is characterized by the (constant-size) truth table of the logical gate.

Therefore, if an efficient prior program samples from some distribution over a collection of binary random variables, then those random variables exhibit many conditional independencies, in the sense that we can introduce a polynomial number of additional boolean random variables (representing intermediate computations) such that there exists a constant-fan-in directed graphical model over all the variables with constant-size conditional probability mass functions.

In Section 5 we return to the question of whether DS is a good model. Here we conclude with a brief discussion of the history of graphical models in AI.

4.6. Graphical models and AI. Graphical models, and, in particular, directed graphical models or Bayesian networks, played a critical role in popularizing probabilistic techniques within AI in the late 1980s and early 1990s. Two developments were central to this shift. First, researchers introduced compact, computer-readable representations of distributions on large (but still finite) collections of random variables, and did so by explicitly representing a graph capturing conditional independencies and exploiting the factorization (10). Second, researchers introduced efficient graph-based algorithms that operated on these representations, exploiting the factorization to compute conditional probabilities. For the first time, a large class of distributions were given a formal representation that enabled the design of general purpose algorithms to compute useful quantities. As a result, the graphical model formalism became a lingua franca between practitioners designing large probabilistic systems, and figures depicting graphical models were commonly used to quickly communicate the essential structure of complex, but structured, distributions.

While there are sophisticated uses of Bayesian networks in cognitive science (see, e.g., [GKT08, §3]), many models are not usefully represented by a Bayesian network. In practice, this often happens when the number of variables or edges is extremely large (or infinite), but there still exists special structure that an algorithm can exploit to perform probabilistic inference efficiently. In the next three sections, we will see examples of models that are not usefully represented by Bayesian networks, but which have concise descriptions as prior programs.

§5. Hierarchical models and learning probabilities from data. The DS program makes a number of implicit assumptions that would deserve scrutiny in a real medical diagnosis setting. For example, DS models the diseases as *a priori* independent, but of course, diseases often arise in clusters, e.g., as the result of an auto-immune condition. In fact, because of the independence and the small marginal probability of each disease, there is an *a priori* bias towards mutually exclusive diagnoses as we saw in the "explaining away" effect in (6). The conditional independence of symptoms given diseases reflects an underlying casual interpretation of DS in terms of diseases *causing* symptoms. In many cases, e.g., a fever or a sore neck, this may be reasonable, while in others, e.g., insulin resistance, it may not.

Real systems that support medical diagnosis must relax the strong assumptions we have made in the simple DS model, while at the same time maintaining enough structure to admit a concise representation. In this and the next section, we show how both the structure and parameters in prior programs like DS *can be learned from data*, providing a clue as to how a mechanical mind could build predictive models of the world simply by experiencing and reacting to it.

5.1. Learning as probabilistic inference. The 95 probabilities in Table 1 eventually parameterize a distribution over 262 144 outcomes. But whence come these 95 numbers? As one might expect by studying the table of numbers, they were designed by hand to elucidate some phenomena and be vaguely plausible. In practice, these parameters would themselves be subject to a great deal of uncertainty, and one might hope to use data from actual diagnostic situations to learn appropriate values.

There are many schools of thought on how to tackle this problem, but a hierarchical Bayesian approach provides a particularly elegant solution that fits entirely within the QUERY framework. The solution is to generalize the DS program in two ways. First, rather than generating one individual's diseases and symptoms, the program will generate data for $n+1$ individuals. Second, rather than using the fixed table of probability values, the program will start by randomly generating a table of probability values, each independent and distributed uniformly at random in the unit interval, and then proceed along the same lines as DS. Let DS' stand for this generalized program.

The second generalization may sound quite surprising, and unlikely to work very well. The key is to consider the combination of the two generalizations. To complete the picture, consider a past record of n individuals and their diagnosis, represented as a (potentially partial) setting of the 18 variables $\{D_1, \ldots, D_{11}; S_1, \ldots, S_7\}$. We define a new predicate OS' that accepts the $n+1$ diagnoses generated by the generalized prior program DS' if and only if the first n agree with the historical records, and the symptoms associated with the $n+1$'st agree with the current patient's symptoms.

What are typical outputs from QUERY(DS', OS')? For very small values of n, we would not expect particularly sensible predictions, as there are many tables of probabilities that could conceivably lead to acceptance by OS'. However, as n grows, some tables are much more likely to lead to acceptance. In particular, for large n, we would expect the hypothesized marginal probability of a disease to be relatively close to the observed frequency of the disease, for otherwise, the probability of acceptance would drop. This effect grows exponentially in n, and so we would expect that the typical accepted sample would quickly correspond with a latent table of probabilities that match the historical record.

We can, in fact, work out the conditional distributions of entries in the table in light of the n historical records. First consider a disease j whose marginal probability, p_j, is modeled as a random variable sampled uniformly at random from the unit interval. The likelihood that the n sampled values of D_j match the historical record is

$$p_j^k \cdot (1 - p_j)^{n-k}, \tag{11}$$

where k stands for the number of records where disease j is present. By Bayes' theorem, in the special case of a uniform prior distribution on p_j, the density of the conditional distribution of p_j given the historical evidence is proportional to the likelihood (11). This implies that, conditionally on the historical record, p_j has a so-called Beta(α_1, α_0) distribution with mean

$$\frac{\alpha_1}{\alpha_1 + \alpha_0} = \frac{k+1}{n+2}$$

and concentration parameter $\alpha_1 + \alpha_0 = n + 2$. Figure 3 illustrates beta distributions under varying parameterizations, highlighting the fact that, as the concentration grows, the distribution begins to concentrate rapidly around its mean. As n grows, predictions made by QUERY(DS', OS') will likely be those of runs where each disease marginals p_j falls near the observed frequency

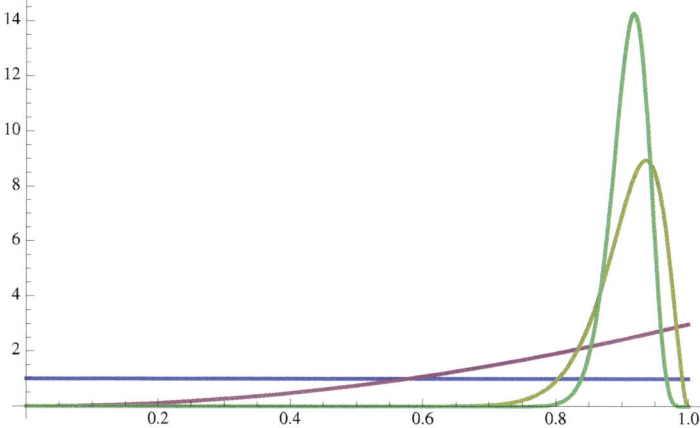

FIGURE 3. Plots of the probability density of Beta(a_1, a_0) distributions with density $f(x; \alpha_1, \alpha_0) = \frac{\Gamma(\alpha_1 + \alpha_0)}{\Gamma(\alpha_1)\Gamma(\alpha_0)} x^{\alpha_1 - 1} (1 - x)^{\alpha_0 - 1}$ for parameters $(1, 1)$, $(3, 1)$, $(30, 3)$, and $(90, 9)$ (respectively, in height). For parameters $\alpha_1, \alpha_0 > 1$, the distribution is unimodal with mean $\alpha_1/(\alpha_1 + \alpha_0)$.

of the jth disease. In effect, the historical record data *determines* the values of the marginals p_j.

A similar analysis can be made of the dynamics of the posterior distribution of the latent parameters ℓ_m and $c_{n,m}$, although this will take us too far beyond the scope of the present article. Abstractly speaking, in finite dimensional Bayesian models like this one satisfying certain regularity conditions, it is possible to show that the predictions of the model converge to those made by the best possible approximation within the model to the distribution of the data. (For a discussion of these issues, see, e.g., [Bar98].)

While the original DS program makes the same inferences in each case, DS′ learns to behave from experience. The key to this power was the introduction of the latent table of probabilities, modeled as random variables. This type of model is referred to as a *hierarchical Bayesian model*. The term "Bayesian" refers to the fact that we have modeled our uncertainty about the unknown probabilities by making them random and specifying a distribution that reflects our subjective uncertainty, rather than a frequency in a large random sample of patients. The term "hierarchy" refers to the fact that in the graphical model representing the program, there is yet another level of random variables (the table of probabilities) sitting above the rest of the original graphical model. More complicated models may have many more layers, or potentially even an infinite number of layers.

An interesting observation is that DS′ is even more compact than DS, as the specification of the distribution of the random table is logarithmic in the size of the table. On the other hand, DS′ relies on data to help it reduce its substantial *a priori* uncertainty regarding these values. This tradeoff—between, on the one hand, the flexibility and complexity of a model and, on the other, the amount of data required in order to make sensible predictions—is seen throughout statistical modeling. We will return to this point in Section 6.3.

Here we have seen how the parameters in prior programs can be modeled as random, and thereby learned from data by conditioning on historical diagnoses. In the next section, we consider the problem of learning not only the parameterization but the structure of the model's conditional independence itself.

§6. **Random structure.** Irrespective of how much historical data we have, DS′ cannot go beyond the conditional independence assumptions implicit in the structure of the prior program. Just as we framed the problem of learning the table of probabilities as a probabilistic inference over a random table, we can frame the problem of identifying the correct structure of the dependence between symptoms and disease as one of probabilistic inference over random conditional independence *structure* among the model variables.

In Section 4.4, we saw that conditional independence relationships among a collection of random variables can be captured by a directed acyclic graph. The approach we will discuss involves treating this graph as a random variable, whose distribution reflects our uncertainty about the statistical dependence among the diseases and symptoms before seeing data, and whose posterior distribution reflects our updated uncertainty about these relationships once the graph is forced to explain any evidence of dependence or independence in the historical data.

The model that we describe in this section introduces several additional layers and many more latent variables. Outside of the Bayesian framework, these latent variables would typically be additional parameters that one would tune to fit the data. Typically, when one adds more parameters to a model, this improves the fit to the data at hand, but introduces a risk of "overfitting", which leads to poor predictive performance on unseen data. However, as we will see in Section 6.3, the problem of overfitting is mitigated in this Bayesian approach, because the latent variables are not optimized, but rather sampled conditionally.

6.1. Learning structure as probabilistic inference. Within AI and machine learning, the problem of learning a probabilistic model from data is a quintessential example of *unsupervised learning*, and the approach of identifying a graph capturing conditional independence relationships among model variables is known as *structure learning*.

In Section 4.4 we saw that every distribution on n binary random variables $X_1, \ldots, X_n$ can be expressed in the form

$$p(X_1, \ldots, X_k) = \prod_{j=1}^{k} p_j(X_j \mid \mathrm{Pa}(X_j)) \tag{12}$$

where G is a directed acyclic graph over the set $\{X_1, \ldots, X_k\}$ of model variables; $\mathrm{Pa}(X_j)$ denotes the parent vertices of X_j; and the $p_j(\cdot \mid \cdot)$ are conditional probability mass functions specifying the distribution of each variable in terms of its parents' values.

From the perspective of this factorization, the tack we took in Section 5.1 was to assume that we knew the graphical structure G (given by DS) and learn (the parameterization of) the conditional mass functions by modeling them as random variables. We will now consider learning both ingredients simultaneously, and later pause to critique this strategy.

6.2. A random probability distribution. Let us return to the setting of medical diagnosis, and in particular the problem of modeling the presence/absence of the 11 diseases and 7 symptoms, represented by the variables $\{D_1, \ldots, D_{11}; S_1, \ldots, S_7\}$.

Towards this end, and with the factorization (12) in mind, consider a prior program, which we will call RPD (for *Random Probability Distribution*), that takes as input two positive integers n and D and produces as output n independent samples from a random probability distribution on $\{0, 1\}^D$.

Intuitively, RPD works in the following way: First, RPD generates a random directed acyclic graph G with D vertices. Next, it generates a *random probability mass function* p, which will specify a distribution over D random variables, $X_1, \ldots, X_D$. The probability mass function will be generated so that it satisfies the conditional independencies implied by the graph G when it is viewed as a directed graphical model. The probability mass function p is generated by choosing random conditional probability mass functions $p(X_j \mid \mathrm{Pa}(X_j))$, one for each variable X_j as in the factorization (12). Specifically, if a variable X_j has k parents $\mathrm{Pa}(X_j)$ (which collectively can take on 2^k possible $\{0, 1\}$-assignments), then we must generate 2^k probabilities, one for each $\{0, 1\}$-assignment v of the parents, indicating the probability $p_{j|v}$ that $X_j = 1$ given that $\mathrm{Pa}(X_j) = v$. In particular, $p_j(1|v) = p_{j|v}$. This fully determines p. RPD then proceeds to generate n samples from p, each a list of D binary values with the same distributions as $X_1, \ldots, X_D$.

More formally, RPD begins by sampling a directed acyclic graph G uniformly at random from the set $\mathcal{G}_D$ of all directed acyclic graphs over the vertex set $\{X_1, \ldots, X_D\}$. For every vertex j and every $\{0, 1\}$-assignment v to X_i's parents $\mathrm{Pa}(X_j)$, we sample a probability value $p_{j|v}$ uniformly at random from $[0, 1]$. Let $j_1, \ldots, j_D$ be a topological ordering of the vertices of G. We then repeat the following procedure n times: First, sample $X_{j_1} \in \{0, 1\}$ with mean $p_{j_1|()}$, and then for $i = 2, \ldots, D$, sample $X_{j_i} \in \{0, 1\}$ with mean $p_{j_i|v}$ where

$v = (X_p \colon p \in \mathrm{Pa}(j_i))$ is the $\{0, 1\}$-assignment of X_j's parents. We then output the variables in order $X_1, \ldots, X_D$, and repeat until we have produced n such samples as output.

With RPD fully specified, let us now consider the output of

$$\mathrm{QUERY}(\mathrm{RPD}(n + 1, 18), \mathrm{OS}') \tag{13}$$

where OS' is defined as in Section 5.1, accepting $n + 1$ diagnoses if and only if the first n agree with historical records, and the symptoms associated with the $n + 1$st agree with the current patient's symptoms. (Note that we are identifying each output $(X_1, \ldots, X_{11}, X_{12}, \ldots, X_{18})$ with a diagnosis $(D_1, \ldots, D_{11}, S_1, \ldots, S_7)$, and have done so in order to highlight the generality of RPD.)

As a first step in understanding RPD, one can show that, conditioned on the graph G, the conditional independence structure of each of its n outputs $(X_1, \ldots, X_D)$ is precisely captured by G, when viewed as a Bayesian network (i.e., the distribution of the X's satisfies the factorization (12)). It is then not hard to see that the probabilities $p_{j|v}$ parameterize the conditional probability mass functions, in the sense that $p(X_j = 1 \mid \mathrm{Pa}(X_j) = v) = p_{j|v}$. Our goal over the remainder of the section will be to elucidate the posterior distribution on the graph and its parameterization, in light of historical data.

To begin, we assume that we know the graph G for a particular output from (13), and then study the likely values of the probabilities $p_{j|v}$ conditioned on the graph G. Given the simple uniform prior distributions, we can in fact derive analytical expressions for the posterior distribution of the probabilities $p_{j|v}$ directly, conditioned on historical data and the particular graph structure G. In much the same was as our analysis in Section 5.1, it is easy to show that the expected value of $p_{j|v}$ on those runs accepted by QUERY is

$$\frac{k_{j|v} + 1}{n_{j|v} + 2}$$

where $n_{j|v}$ is the number of times in the historical data where the pattern $\mathrm{Pa}(X_j) = v$ arises; and $k_{j|v}$ is the number of times when, moreover, $X_j = 1$. This is simply the "smoothed" empirical frequency of the event $X_j = 1$ given $\mathrm{Pa}(X_j) = v$. In fact, the $p_{j|v}$ are conditionally Beta distributed with concentration $n_{j|v} + 2$. Under an assumption that the historical data are conditionally independent and identically distributed according to a measure P, it follows by a law of large numbers argument that these probabilities converge almost surely to the underlying conditional probability $P(X_j = 1 \mid \mathrm{Pa}(X_j) = v)$ as $n \to \infty$.

The variance of these probabilities is one characterization of our uncertainty, and for each probability $p_{j|v}$, the variance is easily shown to scale as $n_{j|v}^{-1}$, i.e., the number of times in the historical data when $\mathrm{Pa}(X_j) = v$.

Informally, this suggests that, the smaller the parental sets (a property of G), the more certain we are likely to be regarding the correct parameterization, and, in terms of QUERY, the smaller the range of values of $p_{j|v}$ we will expect to see on accepted runs. This is our first glimpse at a subtle balance between the simplicity of the graph G and how well it captures hidden structure in the data.

6.3. Aspects of the posterior distribution of the graphical structure. The space of directed acyclic graphs on 18 variables is enormous, and computational hardness results [Coo90, DL93, CSH08] imply there will be no simple way to summarize the structure of the posterior distribution, at least not one that suggests an efficient method in general for choosing structures with high posterior probability. It also goes without saying that one should not expect the PTM defined by (13) to halt within a reasonable time for any appreciable value of n because the probability of generating the structure that fits the data is astronomically small. However it is still instructive to understand the conceptual structure of the posterior distribution of the graph G. On the one hand, there are algorithms that operate quite differently from the naive mechanism of QUERY and work reasonably well in practice at approximating the task defined here, despite hardness results. There are also more restricted, but still interesting, versions of this task for which there exist algorithms that work remarkably well in practice and sometimes provably so [BJ03].

On the other hand, this example is worth studying because it reveals an important aspect of some hierarchical Bayesian models with regard to their ability to avoid "overfitting", and gives some insight into why we might expect "simpler" explanations/theories to win out in the short term over more complex ones.

Consider the set of probability distributions of the form (12) for a particular graph G. We will refer to these simply as the *models in G* when there is no risk of confusion. The first observation to make is that if a graph G is a strict subgraph of another graph G' on the same vertex set, then the set of models in G is a strict subset of those in G'. It follows that, no matter the data set, the best-fitting probability distribution corresponding with G' will be no worse than the best-fitting model in G. Given this observation, one might guess that samples from (13) would be more likely to come from models whose graphs have more edges, as such graphs always contain a model that fits the historical data better.

However, the truth is more subtle. Another key observation is that the posterior probability of a particular graph G does not reflect the best-fitting model in G, but rather reflects the *average* ability of models in G to explain the historical data. In particular, this average is over the random parameterizations $p_{j|v}$ of the conditional probability mass functions. Informally speaking, if a spurious edge exists in a graph G', a typical distribution from G' is less likely to explain the data than a typical distribution from the graph with that edge removed.

In order to characterize the posterior distribution of the graph, we can study the likelihood that a sample from the prior program is accepted, assuming that it begins by sampling a particular graph G. We begin by focusing on the use of each particular probability $p_{j|v}$, and note that every time the pattern $\mathrm{Pa}(X_j) = v$ arises in the historical data, the generative process produces the historical value X_j with probability $p_{j|v}$ if $X_j = 1$ and $1 - p_{j|v}$ if $X_j = 0$. It follows that the probability that the generative process, having chosen graph G and parameters $\{p_{j|v}\}$, proceeds to produce the historical data is

$$\prod_{j=1}^{D} \prod_{v} p_{j|v}^{k_{j|v}} (1 - p_{j|v})^{n_{j|v} - k_{j|v}}, \tag{14}$$

where v ranges over the possible $\{0, 1\}$ assignments to $\mathrm{Pa}(X_j)$ and $k_{j|v}$ and $n_{j|v}$ are defined as above. In order to determine the probability that the generative process produces the historical data (and thus is accepted), assuming only that it has chosen graph G, we must take the expected value of (14) with respect to the uniform probability measure on the parameters, i.e., we must calculate the marginal probability of the historical data conditioned the graph G. Given the independence of the parameters, it is straightforward to show that this expectation is

$$\mathrm{score}(G) := \prod_{j=1}^{D} \prod_{v} (n_{j|v} + 1)^{-1} \binom{n_{j|v}}{k_{j|v}}^{-1}. \tag{15}$$

Because the graph G was chosen uniformly at random, it follows that the posterior probability of a particular graph G is proportional to $\mathrm{score}(G)$.

We can study the preference for one graph G over another G' by studying the ratio of their scores:

$$\frac{\mathrm{score}(G)}{\mathrm{score}(G')}.$$

This score ratio is known as the *Bayes factor*, which Good termed the *Bayes–Jeffreys–Turing factor* [Goo68, Goo75], and which Turing himself called the *factor in favor of a hypothesis* (see [Goo68], [Zab12, §1.4], and [Tur12]). Its logarithm is sometimes known as the *weight of evidence* [Goo68]. The form of (15), a product over the local structure of the graph, reveals that the Bayes factor will depend only on those parts of the graphs G and G' that differ from each other.

Consider the following simplified scenario, which captures several features of learning structure from data: Fix two graphs, G and G', over the same collection of random variables, but assume that in G, two of these random variables, X and Y, have no parents and are thus independent, and in G' there

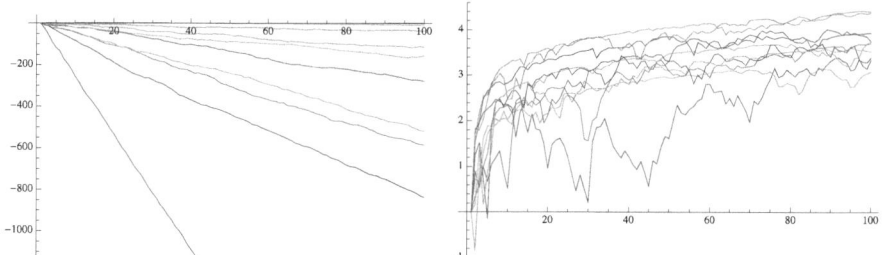

FIGURE 4. Weight of evidence for independence versus dependence (positive values support independence) of a sequence of pairs of random variables sampled from RPD($n, 2$). (left) When presented with data from a distribution where (X, Y) are indeed dependent, the weight of evidence rapidly accumulates for the dependent model, at an asymptotically linear rate in the amount of data. (right) When presented with data from a distribution where (X, Y) are independent, the weight of evidence slowly accumulates for the independent model, at an asymptotic rate that is logarithmic in the amount of data. Note that the dependent model can imitate the independent model, but, on average over random parameterizations of the conditional probability mass functions, the dependent model is worse at modeling independent data.

is an edge from X to Y, and so they are almost surely dependent. From (15), we may deduce that the score ratio is

$$\frac{(n_1 + 1)(n_0 + 1)}{(n + 1)} \frac{\binom{n_1}{k_1}\binom{n_0}{k_0}}{\binom{n}{k}}, \quad (16)$$

where n counts the total number of observations; k counts $Y = 1$; n_1 counts $X = 1$; k_1 counts $X = 1$ and $Y = 1$; n_0 counts $X = 0$; and k_0 counts $X = 0$ and $Y = 1$. In order to understand how the Bayes factor (16) for graph G behaves, let us first consider the case where G' is the true underlying graph, i.e., when Y is indeed dependent on X. Using the law of large numbers, and Stirling's approximation, we can reason that the evidence for G' accumulates rapidly, satisfying

$$\log \frac{\text{score}(G)}{\text{score}(G')} \sim -C \cdot n, \quad \text{a.s.,}$$

for some constant $C > 0$ that depends only on the joint distribution of X and Y. As a concrete example, when X and Y have mean $\frac{1}{2}$, the constant is given by

$$\log \frac{(1-d)^{d-\frac{1}{2}}}{(1+d)^{d+\frac{1}{2}}},$$

where $d = \mathbb{P}\{Y = 1 \mid X = 1\} = 1 - \mathbb{P}\{Y = 1 \mid X = 0\}$. For example, $C \to 0$ as $d \downarrow 0$; $C \approx 0.13$ when $d = 1/2$; and C achieves its maximum, $\log 2$, as $d \uparrow 1$. The first plot in Figure 4 shows the progression of the weight of evidence when data is drawn from distributions generated uniformly at random to satisfy the conditional independencies captured by G'. As predicted, the evidence rapidly accumulates at a linear rate in favor of G'.

On the other hand, when G is the true underlying graph and Y is independent and X, one can show using similar techniques to above that

$$\log \frac{\text{score}(G)}{\text{score}(G')} \sim \frac{1}{2} \log n, \quad \text{a.s.}$$

The second plot in Figure 4 shows the progression of the weight of evidence when data is drawn from distributions generated uniformly at random to satisfy the conditional independencies captured by G. As predicted, the evidence accumulates, but at a much slower logarithmic rate.

In both cases, evidence accumulates for the correct model. In fact, it can be shown that the expected weight of evidence is always non-negative for the true hypothesis, a result due to Turing himself [Goo91, p. 93]. Because the prior probabilities for each graph are fixed and do not vary with the amount of data, the weight of evidence will eventually eclipse any prior information and determine the posterior probability. On the other hand, as we have seen, the evidence accumulates rapidly for dependence and much more slowly for independence and so we might choose our prior distribution to reflect this imbalance, preferring graphs with fewer edges *a priori*.[2]

6.4. Bayes' Occam's razor. In the example above when X and Y are independent, we see that evidence accumulates for the simpler graph over the more complex graph, despite the fact that there is almost always a parameterization of the more complex graph that assigns a higher likelihood to the data than any parametrization of the simpler graph. This phenomenon is known as Bayes' Occam's razor [Mac03, Ch. 28], and it represents a natural way in which hierarchical models like RPD with several layers—a random graph, random conditional probability mass functions generated given the graph, and finally, the random data generated given the graph and the conditional probability mass functions—end up choosing models with intermediate complexity.

One way to understand this phenomenon is to note that, if a model has many degrees of freedom, then each configuration must be assigned, on average, less probability than it would under a simpler model with fewer degrees of freedom. Here, a graph with additional edges has more degrees of freedom, and while it can represent a strictly larger set of distributions than a simpler graph, a distribution with simple graphical structure G is assigned greater probability

---

[2]This analysis in terms of Bayes factors also aligns well with experimental findings on human judgments of evidence for causal structure (see, e.g., [GT05]).

density under G than under a more complex graph. Thus, if fewer degrees of freedom suffice, the simpler model is preferred.

We can apply this same perspective to DS, DS' and RPD: The RPD model has many more degrees of freedom than both DS and DS'. In particular, given enough data, RPD can fit any distribution on a finite collection of binary variables, as opposed to DS', which cannot because it makes strong and immutable assumptions. On the other hand, with only a small amount of training data, one would expect the RPD model to have high posterior uncertainty. Indeed, one would expect much better predictions from DS' versus RPD, if both were fed data generated by DS, especially in the low-data regime.

An important research problem is bridging the gap between RPD and DS'. Whereas DS' makes an immutable choice for one particular structure, RPD assumes *a priori* that every graphical structure is equally likely to explain the data. If, instead, one were uncertain about the structure but expected to find some particular regular pattern in the graphical structure, one could define an alternative model RPD' that placed a non-uniform distribution on the graph, favoring such patterns, and one could then expect better predictions when that pattern was indeed present in the data. However, one often does not know exactly which pattern might arise. But in this case, we can take the same step we took when defining RPD and consider a *random* pattern, drawn from some space of possible patterns. This would constitute an additional level to the hierarchical model. Examples of this idea are described by Mansinghka et al. [MKTG06] and Kemp et al. [KSBT07], and this technique constitutes one aspect of the general approach of "theory-based Bayesian models" [GT06, TGK06, GKT08, KT08, GT09].

Up until this point, we have considered the problem of reasoning and representing our own uncertainty in light of evidence. However, in practice, representations of uncertainty are often useful because they support decision making under uncertainty. In the next section, we show how the QUERY framework can be used to turn models of our uncertainty, including models of the effects of our own actions, into decisions.

§7. Making decisions under uncertainty. Until now, we have discussed how computational processes can represent uncertain knowledge, and how these processes can be transformed using QUERY to reflect our uncertainty after incorporating new evidence. In this section, we consider the problem of making decisions under uncertainty, which will require us to reason not only about the immediate effects of the actions we take, but also about future decisions and the uncertainty we are likely to face when making them. In the end, we will give a recursive characterization of an approximately optimal action, and show how this relates to simple feedback rules that Turing himself proposed.

The solution we describe models decisions as random variables and decision making as sampling. Using PTMs and QUERY, we construct distributions over actions a decision maker might take after reasoning about the effects of those actions and their likely success in achieving some goal or objective. The particular distributions we will construct are based in part on the exponentiated choice rule introduced by Luce [Luc59, Luc77] in the context of modeling human choices.

Our presentation extends that for the "fully observable" case given by Goodman, Mansinghka, Roy, Bonawitz, and Tenenbaum [GMRBT08] and Mansinghka [Man09, §2.2.3]. In particular, recursion plays a fundamental role in our solution, and thus pushes us beyond the representational capacity of many formalisms for expressing complex probabilistic models. Even more so than earlier sections, the computational processes we define with QUERY will not be serious proposals for *algorithms*, although they will define distributions for which we might seek to implement approximate inference. However, those familiar with traditional presentations may be surprised by the ease with which we move between problems often tackled by distinct formalisms and indeed, this is a common feature of the QUERY perspective.

7.1. Diagnosis and Treatment. Returning to our medical diagnosis theme, consider a doctor faced with choosing between one or more treatment plans. What recommendation should they give to the patient and how might we build a system to make similar choices?

In particular, imagine a patient in need of an organ transplant and the question of whether to wait for a human donor or use a synthetic organ. There are a number of sources of uncertainty to contend with: While waiting for a human donor, the patient may become too ill for surgery, risking death. On the other hand, the time before a human organ becomes available would itself be subject to uncertainty. There is also uncertainty involved post-operation: Will the organ be accepted by the patient's immune system without complication? How long should the patient expect to live in both conditions, taking into consideration the deterioration that one should expect if the patient waits quite a while before undergoing a transplant?

This situation is quite a bit more complicated. The decision as to whether to wait changes daily as new evidence accumulates, and how one should act today depends implicitly on the possible states of uncertainty one might face in the future and the decisions one would take in those situations. As we will see, we can use QUERY, along with models of our uncertainty, to make decisions in such complex situations.

7.2. A single decision. We begin with the problem of making a single decision between two alternative treatments. What we have at our disposal are two simulations SIM_x and SIM_y capturing our uncertainty as to the effects of those treatments. These could be arbitrarily complicated computational processes, but in the end we will expect them to produce an output 1 indicating that the resulting simulation of treatment was successful/acceptable and a 0 otherwise.[3] In this way, both simulations act like predicates, and so we will say that a simulation accepts when it outputs 1, and rejects otherwise. In this section we demonstrate the use of PTMs and QUERY to define distributions over *actions*—here, treatments—that are likely to lead to successful outcomes.

Let RT (for *Random Treatment*) be the program that chooses a treatment $Z \in \{x, y\}$ uniformly at random and consider the output of the program

$$\mathsf{CT}^* = \mathsf{QUERY}(\mathsf{RT}, \mathsf{SIM}_Z),$$

where SIM_Z is interpreted as the program that simulates SIM_z when RT outputs $Z = z \in \{x, y\}$. The output of CT^* is a treatment, x or y, and we will proceed by interpreting this output as the treatment chosen by some decision maker.

With a view towards analyzing CT^* (named for *Choose Treatment*), let p_z be the probability that SIM_z accepts (i.e., p_z is the probability that treatment z succeeds), and assume that $p_x > p_y$. Then CT^* "chooses" treatment x with probability

$$\frac{p_x}{p_x + p_y} = \frac{\rho}{\rho + 1}$$

where $\rho := p_x/p_y$ and $\rho > 1$ by assumption. It follows that CT^* is more likely to choose treatment x and that the strength of this preference is controlled by the multiplicative ratio ρ (hence the multiplication symbol $*$ in CT^*). If $\rho \gg 1$, then treatment x is chosen essentially every time.

On the other hand, even if treatment x is twice as likely to succeed, CT^* still chooses treatment y with probability $1/3$. One might think that a person making this decision should always choose treatment x. However, it should not be surprising that CT^* does not represent this behavior because it accepts a proposed action solely on the basis of a single successful simulation of its outcome. With that in mind, let $k \geq 0$ be an integer, and consider the program

$$\mathsf{CT}^*_k = \mathsf{QUERY}(\mathsf{RT}, \mathsf{REPEAT}(k, \mathsf{SIM}_Z)),$$

where the machine REPEAT on input k and SIM_Z accepts if k independent simulations of SIM_Z all accept. The probability that CT^*_k chooses treatment x is

$$\frac{\rho^k}{\rho^k + 1}$$

---

[3] Our discussion is couched in terms of successful/unsuccessful outcomes, rather than in terms of a real-valued loss (as is standard in classical decision theory). However, it is possible to move between these two formalisms with additional hypotheses, e.g., boundedness and continuity of the loss. See [THS06] for one such approach.

and so a small multiplicative difference between p_x and p_y is exponentiated, and CT_k^* chooses the more successful treatment with all but vanishing probability as $k \to \infty$. (See the left plot in Figure 5.) Indeed, in the limit, CT_∞^* would always choose treatment x as we assumed $p_x > p_y$. (For every k, this is the well-known exponentiated Luce choice rule [Luc59, Luc77].)

The behavior of CT_∞^* agrees with that of a classical decision-theoretic approach, where, roughly speaking, one fixes a loss function over possible outcomes and seeks the action minimizing the expected loss. (See [DeG05] for an excellent resource on statistical decision theory.) On the other hand, classical decision theory, at least when applied naively, often fails to explain human performance: It is well-documented that human behavior does not agree with mathematically "optimal" behavior with respect to straightforward formalizations of decision problems that humans face. (See Camerer [Cam11] for a discussion of models of actual human performance in strategic situations.)

While we are not addressing the question of designing efficient algorithms, there is also evidence that seeking optimal solutions leads to computational intractability. (Indeed, PSPACE-hardness [PT87] and even undecidability [MHC03] can arise in the general case of certain standard formulations.)

Some have argued that human behavior is better understood in terms of a large collection of heuristics. For example, Goldstein and Gigerenzer [GG02] propose the "recognition heuristic", which says that when two objects are presented to a human subject and only one is recognized, that the recognized object is deemed to be of greater intrinsic value to the task at hand. The problem with such explanations of human behavior (and approaches to algorithms for AI) is that they often do not explain how these heuristic arise. Indeed, a theory for how such heuristics are learned would be a more concise and explanatory description of the heuristics than the heuristics themselves. A fruitful approach, and one that meshes with our presentation here, is to explain heuristic behavior as arising from approximations to rational behavior, perhaps necessitated by intrinsic computational hardness. Human behavior would then give us clues as to which approximations are often successful in practice and likely to be useful for algorithms.

In a sense, CT_k^* could be such a model for approximately optimal behavior. However, since CT_k^* chooses an action on the basis of the ratio $\rho = p_x/p_y$, one problematic feature is its sensitivity to small differences $|p_x - p_y|$ in the absolute probabilities of success when $p_x, p_y \ll 1$. Clearly, for most decisions, a $1/10\,000$ likelihood of success is not appreciably better than a $1/20\,000$ likelihood. A result by Dagum, Karp, Luby and Ross [DKLR00] on estimating small probabilities to high relative accuracy suggests that this sensitivity in CT_k^* might be a potential source of computational hardness in efforts to design algorithms, not to mention a point of disagreement with

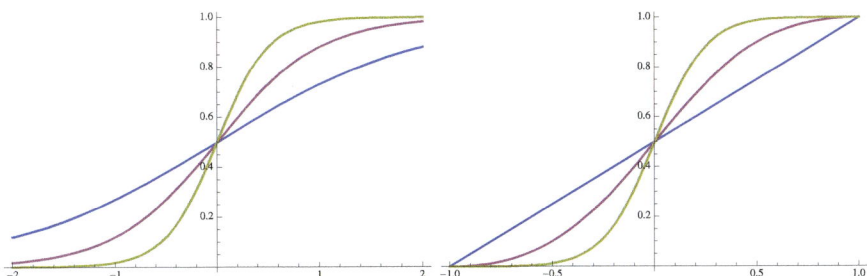

FIGURE 5. Plots of the sigmoidal curves arising from the probability of treatment x under CT^*_k (left) and CT^+_k (right) as a function of the log probability difference $\log \rho = \log p_x - \log p_y$ and probability difference $\alpha = p_x - p_y$, respectively, between the treatments. Here $k \in \{1, 2, 4\}$. The straight line on the right corresponds to $k = 1$, and curvature increases with k.

human behavior. It stands to reason that it may be worthwhile to seek models that do not exhibit this sensitivity.

To this end, let SIM_x and SIM_y be independent simulations and consider the predicate MAJ_x (named for *Majority*) that accepts if SIM_x succeeds and SIM_y fails, rejects in the opposite situation, and chooses to accept or reject uniformly at random otherwise. We then define

$$\mathsf{CT}^+ = \mathsf{QUERY}(\mathsf{RT}, \mathsf{MAJ}_Z).$$

It is straightforward to show that treatment x is chosen with probability

$$\frac{1 + (p_x - p_y)}{2} = \frac{1 + \alpha}{2}$$

and so the output of CT^+ is sensitive to only the additive difference $\alpha = p_x - p_y$ (hence the addition symbol $+$ in CT^+). In particular, when $p_x \approx p_y$, CT^+ chooses an action nearly uniformly at random. Unlike CT^*, it is the case that CT^+ is insensitive to the ratio p_x/p_y when $p_x, p_y \approx 0$.

Similarly to CT^*_k, we may define CT^+_k by

$$\mathsf{CT}^+_k = \mathsf{QUERY}(\mathsf{RT}, \mathsf{REPEAT}(k, \mathsf{MAJ}_Z)),$$

in which case it follows that CT^+_k accepts treatment x with probability

$$\frac{1}{1 + \left(\frac{1-\alpha}{1+\alpha}\right)^k}.$$

Figure 5 shows how this varies as a function of α for several values of k. Again, as $k \to \infty$, the decision concentrates on the treatment with the greatest probability of success, although there is always a region around $\alpha = 0$ where each treatment is almost equally likely to be chosen.

In this section, we have shown how a model of one's uncertainty about the likely success of a single action can be used to produce a distribution over actions that concentrates on actions likely to achieve success. In the next section, we will see how the situation changes when we face multiple decisions. There, the likely success of an action depends on future actions, which in turn depend on the likely success of yet-more-distant actions. Using recursion, we can extend strategies for single decisions to multiple decisions, defining distributions over sequences of actions that, under appropriate limits, agree with notions from classical decision theory, but also suggest notions of approximately optimal behavior.

7.3. Sequences of decisions. How can we make a sequence of good decisions over time, given a model of their effects? Naively, we might proceed along the same lines as we did in the previous section, sampling now a *sequence* of actions conditioned on, e.g., a simulation of these actions leading to a successful outcome. Unfortunately, this does not lead to a sensible notion of approximately optimal behavior, as the first action is chosen on the basis of a fixed sequence of subsequent actions that do not adapt to new observations. Certainly one should react differently when a door leads not to the next room but to a closet!

In order to recover classical notions of optimality under an appropriate limit, we need to evaluate exhaustive plans—called *policies* in the planning literature—that specify how to act in every conceivable situation that could arise. Indeed, the optimal action to take at any moment is that which, when followed by optimal behavior thereafter, maximizes the probability of a successful outcome. As one might expect, the self-referential nature of optimal behavior will lead us to recursive definitions.

Returning to our transplant scenario, each day we are faced with several possible options: Waiting another day for a human donor match; running further diagnostic tests; choosing to go with the synthetic organ; etc. As time passes, observations affect our uncertainty. Observations might include the result of a diagnostic test, the appearance of the patient, how they feel, news about potential donor matches, etc. Underlying these observations (indeed, *generating* these observations) are a network of processes: the dynamics of the patient's organ systems, the biophysical mechanisms underlying the diagnostic tests included in our observations, the sociodynamics of the national donor list, etc.

Observations are the channel through which we can make inferences about the state of the underlying processes and, by reducing our uncertainty, make better decisions. On the other hand, our actions (or inaction) will influence the evolution of the underlying latent processes, and so, in order to choose good actions, we must reason not only about future observations (including eventual success or failure) but our own future actions.

While their may be many details in any particular sequential decision task, we can abstract away nearly all of them. In particular, at any point, the sequence of observations and actions that have transpired constitutes our *belief state*, and our model of the underlying latent processes and the effects of our actions boils down to a description of how our belief state evolves as we take actions and make observations. More concretely, a model for a sequential decision task is captured by a PTM NEXTSTATE, which takes a belief state and an action as input and returns the new, random belief state arising from making an additional observation. Certain belief states are *terminal* and correspond either with a successful or unsuccessful outcome.

The internal details of NEXTSTATE can be arbitrarily complex, potentially representing faithful attempts at simulating the types of processes listed above, e.g., employing detailed models of physiology, diagnostic techniques, the typical progression of the national donor list, success rates of organ transplants, life expectancy under various conditions, etc. Our goal is to transform the computational process NEXTSTATE characterizing the sequential decision task into a computational process representing a *stochastic* policy that maps belief states to *distributions on* actions that have a high probability of success.

To begin, we describe a PTM OUTCOME, which uniformly in a belief state b and index π for a stochastic policy, simulates an outcome resulting from following the policy π, starting from a belief state b. We may describe the behavior of OUTCOME inductively as follows: First, a check is made as to whether b is a terminal belief state. If so, the machine halts and returns 1 (accept) or 0 (reject) depending on whether b represents a successful or unsuccessful outcome, respectively. Otherwise, OUTCOME evaluates $\pi(b)$, producing a random action a, and then performs an independent simulation of NEXTSTATE(b, a) to produce a new belief state b', at which point the process repeats anew. In order to simplify the analysis, we will make the following assumptions: First, we will assume that there is some positive integer M such that, for every belief state b and policy π, we have that OUTCOME(b, π) halts within M iterations. Second, for every non-terminal belief state b and policy π, we will assume that OUTCOME(b, π) accepts with positive probability.

We can now cast the problem of choosing the first of a sequence of actions into the single decision framework as follows: Let SIM$_{b,\pi,z}$ be the PTM that, uniformly in a belief state b, index π for a stochastic policy, and action z, simulates NEXTSTATE(b, z), producing an updated belief state b', and then simulates OUTCOME(b', π), randomly accepting or rejecting depending on whether the simulation of the policy π starting from b' resulted in a successful outcome or not. Let RA be a PTM that samples an action uniformly at random. Then

$$\mathsf{ACT}(b, \pi) := \mathsf{QUERY}(\mathsf{RA}, \mathsf{SIM}_{b,\pi,Z}) \tag{17}$$

represents a distribution on actions that concentrates more mass on the action leading to a higher probability of success under the policy π. Here $\text{SIM}_{b,\pi,z}$ plays a role similar to that played by SIM_z in the single decision framework described earlier. The two additional inputs are needed because we must assign an action (or more accurately, a distribution over actions) to every belief state and policy. As before, we can amplify our preference for the action having the higher probability of success by asking for k simulations to succeed.

In order to determine a complete policy, we must specify the policy π that governs future behavior. The key idea is to choose actions in the future according to (17) as well, and we can implement this idea using recursion. In particular, by Kleene's recursion theorem, there is a PTM POLICY satisfying

$$\text{POLICY}(b) = \text{ACT}(b, \text{POLICY}). \tag{18}$$

The simplifying assumptions we made above are enough to guarantee that POLICY halts with probability one on every belief state. Those familiar with Bellman's "principle of optimality" [Bel57] will notice that (17) and (18) are related to the value iteration algorithm for Markov decision processes [How60].

In order to understand the behavior of POLICY, consider the following simple scenario: A patient may or may not have a particularly serious disease, but if they do, an special injection will save them. On the other hand, if a patient is given the same injection but does not have the condition, there is a good chance of dying from the injection itself. Luckily, there is a diagnostic test that reliably detects the condition. How would POLICY behave?

More concretely, we will assume that the patient is sick with the disease with probability 0.5, and that, if the patient is sick, the injection will succeed in curing the patient with probability 0.95, but will kill them with probability 0.8 if they are not sick with the disease. If the patient is sick, waiting things out is likely to succeed with probability 0.1. Finally, the diagnostic is accurate with probability 0.75. In Figure 6, we present the corresponding belief state transition diagram capturing the behavior of NEXTSTATE.

In order to understand the behavior of POLICY at the initial belief state ε, we begin by studying its behavior after receiving test results. In particular, having received a negative test result, the WAIT action succeeds with probability 0.91 and the INJECT action succeeds with probability 0.275, and so ACT(Negative Test, POLICY) chooses to WAIT with probability ≈ 0.77. (As $k \to \infty$, WAIT is asymptotically almost always chosen.) On the other hand, having received a positive test result, ACT(Positive Test, POLICY) chooses to WAIT with probability ≈ 0.18. (Likewise, as $k \to \infty$, INJECT is asymptotically almost always chosen.) These values imply that $\text{SIM}_{\varepsilon,\text{POLICY},\text{TEST}}$ accepts with probability ≈ 0.76, and so ACT(ε, POLICY) assigns probability 0.40, 0.29, and 0.31 to TEST, WAIT, and INJECT, respectively. As $k \to \infty$, we see that we would asymptotically almost always choose

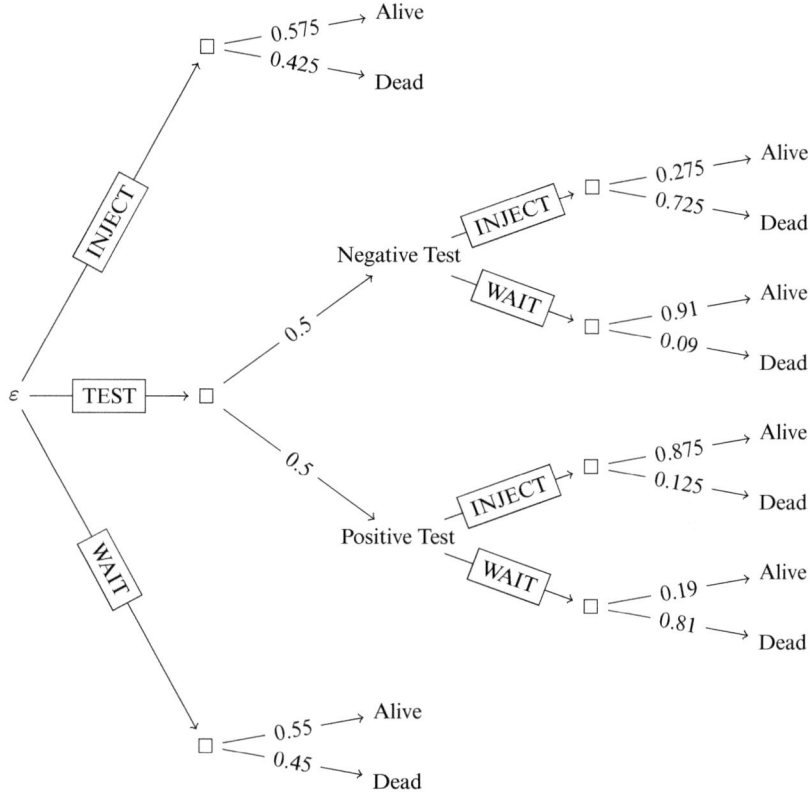

FIGURE 6. A visualization of NEXTSTATE for a simple diagnostic scenario. The initial belief state, ε, corresponds with the distribution assigning equal probability to the patient having the disease and not having the disease. Edges corresponding with actions are labeled with the name of the action in a box. For example, three actions are initially available: *waiting* it out, running a diagnostic *test*, and deciding to administer the *injection*. Square nodes (□) represent the random belief state transitions that follow an action. Edges leaving these nodes are labeled with the probabilities of the transitions.

to run a diagnostic test initially, as it significantly reduces our uncertainty. This simple example only hints at the range of behaviors that can arise from POLICY in complicated sequential decision tasks.

In the definition of ACT, we choose a random action conditioned on future success. Although we can find policies that optimize the probability of success by taking $k \to \infty$, there are good reasons to not use a random action and instead use an initial policy that incorporates one's prior knowledge about the optimal action, much like it is advantageous to use a so-called admissible heuristic in search algorithms like A^*. It is interesting to note that if the initial policy succeeds with very high probability then the QUERY expression above, viewed as algorithms, is even efficient. Returning to a quote in Section 1.1, Turing himself noted that knowing which actions to try first would "make the difference between a brilliant and a footling reasoner" and that this knowledge might be "produced by the machine itself, *e.g.*, by scientific induction" [Tur50, p. 458]. Indeed, recent work in Bayesian reinforcement learning has demonstrated the utility of more informative prior distributions on policies [DWRT10, WGRKT11].

7.4. Turing's insights. There are a wide range of algorithms that have been developed to make decisions under uncertainty. Those familiar with economics and AI may see the connection between our analysis and the Markov Decision Process (MDP) formalism [How60], where the decision maker has full observability, and the Partially Observable MDP (POMDP) formalism, where the agent, like above, has access to only part of the state (see [Mon82] for a classic survey and [KLC98] for an early AI paper introducing the formalism). In AI, these subjects are studied in an area known as *reinforcement learning*, which is in general the study of algorithms that can learn without receiving immediate feedback on their performance. (For a classic survey on reinforcement learning, see [KLM96].)

A popular class of reinforcement learning algorithms are collectively called Q-learning, and were first introduced by Watkins [Wat89]. The simplest variants work by estimating the probability that an action a leads eventually to success, starting from a belief state b and assuming that all subsequent actions are chosen optimally. This function, known as the Q- or action-value function is related to

$$\mathrm{SIM}_{b,\mathrm{POLICY},a}$$

when viewed as a function of belief state b and action a. In particular, the latter is an approximation to the former. In Q-learning, estimates of this function are produced on the basis of experience interacting with the environment. Under certain conditions, the estimate of the Q-function can be shown to converge to its true value [WD92].

It is instructive to compare the Q-learning algorithm to proposals that Turing himself made. In the course of a few pages in his 1950 article, Turing suggests mechanisms that learn by feedback in ways similar to methods in *supervised learning* (immediate feedback) and reinforcement learning (delayed feedback):

> We normally associate punishments and rewards with the teaching process. Some simple child machines can be constructed or programmed on this sort of principle. The machine has to be so

> constructed that events which shortly preceded the occurrence of a punishment signal are unlikely to be repeated, whereas a reward signal increased the probability of repetition of the events which led up to it. [Tur50, p. 457]

This quote describes behavior that, in broad strokes, lines up well with how a reinforcement learning algorithm such as Q-learning chooses which actions to take. Yet such simple approaches to learning from trial and error, or teaching by reward and punishment, do not capture the most powerful ways humans learn and teach each other. The last several decades of research in cognitive development [Car09, Gop12, Sch12] have emphasized the many ways in which children's learning is more driven by intrinsic curiosity and motivation than by immediate external rewards and punishments, more like the discovery and refinement of scientific theories than the mechanisms of Q-learning or the process Turing describes above. Indeed, a close analysis of the recursively defined POLICY would show that its behavior is far more sophisticated than that of Q-learning. In particular, Q-learning does not take advantage of the model NEXTSTATE, essentially assuming that there is no predictable structure in the evolution of the belief state. On the other hand, POLICY will perform information-gathering tasks, which could be construed as curiosity-driven, but are also rational acts that may themselves improve the agent's future ability to act.

Turing himself also realized that more sophisticated learning mechanisms would be required, that a machine would have to learn from "unemotional" channels of communication, which, in the language of this section, would correspond with patterns in the observations themselves not directly linked to eventual success or failure. This type of *unsupervised* learning would be useful if the goals or criteria for success changed, but the environment stayed the same. Turing envisioned that the memory store of a child-machine

> [...] would be largely occupied with definitions and propositions. The propositions would have various kinds of status, *e.g.*, well-established facts, conjectures, mathematically proved theorems, statements given by an authority, expressions having the logical form of proposition but not belief-value. [Tur50, p. 457]

The idea of using a logical language as an underlying representation of knowledge has been studied since the early days of AI, and was even proposed as a means to achieve common-sense reasoning by contemporaries of Turing, such as McCarthy [McC68]. The problem of learning logical formulae from data, especially in domains with complex, discrete structure, is actively pursued today by researchers in Statistical Relational Learning [GT07] and Inductive Logic Programming [Mug91].

Turing imagined that the same collection of logical formulae would also pertain to decision-making:

Certain propositions may be described as 'imperatives'. The machine should be so constructed that as soon as an imperative is classed as 'well-established' the appropriate action automatically takes place. [Tur50, p. 457]

A similar mechanism would later be used in *expert systems*, which first appeared in the 1960s and rose to popularity in the 1980s as they demonstrated their usefulness and commercial viability. (See [LBFL93] for a retrospective on one of the first successful expert system.)

Having seen how QUERY can be used to make decisions under uncertainty, we now conclude with some general thoughts about the use of QUERY in common-sense reasoning.

§8. **Towards common-sense reasoning.** As we have seen, the QUERY framework can be used to model many common-sense reasoning tasks, and the underlying formalism owes much to Turing, as do several details of the approach. In many of the applications we have considered, the key step is providing QUERY with an appropriate model—a generative description of the relevant aspects of nature.

In modeling, too, Turing was a pioneer. As evidenced by his diverse body of work across computation, statistics and even morphogenesis [Tur52], Turing excelled in building simple models of complex natural phenomena. In morphogenesis, in particular, his reaction-diffusion model proposed a particular sort of simplified chemical interactions as a way to understand visible patterns on animals, but also potentially leaf arrangements and even aspects of embryo formation [Cop04, p. 509]. Turing hoped to make progress in understanding these biological phenomena by carefully analyzing simplified mathematical models of these natural systems.

The medical diagnosis model DS that we examined in Section 2 is a crude attempt at the same sort of mechanical/computational description of the natural patterns of co-occurrence of diseases and symptoms. As we have seen, using a generative model of these patterns as an input to QUERY, we can reason about unseen processes, like diseases, from observed ones, like symptoms. We expect the inferences produced by QUERY to be diagnostically more useful when the generative model reflects a deep scientific understanding of the mechanisms underlying the pattern of diseases and symptoms that we find in the human population. But we also expect the inferences produced by QUERY to reflect natural patterns of common-sense reasoning among lay people when fed a model that, like DS, represents a cruder state of uncertainty.

These explanations typify *computational theories* in the sense of Marr [Mar82], and especially the Bayesian accounts developed in Anderson's rational analyses [And90], Shepard's investigations of universal laws [She87], ideal observer models [Gei84, KY03], and the work of Tenenbaum, Griffiths, and colleagues on concept learning and generalization [TG01]; see also Oaksford and Chater's notion of Bayesian rationality [OC98, OC07].

A recent body of literature demonstrates that many human inferences and decisions in *natural* situations are well predicted by probabilistic inference in models defined by simple generative descriptions of the underlying causal structure. One set of examples concerns "inverse planning" in social cognition [BGT07, BST09, GBT09, UBM+09, Bak12] and in language [GS12]. Using the approximate planning/decision-making framework discussed in Section 7 as part of a generative model for human behavior, this research considers situations in which a human reasons about the goals of other agents having only observed their actions—hence the term *inverse* planning—by assuming that the other agents are acting nearly optimally in attaining their goals. These approaches can lead to models that are good at making quantitative predictions of human judgements about the intentions of others. As another example, the "intuitive physics" research by Hamrick and Battaglia [HBT11, Ham12] aims to explain human reasoning about the physical world by positing that we use a crude ability to simulate simple physical models in our minds. Other examples include pragmatics in language [FG12, SG13] and counterfactual reasoning [GGLT12, MUSTT12]. With all of these examples, there is a rather large gap between defining the given problem in that way and being able to computationally solve it. But still there is substantial clarity brought about by the view of using QUERY along with a generative description of underlying causal structure.

Of course, this raises the questions: How do we obtain such models? In particular, how can or should we build them when they are not handed to us? And is there any hope of automating the process by which we, as scientists, invent such models upon mental reflection? These are hard scientific problems, and we have addressed them in only a very narrow sense. In Section 5, we showed how the parameters to the DS could be learned from data by constructing a larger generative process, DS', wherein these parameters are also expressed as being uncertain. We also showed, in Section 6, how the conditional independence structure implicit in the DS model could itself be learned, via inference in the model RPD.

These examples suggest that one possible approach to learning models is via a more abstract version of the sort of inference we have been describing, and this approach is roughly that taken in "theory-based Bayesian models" approach of Griffiths, Kemp, and Tenenbaum [GT06, GKT08, KT08, GT09]. Some examples include attempts to learn structural forms [TGK06], and to learn a theory of causality [GUT11]. There are of course many other proposed approaches to learning models, some with flavors very different from those considered in this paper. Finally, the question of how to approach common-sense reasoning remains. Perhaps common-sense involves knowing how to build one's own models, in a general enough setting to encompass all of experience. It is clearly far too early to tell whether this, or any current approach, will succeed.

Although Turing did not frame his AI work in terms of conditioning, his generative models for morphogenesis did capture one of the key ideas presented here—that of explaining a natural phenomenon via a detailed stochastic model of the underlying causal process. More generally, given the wide range of Turing's ideas that appear together in the approach to AI we have described, it is fascinating to speculate on what sort of synthesis Turing might have made, if he had had the opportunity.

A tantalizing clue is offered by his wartime colleague I. J. Good: Near the end of his life, Turing was a member, along with several prominent neurologists, statisticians, and physicists, of a small exclusive discussion group known as the *Ratio Club*, named in part because of "the dependence of perception on the judging of ratios" [Goo91, p. 101].

One can only guess at how Turing might have combined his computational insight, statistical brilliance, and passion for modeling natural phenomena into still further pursuits in AI.

Acknowledgements. The authors would like to thank Nate Ackerman, Chris Baker, Owain Evans, Leslie Kaelbling, Jonathan Malmaud, Vikash Mansinghka, Timothy O'Donnell, and Juliet Wagner for very helpful discussions and critical feedback on drafts, and Noah Goodman, Susan Holmes, Max Siegel, Andreas Stuhlmüller, and Sandy Zabell for useful conversations. This publication was made possible through the support of grants from the John Templeton Foundation and Google. The opinions expressed in this publication are those of the authors and do not necessarily reflect the views of the John Templeton Foundation. This paper was partially written while C.E.F. and D.M.R. were participants in the program *Semantics and Syntax: A Legacy of Alan Turing* at the Isaac Newton Institute for the Mathematical Sciences. D.M.R. is supported by a Newton International Fellowship and Emmanuel College.

REFERENCES

[AFR11] N. L. ACKERMAN, C. E. FREER, and D. M. ROY, *Noncomputable conditional distributions*, **Proceedings of the 26th annual IEEE symposium on Logic in Computer Science (*LICS 2011*)**, 2011, pp. 107–116.

[And90] J. R. ANDERSON, *The adaptive character of thought*, Erlbaum, Hillsdale, NJ, 1990.

[AB13] J. AVIGAD and V. BRATTKA, Computability and analysis: The legacy of Alan Turing, in this volume.

[BJ03] F. R. BACH and M. I. JORDAN, *Learning graphical models with Mercer kernels*, **Advances in Neural Information Processing Systems 15 (*NIPS 2002*)** (S. Becker, S. Thrun, and K. Obermayer, editors), The MIT Press, Cambridge, MA, 2003, pp. 1009–1016.

[Bak12] C. L. BAKER, *Bayesian theory of mind: Modeling human reasoning about beliefs, desires, goals, and social relations*, Ph.D. thesis, Massachusetts Institute of Technology, 2012.

[BGT07] C. L. BAKER, N. D. GOODMAN, and J. B. TENENBAUM, *Theory-based social goal inference*, **Proceedings of the 30th annual conference of the Cognitive Science Society**, 2007, pp. 1447–1452.

[BST09] C. L. BAKER, R. SAXE, and J. B. TENENBAUM, *Action understanding as inverse planning*, **Cognition**, vol. 113 (2009), no. 3, pp. 329–349.

[Bar98] A. R. BARRON, *Information-theoretic characterization of Bayes performance and the choice of priors in parametric and nonparametric problems*, **Bayesian statistics 6: Proceedings of the sixth Valencia international meeting** (J. M. Bernardo, J. O. Berger, A. P. Dawid, and A. F. M. Smith, editors), 1998, pp. 27–52.

[Bel57] R. BELLMAN, **Dynamic programming**, Princeton University Press, Princeton, NJ, 1957.

[Bla97] J. BLANCK, *Domain representability of metric spaces*, **Annals of Pure and Applied Logic**, vol. 83 (1997), no. 3, pp. 225–247.

[Cam11] C. F. CAMERER, **Behavioral game theory: Experiments in strategic interaction**, The Roundtable Series in Behavioral Economics, Princeton University Press, 2011.

[Car09] S. CAREY, **The origin of concepts**, Oxford University Press, New York, 2009.

[CSH08] V. CHANDRASEKARAN, N. SREBRO, and P. HARSHA, *Complexity of inference in graphical models*, **Proceedings of the twenty fourth conference on Uncertainty in Artificial Intelligence (UAI 2008)** (Corvalis, Oregon), AUAI Press, 2008, pp. 70–78.

[Coo90] G. F. COOPER, *The computational complexity of probabilistic inference using Bayesian belief networks*, **Artificial Intelligence**, vol. 42 (1990), no. 2–3, pp. 393–405.

[Cop04] B. J. Copeland (editor), **The essential Turing: Seminal writings in computing, logic, philosophy, artificial intelligence, and artificial life: Plus the secrets of enigma**, Oxford University Press, Oxford, 2004.

[CP96] B. J. COPELAND and D. PROUDFOOT, *On Alan Turing's anticipation of connectionism*, **Synthese**, vol. 108 (1996), no. 3, pp. 361–377.

[DKLR00] P. DAGUM, R. KARP, M. LUBY, and S. ROSS, *An optimal algorithm for Monte Carlo estimation*, **SIAM Journal on Computing**, vol. 29 (2000), no. 5, pp. 1484–1496.

[DL93] P. DAGUM and M. LUBY, *Approximating probabilistic inference in Bayesian belief networks is NP-hard*, **Artificial Intelligence**, vol. 60 (1993), no. 1, pp. 141–153.

[dMSS56] K. DE LEEUW, E. F. MOORE, C. E. SHANNON, and N. SHAPIRO, *Computability by probabilistic machines*, **Automata Studies**, Annals of Mathematical Studies, no. 34, Princeton University Press, Princeton, NJ, 1956, pp. 183–212.

[DeG05] M. H. DEGROOT, **Optimal statistical decisions**, Wiley Classics Library, Wiley, 2005.

[DWRT10] F. DOSHI-VELEZ, D. WINGATE, N. ROY, and J. TENENBAUM, *Nonparametric Bayesian policy priors for reinforcement learning*, **Advances in Neural Information Processing Systems 23 (NIPS 2010)** (J. Lafferty, C. K. I. Williams, J. Shawe-Taylor, R. S. Zemel, and A. Culotta, editors), 2010, pp. 532–540.

[Eda96] A. EDALAT, *The Scott topology induces the weak topology*, **11th annual IEEE symposium on Logic in Computer Science (LICS 1996)**, IEEE Computer Society Press, Los Alamitos, CA, 1996, pp. 372–381.

[EH98] A. EDALAT and R. HECKMANN, *A computational model for metric spaces*, **Theoretical Computer Science**, vol. 193 (1998), no. 1–2, pp. 53–73.

[FG12] M. C. FRANK and N. D. GOODMAN, *Predicting pragmatic reasoning in language games*, **Science**, vol. 336 (2012), no. 6084, p. 998.

[Gac05] P. GÁCS, *Uniform test of algorithmic randomness over a general space*, **Theoretical Computer Science**, vol. 341 (2005), no. 1–3, pp. 91–137.

[GHR10] S. GALATOLO, M. HOYRUP, and C. ROJAS, *Effective symbolic dynamics, random points, statistical behavior, complexity and entropy*, **Information and Computation**, vol. 208 (2010), no. 1, pp. 23–41.

[Gei84] W. S. Geisler, *Physical limits of acuity and hyperacuity*, **Journal of the Optical Society of America A**, vol. 1 (1984), no. 7, pp. 775–782.

[GG12] T. Gerstenberg and N. D. Goodman, *Ping pong in Church: Productive use of concepts in human probabilistic inference*, **Proceedings of the thirty-fourth annual conference of the Cognitive Science Society** (Austin, TX) (N. Miyake, D. Peebles, and R. P. Cooper, editors), Cognitive Science Society, 2012.

[GGLT12] T. Gerstenberg, N. D. Goodman, D. A. Lagnado, and J. B. Tenenbaum, *Noisy Newtons: Unifying process and dependency accounts of causal attribution*, **Proceedings of the thirty-fourth annual conference of the Cognitive Science Society** (Austin, TX) (N. Miyake, D. Peebles, and R. P. Cooper, editors), Cognitive Science Society, 2012.

[GT07] L. Getoor and B. Taskar, *Introduction to statistical relational learning*, The MIT Press, 2007.

[GG02] D. G. Goldstein and G. Gigerenzer, *Models of ecological rationality: The recognition heuristic*, **Psychological Review**, vol. 109 (2002), no. 1, pp. 75–90.

[Goo61] I. J. Good, *A causal calculus. I*, **The British Journal for the Philosophy of Science**, vol. 11 (1961), pp. 305–318.

[Goo68] ———, *Corroboration, explanation, evolving probability, simplicity and a sharpened razor*, **The British Journal for the Philosophy of Science**, vol. 19 (1968), no. 2, pp. 123–143.

[Goo75] ———, *Explicativity, corroboration, and the relative odds of hypotheses*, **Synthese**, vol. 30 (1975), no. 1, pp. 39–73.

[Goo79] ———, *A. M. Turing's statistical work in World War II*, **Biometrika**, vol. 66 (1979), no. 2, pp. 393–396, Studies in the history of probability and statistics. XXXVII.

[Goo91] ———, *Weight of evidence and the Bayesian likelihood ratio*, **The use of statistics in forensic science** (C. G. G. Aitken and D. A. Stoney, editors), Ellis Horwood, Chichester, 1991.

[Goo00] ———, *Turing's anticipation of empirical Bayes in connection with the cryptanalysis of the naval Enigma*, **Journal of Statistical Computation and Simulation**, vol. 66 (2000), no. 2, pp. 101–111.

[GBT09] N. D. Goodman, C. L. Baker, and J. B. Tenenbaum, *Cause and intent: Social reasoning in causal learning*, **Proceedings of the 31st annual conference of the Cognitive Science Society**, 2009, pp. 2759–2764.

[GMRBT08] N. D. Goodman, V. K. Mansinghka, D. M. Roy, K. Bonawitz, and J. B. Tenenbaum, *Church: A language for generative models*, **Proceedings of the twenty-fourth conference on Uncertainty in Artificial Intelligence (UAI 2008)** (Corvalis, Oregon), AUAI Press, 2008, pp. 220–229.

[GS12] N. D. Goodman and A. Stuhlmüller, *Knowledge and implicature: Modeling language understanding as social cognition*, **Proceedings of the thirty-fourth annual Conference of the Cognitive Science Society** (Austin, TX) (N. Miyake, D. Peebles, and R. P. Cooper, editors), Cognitive Science Society, 2012.

[GT12] N. D. Goodman and J. B. Tenenbaum, *The probabilistic language of thought*, in preparation, 2012.

[GTFG08] N. D. Goodman, J. B. Tenenbaum, J. Feldman, and T. L. Griffiths, *A rational analysis of rule-based concept learning*, **Cognitive Science**, vol. 32 (2008), no. 1, pp. 108–154.

[GTO11] N. D. Goodman, J. B. Tenenbaum, and T. J. O'Donnell, *Probabilistic models of cognition*, **Church wiki**, (2011), http://projects.csail.mit.edu/church/wiki/Probabilistic_Models_of_Cognition.

[GUT11] N. D. Goodman, T. D. Ullman, and J. B. Tenenbaum, *Learning a theory of causality*, **Psychological Review**, vol. 118 (2011), no. 1, pp. 110–119.

[Gop12] A. Gopnik, *Scientific thinking in young children: Theoretical advances, empirical research, and policy implications*, **Science**, vol. 337 (2012), no. 6102, pp. 1623–1627.

[GKT08] T. L. Griffiths, C. Kemp, and J. B. Tenenbaum, *Bayesian models of cognition*, **Cambridge handbook of computational cognitive modeling**, Cambridge University Press, 2008.

[GT05] T. L. GRIFFITHS and J. B. TENENBAUM, *Structure and strength in causal induction*, **Cognitive Psychology**, vol. 51 (2005), no. 4, pp. 334–384.

[GT06] ———, *Optimal predictions in everyday cognition*, **Psychological Science**, vol. 17 (2006), no. 9, pp. 767–773.

[GT09] ———, *Theory-based causal induction*, **Psychological Review**, vol. 116 (2009), no. 4, pp. 661–716.

[GSW07] T. GRUBBA, M. SCHRÖDER, and K. WEIHRAUCH, *Computable metrization*, **Mathematical Logic Quarterly**, vol. 53 (2007), no. 4–5, pp. 381–395.

[Ham12] J. HAMRICK, **Physical reasoning in complex scenes is sensitive to mass**, Master of Engineering thesis, Massachusetts Institute of Technology, Cambridge, MA, 2012.

[HBT11] J. HAMRICK, P. W. BATTAGLIA, and J. B. TENENBAUM, *Internal physics models guide probabilistic judgments about object dynamics*, **Proceedings of the thirty-third annual Conference of the Cognitive Science Society** (Austin, TX) (C. Carlson, C. Hölscher, and T. Shipley, editors), Cognitive Science Society, 2011, pp. 1545–1550.

[Hem02] A. HEMMERLING, *Effective metric spaces and representations of the reals*, **Theoretical Computer Science**, vol. 284 (2002), no. 2, pp. 347–372.

[Hod97] A. HODGES, **Turing: A natural philosopher**, Phoenix, London, 1997.

[How60] R. A. HOWARD, **Dynamic programming and Markov processes**, The MIT Press, Cambridge, MA, 1960.

[KLC98] L. P. KAELBLING, M. L. LITTMAN, and A. R. CASSANDRA, *Planning and acting in partially observable stochastic domains*, **Artificial Intelligence**, vol. 101 (1998), pp. 99–134.

[KLM96] L. P. KAELBLING, M. L. LITTMAN, and A. W. MOORE, *Reinforcement learning: A survey*, **Journal of Artificial Intelligence Research**, vol. 4 (1996), pp. 237–285.

[Kal02] O. KALLENBERG, **Foundations of modern probability**, 2nd ed., Probability and its Applications, Springer, New York, 2002.

[KGT08] C. KEMP, N. D. GOODMAN, and J. B. TENENBAUM, *Learning and using relational theories*, **Advances in Neural Information Processing Systems 20 (NIPS 2007)**, 2008.

[KSBT07] C. KEMP, P. SHAFTO, A. BERKE, and J. B. TENENBAUM, *Combining causal and similarity-based reasoning*, **Advances in Neural Information Processing Systems 19 (NIPS 2006)** (B. Schölkopf, J. Platt, and T. Hoffman, editors), The MIT Press, Cambridge, MA, 2007, pp. 681–688.

[KT08] C. KEMP and J. B. TENENBAUM, *The discovery of structural form*, **Proceedings of the National Academy of Sciences**, vol. 105 (2008), no. 31, pp. 10687–10692.

[KY03] D. KERSTEN and A. YUILLE, *Bayesian models of object perception*, **Current Opinion in Neurobiology**, vol. 13 (2003), no. 2, pp. 150–158.

[LBFL93] R. K. LINDSAY, B. G. BUCHANAN, E. A. FEIGENBAUM, and J. LEDERBERG, *DENDRAL: A case study of the first expert system for scientific hypothesis formation*, **Artificial Intelligence**, vol. 61 (1993), no. 2, pp. 209–261.

[Luc59] R. D. LUCE, **Individual choice behavior**, John Wiley, New York, 1959.

[Luc77] ———, *The choice axiom after twenty years*, **Journal of Mathematical Psychology**, vol. 15 (1977), no. 3, pp. 215–233.

[Mac03] D. J. C. MACKAY, **Information theory, inference, and learning algorithms**, Cambridge University Press, Cambridge, UK, 2003.

[MHC03] O. MADANI, S. HANKS, and A. CONDON, *On the undecidability of probabilistic planning and related stochastic optimization problems*, **Artificial Intelligence**, vol. 147 (2003), no. 1–2, pp. 5–34.

[Man09] V. K. MANSINGHKA, **Natively probabilistic computation**, Ph.D. thesis, Massachusetts Institute of Technology, 2009.

[Man11] ———, *Beyond calculation: Probabilistic computing machines and universal stochastic inference*, **NIPS Philosophy and Machine Learning Workshop**, (2011).

[MJT08] V. K. MANSINGHKA, E. JONAS, and J. B. TENENBAUM, *Stochastic digital circuits for probabilistic inference*, **Technical Report MIT-CSAIL-TR-2008-069**, Massachusetts Institute of Technology, 2008.

[MKTG06] V. K. MANSINGHKA, C. KEMP, J. B. TENENBAUM, and T. L. GRIFFITHS, *Structured priors for structure learning*, **Proceedings of the twenty-second conference on Uncertainty in Artificial Intelligence (UAI 2006)** (Arlington, Virginia), AUAI Press, 2006, pp. 324–331.

[MR13] V. K. MANSINGHKA and D. M. ROY, *Stochastic inference machines*, in preparation.

[Mar82] D. MARR, **Vision**, Freeman, San Francisco, 1982.

[McC68] JOHN MCCARTHY, *Programs with common sense*, **Semantic information processing**, The MIT Press, 1968, pp. 403–418.

[MUSTT12] J. MCCOY, T. D. ULLMAN, A. STUHLMÜLLER, T. GERSTENBERG, and J. B. TENENBAUM, *Why blame Bob? Probabilistic generative models, counterfactual reasoning, and blame attribution*, **Proceedings of the thirty-fourth annual conference of the Cognitive Science Society** (Austin, TX) (N. Miyake, D. Peebles, and R. P. Cooper, editors), Cognitive Science Society, 2012.

[MP43] W. S. MCCULLOCH and W. PITTS, *A logical calculus of the ideas immanent in nervous activity*, **Bulletin of Mathematical Biology**, vol. 5 (1943), no. 4, pp. 115–133.

[Mon82] GEORGE E. MONAHAN, *A survey of partially observable Markov Decision Processes: Theory, models, and algorithms*, **Management Science**, vol. 28 (1982), no. 1, pp. 1–16.

[Mug91] S. MUGGLETON, *Inductive logic programming*, **New Generation Computing**, vol. 8 (1991), no. 4, pp. 295–318.

[OC98] M. Oaksford and N. Chater (editors), **Rational models of cognition**, Oxford University Press, Oxford, 1998.

[OC07] ———, **Bayesian rationality: The probabilistic approach to human reasoning**, Oxford University Press, New York, 2007.

[PT87] C. H. PAPADIMITRIOU and J. N. TSITSIKLIS, *The complexity of Markov Decision Processes*, **Mathematics of Operations Research**, vol. 12 (1987), no. 3, pp. 441–450.

[Pea88] J. PEARL, **Probabilistic reasoning in intelligent systems: Networks of plausible inference**, Morgan Kaufmann, San Francisco, 1988.

[Pea04] ———, *Graphical models for probabilistic and causal reasoning*, **Computer science handbook** (A. B. Tucker, editor), CRC Press, 2nd ed., 2004.

[Pfa79] J. PFANZAGL, *Conditional distributions as derivatives*, **The Annals of Probability**, vol. 7 (1979), no. 6, pp. 1046–1050.

[Rao88] M. M. RAO, *Paradoxes in conditional probability*, **Journal of Multivariate Analysis**, vol. 27 (1988), no. 2, pp. 434–446.

[Rao05] ———, **Conditional measures and applications**, 2nd ed., Pure and Applied Mathematics, vol. 271, Chapman & Hall/CRC, Boca Raton, FL, 2005.

[RH11] S. RATHMANNER and M. HUTTER, *A philosophical treatise of universal induction*, **Entropy**, vol. 13 (2011), no. 6, pp. 1076–1136.

[Roy11] D. M. ROY, **Computability, inference and modeling in probabilistic programming**, Ph.D. thesis, Massachusetts Institute of Technology, 2011.

[Sch07] M. SCHRÖDER, *Admissible representations for probability measures*, **Mathematical Logic Quarterly**, vol. 53 (2007), no. 4–5, pp. 431–445.

[Sch12] L. SCHULZ, *The origins of inquiry: Inductive inference and exploration in early childhood*, **Trends in Cognitive Sciences**, vol. 16 (2012), no. 7, pp. 382–389.

[She87] R. N. SHEPARD, *Toward a universal law of generalization for psychological science*, **Science**, vol. 237 (1987), no. 4820, pp. 1317–1323.

[SMH+91] M. A. SHWE, B. MIDDLETON, D. E. HECKERMAN, M. HENRION, E. J. HORVITZ, H. P. LEHMANN, and G. F. COOPER, *Probabilistic diagnosis using a reformulation of the INTERNIST-1/QMR knowledge base*, **Methods of Information in Medicine**, vol. 30 (1991), pp. 241–255.

[SG92] A. F. M. SMITH and A. E. GELFAND, *Bayesian statistics without tears: A sampling-resampling perspective*, **The American Statistician**, vol. 46 (1992), no. 2, pp. 84–88.

[Sol64] R. J. SOLOMONOFF, *A formal theory of inductive inference: Parts I and II*, **Information and Control**, vol. 7 (1964), no. 1, pp. 1–22 and 224–254.

[SG13] A. STUHLMÜLLER and N. D. GOODMAN, *Reasoning about reasoning by nested conditioning: Modeling theory of mind with probabilistic programs*, submitted.

[SG12] ———, *A dynamic programming algorithm for inference in recursive probabilistic programs*, **Second Statistical Relational AI workshop at UAI 2012** (*StaRAI-12*), (2012).

[TG01] J. B. TENENBAUM and T. L. GRIFFITHS, *Generalization, similarity, and Bayesian inference*, **Behavioral and Brain Sciences**, vol. 24 (2001), no. 4, pp. 629–640.

[TGK06] J. B. TENENBAUM, T. L. GRIFFITHS, and C. KEMP, *Theory-based Bayesian models of inductive learning and reasoning*, **Trends in Cognitive Sciences**, vol. 10 (2006), no. 7, pp. 309–318.

[TKGG11] J. B. TENENBAUM, C. KEMP, T. L. GRIFFITHS, and N. D. GOODMAN, *How to grow a mind: Statistics, structure, and abstraction*, **Science**, vol. 331 (2011), no. 6022, pp. 1279–1285.

[Teu02] C. TEUSCHER, **Turing's connectionism: An investigation of neural network architectures**, Springer-Verlag, London, 2002.

[Tju74] T. TJUR, **Conditional probability distributions**, Lecture Notes, no. 2, Institute of Mathematical Statistics, University of Copenhagen, Copenhagen, 1974.

[Tju75] ———, *A constructive definition of conditional distributions*, Preprint 13, Institute of Mathematical Statistics, University of Copenhagen, Copenhagen, 1975.

[Tju80] ———, **Probability based on Radon measures**, Wiley Series in Probability and Mathematical Statistics, John Wiley & Sons Ltd., Chichester, 1980.

[THS06] M. TOUSSAINT, S. HARMELING, and A. STORKEY, *Probabilistic inference for solving (PO)MDPs*, **Technical Report EDI-INF-RR-0934**, University of Edinburgh, School of Informatics, 2006.

[Tur36] A. M. TURING, *On computable numbers, with an application to the Entscheidungsproblem*, **Proceedings of the London Mathematical Society. Second Series**, vol. 42 (1936), no. 1, pp. 230–265.

[Tur39] ———, *Systems of logic based on ordinals*, **Proceedings of the London Mathematical Society. Second Series**, vol. 45 (1939), no. 1, pp. 161–228.

[Tur48] ———, **Intelligent machinery**, National Physical Laboratory Report, 1948.

[Tur50] ———, *Computing machinery and intelligence*, **Mind**, vol. 59 (1950), pp. 433–460.

[Tur52] ———, *The chemical basis of morphogenesis*, **Philosophical Transactions of the Royal Society of London. Series B, Biological Sciences**, vol. 237 (1952), no. 641, pp. 37–72.

[Tur96] ———, *Intelligent machinery, a heretical theory*, **Philosophia Mathematica. Philosophy of Mathematics, its Learning, and its Applications. Series III**, vol. 4 (1996), no. 3, pp. 256–260, Originally a radio presentation, 1951.

[Tur12] ———, *The applications of probability to cryptography, c. 1941*, UK National Archives, HW 25/37, 2012.

[UBM+09] T. D. ULLMAN, C. L. BAKER, O. MACINDOE, O. EVANS, N. D. GOODMAN, and J. B. TENENBAUM, *Help or hinder: Bayesian models of social goal inference*, **Advances in Neural Information Processing Systems 22** (*NIPS 2009*), 2009, pp. 1874–1882.

[Wat89] C. J. C. H. WATKINS, **Learning from delayed rewards**, Ph.D. thesis, King's College, University of Cambridge, 1989.

[WD92] C. J. C. H. WATKINS and P. DAYAN, *Q-Learning*, **Machine Learning**, vol. 8 (1992), pp. 279–292.

[Wei93] K. WEIHRAUCH, *Computability on computable metric spaces*, **Theoretical Computer Science**, vol. 113 (1993), no. 2, pp. 191–210.

[Wei99] ———, *Computability on the probability measures on the Borel sets of the unit interval*, **Theoretical Computer Science**, vol. 219 (1999), no. 1–2, pp. 421–437.

[Wei00] ———, *Computable analysis: An introduction*, Texts in Theoretical Computer Science, An EATCS Series, Springer-Verlag, Berlin, 2000.

[WGRKT11] D. WINGATE, N. D. GOODMAN, D. M. ROY, L. P. KAELBLING, and J. B. TENENBAUM, *Bayesian policy search with policy priors*, **Proceedings of the twenty-second International Joint Conference on Artificial Intelligence (IJCAI)** (Menlo Park, CA) (T. Walsh, editor), AAAI Press, 2011.

[WGSS11] D. WINGATE, N. D. GOODMAN, A. STUHLMÜLLER, and J. M. SISKIND, *Nonstandard interpretations of probabilistic programs for efficient inference*, **Advances in Neural Information Processing Systems 24 (NIPS 2011)**, 2011.

[WSG11] D. WINGATE, A. STUHLMÜLLER, and N. D. GOODMAN, *Lightweight implementations of probabilistic programming languages via transformational compilation*, **Proceedings of the fourteenth international conference on Artificial Intelligence and Statistics (AISTATS)**, Journal of Machine Learning Research: Workshop and Conference Proceedings, vol. 15, 2011, pp. 770–778.

[Yam99] T. YAMAKAMI, *Polynomial time samplable distributions*, **Journal of Complexity**, vol. 15 (1999), no. 4, pp. 557–574.

[Zab95] S. L. ZABELL, *Alan Turing and the central limit theorem*, **American Mathematical Monthly**, vol. 102 (1995), no. 6, pp. 483–494.

[Zab12] ———, *Commentary on Alan M. Turing: The applications of probability to cryptography*, **Cryptologia**, vol. 36 (2012), no. 3, pp. 191–214.

MASSACHUSETTS INSTITUTE OF TECHNOLOGY
COMPUTER SCIENCE AND ARTIFICIAL INTELLIGENCE LABORATORY, USA
E-mail: freer@math.mit.edu

UNIVERSITY OF CAMBRIDGE
DEPARTMENT OF ENGINEERING, UK
E-mail: d.roy@eng.cam.ac.uk

MASSACHUSETTS INSTITUTE OF TECHNOLOGY
DEPARTMENT OF BRAIN AND COGNITIVE SCIENCES, USA
E-mail: jbt@mit.edu

MATHEMATICS IN THE AGE OF THE TURING MACHINE

THOMAS C. HALES

FIGURE 1. Alan Turing (image source [Ima13])

"*And when it comes to mathematics, you must realize that this is the human mind at the extreme limit of its capacity.*" (H. Robbins)

"*... so reduce the use of the brain and calculate!*" (E. W. Dijkstra)

"*The fact that a brain* can *do it seems to suggest that the difficulties [of trying with a machine] may not really be so bad as they now seem.*" (A. Turing)

Research supported in part by NSF grant 0804189 and the Benter Foundation.

§1. Computer calculation.

1.1. A panorama of the status quo. Where stands the mathematical endeavor?

In 2012, many mathematical utilities are reaching consolidation. It is an age of large aggregates and large repositories of mathematics: the arXiv, Math Reviews, and euDML, which promises to aggregate the many European archives such as Zentralblatt Math and Numdam. Sage aggregates dozens of mathematically oriented computer programs under a single Python-scripted front-end.

Book sales in the U.S. have been dropping for the past several years. Instead, online sources such as Wikipedia and Math Overflow are rapidly becoming students' preferred math references. The Polymath blog organizes massive mathematical collaborations. Other blogs organize previously isolated researchers into new fields of research. The slow, methodical deliberations of referees in the old school are giving way; now in a single stroke, Tao blogs, gets feedback, and publishes.

Machine Learning is in its ascendancy. *LogAnswer* and *Wolfram Alpha* answer our elementary questions about the quantitative world; *Watson* our *Jeopardy* questions. *Google Page* ranks our searches by calculating the largest eigenvalue of the largest matrix the world has ever known. *Deep Blue* plays our chess games. The million-dollar-prize-winning *Pragmatic Chaos* algorithm enhances our *Netflix searches*. The major proof assistants now contain tens of thousands of formal proofs that are being mined for hints about how to prove the next generation of theorems.

Mathematical models and algorithms rule the quantitative world. Without applied mathematics, we would be bereft of Shor's factorization algorithm for quantum computers, Yang–Mills theories of strong interactions in physics, invisibility cloaks, Radon transforms for medical imaging, models of epidemiology, risk analysis in insurance, stochastic pricing models of financial derivatives, RSA encryption of sensitive data, Navier–Stokes modeling of fluids, and models of climate change. Without it, entire fields of engineering from Control Engineering to Operations Research would close their doors. The early icon of mathematical computing, Von Neumann, divided his final years between meteorology and hydrogen bomb calculations. Today, applications fuel the economy: in 2011 rankings, the first five of the "10 best jobs" are math or computer related: software engineer, mathematician, actuary, statistician, and computer systems analyst [CC11].

Computers have rapidly become so pervasive in mathematics that future generations may look back to this day as a golden dawn. A comprehensive survey is out of the question. It would almost be like asking for a summary of applications of symmetry to mathematics. Computability—like symmetry— is a wonderful structural property that some mathematical objects possess that makes answers flow more readily wherever it is found. This section gives many examples that give a composite picture of computers in mathematical research, showing that computers are neither the panacea that the public at large might imagine, nor the evil that the mathematical purist might fear. I have deliberately selected many examples from pure mathematics, partly because of my own background and partly to correct the conventional wisdom that couples computers with applied mathematics and blackboards with pure mathematics.

1.2. Birch and Swinnerton-Dyer conjecture. I believe that the Birch and Swinnerton-Dyer conjecture is the deepest conjecture ever to be formulated with the help of a computer [BSD65]. The Clay Institute has offered a one-million dollar prize to anyone who settles it.

Let E be an elliptic curve defined by an equation $y^2 = x^3 + ax + b$ over the field of rational numbers. Motivated by related quantities in Siegel's work on quadratic forms, Birch and Swinnerton-Dyer set out to estimate the quantity

$$\prod N_p/p, \tag{1}$$

where N_p is the number of rational points on E modulo p, and the product extends over primes $p \leq P$ [Bir02]. Performing experiments on the EDSAC II computer at the Computer laboratory at Cambridge University during the years 1958–1962, they observed that as P increases, the products (1) grow asymptotically in P as

$$c(E) \log^r P,$$

for some constant c, where r is the Mordell–Weil rank of E; that is, the maximum number of independent points of infinite order in the group $E(\mathbb{Q})$ of rational points. Following the suggestions of Cassels and Davenport, they reformulated this numerical asymptotic law in terms of the zeta function $L(E, s)$ of the elliptic curve. Thanks to the work of Wiles and subsequent extensions of that work, it is known that $L(E, s)$ is an entire function of the complex variable s. The Birch and Swinnerton-Dyer conjecture asserts that the rank r of an elliptic curve over $\mathbb{Q}$ is equal to the order of the zero of $L(E, s)$ at $s = 1$.

A major (computer-free) recent theorem establishes that the Birch and Swinnerton-Dyer conjecture holds for a positive proportion of all elliptic curves over $\mathbb{Q}$ [BS10]. This result, although truly spectacular, is mildly misleading in the sense that the elliptic curves of high rank rarely occur but pose the greatest difficulties.

1.3. Sato–Tate. The Sato–Tate conjecture is another major conjecture about elliptic curves that was discovered by computer. If E is an elliptic curve with rational coefficients

$$y^2 = x^3 + ax + b,$$

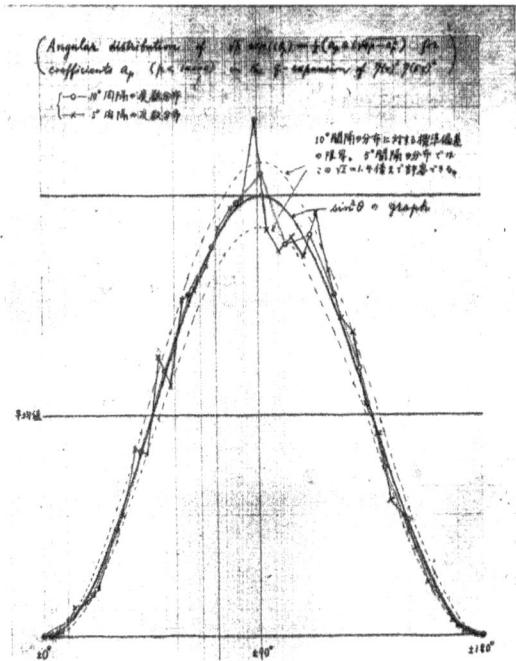

FIGURE 2. Data leading to the Sato–Tate conjecture (image source [SN63]).

then the number of solutions modulo a prime number p (including the point at infinity) has the form

$$1 + p - 2\sqrt{p}\cos\theta_p,$$

for some real number $0 \leq \theta_p \leq \pi$. In 1962, Sato, Nagashima, and Namba made calculations of θ_p on a Hitachi HIPAC 103 computer to understand how these numbers are distributed as p varies for a fixed elliptic curve E [Sch]. By the spring of 1963, the evidence suggested $\sin^2\theta$ as a good fit of the data (Figure 2). That is, if $P(n)$ is the set of the first n primes, and $f\colon [0,\pi] \to \mathbb{R}$ is any smooth test function, then for large n,

$$\frac{1}{n}\sum_{p\in P(n)} f(\theta_p) \quad\text{tends to}\quad \frac{2}{\pi}\int_0^\pi f(\theta)\sin^2\theta\,d\theta.$$

The Sato–Tate conjecture (1963) predicts that this same distribution is obtained, no matter the elliptic curve, provided the curve does not have complex multiplication. Tate, who arrived at the conjecture independently, did so without computer calculations.

Serre interpreted Sato–Tate as a generalization of Dirichlet's theorem on primes in arithmetic progression, and gave a proof strategy of generalizing the analytic properties of L-functions used in the proof of Dirichlet's theorem [Ser68]. Indeed, a complete proof of Sato–Tate conjecture has now been found and is based on extremely deep analytic properties of L-functions [Car07]. The proof of the Sato–Tate conjecture and its generalizations has been one of the most significant recent advances in number theory.

1.4. Transient uses of computers. It has become common for problems in mathematics to be first verified by computer and later confirmed without them. Some examples are the construction of sporadic groups, counterexamples to a conjecture of Euler, the proof of the Catalan conjecture, and the discovery of a formula for the binary digits of π.

Perhaps the best known example is the construction of sporadic groups as part of the monumental classification of finite simple groups. The sporadic groups are the 26 finite simple groups that do not fall into natural infinite families. For example, Lyons (1972) predicted the existence of a sporadic group of order

$$2^8 \cdot 3^7 \cdot 5^6 \cdot 7 \cdot 11 \cdot 31 \cdot 37 \cdot 67.$$

In 1973, Sims proved the existence of this group in a long unpublished manuscript that relied on many specialized computer programs. By 1999, the calculations had become standardized in group theory packages, such as GAP and Magma [HS99]. Eventually, computer-free existence and uniqueness proofs were found [MC02], [AS92].

Another problem in finite group theory with a computational slant is the inverse Galois problem: is every subgroup of the symmetric group S_n the Galois group of a polynomial of degree n with rational coefficients? In the 1980s Malle and Matzat used computers to realize many groups as Galois groups [MM99], but with an infinite list of finite groups to choose from, non-computational ideas have been more fruitful, such as Hilbert irreducibility, rigidity, and automorphic representations [KLS08].

Euler conjectured (1769) that a fourth power cannot be the sum of three positive fourth powers, that a fifth power cannot be the sum of four positive fifth powers, and so forth. In 1966, a computer search [LP66] on a CDC 6600 mainframe uncovered a counterexample

$$27^5 + 84^5 + 110^5 + 133^5 = 144^5,$$

which can be checked by hand (I dare you). The two-sentence announcement of this counterexample qualifies as one of the shortest mathematical publications of all times. Twenty years later, a more subtle computer search gave another counterexample [Elk88]:

$$2682440^4 + 15365639^4 + 18796760^4 = 20615673^4.$$

The Catalan conjecture (1844) asserts that the only solution to the equation

$$x^m - y^n = 1,$$

in positive integers x, y, m, n with exponents m, n greater than 1 is the obvious

$$3^2 - 2^3 = 1.$$

That is, 8 and 9 are the only consecutive positive perfect powers. By the late 1970s, Baker's methods in diophantine analysis had reduced the problem to an astronomically large and hopelessly infeasible finite computer search. Mihăilescu's proof (2002) of the Catalan conjecture made light use of computers (a one-minute calculation), and later the computer calculations were entirely eliminated [Mih04], [Met03].

Bailey, Borwein, and Plouffe found an algorithm for calculating the nth binary digit of π directly: it jumps straight to the nth digit without first calculating any of the earlier digits. They understood that to design such an algorithm, they would need an infinite series for π in which powers of 2 controlled the denominators. They did not know of any such formula, and made a computer search (using the PSLQ lattice reduction algorithm) for any series of the desired form. Their search unearthed a numerical identity

$$\pi = \sum_{n=0}^{\infty} \left(\frac{4}{8n+1} - \frac{2}{8n+4} - \frac{1}{8n+5} - \frac{1}{8n+6} \right) \left(\frac{1}{16} \right)^n,$$

which was then rigorously proved and used to implement their binary-digits algorithm.

1.5. Rogers–Ramanujan identities. The famous Rogers–Ramanujan identities

$$1 + \sum_{k=1}^{\infty} \frac{q^{k^2+ak}}{(1-q)(1-q^2)\cdots(1-q^k)} = \prod_{j=0}^{\infty} \frac{1}{(1-q^{5j+a+1})(1-q^{5j-a+4})},$$

$$a = 0, 1.$$

can now be proved by an almost entirely mechanical procedure from Jacobi's triple product identity and the q-WZ algorithm of Wilf and Zeilberger that checks identities of q-hypergeometric finite sums [Pau94]. Knuth's foreword to a book on the WZ method opens, "Science is what we understand well enough to explain to a computer. Art is everything else we do." Through the WZ method, many summation identities have become a science [PWZ96].

1.6. Packing tetrahedra. Aristotle erroneously believed that regular tetrahedra tile space: "It is agreed that there are only three plane figures which can fill a space, the triangle, the square, and the hexagon, and only two solids, the pyramid and the cube" [AriBC]. However, centuries later, when the dihedral angle of the regular tetrahedron was calculated:

$$\arccos(1/3) \approx 1.23 < 1.25664 \approx 2\pi/5,$$

it was realized that a small gap is left when five regular tetrahedra are grouped around a common edge (Figure 3). In 1900, in his famous list of problems, Hilbert asked "How can one arrange most densely in space an infinite number of equal solids of given form, e.g., spheres with given radii or regular tetrahedra ... ?"

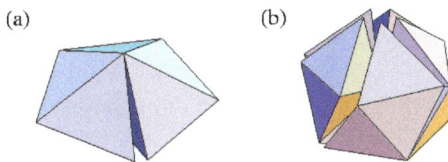

FIGURE 3. Regular tetrahedra fail to tile space (image source [Doy11]).

Aristotle notwithstanding, until recently, no arrangements of regular tetrahedra with high density were known to exist. In 2000, Betke and Henk developed an efficient computer algorithm to find the densest lattice packing of a general convex body [BH00]. This opened the door to experimentation [CT06]. For example, the algorithm can determine the best lattice packing of the convex hull of the cluster of tetrahedra in Figure 3. In rapid succession came new record-breaking arrangements of tetrahedra, culminating in what is now conjectured to be the best possible [CEG10]. (See Figure 4.) Although

FIGURE 4. The best packing of tetrahedra is believed to be the Chen–Engel–Glotzer arrangement with density $4000/4671 \approx 0.856$ (image source [CEG10]).

Chen had the panache to hand out Dungeons and Dragons tetrahedral dice to the audience for a hands-on modeling session during her thesis defense, the best arrangement was found using Monte Carlo experiments. In the numerical simulations, a finite number of tetrahedra are randomly placed in a box of variable shape. The tetrahedra are jiggled as the box slowly shrinks until no further improvement is possible. Now that a precise conjecture has been formulated, the hardest part still remains: to give a proof.

1.7. The Kepler conjecture. Hilbert's 18th problem asks to find dense packings of both spheres and regular tetrahedra. The problem of determining the best sphere packing in three dimensions is the Kepler conjecture. Kepler was led to the idea of density as an organizing principle in nature by observing the tightly packed seeds in a pomegranate. Reflecting on the hexagonal symmetry of snowflakes and honeycombs, by capping each honeycomb cell with a lid of the same shape as the base of the cell, he constructed a closed twelve-sided cell that tiles space. Kepler observed that the familiar pyramidal cannonball arrangement is obtained when a sphere is placed in each capped honeycomb cell (Figure 5). This he believed to be the densest packing.

FIGURE 5. An optimal sphere packing is obtained by placing one sphere in each three-dimensional honeycomb cell (image source [RhD11]).

L. Fejes Tóth proposed a strategy to prove Kepler's conjecture in the 1950s, and later he suggested that computers might be used. The proof, finally obtained by Ferguson and me in 1998, is one of the most difficult nonlinear optimization problems ever rigorously solved by computer [Hal05b]. The computers calculations originally took about 2000 hours to run on Sparc workstations. Recent simplifications in the proof have reduced the runtime to about 20 hours and have reduced the amount of customized code by a factor of more than 10.

1.8. The four-color theorem. The four-color theorem is the most celebrated computer proof in the history of mathematics. The problem asserts that it is possible to color the countries of any map with at most four colors in such a way that contiguous countries receive different colors. The proof of this theorem required about 1200 hours on an IBM 370-168 in 1976. So much has been written about Appel and Haken's computer solution to this problem that it is pointless to repeat it here [AHK77]. Let it suffice to cite a popular account [Wil02], a sociological perspective [Mac01], the second generation proof [RSST97], and the culminating formal verification [Gon08].

1.9. Projective planes. A finite projective plane of order $n > 1$ is defined to be a set of $n^2 + n + 1$ lines and $n^2 + n + 1$ points with the following properties:

1. Every line contains $n + 1$ points;
2. Every point is on $n + 1$ lines;
3. Every two distinct lines have exactly one point of intersection; and
4. Every two distinct points lie on exactly one line.

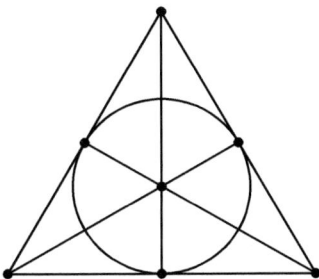

FIGURE 6. The Fano plane is a finite projective plane of order 2.

The definition is an abstraction of properties that evidently hold for $\mathbb{P}^2(\mathbb{F}_q)$, the projective plane over a finite field $\mathbb{F}_q$, with $q = n$, for any prime power q. In particular, a finite projective plane exists whenever n is a positive power of a prime number (Figure 6).

The conjecture is that every finite projective plane of order $n > 1$ is a prime power. The smallest integers $n > 1$ that are *not* prime powers are

$$6, 10, 12, 14, 15, \ldots.$$

The brute force approach to this conjecture is to eliminate each of these possibilities in turn. The case $n = 6$ was settled in 1938. Building on a number of theoretical advances [MST73], Lam eliminated the case $n = 10$ in 1989, in one of the most difficult computer proofs in history [LTS89]. This calculation was executed over a period of years on multiple machines and eventually totaled about 2000 hours of Cray-1A time.

Unlike the computer proof of the four-color theorem, the projective plane proof has never received independent verification. Because of the possibilities of programming errors and soft errors (see Section 3.5), Lam is unwilling to call his result a proof. He writes, "From personal experience, it is extremely easy to make programming mistakes. We have taken many precautions, ... Yet, I want to emphasize that this is only an experimental result and it desperately needs an independent verification, or better still, a theoretical explanation" [Lam91].

Recent speculation at *Math Overflow* holds that the next case, $n = 12$, remains solidly out of computational reach [Hor10].

1.10. Hyperbolic manifolds. Computers have helped to resolve a number of open conjectures about hyperbolic manifolds (defined as complete Riemannian manifolds with constant negative sectional curvature -1), including the proof that the space of hyperbolic metrics on a closed hyperbolic 3-manifold is contractible [GMT03], [Gab10].

1.11. Chaos theory and strange attractors. The theory of chaos has been one of the great success stories of twentieth century mathematics and science. Turing[1] expressed the notion of chaos with these words, "quite small errors in the initial conditions can have an overwhelming effect at a later time. The displacement of a single electron by a billionth of a centimetre at one moment might make the difference between a man being killed by an avalanche a year later, or escaping" [Tur50]. Later, the metaphor became a butterfly that stirs up a tornado in Texas by flapping its wings in Brazil.

Thirteen years later, Lorenz encountered chaos as he ran weather simulations on a Royal McBee LGP-30 computer [Lor63]. When he reran an earlier numerical solution with what he thought to be identical initial data, he obtained wildly different results. He eventually traced the divergent results to a slight discrepancy in initial conditions caused by rounding in the printout. The *Lorenz oscillator* is the simplified form of Lorenz's original ordinary differential equations.

A set A is *attracting* if it has a neighborhood U such that

$$A = \bigcap_{t \geq 0} f_t(U),$$

where $f_t(x)$ is the solution of the dynamical system (in present case the Lorenz oscillator) at time t with initial condition x. That is, U flows towards the attracting set A. Simulations have discovered attracting sets with strange properties such as non-integral Hausdorff dimension and the tendency for a small slab of volume to quickly spread throughout the attractor.

Lorenz conjectured in 1963 that his oscillator has a strange attractor (Figure 7). In 1982, the Lax report cited soliton theory and strange attractors as two prime examples of the "discovery of new phenomena through numerical experimentation," and calls such discovery perhaps the most "significant

---

[1] For early history, see [Wol02, p. 971]. Turing vainly hoped that digital computers might be insulated from the effects of chaos.

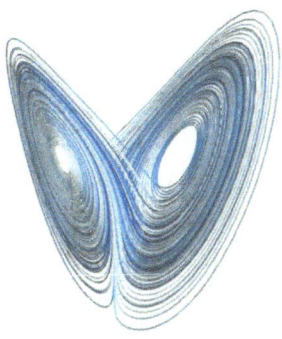

FIGURE 7. The Lorenz oscillator gives one of the most famous images of mathematics, a *strange attractor* in dynamical systems (image source [Aga13]).

application of scientific computing" [Lax82]. Smale, in his list of 18 "Mathematical Problems for the Next Century" made the fourteenth problem to present a rigorous proof that the dynamics of the Lorenz oscillator is a strange attractor [Sma98] with various additional properties that make it a "geometric Lorenz attractor."

Tucker has solved Smale's fourteenth problem by computer [Tuc02] [Ste00]. One particularly noteworthy aspect of this work is that chaotic systems, by their very nature, pose particular hardships for rigorous computer analysis. Nevertheless, Tucker implemented the classical Euler method for solving ordinary differential equations with particular care, using interval arithmetic to give mathematically rigorous error bounds. Tucker has been awarded numerous prizes for this work, including the Moore Prize (2002) and the EMS Prize (2004).

Smale's list in general envisions a coming century in which computer science, especially computational complexity, plays a much larger role than during the past century. He finishes the list with the open-ended philosophical problem that echoes Turing: *"What are the limits of intelligence, both artificial and human?"*

1.12. 4/3. Mandelbrot's conjectures in fractal geometry have resulted in two Fields Medals. Here he describes the discovery of the 4/3-conjecture made in [Man82]. "The notion that these conjectures might have been reached by pure thought—with no picture—is simply inconceivable. ... I had my programmer draw a very big sample [Brownian] motion and proceeded to play with it." He goes on to describe computer experiments that led him to enclose the Brownian motion into black clusters that looked to him like islands with jagged coastlines (Figure 8). "[I]nstantly, my long previous experience with the coastlines of actual islands on Earth came handy and made me suspect that the boundary of Brownian motion has a fractal dimension equal to 4/3" [Man04].

This conjecture, which Mandelbrot's trained eye spotted in an instant, took 18 years to prove [LSW01].

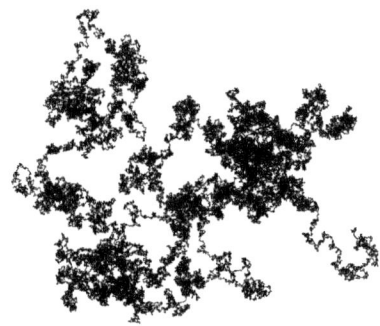

FIGURE 8. A simulation of planar Brownian motion. Mandelbrot used "visual inspection supported by computer experiments" to formulate deep conjectures in fractal geometry (image generated from source code at [LSW01]).

1.13. Sphere eversion visualization. Smale (1958) proved that it is possible to turn a sphere inside out without introducing any creases.[2] For a long time, this paradoxical result defied the intuition of experts. R. Bott, who had been Smale's graduate advisor, refused to believe it at first. Levy writes that trying to visualize Smale's mathematical argument "is akin to describing what happens to the ingredients of a soufflé in minute detail, down to the molecular chemistry, and expecting someone who has never seen a soufflé to follow this 'recipe' in preparing the dish" [Lev95].

It is better to see and taste a soufflé first. The computer videos of this theorem are spectacular. Watch them on YouTube! As we watch the sphere turn inside out, our intuition grows. The computer calculations behind the animations of the first video (the Optiverse) start with a sphere, half inverted and half right-side out [SFL]. From halfway position, the path of steepest descent of an energy functional is used to calculate the unfolding in both directions to the round spheres, with one fully inverted (Figure 9). The second video is based on Thurston's "corrugations" [LMM94]. As the name suggests, this sphere eversion has undulating ruffles that dance like a jellyfish, but avoids sharp creases. Through computers, understanding.

---

[2] I am fond of this example, because The Scientific American article [Phi66] about this theorem was my first exposure to "real mathematics" as a child.

FIGURE 9. Computer-generated stages of a sphere eversion (image source [Op111]).

Speaking of Thurston, he contrasts "our amazingly rich abilities to absorb geometric information and the weakness of our innate abilities to convey spatial ideas.... We effortlessly look at a two-dimensional picture and reconstruct a three-dimensional scene, but we can hardly draw them accurately" [Pit11]. As more and more mathematics migrates to the computer, there is a danger that geometrical intuition becomes buried under a logical symbolism.

1.14. Minimal surface visualization. Weber and Wolf [WW11] report that the use of computer visualization has become "commonplace" in minimal surface research, "a conversation between visual aspects of the minimal surfaces and advancing theory, each supporting the other." This started when computer illustrations of the Costa surface (a particular minimal surface, Figure 10) in the 1980s revealed dihedral symmetries of the surface that were not seen directly from its defining equations. The observation of symmetry turned out to be the key to the proof that the Costa surface is an embedding. The symmetries further led to a conjecture and then proof of the existence of other minimal surfaces of higher genus with similar dihedral symmetries. As Hoffman wrote about his discoveries, "The images produced along the way were the objects that we used to make discoveries. They are an integral part of the process of doing mathematics, not just a way to convey a discovery made without their use" [Hof87].

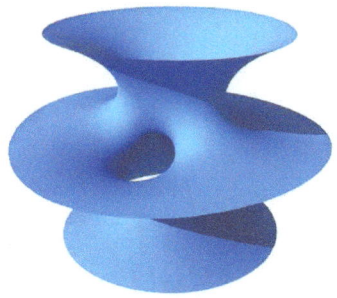

FIGURE 10. The Costa surface launched an era of computer exploration in minimal surface theory (image source [San12]).

1.15. Double bubble conjecture. Closely related to minimal surfaces are surfaces of constant mean curvature. The mean curvature of a minimal surface is zero; surfaces whose mean curvature is constant are a slight generalization. They arise as surfaces that are minimal subject to the constraint that they enclose a region of fixed volume. Soap bubble films are surfaces of constant mean curvature.

The isoperimetric inequality asserts that the sphere minimizes the surface area among all surfaces that enclose a region of fixed volume. The double bubble problem is the generalization of the isoperimetric inequality to two enclosed volumes. What is the surface minimizing way to enclose two separate regions of fixed volume? In the nineteenth century, Boys [Boy90] and Plateau observed experimentally that the answer should be two partial spherical bubbles joined along a shared flat disk (Figure 11). The size of the shared disk is determined by the condition that angles should be 120° where the three surfaces meet. This is the *double bubble conjecture*.

The first case of double bubble conjecture to be established was that of two equal volumes [HHS95]. The proof was a combination of conventional analysis and computer proof. Conventional analysis (geometric measure theory) established the existence of a minimizer and reduced the possibilities to a small number of figures of revolution, and computers were to analyze each of the cases, showing in each case by interval analysis either that the case was not a local minimizer or that its area was strictly larger than the double bubble. Later theorems proved the double bubble conjecture in the general unequal volume case without the use of computers [HMRR00].

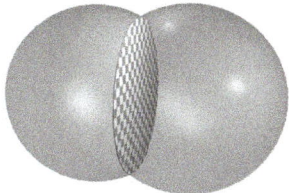

FIGURE 11. The optimality of a double bubble was first established by computer, using interval analysis (image source [Tsi13]).

The natural extension of the double bubble conjecture from two bubbles to an infinite bubbly foam is the Kelvin problem. The problem asks for the surface area minimizing partition of Euclidean space into cells of equal volume. Kelvin's conjecture—a tiling by slight perturbations of truncated octahedra—remained the best known partition until a counterexample was constructed by two physicists, Phelan and Weaire in 1993 (Figure 12). The

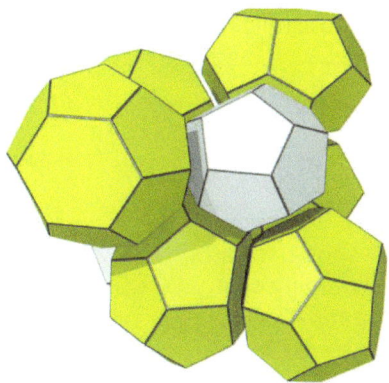

FIGURE 12. The Phelan–Weaire foam, giving the best known partition of Euclidean space into cells of equal volume, was constructed with Surface Evolver software. This foam inspired the bubble design of the Water Cube building in the 2008 Beijing Olympics (image source [PW11]).

counterexample exists not as a physical model, nor as an exact mathematical formula, but only as an image generated from a triangular mesh in the *Surface*

Evolver computer program. By default, the counterexample has become the new conjectural answer to the Kelvin problem, which I fully expect to be proved someday by computer.

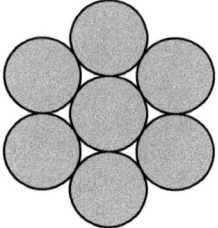

FIGURE 13. In two dimensions, the kissing number is 6. In eight dimensions, the answer is 240. The proof certificate was found by linear programming.

1.16. Kissing numbers. In the plane, at most six pennies can be arranged in a hexagon so that they all touch one more penny placed at the center of the hexagon (Figure 13). Odlyzko and Sloane, solved the corresponding problem in dimension 8: at most 240 nonoverlapping congruent balls can be arranged so that they all touch one more at the center.

Up to rotation, a unique arrangement of 240 exists. To the cognoscenti, the proof of this fact is expressed as one-line certificate:

$$\left(t - \frac{1}{2}\right) t^2 \left(t + \frac{1}{2}\right)^2 (t + 1).$$

(For an explanation of the certificates, see [PZ04].) The certificate was produced by a linear programming computer search, but once the certificate is in hand, the proof is computer-free.

As explained above, six is the *kissing number* in two dimensions, 240 is the kissing number in eight dimensions. In three dimensions, the kissing number is 12. This three-dimensional problem goes back to a discussion between Newton and Gregory in 1694, but was not settled until the 1950s. A recent computer proof makes an exhaustive search through nearly 100 million combinatorial possibilities to determine exactly how much the twelve spheres must shrink to accommodate a thirteenth [MT10]. Bachoc and Vallentin were recently awarded the SIAG/Optimization prize for their use of semi-definite programming algorithms to establish new proofs of the kissing number in dimensions 3, 4, 8 and new bounds on the kissing number in various other dimensions [BV08].

1.17. Digression on E_8. It is no coincidence that the calculation of Odlyzko and Sloane works in dimension 8. Wonderful things happen in eight dimensional space and again in 24 dimensions.

Having mentioned the 240 balls in eight dimensions, I cannot resist mentioning some further computer proofs. The centers of the 240 balls are vectors whose integral linear combinations generate a lattice in $\mathbb{R}^8$, known as the E_8 lattice (Figure 14).

There is a packing of congruent balls in eight dimensions that is obtained by centering one ball at each vector in the E_8 lattice, making the balls as large as possible without overlap. Everyone believes that this packing in eight dimensions is the densest possible, but this fact currently defies proof. If the center of the balls are the points of a lattice, then the packing is called a *lattice packing*. Cohn and Kumar have a beautiful computer assisted proof that the E_8 packing is the densest of all lattice packings in $\mathbb{R}^8$ (and the corresponding result in dimension 24 for the Leech lattice). The proof is based on the Poisson summation formula. Pfender and Ziegler's account of this computer-assisted proof won the Chauvenet Prize of the MAA for writing [PZ04].

FIGURE 14. The E_8 lattice is generated by eight vectors in $\mathbb{R}^8$ whose mutual angles are 120° or 90° depending on whether the corresponding dots are joined by a segment are not.

The 240 vectors that generate the E_8 lattice are the *roots* of a 240 + 8 dimensional Lie group (also called E_8); that is, a differentiable manifold that has the analytic structure of a group. All simple Lie groups were classified in the nineteenth century.[3] They fall into infinite families named alphabetically, A_n, B_n, C_n, D_n, with 5 more exceptional cases that do not fall into infinite families E_6, E_7 E_8, F_4, G_2. The exceptional Lie group[4] of highest dimension is E_8.

The long-term *Atlas Project* aims to use computers to determine all unitary representations of real reductive Lie groups [Atl]. The 19-member team focused on E_8 first, because everyone respects the formidable E_8. By 2007, a computer had completed the character table of E_8. Since there are infinitely

---

[3] I describe the families over $\mathbb{C}$. Each complex Lie group has a finite number of further real forms.

[4] For decades, E_8 has stood for the ultimate in speculative physics, whether in heterotic string theory or a "theory of everything." Last year, E_8 took a turn toward the real world, when E_8 calculations predicted neutron scattering experiments with a cobalt niobate magnet [BG11].

many irreducible characters and each character is an analytic function on (a dense open subset of) the group, it is not clear without much further explanation what it might even mean for a computer to output the full character table as a 60 gigabyte file [Ada11]. What is significant about this work is that it brings the computer to bear on some abstract parts of mathematics that have been traditionally largely beyond the reach of concrete computational description, including infinite dimensional representations of Lie groups, intersection cohomology and perverse sheaves. Vogan's account of this computational project was awarded the 2011 Conant Prize of the AMS [Vog07].

While on the topic of computation and representation theory, I cannot resist a digression into the P versus NP problem, the most fundamental unsolved problem in mathematics. In my opinion, attempts to settle P versus NP from the axioms of ZFC are ultimately as ill-fated as Hilbert's program in the foundations of math (which nonetheless spurred valuable partial results such as the decision procedures of Presburger and Tarski), but if I were to place faith anywhere, it would be in Mulmuley's program in *geometric complexity theory*. The program invokes geometric invariant theory and representation theoretic invariants to tease apart complexity classes: if the irreducible constituents of modules canonically associated with two complexity classes are different, then the two complexity classes are distinct. In this approach, the determinant and permanent of a matrix are chosen as the paradigms of what is easy and what is hard to compute, opening up complexity theory to a rich algebro-geometric structure [Mul11], [For09].

1.18. Future computer proofs. Certain problems are natural candidates for computer proof: the Kelvin problem by the enumeration of the combinatorial topology of possible counterexamples; the search for a counterexample to the two-dimensional Jacobian conjecture through the minimal model program [Bor09]; resolution of singularities in positive characteristic through an automated search for numerical quantities that decrease under suitable blowup; existence of a projective plane of order 12 by constraint satisfaction programming; the optimality proof of the best known packing of tetrahedra in three dimensions [CEG10]; Steiner's isoperimetric conjecture (1841) for the icosahedron [Ste41]; and the Reinhardt conjecture through nonlinear optimization [Hal11]. But proceed with caution! Checking on our zeal for brute computation, computer-generated patterns can sometimes fail miserably. For example, the sequence:

$$\left\lceil \frac{2}{2^{1/n}-1} \right\rceil - \left\lfloor \frac{2n}{\log 2} \right\rfloor, \quad n = 1, 2, 3, \ldots$$

starts out as the zero sequence, but remarkably first gives a nonzero value when n reaches $777, 451, 915, 729, 368$ and then again when $n = 140, 894, 092, 055, 857, 794$. See [Sta07].

At the close of this first section, we confess that a survey of mathematics in the age of the Turing machine is a reckless undertaking, particularly if it almost completely neglects software products and essential mathematical algorithms—the Euclidean algorithm, Newton's method, Gaussian elimination, fast Fourier transform, simplex algorithm, sorting, Schönhage–Strassen, and many more. A starting point for the exploration of mathematical software is KNOPPIX/Math, a bootable DVD with over a hundred free mathematical software products (Figure 15) [Ham08]. Sage alone has involved over 200 developers and includes dozens of other packages, providing an open-source Python scripted alternative to computer algebra systems such as Maple and Mathematica.

TeX	Active-DVI, AUCTeX, TeXmacs, Kile, WhizzyTeX
computer algebra	Axiom, CoCoA4, GAP, Macaulay2, Maxima, PARI/GP, Risa/Asir, Sage, Singular, Yacas
numerical calc	Octave, Scilab, FreeFem++, Yorick
visualization	3D-XplorMath-J, Dynagraph, GANG, Geomview, gnuplot, JavaView, K3DSurf
geometry	C.a.R, Dr.Geo, GeoGebra, GEONExT, KidsCindy, KSEG
programming	CLISP, Eclipse, FASM, Gauche, GCC, Haskell, Lisp Prolog, Guile, Lazarus, NASM, Objective Caml, Perl, Python, Ruby, Squeak

FIGURE 15. Some free mathematical programs on the Knoppix/Math DVD [Ham08].

§2. Computer proof. Proof assistants represent the best effort of logicians, computer scientists, and mathematicians to obtain complete mathematical rigor by computer. This section gives a brief introduction to proof assistants and describes various recent projects that use them.

The first section described various computer calculations in math, and this section turns to computer reasoning. I have never been able to get used to it being the mathematicians who use computers for calculation and the computers scientists who use computers for proofs!

2.1. Design of proof assistants. A formal proof is a proof that has been checked at the level of the primitive rules of inference all the way back to the fundamental axioms of mathematics. The number of primitive inferences is generally so large that it is quite hopeless to construct a formal proof by hand of anything but theorems of the most trivial nature. McCarthy and de Bruijn suggested that we program computers to generate formal proofs from high-level descriptions of the proof. This suggestion has led to the development of proof assistants.

A *proof assistant* is an interactive computer program that enables a user to generate a formal proof from its high-level description. Some examples of theorems that have been formally verified by proof assistants appear in Figure 16. The computer code that implements a proof assistant lists the fundamental axioms of mathematics and gives procedures that implement each of the rules of logical inference. Within this general framework, there are enormous variations from one proof assistant to the next. The feature table in Figure 16 is reproduced from [Wie06]. The columns list different proof assistants, HOL, Mizar, etc.

Since it is the one that I am most familiar with, my discussion will focus largely on a particular proof assistant, *HOL Light*, which belongs to the *HOL* family of proof assistants. *HOL* is an acronym for Higher-Order Logic, which is the underlying logic of these proof assistants. A fascinating account of the history of HOL appears in [Gor00]. In 1972, R. Milner developed a proof-checking program based on a deductive system LCF (for Logic of Computable Functions) that had been designed by Dana Scott a few years earlier. A long series of innovations (such as goal-directed proofs and tactics, the ML language, enforcing proof integrity through the type system, conversions and theorem continuations, rewriting with discrimination nets, and higher-order features) have led from LCF to HOL.

Without going into full detail, I will make a few comments about what some of the features mean. Different systems can be commended in different ways: HOL Light for its small trustworthy kernel, Coq for its powerful type system, Mizar for its extensive libraries, and Isabelle/HOL for its support and usability.

Small proof kernel. If a proof assistant is used to check the correctness of proofs, who checks the correctness of the proof assistant itself? De Bruijn proposed that the proofs of a proof assistant should be capable of being checked by a short piece of computer code—something short enough to be checked by hand. For example, the kernel of the proof assistant HOL Light is just 430 lines of very readable computer code. The architecture of the system is such that if these 430 lines are bug free then it is incapable[5] of generating a theorem that hasn't been properly proved.

Automating calculations. Mathematical argument involves both calculation and proof. The foundations of logic often specify in detail what constitutes a mathematical proof (a sequence of logical inferences from the axioms), but

---

[5]I exaggerate. Section 3 goes into detail about trust in computers.

Year	Theorem	Proof System	Formalizer	Traditional Proof
1986	First Incompleteness	Boyer–Moore	Shankar	Gödel
1990	Quadratic Reciprocity	Boyer–Moore	Russinoff	Eisenstein
1996	Fundamental – of Calculus	HOL Light	Harrison	Henstock
2000	Fundamental – of Algebra	Mizar	Milewski	Brynski
2000	Fundamental – of Algebra	Coq	Geuvers et al.	Kneser
2004	Four Color	Coq	Gonthier	Robertson et al.
2004	Prime Number	Isabelle	Avigad et al.	Selberg–Erdös
2005	Jordan Curve	HOL Light	Hales	Thomassen
2005	Brouwer Fixed Point	HOL Light	Harrison	Kuhn
2006	Flyspeck I	Isabelle	Bauer–Nipkow	Hales
2007	Cauchy Residue	HOL Light	Harrison	classical
2008	Prime Number	HOL Light	Harrison	analytic proof
2012	Odd Order Theorem	Coq	Gonthier	Feit–Thompson

FIGURE 16. Examples of Formal Proofs, adapted from [Hal08].

downgrade calculation to second-class status, requiring every single calculation to undergo a cumbersome translation into logic. Some proof assistants allow *reflection* (sometimes implausibly attributed to *Poincaré*), which admits as proof the output from a verified algorithm (bypassing the expansive translation into logic of each separate execution of the algorithm) [Poi52, p. 4], [Bar07].

Constructive logic. The law of excluded middle $\phi \vee \neg \phi$ is accepted in classical logic, but rejected in constructive logic. A proof assistant may be constructive or classical. A box (*A Mathematical Gem*, see p. 275) shows how HOL Light becomes classical through the introduction of an axiom of choice.

Logical framework. Many different systems of logic arise in computer science. In some proof assistants the logic is fixed. Other proof assistants are more flexible, allowing different logics to be plugged in and played with. The more flexible systems implement a meta-language, a *logical framework*, that gives support for the implementation of multiple logics. Within a logical framework, the logic and axioms of a proof assistant can themselves be formalized, and machine translations[6] can be constructed between different foundations of mathematics [IR11].

---

[6]My long term Flyspeck project seeks to give a formal proof of the Kepler conjecture [Hal05a]. This project is now scattered between different proof assistants. Logical framework based translations between proof assistants gives me hope that an automated tool may assemble the scattered parts of the project.

proof assistant	HOL	Mizar	PVS	Coq	Otter/Ivy	Isabelle/Isar	Alfa/Agda	ACL2	PhoX	IMPS	Metamath	Theorema	Lego	Nuprl	Ωmega	B method	Minilog
small proof kernel ('proof objects')	+	-	-	+	+	+	+	-	+	-	+	-	+	-	+	-	+
calculations can be proved automatically	+	-	+	+	+	+	-	+	+	+	-	+	+	+	+	+	+
extensible/programmable by the user	+	-	+	+	-	+	-	-	-	-	-	-	-	+	+	-	+
powerful automation	+	-	+	-	+	+	-	+	-	+	-	+	-	-	+	+	-
readable proof input files	-	+	-	-	-	+	-	+	-	-	-	+	-	-	-	-	-
constructive logic supported	-	-	-	+	-	+	+	-	-	-	+	-	+	+	-	-	+
logical framework	-	-	-	-	-	+	-	-	-	-	+	-	-	-	-	-	-
typed	+	+	+	+	-	+	+	-	+	+	-	-	+	+	+	-	+
decidable types	+	+	-	+	-	+	+	-	+	+	-	-	+	-	+	-	+
dependent types	-	+	+	+	-	-	+	-	-	-	-	-	+	+	-	-	-
based on higher order logic	+	-	+	+	-	+	+	-	+	+	-	+	+	+	+	-	-
based on ZFC set theory	-	+	-	-	-	+	-	-	-	-	+	-	-	-	-	+	-
large mathematical standard library	+	+	+	+	-	+	-	-	-	+	-	-	-	+	-	-	-

FIGURE 17. Features of proof assistants [Wie06]. The table is published by permission from Springer Science Business Media B.V.

Type theory. Approaching the subject of formal proofs as a mathematician whose practice was shaped by Zermelo–Fraenkel set theory, I first treated types as nothing more than convenient identifying labels (such real number, natural number, list of integers, or boolean) attached to terms, like the PLU stickers on fruit that get peeled away before consumption. Types are familiar from programming languages as a way of identifying what data structure is what. In the simple type system of HOL Light, to each term is affixed a unique type, which is either a primitive type (such as the boolean type *bool*), a type variable $(A, B, C, \ldots)$, or inductively constructed from other types with the arrow constructor $(A \to B, A \to (bool \to C)$, etc.). There is also a way to create subtypes of existing types. If the types are interpreted naively as sets, then $x{:}A$ asserts that the term x is a member of A, and $f\colon A \to B$ asserts that f is a member of $A \to B$, the set of functions from A to B.

In untyped set theory, it is possible to ask ridiculous questions such as whether the real number $\pi = 3.14\ldots$, when viewed as a raw set, is a finite group. In fact, in a random exploration of set theory, like a monkey composing sonnets at the keyboard, ridiculous questions completely overwhelm all serious

> *A Mathematical Gem—Proving the Excluded Middle*
>
> The logic of HOL Light is intuitionistic until the axiom of choice is introduced and classical afterwards. By a result of Diononescu [Bee85], choice and extensionality imply the law of excluded middle:
>
> $$\phi \vee \neg\phi.$$
>
> The proof is such a gem that I have chosen to include it as the only complete proof in this survey article. Consider the two sets of booleans
>
> $$P_1 = \{x \mid (x = \text{false}) \vee ((x = \text{true}) \wedge \phi)\} \quad \text{and}$$
> $$P_2 = \{x \mid (x = \text{true}) \vee ((x = \text{false}) \wedge \phi))\}.$$
>
> The sets are evidently nonempty, because false $\in P_1$ and true $\in P_2$. By choice, we may pick $x_1 \in P_1$ and $x_2 \in P_2$; and by the definition of P_1 and P_2:
>
> $$(x_1 = \text{false}) \vee (x_1 = \text{true}), \quad (x_2 = \text{false}) \vee (x_2 = \text{true}).$$
>
> We may break the proof of the excluded middle into four cases, depending on the two possible truth values of each of x_1 and x_2.
>
> **Cases** $(x_1, x_2) = (\text{true}, \text{true})$, $(x_1, x_2) = (\text{true}, \text{false})$: By the definition of P_1, if $x_1 = \text{true}$, then ϕ, so $\phi \vee \neg\phi$.
>
> **Case** $(x_1, x_2) = (\text{false}, \text{false})$: Similarly, by the definition of P_2, if $x_2 = \text{false}$, then ϕ, so also $\phi \vee \neg\phi$.
>
> **Case** $(x_1, x_2) = (\text{false}, \text{true})$: If ϕ, then $P_1 = P_2$, and the choices x_1 and x_2 reduce to a single choice $x_1 = x_2$, which contradicts $(x_1, x_2) = (\text{false}, \text{true})$. Hence ϕ implies false; which by the definition of negation gives $\neg\phi$, so also $\phi \vee \neg\phi$. *Q.E.D.*

content. Types organize data on the computer in meaningful ways to cut down on the static noise in the system. The question about π and groups is not well-typed and cannot be asked. Russell's paradox also disappears: $X \notin X$ is not well-typed. For historical reasons, this is not surprising: Russell and Whitehead first introduced types to overcome the paradoxes of set theory, and from there, through Church, they passed into computer science.

Only gradually have I come to appreciate the significance of a comprehensive *theory of types*. The type system used by a proof assistant determines to a large degree how much of a proof the user must contribute and how much the computer automates behind the scenes. The type system is *decidable* if there is a decision procedure to determine the type of each term.

A type system is *dependent* if a type can depend on another term. For example, Euclidean space $\mathbb{R}^n$, depends on its dimension n. For this reason, Euclidean space is most naturally implemented in a proof assistant as a dependent type. In a proof assistant such as HOL Light that does not have dependent types, extra work is required to develop a Euclidean space library.

2.2. Propositions as types. I mentioned the naive interpretation of each type A as a set and a term $x{:}A$ as a member of the set. A quite different interpretation of types has had considerable influence in the design of proof assistants. In this "terms-as-proofs" view, a type A represents a proposition and a term $x{:}A$ represents a proof of the proposition A. A term with an arrow type, $f\colon A \to B$, can be used to to construct a proof $f(x)$ of B from a proof x of A. In this interpretation, the arrow is logical implication.

A further interpretation of types comes from programming languages. In this "terms-as-computer-programs" view, a term is a program and the type is its specification. For example, $f\colon A \to B$ is a program f that takes input of type A and returns a value of type B.

By combining the "terms as proofs" with the "terms as computer programs" interpretations, we get the famous *Curry–Howard correspondence* that identifies proofs with computer programs and identifies each proposition with the type of a computer program. For example, the most fundamental rule of logic,

$$\frac{A,\quad A\to B}{B}, \quad (modus\ ponens)$$

(from A and A-implies-B follows B) is identified with the function application in a computer program; from $x{:}A$ and $f\colon A \to B$ we get $f(x){:}B$. To follow the correspondence is to extract an executable computer program from a mathematical proof. The Curry–Howard correspondence has been extremely fruitful, with a multitude of variations, running through a gamut of proof systems in logic and identifying each with a suitable programming domain.

2.3. Proof tactics. In some proof assistants, the predominant proof style is a backward style proof. The user starts with a *goal*, which is a statement to be proved. In interactive steps, the user reduces the goal to successively simpler goals until there is nothing left to prove.

Each command that reduces a goal to simpler goals is called a *tactic*. For example, in the proof assistant HOL Light, there are about 100 different commands that are tactics or higher-order operators on tactics (called tacticals). Figure 18 shows the most commonly used proof commands in HOL Light. The most common tactic is *rewriting*, which takes a theorem of the form $a = b$ and substitutes b for an occurrence of a in the goal.

In the Coq proof assistant, the tactic system has been streamlined to an extraordinary degree by the *SSReflect* package, becoming a model of efficiency for other proof assistants to emulate, with an extremely small number of tactics such as the *move* tactic for bookkeeping, one for rewriting, ones for forward and backward reasoning, and another for case analysis [GM11], [GMT11].

name	purpose	usage
THEN	combine two tactics into one	37.2%
REWRITE	use $a = b$ to replace a with b in goal	14.5%
MP_TAC	introduce a previously proved theorem	4.0%
SIMP_TAC	rewriting with conditionals	3.1%
MATCH_MP_TAC	reduce a goal b to a, given a theorem $a \implies b$	3.0%
STRIP_TAC	(bookkeeping) unpackage a bundled goal	2.9%
MESON_TAC	apply first-order reasoning to solve the goal	2.6%
REPEAT	repeat a tactic as many times as possible	2.5%
DISCH_TAC	(bookkeeping) move hypothesis	2.3%
EXISTS_TAC	instantiate an existential goal $\exists x \ldots$	2.3%
GEN_TAC	instantiate a universal goal $\forall x \ldots$	1.4%

FIGURE 18. A few of the most common proof commands in the HOL Light proof assistant.

The package also provides support for exploiting the computational content of proofs, by integrating logical reasoning with efficient computational algorithms.

2.4. First-order automated reasoning. Many proof assistants support some form of automated reasoning to relieve the user of doing rote logic by hand. For example, Table 18 lists *meson* (an acronym for Loveland's Model Elimination procedure), which is HOL Light's tactic for automated reasoning [Har09, Sec. 3.15], [Har96]. The various automated reasoning tools are generally *first-order* theorem provers. The classic resolution algorithm for first-order reasoning is illustrated in a box (*Proof by Resolution*, see page 278).

Writing about first-order automated reasoning, Huet and Paulin–Mohring [BC04] describe the situation in the early 1970s as a "catastrophic state of the art." "The standard mode of use was to enter a conjecture and wait for the computer's memory to exhaust its capacity. Answers were obtained only in exceptionally trivial cases." They go on to describe numerous developments (Knuth–Bendix, LISP, rewriting technologies, LCF, ML, Martin-Löf type theory, NuPrl, Curry–Howard correspondence, dependent types, etc.) that led up to the Coq proof assistant. These developments led away from first-order theorem proving with its "thousands of unreadable logical consequences" to a highly structured approach to theorem proving in Coq.

First-order theorem proving has developed significantly over the years into sophisticated software products. They are no longer limited to "exceptionally limited cases." Many different software products compete in an annual competition (CASC), to see which can solve difficult first-order problems the fastest. The LTB (large theory batch) division of the competition includes problems with thousands of axioms [PSST08]. Significantly, this is the same

Proof by Resolution

Resolution is the granddaddy of automated reasoning in first-order logic. The resolution rule takes two disjunctions

$$P \vee A \quad \text{and} \quad \neg P' \vee B$$

and concludes

$$A' \vee B',$$

where A' and B' are the specializations of A and B, respectively, under the *most general unifier* of P and P'. (Examples of this in practice appear below.)

This box presents a rather trivial example of proof by resolution, to deduce the easy theorem asserting that every infinite set has a member. The example will use the following notation. Let $\emptyset$ be a constant representing the empty set and constant c representing a given infinite set. We use three unary predicates e, f, i that have interpretations

$$e(X) \text{ "}X \text{ is empty"}, \quad f(X) \text{ "}X \text{ is finite"}, \quad i(X) \text{ "}X \text{ is infinite."}$$

The binary predicate ($\in$) denotes set membership. We prove

$$i(c) \Rightarrow (\exists z. z \in c)$$

"an infinite set has a member" by resolution.

To argue by contradiction, we introduce the hypothesis $i(c)$ and the negated conclusion $\neg(Z \in c)$ as axioms. Here are the axioms that we allow in the deduction. The axioms have been preprocessed, stripped of quantifiers, and written as a disjunction of literals. Upper case letters are variables.

Axiom	Informal Description
1. $i(c)$	Assumption of desired theorem.
2. $\neg(Z \in c)$	Negation of conclusion of desired theorem.
3. $e(X) \vee (u(X) \in X)$	A nonempty set has a member.
4. $e(\emptyset)$	The empty set is empty.
5. $f(\emptyset)$	The empty set is finite.
6. $\neg i(Y) \vee \neg f(Y)$	A set is not both finite and infinite.
7. $\neg e(U) \vee \neg e(V) \vee \neg i(U) \vee i(V)$	Weak indistinguishability of empty sets.

Here are the resolution inferences from this list of axioms. The final step obtains the desired contradiction.

Inference	Resolvant
8. (resolving 2,3, unifying X with c and $u(X)$ with Z)	$e(c)$
9. (resolving 7,8, unifying U with c)	$\neg e(V) \vee \neg i(c) \vee i(V)$
10. (resolving 1,9)	$\neg e(V) \vee i(V)$
11. (resolving 4,10, unifying V with $\emptyset$)	$i(\emptyset)$
12. (resolving 6,11, unifying Y with $\emptyset$)	$\neg f(\emptyset)$
13. (resolving 12,5)	$\bot$.

Q.E.D.

order of magnitude as the total number of theorems in a proof assistant. What this means is that a first-order theorem provers have reached the stage of development that they might be able to give fully automated proofs of new theorems in a proof assistant, working from the full library of previously proved theorems.

Sledgehammer. The Sledgehammer tactic is Paulson's implementation of this idea of full automation in the Isabelle/HOL proof assistant [Pau10]. As the name 'Sledgehammer' suggests, the tactic is all-purpose and powerful, but demolishes all higher mathematical structure, treating every goal as a massive unstructured problem in first-order logic. If L is the set of all theorems in the Isabelle/HOL library, and g is a goal, it would be possible to hand off the problem $L \implies g$ to a first-order theorem prover. However, success rates are dramatically improved, when the theorems in L are first assessed by heuristic rules for their likely relevance for the goal g, in a process called *relevance filtering*. This filtering is used to reduce L to an axiom set L' of a few hundred theorems that are deemed most likely to prove g.

The problem $L' \implies g$ is stripped of type information, converted to a first-order, and fed to first-order theorem provers. Experiments indicate that it is more effective to feed a problem in parallel into multiple first-order provers for a five-second burst than to hand the problem to the best prover (Vampire) for a prolonged attack [Pau10], [BN10]. When luck runs in your favor, one of the first-order theorem provers finds a proof.

The reconstruction of a formal proof from a first-order proof can encounter hurdles. For one thing, when type information is stripped from the problem (which is done to improve performance), soundness is lost. "In unpublished work by Urban, MaLARea [a machine learning program for relevance ranking] easily proved the full Sledgehammer test suite by identifying an inconsistency in the translated lemma library; once MaLARea had found the inconsistency in one proof, it easily found it in all the others" [Pau10], [Urb07]. Good results have been obtained in calling the first-order prover repeatedly to find a smaller set of axioms $L'' \subset L'$ that imply the goal g. A manageably sized set L'' is then passed to the Metis tactic[7] in Isabelle/HOL, which constructs a formal proof $L'' \implies g$ from scratch.

Böhme and Nipkow took 1240 proof goals that appear in several diverse theories of the Isabelle/HOL system and ran sledgehammer on all of them [BN10]. The results are astounding. The success rate (of obtaining fully reconstructed formal proofs) when three different first-order provers run for two-minutes each was 48%. The proofs of these same goals by hand might represent years of human labor, now fully automated through a single new tool.

Sledgehammer has led to a new style of theorem proving, in which the user is primarily responsible for stating the goals. In the final proof script, there is no

---
[7]Metis is a program that automates first-order reasoning [Met].

explicit mention of sledgehammer. Metis proves the goals, with sledgehammer operating silently in the background to feed Metis with whatever theorems it needs. For example, a typical proof script might contain lines such as [Pau10]:

> **hence** "$x \subseteq$ space M"
> **by** (Metis sets into space lambda system sets)

The first line is the goal that the user types. The second line has been automatically inserted into the proof script by the system, with the relevant theorems sets, into etc. selected by Sledgehammer.

2.5. Computation in proof assistants. One annoyance of formal proof systems is the difficulty in locating the relevant theorems. At last count, HOL Light had about 14,000 theorems and nearly a thousand procedures for proof construction. Larger developments, such as Mizar, have about twice as many theorems. Good search tools have somewhat relieved the burden of locating theorems in the libraries. However, as the formal proof systems continue to grow, it becomes ever more important to find ways to use theorems without mentioning them by name.

As an example of a feature which commendably reduces the burden of memorizing long lists of theorem names, I mention the REAL_RING command in HOL Light, which is capable of proving any system of equalities and inequalities that holds over an arbitrary integral domain. For example, I can give a one-line formal proof of an isogeny $(x_1, y_1) \mapsto (x_2, y_2)$ of elliptic curves: if we have a point on the first elliptic curve:

$$y_1^2 = 1 + ax_1^2 + bx_1^4,$$
$$x_2 y_1 = x_1,$$
$$y_2 y_1^2 = (1 - bx_1^4),$$
$$y_1 \neq 0$$

then (x_2, y_2) lies on a second elliptic curve

$$y_2^2 = 1 + a' x_2^2 + b' x_2^4,$$

where $a' = -2a$ and $b' = a^2 - 4b$. In the proof assistant, the input of the statement is as economical as what I have written here. We expect computer algebra systems to be capable of checking identities like this, but to my amazement, I found it *easier* to check this isogeny in HOL Light than to check it in *Mathematica*.

The algorithm works in the following manner. A universally quantified system of equalities and inequalities holds over all integral domains if and only if it holds over all fields. By putting the formula in conjunctive normal form, it is enough to prove a finite number of polynomial identities of the form:

$$(p_1 = 0) \vee \cdots \vee (p_n = 0) \vee (q_1 \neq 0) \vee \cdots \vee (q_k \neq 0). \quad (2)$$

An element in a field is zero, if and only if it is not a unit. Thus we may rewrite each polynomial equality $p_i = 0$ as an equivalent inequality $1 - p_i z_i \neq 0$. Thus, without loss of generality, we may assume that $n = 0$; so that all disjuncts are inequalities. The formula (2) is logically equivalent to

$$(q_1 = 0) \wedge \cdots \wedge (q_k = 0) \implies \text{false}.$$

In other words, it is enough to prove that the zero set of the ideal $I = (q_1, \ldots, q_n)$ is empty. For this, we may use Gröbner bases[8] to prove that $1 \in I$, to certify that the zero set is empty.

Gröbner basis algorithms give an example of a *certificate-producing procedure*. A formal proof is obtained in two stages. In the first stage an unverified algorithm produces a certificate. In the second stage the proof assistant analyzes the certificate to confirm the results. Certificate-producing procedures open the door to external tools, which tremendously augment the power of the proof assistant. The meson is procedure implemented this way, as a search followed by verification. Other certificate-producing procedures in use in proof assistants are linear programming, SAT, and SMT.

Another praiseworthy project is Kaliszyk and Wiedijk's implementation of a computer algebra system on top of the proof assistant HOL Light. It combines the ease of use of computer algebra with the rigor of formal proof [KW07]. Even with its notational idiosyncrasies (`&` and `#` as a markers of real numbers, `Cx` as a marker of complex numbers, `ii` for $\sqrt{-1}$, and `--` for unary negation), it is the kind of product that I can imagine finding widespread adoption by mathematicians. Some of the features of the system are shown in Figure 19.

2.6. Formalization of finite group theory. The Feit–Thompson theorem, or odd-order theorem, is one of the most significant theorems of the twentieth century. (For his work, Thompson was awarded the three highest honors in the mathematical world: the Fields Medal, the Abel Prize, and the Wolf Prize.) The Feit–Thompson theorem states that every finite simple group has even order, except for cyclic groups of prime order. The proof, which runs about 250 pages, is extremely technical. The Feit–Thompson theorem launched the endeavor to classify all finite simple groups, a monumental undertaking that consumed an entire generation of group theorists.

Gonthier's team has formalized the proof of the Feit–Thompson theorem [Gon12]. To me as a mathematician, nothing else that has been done by the formal proof community compares in splendor to the formalization of this theorem. Finally, we are doing real mathematics! The project formalized

---

[8] Kaliszyk's benchmarks suggest that the Gröbner basis algorithm in the proof assistant Isabelle runs about twenty times faster than that of HOL Light.

```
In1  := (3 + 4 DIV 2) EXP 3 * 5 MOD 3
Out1 := 250
In2  := vector [&2; &2] - vector [&1; &0] + vec 1
Out2 := vector [&2; &3]
In3  := diff (diff (\x. &3 * sin (&2 * x) + &7 + exp (exp x)))
Out3 := \x. exp x pow 2 * exp (exp x) + exp x * exp (exp x) +
             -- &12 * sin (&2 * x)
In4  := N (exp (&1)) 10
Out4 := #2.7182818284 + ... (exp (&1)) 10 F
In5  := 3 divides 6 /\ EVEN 12
Out5 := T
In6  := Re ((Cx (&3) + Cx (&2) * ii) /
            (Cx (-- &2) + Cx (&7) * ii))
Out6 := &8 / &53
```

FIGURE 19. Interaction with a formally verified computer algebra system [KW07].

two books, [BG94] and [Pet00], as well as a significant body of background material.

The structures of abstract algebra—groups, rings, modules, algebras, algebraically closed fields and so forth—have all been laid out formally in the Coq proof assistant. Analogous algebraic hierarchies appear in systems such as OpenAxiom, MathScheme, Mizar, and Isabelle; and while some of these hierarchies are elaborate, none have delved so deeply as the development for Feit–Thompson. It gets multiple abstract structures to work coherently together in a formal setting. "The problem is not so much in capturing the semantics of each individual construct but rather in having all the concepts working together well" [GMR07].

```
Structure finGroupType Type := FinGroupType {
   element :> finType;
        1 : element;
       -1 : element → element;
        * : element → element → element;
    unitP : ∀x, 1 * x = x;
     invP : ∀x, x⁻¹ * x = 1;
     mulP : ∀x₁ x₂ x₃, x₁ * (x₂ * x₃) = (x₁ * x₂) * x₃
}.
```

FIGURE 20. The structure of a finite group [GMR07].

The definition of a finite group in Coq is similar to the textbook definition, expressed in types and structures (Figure 20). It declares a finite type called element that is the group carrier or domain. The rest of the structure specifies a left-unit element 1, a left-inverse $^{-1}$ and an associative binary operation $(*)$.

Other aspects of Gonthier's recent work can be found at [Gon11], [GGMR09], [BGBP08]. Along different lines, a particularly elegant organization of abstract algebra and category theory is obtained with type classes [SvdW11].

2.7. Homotopy type theory. The simple type theory of HOL Light is adequate for real analysis, where relatively few types are needed—one can go quite far with natural numbers, real numbers, booleans, functions between these types, and a few functionals. However, the dependent type theory of Coq is better equipped than HOL Light for the hierarchy of structures from groups to rings of abstract algebra. But even Coq's type theory is showing signs of strain in dealing with abstract algebra. For instance, an unpleasant limitation of Coq's theory of types is that it lacks the theorem of extensionality for functions: if two functions take the same value for every argument, it *does not* follow that the two functions are equal.[9] The gymnastics to solve the problem of function extensionality in the context of the Feit–Thompson theorem are found in [GMR07].

A lack of function extensionality is an indication that equality in type theory may be misconceived. Recently, *homotopy type theory* has exploded onto the scene, which turns to homotopy theory and higher categories as models of type theory [HTT11]. It is quite natural to interpret a dependent type (viewed as a family of types parametrized by a second type) topologically as a fibration (viewed as a family of fibers parametrized by a base space) [AW09]. Voevodsky took the homotopical notions of equality and equivalence and translated them back into type theory, obtaining the *univalence axiom* of type theory, which posits what types are equivalent [Voe11], [PW12], [KLV12b], [KLV12a]. One consequence of the univalence axiom is the theorem of extensionality for functions. Another promising sign for computer theorem-proving applications is that the univalence axiom appears to preserve the computable aspects of type theory (unlike for instance, the axiom of choice which makes non-computable choices) [LH11]. We may hope that some day there may be a back infusion of type-theoretic proofs into homotopy theory.

2.8. Language of mathematics. Ganesalingam's thesis is the most significant linguistic study of the language of mathematics to date [Gan09], [Gan10]. Ganesalingam was awarded the 2011 Beth Prize for the best dissertation in Logic, Language, or Information. Although this research is still at an early stage, it suggests that the mechanical translation of mathematical prose into formal computer syntax that faithfully represents the semantics is a realistic hope for the not-to-distant future.

---

[9]HOL Light avoids this problem by positing extensionality as a mathematical axiom.

The linguistic problems surrounding the language of mathematics differ in various ways from those of say standard English. A mathematical text introduces new definitions and notations as it progresses, whereas in English, the meaning of words is generally fixed from the outset. Mathematical writing freely mixes English with symbolic expressions. At the same time, mathematics is self-contained in a way that English can never be; to understand English is to understand the world. By contrast, the meaning in a carefully written mathematical text is determined by Zermelo–Fraenkel set theory (or your favorite foundational system).

Ganesalingam's analysis of notational syntax is general enough to treat quite general mixfix operations generalizing infix (e.g. $+$), postfix (e.g. factorial !), and prefix (cos). He analyzes subscripted infix operators (such as a semidirect product $H \rtimes_\alpha N$), multi-symboled operators (such as the three-symboled [:] operator for the degree $[K : k]$ of a field-extension), prefixed words (R-module), text within formulas $\{(a,b) \mid a \text{ is a factor of } b\}$, unusual script placement $^L G$, chained relations $a < b < c$, ellipses $1+2+\cdots+n$, contracted forms $x, y \in \mathbb{N}$, and exposed formulas (such as "for all $x > 0, \ldots$" to mean "for all x, if $x > 0$, then $\ldots$").

The thesis treats what is called the formal mode of mathematics—the language divested of all the informal side-remarks. The syntax is treated as a context-free grammar, and the semantics are analyzed with a variant of *discourse representation theory*, which in my limited understanding is something very similar to first-order logic; but different in one significant aspect: it provides a theory of pronoun references; or put more precisely, a theory of what may be the "legitimate antecedent for anaphor."

A major issue in Ganesalingam's thesis is the resolution of ambiguity. For example, in the statement

$$P \text{ is prime} \qquad (3)$$

the term 'prime' may mean prime number, prime ideal, or prime manifold. His solution is to attach type information to terms (in the sense of types as discussed above). The reading of (3) depends on the type of P, variously a number, a subset of a ring, or a manifold. In this analysis, resolution of ambiguity becomes a task of a type inference engine.

Because of the need for type information, Ganesalingam raises questions about the suitability of Zermelo–Fraenkel set theory as the ultimate semantics of mathematics. A number of formal-proof researchers have been arguing in favor of typed foundational systems for many years. It is encouraging that there is remarkable convergence between Ganesalingam's linguistic analysis, innovations in the Mizar proof assistant, and the development of abstract algebra in Coq. For example, in various camps we find ellipses (aka big operators), mixfix operators, type inference, missing argument inference mechanisms, and so forth. Also see [Hoe11] and [Pas07]. Mathematical abuses of notation have turned out to be rationally construed after all!

2.9. Looking forward. Let's take the long term view that the longest proofs of the last century are of insignificant complexity compared to what awaits. Why would we limit our creative endeavors to 10,000 page proofs when we have tools that allow us to go to a million pages or more? So far it is rare for a computer proof has defied human understanding. No human has been able to make sense of an unpublished 1500 page computer-generated proof about Bruck loops[10] [PS08]. Eventually, we will have to content ourselves with fables that approximate the content of a computer proof in terms that humans can comprehend.

Turing's great theoretical achievements were to delineate what a computer can do in the concept of a universal Turing machine, to establish limits to what a computer can do in his solution to the *Entscheidungsproblem*, and yet to advocate nonetheless that computers might imitate all intelligent activity. It remains a challenging research program: to show that one limited branch of mathematics, computation, might stand for all mathematical activity.

In the century since Turing's birth, the computer has become so ubiquitous and the idea of computer as brain so commonplace that it bears repeating that we must still think very long and hard about how to construct a computer that can imitate a living, thinking mathematician.

Proof assistant technology is still under development in labs; far more is needed before it finds widespread adoption. Ask any proof assistant researcher, and you will get a sizable list of features to implement: more automation, better libraries, and better user interfaces! Wiedijk discusses ten design questions for the next generation of proof assistants, including the type system, which axiomatic foundations of mathematics to use, and the language of proof scripts [Wie10b].

Everyone actively involved in proof formalization experiences the incessant barrage of problems that have been solved multiple times before and that other users will have to solve multiple times again, because the solutions are not systematic. To counter this, the DRY "Don't Repeat Yourself" principle of programming, formulated in [HT00], has been carried to a refreshing extreme by Carette in his proof assistant design. For example, in his designs, a morphism is defined only once, eliminating the need for separate definitions of a morphism of modules, of algebras, of varieties, and so forth. Carette's other design maxims include "math has a lot of structure; use it" and "abstract mathematical structures produce the best code" [CES11]. Indeed, mathematicians turn to abstraction to bring out relevant structure. This applies to computer code and mathematical reasoning alike. American Math Society

---

[10]The theorem states that Bruck loops with abelian inner mapping group are centrally nilpotent of class two.

guidelines for mathematical writing apply directly to the computer: "omit any computation which is routine.... Merely indicate the starting point, describe the procedure, and state the outcome" [DCF+62] (except that computations should be automated rather than entirely omitted).

We need to separate the concerns of construction, maintenance, and presentation of proofs. The construction of formal proofs from a mathematical text is an extremely arduous process, and yet I often hear proposals that would increase the labor needed to formalize a proof, backed by secondary goals such as ease of maintenance, elegance of presentation, fidelity to printed texts, and pedagogy.[11] Better to avail ourselves of automation that was not available in the day of paper proofs, and to create new mathematical styles suited to the medium, with proofs that variously look like a computer-aided design session, a functional program, or a list of hypotheses as messages in gmail. The most pressing concern is to reduce the skilled labor it takes a user to construct a formal proof from a pristine mathematical text.

The other concerns of proof transformation should be spun off as separate research activities: refactored proofs, proof scripts optimized for execution time, translations into other proof assistants, natural language translations, natural language abstracts, probabilistically checkable proofs, searchable metadata extracts, and proof mining.

For a long time, proof formalization technology was unable to advance beyond the mathematics of the 19th century, picking classical gems such as the Jordan curve theorem, the prime number theorem, or Dirichlet's theorem on primes in arithmetic progressions. With the Feit–Thompson theorem, formalization has risen to a new level, by taking on the work of a Fields medalist.

At this level, there is an abundant supply of mathematical theorems to choose from. A Dutch research agenda lists the formalization of Fermat's Last Theorem as the first in a list of "Ten Challenging Research Problems for Computer Science" [Ber05]. Hesselink predicts that this one formalization project alone will take about "fifty years, with a very wide margin." Small pieces of the proof of Fermat, such as class field theory, the Langlands–Tunnell theorem, or the arithmetic theory of elliptic curves would be a fitting starting point. The aim is to develop technologies until formal verification of theorems becomes routine at the level of Atiyah–Singer index theorem, Perelman's proof of the Poincaré conjecture, the Green–Tao theorem on primes in arithmetic progression, or Ngô's proof of the fundamental lemma.

---

[11] To explain the concept of *separation of concerns*, Dijkstra tells the story of an old initiative to create a new programming language that failed miserably because the designers felt that the new language had to look just like *FORTRAN* to gain broad acceptance. "The proper technique is clearly to postpone the concerns for general acceptance until you have reached a result of such a quality that it deserves acceptance" [Dij82].

Starting from the early days of Newell, Shaw, and Simon's experiments, researchers have dreamed of a general-purpose mechanical problem solver. Generations later, after untold trials, it remains an unwavering dream. I will end this section with one of the many proposals for a general problem solving algorithm. Kurzweil breaks general problem solving into three phases:
1. State your problem in precise terms.
2. Map out the contours of the solution space by traversing it recursively, within the limits of available computational resources.
3. Unleash an evolutionary algorithm to configure a neural net to tackle the remaining leaves of the tree.

He concludes, "And if all of this doesn't work, then you have a difficult problem indeed" [Kur99]. Yes, indeed we do! Some day, energy and persistence will conquer.

§3. Issues of trust. We all have first-hand experience of the bugs and glitches of software. We exchange stories when computers run amok. Science recently reported the story of a textbook "The Making of a Fly" that was on sale at Amazon for more than 23 million dollars [Sci11]. The skyrocketing price was triggered by an automated bidding war between two sellers, who let their algorithms run unsupervised. The textbook's author, Berkeley professor Peter Lawrence, said he hoped that the price would reach "a billion." An overpriced textbook on the fly is harmless, except for students who have it as a required text.

But what about the Flash Crash on Wall Street that brought a 600 point plunge in the Dow Jones in just 5 minutes at 2:41 pm on May 6, 2010? According to the New York Times [NYT10], the flash crash started when a mutual fund used a computer algorithm "to sell $4.1 billion in futures contracts." The algorithm was designed to sell "without regard to price or time. . . . [A]s the computers of the high-frequency traders traded [futures] contracts back and forth, a 'hot potato' effect was created." When computerized traders backed away from the unstable markets, share prices of major companies fluctuated even more wildly. "Over 20,000 trades across more than 300 securities were executed at prices more than 60% away from their values just moments before" [SEC10]. Throughout the crash, computers followed algorithms to a T, to the havoc of the global economy.

3.1. Mathematical error. Why use computers to verify mathematics? The simple answer is that carefully implemented proof checkers make fewer errors than mathematicians (except J.-P. Serre).

Incorrect proofs of correct statements are so abundant that they are impossible to catalogue. Ralph Boas, former executive editor of Math Reviews, once remarked that proofs are wrong "half the time" [Aus08]. Kempe's claimed proof of the four-color theorem stood for more than a decade before Heawood refuted it [Mac01, p. 115]. "More than a thousand false proofs [of Fermat's Last Theorem] were published between 1908 and 1912 alone" [Cor10]. Many published theorems are like the hanging chad ballots of the 2000 U.S. presidential election, with scrawls too ambivalent for a clear yea or nay. One mathematician even proposed to me that a new journal is needed that unlike the others only publishes reliable results. Euclid gave us a method, but even he erred in the proof of the very first proposition of the Elements when he assumed without proof that two circles, each passing through the other's center, must intersect. The concept that is needed to repair the gap in Euclid's reasoning is an intermediate value theorem. This defect was not remedied until Hilbert's 'Foundations of Geometry.'

Examples of widely accepted proofs of false or unprovable statements show that our methods of proof-checking are far from perfect. Lagrange thought he had a proof of the parallel postulate, but had enough doubt in his argument to withhold it from publication. In some cases, entire schools have become sloppy, such as the Italian school of algebraic geometry or real analysis before the revolution in rigor towards the end of the nineteenth century. Plemelj's 1908 accepted solution to Hilbert's 21st problem on the monodromy of linear differential equations was refuted in 1989 by Bolibruch. Auslander gives the example of a theorem[12] published by Waraskiewicz in 1937, generalized by Choquet in 1944, then refuted with a counterexample by Bing in 1948 [Aus08]. Another example is the approximation problem for Sobolev maps between two manifolds [Bet91], which contains a faulty proof of an incorrect statement. The corrected theorem appears in [HL03]. Such examples are so plentiful that a Wiki page has been set up to classify them, with references to longer discussions at Math Overflow [Wik11], [Ove09], [Ove10].

Theorems that are calculations or enumerations are especially prone to error. Feynman laments, "I don't notice in the morass of things that something, a little limit or sign, goes wrong. ... I have mathematically proven to myself so many things that aren't true" [Fey00, p. 885]. Elsewhere, Feynman describes two teams of physicists who carried out a two-year calculation of the electron magnetic moment and independently arrived at the same predicted value. When experiment disagreed with prediction, the discrepancy was eventually traced to an arithmetic error made by the physicists, whose calculations were not so independent as originally believed [Fey85, p. 117]. Pontryagin and Rokhlin erred in computing stable homotopy groups of spheres. Little's tables of knots from 1885 contains duplicate entries that went undetected until 1974. In enumerative geometry, in 1848, Steiner counted 7776 plane conics tangent to 5 general plane conics, when there are actually only 3264. One of the most persistent blunders in the history of mathematics has been the misclassification (or misdefinition) of convex Archimedean polyhedra. Time and again, the

---

[12] The claim was that every homogeneous plane continuum is a simple closed curve.

pseudo-rhombic cuboctahedron has been overlooked or illogically excluded from the classification (Figure 21) [Gru11].

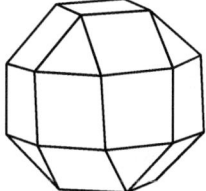

FIGURE 21. Throughout history, the pseudo-rhombic cuboctahedron has been overlooked or misclassified.

3.2. In HOL Light we trust. To what extent can we trust theorems certified by a proof assistant such as HOL Light? There are various aspects to this question. Is the underlying logic of the system consistent? Are there any programming errors in the implementation of the system? Can a devious user find ways to create bogus theorems that circumvent logic? Are the underlying compilers, operating system, and hardware reliable?

As mentioned above, formal methods represent the best cumulative effort of logicians, computer scientists and mathematicians over the decades and even over the centuries to create a trustworthy foundation for the practice of mathematics, and by extension, the practice of science and engineering.

3.3. A network of mutual verification. John Harrison repeats the classical question *"Quis custodiet ipsos custodes"*—who guards the guards [Har06]? How do we prove the correctness of the prover itself? In that article, he proves the consistency of the HOL Light logic and the correctness of its implementation in computer code. He makes this verification in HOL Light itself! To skirt Gödel's theorem, which implies that HOL Light—if consistent—cannot prove its own consistency, he gives two versions of his proof. The first uses HOL Light to verify a weakened version of HOL Light that does not have the axiom of infinity. The second uses a HOL Light with a strengthened axiom of infinity to verify standard HOL Light.

Recently, Adams has implemented a version of HOL called HOL Zero. His system has the ability to import mechanically proofs that were developed in HOL Light [Ada09]. He imported the self-verification of HOL Light, to obtain an external verification. You see where this is going. As mechanical translation capabilities are developed for proof assistants, it becomes possible for different proof assistants to share consistency proofs, similar to the way that different axiomatic systems give relative consistency proofs of one another. We are headed in the direction of knowing that if the logic or implementation of one proof assistant has an error, then all other major proof assistants must fail in tandem. Other self-verification projects are Coq in Coq (Coc) and ACL2 in ACL2 (Milawa) [Bar98], [Dav09].

3.4. Hacking HOL. Of course, every formal verification project is a verification of an abstract model of the computer code, the computer language, and its semantics. In practice, there are gaps between the abstract model and implementation.

This leaves open the possibility that a hacker might find ways to create an unauthorized theorem; that is, a theorem generated by some means other than the rules of inference of HOL Logic. Indeed, there are small openings that a hacker can exploit.[13] Adams maintains a webpage of known vulnerabilities in his system and offers a cash bounty to anyone who uncovers a new vulnerability.

These documented vulnerabilities need to be kept in perspective. They lie at the fringe of the most reliable software products ever designed. Proof assistants are used to verify the correctness of chips and microcode [Fox03], operating system kernels [KAE+10], compilers [Ler06], safety-critical software such as aircraft guidance systems, security protocols, and mathematical theorems that defeat the usual refereeing process.

Some take the view that nothing short of absolute certainty in mathematics gives an adequate basis for science. Poincaré was less exacting,[14] only demanding the imprecision of calculation not to exceed experimental error. As Harrison reminds us, "a foundational death spiral adds little value" [Har10].

3.5. Soft errors. Mathematicians often bring up the "cosmic ray argument" against the use of computers in math. Let's look at the underlying science.

A soft error in a computer is a transient error that cannot be attributed to permanent hardware defects nor to bugs in software. Hard errors—errors that can be attributed to a lasting hardware failure—also occur, but at rates that are ten times smaller than soft errors [MW04]. Soft errors come from many sources. A typical soft error is caused by cosmic rays, or rather by the shower of energetic neutrons they produce through interactions in the earth's atmosphere. A nucleus of an atom in the hardware can capture one of these energetic neutrons and throw off an alpha particle, which strikes a memory circuit and changes the value stored in memory. To the end user, a soft error

---

[13]For example, strings are mutable in HOL Light's source language, Objective CAML, allowing theorems to be maliciously altered. Also, Objective CAML has *object magic*, which is a way to defeat the type system. These vulnerabilities and all other vulnerabilities that I know would be detected during translation of the proof from HOL Light to HOL Zero. A stricter standard is Pollack consistency, which requires a proof assistant to avoid the appearance of inconsistency [Ada09], [Wie10a]. For example, some proof assistants allow the substitution of a variable whose name is a meaningless sequence of characters 'n<0 ∧ 0' for t in $\exists n. \, t < n$ to obtain a Pollack-inconsistency $\exists n. \, n < 0 \wedge 0 < n$.

[14]"Il est donc inutile de demander au calcul plus de précision qu'aux observations; mais on ne doit pas non plus lui en demander moins" [Poi92].

appears as a gremlin, a seemingly inexplicable random error that disappears when the computer is rebooted and the program runs again.

As an example, we will calculate the expected number of soft errors in one of the mathematical calculations of Section 1.17. The Atlas Project calculation of the E_8 character table was a 77 hour calculation that required 64 gigabytes RAM [Ada07]. Soft errors rates are generally measured in units of failures-in-time (FIT). One FIT is defined as one error per 10^9 hours of operation. If we assume a soft error rate of 10^3 FIT per Mbit, (which is a typical rate for a modern memory device operating at sea level[15] [Tez04]), then we would expect there to be about 40 soft errors in memory during the calculation:

$$\frac{10^3 \text{ FIT}}{1 \text{ Mbit}} \cdot 64 \text{ GB} \cdot 77 \text{ hours} = \frac{10^3 \text{ errors}}{10^9 \text{ hours Mbit}} \cdot (64 \cdot 8 \cdot 10^3 \text{ Mbit}) \cdot 77 \text{ hours}$$

$$\approx 39.4 \text{ errors.}$$

This example shows that soft errors can be a realistic concern in mathematical calculations. (As added confirmation, the E_8 calculation has now been repeated about 5 times with identical results.)

In software that has been thoroughly debugged, soft errors become the most significant source of error in computation. Although there are numerous ways to protect against soft errors with methods such as repeated calculations and error-correcting codes, hardware redesign carries an economic cost. In fact, soft errors are on the rise through miniaturization: a smaller circuit generally has a lower capacitance and responds to less energetic alpha particles than a larger circuit.

Soft errors are depressing news in the ultra-reliable world of proof assistants. Alpha particles rain on perfect and imperfect software alike. In fact, because the number of soft errors is proportional to the execution time of a calculation, by being slow and methodical, the probability of a soft error during a calculation inside a proof assistant can be much higher than the probability when done outside.

Soft errors and susceptibility to hacking have come to be more than a nuisance to me. They alter my philosophical views of the foundations of mathematics. I am a computational formalist—a formalist who admits physical limits to the reliability of any verification process, whether by hand or machine. These limits taint even the simplest theorems, such as our ability to verify that $1 + 1 = 2$ is a consequence of a set of axioms. One rogue alpha particle brings all my schemes of perfection to nought. The rigor of mathematics and the reliability of technology are mutually dependent; math to provide ever more accurate models of science, and technology to provide ever more reliable execution of mathematical proofs.

---

[15]The soft error rate is remarkably sensitive to elevation; a calculation in Denver produces about three times more soft errors than the same calculation on identical hardware in Boston.

§4. **Concluding remarks.** To everyone who has made it this far in this essay, I highly recommend MacKenzie's book [Mac01]. It written by a sociologist with a fine sensitivity to mathematics. The author received the Robert K. Merton Award of the American Sociological Association in 2003 for this book.

A few years ago, a special issue of the Notices of the AMS presented a general introduction to formal proofs [Hal08], [Har08], [Gon08], [Wie08]. I also particularly recommend the body of research articles by Harrison, Gonthier, and Carette.

I thank Adams (both Jeff and Mark), Urban, Carette, Kapulkin, Harrison, and Manfredi for conversations about ideas in this article.

REFERENCES

[Aga13] http://en.wikipedia.org/wiki/Lorenz_attractor, accessed 2013.

[Ada07] J. ADAMS, *Atlas of Lie groups and representations*, MIT colloquium slides, http://atlas.math.umd.edu/talks/boston.pdf, 2007.

[Ada11] ———, *Computing global characters*, http://www.liegroups.org/papers/characters.pdf, 2011.

[Ada09] M. ADAMS, *Importing proofs into HOL Zero*, private communication, 2009.

[AHK77] K. APPEL, W. HAKEN, and J. KOCH, *Every planar map is four colorable*, **Illinois Journal of Mathematics**, vol. 21 (1977), pp. 439–567.

[AriBC] ARISTOTLE, *On the heavens*, http://classics.mit.edu/Aristotle/heavens.html, 350BC, translated by J. L. Stocks.

[AS92] M. ASCHBACHER and Y. SEGEV, *The uniqueness of groups of Lyons type*, **Journal of the American Mathematical Society**, vol. 5 (1992), no. 1, pp. 75–98.

[Atl] *Atlas of Lie groups and representations*, http://www.liegroups.org/.

[Aus08] JOSEPH AUSLANDER, *On the roles of proof in mathematics*, **Proof and other dilemmas** (Bonnie Gold and Roger A. Simons, editors), Mathematical Association of America, 2008.

[AW09] S. AWODEY and M. WARREN, *Homotopy theoretic models of identity types*, **Mathematical Proceedings of the Cambridge Philosophical Society**, vol. 146 (2009), no. 1, pp. 45–55.

[BV08] C. BACHOC and F. VALLENTIN, *New upper bounds for kissing numbers from semidefinite programming*, **Journal of the American Mathematical Society**, vol. 21 (2008), pp. 909–924.

[Bar07] H. BARENDREGT, *Foundations of mathematics from the perspective of computer verification*, **Mathematics, computer science, logic—a never ending story**, Springer-Verlag, 2007.

[Bar98] B. BARRAS, *Verification of the interface of a small proof system in Coq*, **Types for proofs and programs**, Lecture Notes in Computer Science, vol. 1512, 1998, pp. 28–45.

[Bee85] M. BEESON, **Foundations of constructive mathematics: metamathematical studies**, Springer-Verlag, 1985.

[BG94] H. BENDER and G. GLAUBERMAN, *Local analysis for the odd order theorem*, LMS, vol. 188, Cambridge University Press, 1994.

[Ber05] J. BERGSTRA, *Nationale Onderzoeksagenda Informatie en Communicatietechnologie* (*NOAG-ict*) *2005–2010*, Albani drukkers, Den Haag, 2005.

[BC04] Y. BERTOT and P. CASTÉRAN, *Interactive theorem proving and program development Coq'Art: The calculus of inductive constructions*, Springer-Verlag, 2004.

[BGBP08] Y. BERTOT, G. GONTHIER, S. OULD BIHA, and I. PASCA, *Canonical big operators*, **Theorem proving in higher order logics**, Lecture Notes in Computer Science, vol. 5170, Springer-Verlag, 2008, pp. 86–101.

[Bet91] FABRICE BETHUEL, *The approximation problem for Sobolev maps between two manifolds*, **Acta Mathematica**, vol. 167 (1991), no. 3–4, pp. 153–206.
[BH00] U. BETKE and M. HENK, *Densest lattice packings of 3-polytopes*, **Computational Geometry**, vol. 16 (2000), pp. 157–186.
[BS10] M. BHARGAVA and A. SHANKAR, *Ternary cubic forms having bounded invariants, and the existence of a positive proportion of elliptic curves having rank 0*, http://arxiv.org/abs/1007.0052v1 [math.NT], 2010.
[Bir02] B. J. BIRCH, *In lieu of birthday greetings*, **Number theory and algebraic geometry**, Cambridge Univ. Press, 2002, pp. 1–30.
[BSD65] B. J. BIRCH and H. P. F. SWINNERTON-DYER, *Notes on elliptic curves, II*, **Journal für die Reine und Angewandte Mathematik**, vol. 218 (1965), pp. 79–108.
[BN10] SASCHA BÖHME and TOBIAS NIPKOW, *Sledgehammer: Judgement day*, **Automated reasoning (iJCAR 2010)** (J. Giesl and R. Hähnle, editors), Lecture Notes in Computer Science, vol. 6173, Springer-Verlag, 2010, pp. 107–121.
[Bor09] A. BORISOV, *A geometric approach to the two-dimensional Jacobian conjecture*, http://arxiv.org/abs/0912.4803, 2009.
[BG11] D. BORTHWICK and S. GARIBALDI, *Did a 1-dimensional magnet detect a 248-dimensional Lie algebra?*, **Notices of the American Mathematical Society**, vol. 58 (2011), no. 8, pp. 1055–1066.
[Boy90] C. V. BOYS, *Soap bubbles*, (Dover reprint), 1890.
[Car07] H. CARAYOL, *La conjecture de Sato–Tate [d'après Clozel, Harris, Shepherd-Barron, Taylor]*, **Séminaire Bourbaki**, vol. 59 (2006–07), pp. 345–391.
[CES11] J. CARETTE, M. ELSHEIKH, and S. SMITH, *A generative geometric kernel*, **Proceedings of the 20th ACM SIGPLAN workshop on partial evaluation and program manipulation**, PEPM '11, 2011, pp. 53–62.
[CC11] *10 best jobs 2011*, http://www.careercast.com/jobs-rated/10-best-jobs-2011
[CEG10] E. R. CHEN, MICHAEL ENGEL, and SHARON C. GLOTZER, *Dense crystalline dimer packings of regular tetrahedra*, **Discrete & Computational Geometry**, vol. 44 (2010), pp. 253–280, http://arxiv.org/abs/1001.0586.
[CT06] J. H. CONWAY and S. TORQUATO, *Packing, tiling, and covering with tetrahedra*, **Proceedings of the National Academy of Sciences**, vol. 103 (July 11, 2006), no. 28, pp. 10612–10617.
[Cor10] LEO CORRY, *On the history of Fermat's last theorem: fresh views on an old tale*, **Mathematische Semesterberichte**, vol. 57 (2010), no. 1, pp. 123–138.
[Dav09] J. DAVIS, *A self-verifying theorem prover*, Ph.D. thesis, University of Texas at Austin, 2009.
[Dij82] E. W. DIJKSTRA, *On the role of scientific thought*, **Edsger W. Dijkstra: Selected writings on computing: A personal perspective**, Springer-Verlag, 1982, http://www.cs.utexas.edu/users/EWD/ewd04xx/EWD447.PDF.
[DCF+62] J. L. DOOB, L. CARLITZ, F. A. FICKEN, G. PARANIAN, and N. E. STEENROD, **Manual for authors of mathematical papers**, vol. 68, 1962.
[Doy11] J. DOYE and D. WALES, http://arxiv.org/abs/cond-mat/9709201, accessed 2011.
[Elk88] N. ELKIES, *On $A^4 + B^4 + C^4 = D^4$*, **Mathematics of Computation**, vol. 51 (1988), no. 184, pp. 825–835.
[Fey85] R. P. FEYNMAN, **QED: the strange theory of light and matter**, Princeton University Press, 1985.
[Fey00] ———, *Selected papers of Richard Feynman*, World Scientific, 2000.
[For09] L. FORTNOW, *The status of the P versus NP problem*, **Communications of the ACM**, vol. 52 (2009), no. 9, pp. 78–86.
[Fox03] A. FOX, *Formal specification and verification of ARM6*, **Theorem proving in higher order logics**, Lecture Notes in Computer Science, vol. 2758, Springer-Verlag, 2003, pp. 25–40.

[Gab10] D. GABAI, *Hyperbolic 3-manifolds in the 2000's*, **Proceedings of the International Congress of Mathematicians**, 2010.

[GMT03] D. GABAI, R. MEYERHOFF, and N. THURSTON, *Homotopy hyperbolic 3-manifolds are hyperbolic*, **Annals of Mathematics**, vol. 157 (2003), pp. 335–431.

[Gan09] M. GANESALINGAM, *The language of mathematics*, Ph.D. thesis, Cambridge University, 2009, http://people.pwf.cam.ac.uk/mg262/GanesalingamMdis.pdf.

[Gan10] ———, *The language of mathematics*, slides, http://www.srcf.ucam.org/principia/files/ganesalingam.pdf, 2010.

[GGMR09] F. GARILLOT, G. GONTHIER, A. MAHBOUBI, and L. RIDEAU, *Packaging mathematical structures*, **Theorem proving in higher order logics**, Lecture Notes in Computer Science, vol. 5674, Springer-Verlag, 2009, pp. 327–342.

[Gon08] ———, *Formal proof—the four colour theorem*, **Notices of the American Mathematical Society**, vol. 55 (December 2008), no. 11, pp. 1382–1393.

[Gon11] G. GONTHIER, *Point-free, set-free concrete linear algebra*, **Interactive theorem proving**, Lecture Notes in Computer Science, vol. 6898, Springer-Verlag, 2011, pp. 103–118.

[Gon12] ———, http://www.msr-inria.inria.fr/events-news/feit-thompson-proved-in-coq, 2012.

[GM11] G. GONTHIER and A. MAHBOUBI, *An introduction to small scale reflection in Coq*, http://hal.inria.fr/inria-00515548/, 2011.

[GMR07] G. GONTHIER, A. MAHBOUBI, and L. RIDEAU, *A modular formalisation of finite group theory*, **Theorem proving in higher order logics**, Lecture Notes in Computer Science, vol. 4732, Springer-Verlag, 2007, pp. 86–101.

[GMT11] G. GONTHIER, A. MAHBOUBI, and E. TASSI, *A small scale reflection extension for the Coq system*, http://hal.archives-ouvertes.fr/inria-00258384/, 2011.

[Gor00] M. GORDON, *From LCF to HOL: a short history*, **Proof, language, and interaction: essays in honour of Robin Milner**, MIT, 2000, pp. 169–185.

[Gru11] B. GRÜNBAUM, *An enduring error*, **The best writing on mathematics 2010** (M. Pitici, editor), Princeton University Press, 2011.

[Hal05a] T. C. HALES, *The Flyspeck project*, **Dagstuhl seminar proceedings** (T. Coquand and H. Lombardi, editors), vol. 25, IBFI, 2005, pp. 489–507.

[Hal05b] ———, *A proof of the Kepler conjecture*, **Annals of Mathematics**, vol. 162 (2005), pp. 1065–1185.

[Hal08] ———, *Formal proof*, **Notices of the American Mathematical Society**, vol. 55 (December 2008), no. 11, pp. 1370–1380.

[Hal11] ———, *On the Reinhardt conjecture*, http://arxiv.org/abs/1103.4518, 2011.

[Ham08] T. HAMADA, *KNOPPIX/math: A live system for enjoying mathematics with computer*, **ACM Communications in Computer Algebra**, vol. 42 (2008), no. 3, pp. 175–176, http://www.knoppix-math.org/.

[HL03] FENGBO HANG and FANGUA LIN, *Topology of Sobolev mappings. II*, **Acta Mathematica**, vol. 191 (2003), no. 1, pp. 55–107.

[Har96] J. HARRISON, *Optimizing proof search in model elimination*, **Proceedings of the 13th international Conference on Automated Deduction (CADE-13)**, Lecture Notes in Computer Science, vol. 1104, Springer-Verlag, 1996, pp. 313–327.

[Har06] ———, *Towards self-verification of HOL light*, **Proceedings of IJCAR 2006**, Lecture Notes in Computer Science, vol. 4130, Springer-Verlag, 2006, pp. 177–191.

[Har08] ———, *Formal proof—theory and practice*, **Notices of the American Mathematical Society**, vol. 55 (December 2008), no. 11, pp. 1395–1406.

[Har09] ———, *Handbook of practical logic and automated reasoning*, Cambridge University Press, 2009.

[Har10] ———, *On the cruelty of really doing formal proofs*, www.srcf.ucam.org/principia/files/jrhslides.pdf, 2010, Principia Mathematica anniversary symposium.

[HHS95] J. HASS, M. HUTCHINGS, and R. SCHLAFLY, *The double bubble conjecture*, **Electronic Research Announcements of the American Mathematical Society**, vol. 1 (1995), no. 3.

[HS99] G. HAVAS and C. C. SIMS, *A presentation for the Lyons simple group*, http://dimacs.rutgers.edu/~havas/TR0416.pdf, 1999.

[Hoe11] J. VAN DER HOEVEN, *Towards semantic mathematical editing*, http://hal.archives-ouvertes.fr/hal-00569351/fr/, 2011.

[Hof87] D. HOFFMAN, *The computer-aided discovery of new embedded minimal surfaces*, **The Mathematical Intelligencer**, vol. 9 (1987), no. 3, pp. 8–81.

[Hor10] M. HORN, *Projective plane of order* 12, http://mathoverflow.net/questions/38632/projective-plane-of-order-12, 2010, Math Overflow.

[HTT11] *Homotopy type theory*, http://homotopytypetheory.org/, 2011.

[HT00] A. HUNT and D. THOMAS, *The pragmatic programmer*, Addison Wesley, 2000.

[HMRR00] M. HUTCHINGS, F. MORGAN, M. RITORÉ, and A. ROS, *Proof of the double bubble conjecture*, **Electronic Research Announcements of the American Mathematical Society**, vol. 6 (2000), pp. 45–49.

[IR11] M. IANCU and F. RABE, *Formalizing foundations of mathematics*, **Mathematical Structures in Computer Science**, vol. 21 (2011), no. 4, pp. 883–911.

[Ima13] Wellcome Images, http://images.wellcome.ac.uk, London, accessed 2013.

[KW07] C. KALISZYK and F. WIEDIJK, *Certified computer algebra on top of an interactive theorem prover*, **Proceedings of the 14th symposium on the integration of symbolic computation and mechanised reasoning** (*Calculemus'07*) (Manuel Kauers, Manfred Kerber, Robert Miner, and Wolfgang Windsteiger, editors), Lecture Notes in Computer Science, vol. 4573, Springer-Verlag, 2007, http://score.cs.tsukuba.ac.jp/~kaliszyk/docs/kaliszyk_p04_calc.pdf, pp. 94–105.

[KLV12a] C. KAPULKIN, P. F. LUMSDAINE, and V. VOEVODSKY, *The simplicial model of univalent foundations*, http://arxiv.org/abs/1211.2851, 2012.

[KLV12b] ———, *Univalence in simplicial sets*, http://arxiv.org/abs/1203.2553, 2012.

[KLS08] C. KHARE, M. LARSEN, and G. SAVIN, *Functoriality and the inverse Galois problem*, **Compositio Mathematica**, vol. 144 (2008), pp. 541–564.

[KAE+10] G. KLEIN, J. ANDRONICK, K. ELPHINSTONE, G. HEISER, D. COCK, P. DERRIN, D. ELKADUWE, K. ENGELHARDT, R. KOLANSKI, M. NORRISH, T. SEWELL, H. TUCH, and S. WINWOOD, *seL4: Formal verification of an operating-system kernel*, **Communications of the ACM**, vol. 53 (2010), no. 6, pp. 107–115.

[Kur99] R. KURZWEIL, *The age of spiritual machines*, Penguin, 1999.

[Lam91] C. W. H. LAM, *The search for a finite projective plane of order* 10, **American Mathematical Monthly**, vol. 98 (1991), no. 4, pp. 305–318.

[LTS89] C. W. H. LAM, L. THIEL, and S. SWIEREZ, *The non-existence of finite projective planes of order* 10, **Canadian Journal of Mathematics**, vol. 41 (1989), no. 6, pp. 1117–1123.

[LP66] L. J. LANDER and T. R. PARKIN, *Counterexample to Euler's conjecture on sums of like powers*, **American Mathematical Society. Bulletin**, (1966), p. 1079.

[LSW01] G. LAWLER, O. SCHRAMM, and W. WERNER, *The dimension of the planar Brownian frontier is 4/3*, **Mathematical Research Letters**, vol. 8 (2001), pp. 401–411, http://www.mrlonline.org/mrl/2001-008-004/2001-008-004-001.pdf.

[Lax82] P. LAX, *Report of the panel on large scale computing in science and engineering*, www.pnl.gov/scales/docs/lax_report1982.pdf, 1982.

[Ler06] X. LEROY, *Formal certification of a compiler back-end, or: programming a compiler with a proof assistant*, **33rd acm symposium on Principles of Programming Languages**, ACM Press, 2006, http://compcert.inria.fr/, pp. 42–54.

[Lev95] S. LEVY, *Making waves: A guide to the ideas behind "outside in"*, AK Peters, 1995, extract at http://www.math.sunysb.edu/CDproject/OvUM/cool-links/www.geom.umn.edu/docs/outreach/oi/history.html.

[LMM94] S. LEVY, D. MAXWELL, and T. MUNZNER, *Outside in*, AK Peters, 1994, video, The Geometry Center (also on YouTube).

[LH11] D. LICATA and R. HARPER, *Canonicity for 2-dimensional type theory*, preprint 2011, http://www.cs.cmu.edu/~drl/pubs/lh112tt/lh112tt.pdf.

[Lor63] E. N LORENZ, *Deterministic nonperiodic flow*, **Journal of the Atmospheric Sciences**, vol. 20 (1963), no. 2, pp. 130–141.

[Mac01] D. MACKENZIE, *Mechanizing proof*, MIT Press, Cambridge, MA, 2001.

[MST73] F. J. MACWILLIAMS, N. J. A. SLOANE, and J. G. THOMPSON, *On the existence of a projective plane of order 10*, **Journal of Combinatorial Theory. Series A**, vol. 14 (1973), pp. 66–78.

[MM99] G. MALLE and MATZAT, *Inverse Galois theory*, Springer-Verlag, 1999.

[Man82] B. MANDELBROT, *The fractal geometry of nature*, Freeman and Co., 1982.

[Man04] ———, *A theory of roughness*, http://www.edge.org/3rd_culture/mandelbrot04/mandelbrot04_index.html, 2004.

[MW04] R. MASTIPURAM and E. C. WEE, *Soft errors' impact on system reliability*, **EDN**, (Sept. 30, 2004).

[MC02] U. MEIERFRANKENFELD and C. W. PARKER, *A computer-free construction of the Lyons group*, to appear, http://www.math.msu.edu/~meier/Preprints/Ly/ly.pdf, 2002.

[Met] *Metis*, http://www.gilith.com/software/metis/index.html.

[Met03] T. METSÄNKYLÄ, *Catalan's conjecture: another old Diophantine problem solved*, **American Mathematical Society. Bulletin**, vol. 41 (2003), no. 1, pp. 43–57.

[Mih04] P. MIHĂILESCU, *Primary cyclotomic units and a proof of Catalan's conjecture*, **Journal für die Reine und Angewandte Mathematik**, vol. 572 (2004), pp. 167–195.

[Mul11] K. MULMULEY, *On P vs. NP, and geometric complexity theory*, **Journal of the ACM**, vol. 58 (April 2011), no. 2.

[MT10] O. R. MUSIN and A. S. TARASOV, *The strong thirteen spheres problem*, preprint http://arxiv.org/abs/1002.1439, February 2010.

[NYT10] *Lone $4.1 billion sale led to 'flash crash' in May*, New York Times http://www.nytimes.com/2010/10/02/business/02flash.html, October 10 2010.

[Op111] http://new.math.uiuc.edu/optiverse/images.html, accessed 2011.

[Ove09] *Most interesting mathematics mistake?*, http://mathoverflow.net/questions/879, 2009.

[Ove10] *Widely accepted mathematical results that were later shown wrong?*, http://mathoverflow.net/questions/35468, 2010.

[Pas07] A. PASKEVICH, *The syntax and semantics of the ForTheL language*, http://nevidal.org/download/forthel.pdf, 2007.

[Pau94] P. PAULE, *Short and easy computer proofs of the Rogers–Ramanujan identities and of identities of similar type*, **Electronic Journal of Combinatorics**, vol. 1 (1994), pp. 1–9.

[Pau10] L. PAULSON, *Three years of experience with sledgehammer, a practical link between automatic and interactive theorem provers*, 2010, **PAAR-2010, Practical Aspects of Automated Reasoning**.

[PSST08] A. PEASE, G. SUTCLIFFE, N. SIEGEL, and S. TRAC, *The annual SUMO reasoning prizes at CASC*, **First international workshop on Practical Aspects of Automated Reasoning**, CEUR Workshop Proceedings, vol. 373, 2008.

[PW12] A. PELAYO and M. A. WARREN, *Homotopy type theory and Voevodsky's univalent foundations*, http://arxiv.org/abs/1210.5658, 2012.

[Pet00] T. PETERFALVI, *Character theory for the odd order theorem*, LMS, vol. 272, Cambridge University Press, 2000.

[PWZ96] M. PETKOVŠEK, H. S. WILF, and D. ZEILBERGER, $A = B$, A. K. Peters, 1996.

[PZ04] F. PFENDER and G. ZIEGLER, *Kissing numbers, sphere packings, and some unexpected proofs*, **Notices of the American Mathematical Society**, (2004), pp. 873–883.

[Phi66] A. PHILLIPS, *Turning a surface inside out*, **Scientific American**, (1966), pp. 112–120.

[PS08] J. D. PHILLIPS and D. STANOVSKÝ, *Using automated theorem provers in nonassociative algebra*, http://www.karlin.mff.cuni.cz/~stanovsk/math/lpar08.pdf, 2008.

[Pit11] M. Pitici (editor), *The best writing on mathematics 2010*, Princeton Univ. Press, 2011, forward by W. Thurston.

[Poi92] H. POINCARÉ, *Les méthodes nouvelles de la mécanique céleste*, Dover reprint, 1892.

[Poi52] ———, *Science and hypothesis*, Dover reprint, 1952.

[PW11] http://it.wikipedia.org/wiki/File:Foam_-_Weaire-Phelan_structure.png, accessed 2011.

[RhD11] http://upload.wikimedia.org/wikipedia/commons/8/86/Rhombic_dodecahedra.jpg, accessed 2011.

[RSST97] NEIL ROBERTSON, DANIEL SANDERS, PAUL SEYMOUR, and ROBIN THOMAS, *The four-colour theorem*, **Journal of Combinatorial Theory, Series B**, vol. 70 (1997), pp. 2–44.

[San12] ANDERS SANDBERG, http://www.flickr.com/photos/arenamontanus/8059864268/, accessed 2012.

[SN63] M. SATO and K. NAMBA, unpublished data, April 3, 1963.

[Sch] R. SCHMIDT, *The Sato–Tate conjecture*, www.math.ou.edu/~rschmidt/satotate/page5.html.

[Sci11] *The \$23 million textbook*, **Science**, vol. 332 (2011), pp. 647–648.

[SEC10] *Findings regarding the market events of May 6, 2010*, http://sec.gov/news/studies/2010/marketevents-report.pdf, September 10, 2010, Report of the Staffs of the Cftc and Sec to the Joint Advisory Committee on Emerging Regulatory Issues.

[Ser68] J.-P. SERRE, *Abelian ℓ-adic representations and elliptic curves*, WA Benjamin, 1968.

[Sma98] S. SMALE, *Mathematical problems for the next century*, **The Mathematical Intelligencer**, vol. 20 (1998), no. 2, pp. 7–15.

[SvdW11] B. SPITTERS and E. VAN DER WEEGEN, *Type classes for mathematics in type theory*, http://arxiv.org/abs/1102.1323, 2011.

[Sta07] R. STANLEY, *Sequence A129935 in the on-line encyclopedia of integer sequences*, published electronically at http://oeis.org/A129935, 2007.

[Ste41] J. STEINER, *Über Maximum und Minimum bei den Figuren in der Ebene, auf der Kugelfläche und in Raume überhaupt*, **Comptes Rendu de l'Académie des Sciences. Paris**, vol. 12 (1841), pp. 177–308.

[Ste00] I. STEWART, *The Lorenz attractor exists*, **Nature**, vol. 406 (31 Aug 2000), pp. 948–949.

[SFL] J. M. SULLIVAN, G. FRANCIS, and S. LEVY, *The Optiverse*, video http://new.math.uiuc.edu/optiverse/ or YouTube.

[Tez04] *Tezzaron semiconductor*, http://www.tezzaron.com/about/papers/soft_errors_1_1_secure.pdf, 2004.

[Tsi13] MICHAEL TSIRELSON, http://en.wikipedia.org/wiki/File:Double_bubble.png, accessed 2013.

[Tuc02] W. TUCKER, *A rigorous ODE solver and Smale's 14th problem*, **Foundations of Computational Mathematics**, vol. 2 (2002), pp. 53–117.

[Tur50] A. TURING, *Computing machinery and intelligence*, **Mind**, vol. 59 (1950), pp. 433–460.

[Urb07] J. URBAN, *MaLARea: A metasystem for automated reasoning in large theories*, **Proceedings of the CADE-21 workshop on empirically successful automated reasoning in large theories** (J. Urban, G. Sutcliffe, and S. Schultz, editors), 2007, pp. 45–58.

[Voe11] V. VOEVODSKY, *Univalent foundations*, http://hottheory.files.wordpress.com/2011/06/report-11_2011.pdf, 2011, Mathematisches Forschungsinstitut Oberwolfach, pp. 7–9.

[Vog07] D. VOGAN, *The character table for E_8*, **Notices of the American Mathematical Society**, vol. 54 (2007), no. 9, pp. 1022–1034.

[WW11] M. WEBER and M. WOLF, *About the cover: early images of minimal surfaces*, **American Mathematical Society. Bulletin**, vol. 48 (2011), no. 3, pp. 457–460.

[Wie06] F. Wiedijk (editor), *The seventeen provers of the world*, Lecture Notes in Artificial Intelligence, Springer-Verlag, 2006.

[Wie08] ———, *Formal proof—getting started*, **Notices of the American Mathematical Society**, vol. 55 (December 2008), no. 11, pp. 1408–1414.

[Wie10a] ———, *Pollack inconsistency*, **9th international workshop on user interfaces for theorem provers** (C. Sacerdoti Coen and D. Aspinall, editors), ENTCS, 2010.

[Wie10b] ———, *The next generation of proof assistants*, http://www.cs.ru.nl/F.Wiedijk/talks/lsfa.ps.gz, slides, 2010.

[Wik11] *List of published incomplete proofs*, http://en.wikipedia.org/wiki/List_of_published_incomplete_proofs, accessed 9/2011.

[Wil02] R. WILSON, *Four colors suffice: How the map problem was solved*, Princeton University Press, 2002.

[Wol02] S. WOLFRAM, *A new kind of science*, Wolfram Media, 2002, http://www.wolframscience.com/reference/notes/971c.

UNIVERSITY OF PITTSBURGH
E-mail: hales@pitt.edu

TURING AND THE DEVELOPMENT OF COMPUTATIONAL COMPLEXITY

STEVEN HOMER AND ALAN L. SELMAN

Abstract. Turing's beautiful capture of the concept of computability by the "Turing machine" linked computability to a device with explicit steps of operations and use of resources. This invention led in a most natural way to build the foundations for computational complexity.

§1. Introduction. Computational complexity provides mechanisms for classifying combinatorial problems and measuring the computational resources necessary to solve them. The discipline provides explanations of why no practical solutions to certain problems have been found, and provides a way of anticipating difficulties involved in solving these problems. The classification is quantitative and is intended to investigate what resources are necessary, lower bounds, and what resources are sufficient, upper bounds, to solve various problems.

This classification should not depend on a particular computational model but rather should measure the intrinsic difficulty of a problem. Precisely for this reason, as we will explain, the basic model of computation for our study is the multitape Turing machine.

Computational complexity theory today addresses issues of contemporary concern, for example, parallel computation, circuit design, computations that depend on random number generators, and development of efficient algorithms. Above all, computational complexity is interested in distinguishing problems that are *efficiently* computable. Algorithms whose running times are n^2 in the size of their inputs can be implemented to execute efficiently even for fairly large values of n, but algorithms that require an exponential running time can be executed only for small values of n. It is common to identify efficiently computable problems with those that have polynomial time algorithms.

A complexity measure quantifies the use of a particular computational resource during execution of a computation. The two most important and most common measures are *time*, the time it takes a program to execute, and *space*, the amount of memory used during a computation. However, other

Turing's Legacy: Developments from Turing's Ideas in Logic
Edited by Rod Downey
Lecture Notes in Logic, 42
© 2014, Association for Symbolic Logic

measures are considered as well, and we will introduce other resources as we proceed through this exposition.

Computational complexity relies on an expanded version of the Church–Turing Thesis[1] [Chu36, Tur36], one that is even stronger than the original thesis. This expanded version asserts that any two general and reasonable models of sequential computation are polynomial related. That is, a problem that has time complexity t on some general and reasonable model of computation has time complexity $p(t)$, for some polynomial p, in the multitape Turing machine model. The assertion has been proven for all known reasonable models of sequential computation including random access machines (RAMS). This thesis is particularly fortunate because of another assertion known as the Cobham–Edmonds Thesis [Cob64, Edm65]. The Cobham–Edmonds Thesis asserts that computational problems can be feasibly computed on some computational device only if they can be computed in polynomial time. (Truth be told, an n^{100} time algorithm is not a useful algorithm. It is a remarkable phenomenon though, that problems for which polynomial algorithms are found have such algorithms with small exponents and with small coefficients.) Combining these two theses, a problem can be feasibly computed only if it can be computed in polynomial time on some multitape Turing machine.

Computational complexity forms a basis for the classification and analysis of combinatorial problems. To illustrate this, consider the problem of determining whether an arbitrary graph possesses a Hamiltonian Circuit (i.e., a simple cycle that contains every vertex). Currently it is not known whether this problem has a feasible solution, and all known solutions are equivalent to a sequential search of all paths through the graph, testing each in turn for the Hamiltonian property. Recalling that input to a Turing machine is a word over some finite input alphabet, it must be possible to encode data structures such as graphs into words so that what is intuitively the size of the graph differs from the length of the input by no more than a polynomial. In fact, it is easy to do so. Then, we demand that the theory is capable of classifying the intrinsic complexity of this problem in a precise way, and is capable of elucidating the difficulty in finding an efficient solution to this problem.

Finally we should mention the seminal paper of Hartmanis and Stearns, "On the Computational Complexity of Algorithms" [HS65], from which the discipline takes its name. This paper formulated definitions of time and space complexity on multitape Turing machines and proved results demonstrating that with more time, more problems can be computed. It was a fundamental step to make complexity classes, defined by functions that bound the amount of resources use, the main subject of study. More abstract definitions of computability could not have offered natural guidance to an intuitively satisfying and practical formulation of computational complexity.

---

[1] Some scholars question the correctness of the Church–Turing Thesis. For a discussion of these issues and relevant citations we refer the reader to Davis [Dav06a, Dav06b].

The Turing machine beautifully captured the discrete step-by-step nature of computation. Furthermore, this machine enabled the definition and measurement of time and space complexity in a natural and precise manner amenable to quantifying the resources used by a computation. This natural precision of Turing machines was crucial in guiding the originators of this field in their fundamental definitions and first results which set the tenor for this research area up to the present day. Other, more abstract definitions of computability, while having advantages of brevity and elegance, could not and did not offer the needed precision or guidance toward this intuitively satisfying and practical formulation of complexity theory.

§2. Modes of computation. A computation of a Turing machine is a sequence of moves, as determined by its transition function. Ordinarily, the transition function is single-valued. Such Turing machines are called *deterministic* and their computations are sequential.

A *nondeterministic* Turing machine is one that allows for a choice of next moves; in this case the transition function is multivalued. If M is a nondeterministic Turing machine and x is an input word, then M and x specify a *computation tree*. The root of the tree is the *initial configuration* of M. The children of a node are the configurations that may follow in one move. A path from the root to a leaf is a computation, and a computation is accepting if the leaf is an accepting configuration.

Given a Turing machine M, the language $L(M)$ *accepted* by M is the set of words such that *some* computation of M on x is accepting. (Observe that this definition is meaningful for both the deterministic and the nondeterministic modes.) It is easy to see, by a breadth-first search of the computation tree, that for every nondeterministic Turing machine, there is a deterministic Turing machine that accepts the same language. The difficulty for computational complexity is that this search technique results in a deterministic machine that is exponentially slower than the original nondeterministic one.

The advantage of nondeterministic Turing machines is that they are a very useful mode for classification of computational problems. For example, whereas it is not known whether there is a deterministic polynomial time-bounded Turing machine to solve (an encoding of) the Hamiltonian Circuit problem, it is easy to design a nondeterministic polynomial time-bounded Turing machine that does solve this problem. This is what makes it possible to give an exact classification of the Hamiltonian Circuit problem. Indeed, this problem is known to be NP-complete, which places it among hundreds of other important computational problems whose deterministic complexity still remain open.

Nondeterminism was first considered in the classic paper of Rabin and Scott on finite automata [RS59]. We will return to nondeterminism and to a precise definition of NP-completeness in a later section.

§3. Complexity classes and complexity measures.

We assume that a multi-tape Turing machine has its input written on one of the work tapes, which can be rewritten and used an as an ordinary work tape. The machine may be either deterministic or nondeterministic. Let M be such a Turing machine, and let T be a function defined on the set of natural numbers. M is a $T(n)$ *time-bounded* Turing machine if for every input of length n, M makes at most $T(n)$ moves before halting. If M is nondeterministic, then every computation of M on words of length n must take at most $T(n)$ steps. The language $L(M)$ that is accepted by a deterministic $T(n)$ time-bounded M has *time complexity* $T(n)$. By convention, the time it takes to read the input is counted, and every machine is entitled to read its input.

Denote the length of a word x by $|x|$. We might be tempted to say that a nondeterministic Turing machine is $T(n)$ time-bounded if for every input word $x \in L(M)$, the number of steps of the shortest accepting computation of M on x is at most $T(|x|)$. It turns out that the formulations are equivalent for the specific time bounds that we will write about. But, they are not known to be equivalent for arbitrary time bounds.

A complexity class is a collection of sets that can be accepted by Turing machines with the same resources. Now we define the time-bounded complexity classes:

Define DTIME($T(n)$) to be the set of all languages having time-complexity $T(n)$.
Define NTIME($T(n)$) to be the set of all languages accepted by nondeterministic $T(n)$ time-bounded Turing machines.

In order to define space complexity, we need to use off-line Turing machines. An *off-line* Turing machine is a multitape Turing machine with a separate read-only input tape. The Turing machine can read the input, but cannot write over the input. Let M be an off-line multitape Turing machine and let S be a function defined on the set of natural numbers. M is an $S(n)$ *space-bounded* Turing machine if for every word of length n, M scans at most $S(n)$ cells on any storage tape. If M is nondeterministic, then every computation must scan no more than $S(n)$ cells on any storage tape. The language $L(M)$ that is accepted by an $S(n)$ deterministic space-bounded Turing machine has *space-complexity* $S(n)$.

Observe that the space taken by the input is not counted. So space-complexity might be less than the length of the input or substantially more. One might be tempted to say that a nondeterministic Turing machine is $S(n)$ space-bounded if for every word of length n that belongs to $L(M)$, there is an accepting computation that uses no more than $S(n)$ work cells on any work tape. The comment made for time-bounds applies here as well.

Now, we define the space-bounded complexity classes:

Define DSPACE($S(n)$) to be the set of all languages having space-complexity $S(n)$.
Define NSPACE($S(n)$) to be the set of all languages accepted by nondeterministic $S(n)$ space-bounded Turing machines.

We note that the study of time complexity begins properly with the paper [HS65] of Hartmanis and Stearns and the study of space complexity begins with the paper [HLS65] of Hartmanis, Lewis and Stearns. These seminal papers introduced some of the issues that remain of concern even today. These include time/space trade-offs, inclusion relations, hierarchy results, and efficient simulation of nondeterministic computations.

We will be primarily concerned with classes defined by logarithmic, polynomial and exponential functions. As we proceed to relate and discuss various facts about complexity classes in general, we will see what impact they have on the following list of *standard* complexity classes. These classes are well-studied in the literature and each contains important computational problems. The classes are introduced with their common notations.

1. L = DSPACE($\log(n)$);
2. NL = NSPACE($\log(n)$);
3. POLYLOGSPACE = $\bigcup \{$DSPACE($\log(n)^k$) $\mid k \geq 1\}$;
4. DLBA = $\bigcup \{$DSPACE(kn) $\mid k \geq 1\}$;
5. LBA = $\bigcup \{$NSPACE(kn) $\mid k \geq 1\}$;
6. P = $\bigcup \{$DTIME(n^k) $\mid k \geq 1\}$;
7. NP = $\bigcup \{$NTIME(n^k) $\mid k \geq 1\}$;
8. PSPACE = $\bigcup \{$DSPACE(n^k) $\mid k \geq 1\}$;
9. E = $\bigcup \{$DTIME(k^n) $\mid k \geq 1\}$;
10. NE = $\bigcup \{$NTIME(k^n) $\mid k \geq 1\}$;
11. EXP = $\bigcup \{$DTIME($2^{p(n)}$) $\mid p$ is a polynomial$\}$;
12. NEXP = $\bigcup \{$NTIME($2^{p(n)}$) $\mid p$ is a polynomial$\}$.

The class NL contains the *directed reachability problem*, of determining for arbitrary directed graphs G and vertices u and v, whether there is a path from u to v. This problem is not known to belong to L. The latter class contains the reachability problem for undirected graphs, as well as restriction of the directed reachability problem to the case that no vertex has more than one directed edge leading from it. [Jon73, Jon75]. The famous class P is identified with the class of feasibly computed problems. The corresponding nondeterministic class NP will be discussed in a later section.

The class denoted by LBA is so named because it is known to be identical to the class of languages accepted by *linear-bounded automata*, otherwise known as the context-sensitive languages [Myh60, Kur64]. This comment explains the notation for the corresponding deterministic class as well.

E characterizes the complexity of languages accepted by writing push-down automata [Mag69], and NE characterizes the complexity of the spectrum problem in finite model theory [JS74]. PSPACE contains many computational games, such as HEX [ET76].

§4. **Basic results.** Now we survey several different types of results that apply to all complexity classes. These results demonstrate that the definitions of these classes are invariant under small changes and they prove a variety of relationships between these classes. These are for the most part early results, and their proofs typically involve intricate Turing machine simulations.

The concepts of reducibility used in comparing combinatorial problems, including in definitions of completeness, and in related proof methods such as simple diagonalization and simulation common to complexity theory, arose as generalizations of fundamental concepts of computability theory. These ideas originated with the founders of computability theory prominently including Gödel, Church, Turing, Kleene, Post and others. Alan Turing in particular defined his eponymous machines, embodying the specific mode of computation most amenable to the application of these ideas (as mentioned above), and also introduced the oracle Turing machine [Tur50], which enabled a machine-based definition of the most general notion of relative computability. While other, machine-independent concepts give rise to the same general results, the use of Turing's machines allow for a precise quantified measurement of time, space, nondeterminism, randomness, etc., which provides the more exact and fine-grained notions of computation most applicable to computer science.

4.1. Linear compression and speedup. The first results are of the form: if a language can be accepted with resource $f(n)$, then it can be accepted with resource $cf(n)$, for any $c > 0$. These results justify use of "big-oh" notation for complexity functions.

The Space Compression Theorem [HLS65] asserts that if L is accepted by a k-tape $S(n)$ space-bounded Turing machine, then for any $c > 0$, L is accepted by a k-tape $cS(n)$ space-bounded Turing machine. If the $S(n)$ space-bounded Turing machine is nondeterministic, then so is the $cS(n)$ space-bounded Turing machine. A simple machine simulation proves the result. As a corollary, it follows that $\mathrm{DSPACE}(S(n)) = \mathrm{DSPACE}(cS(n))$ and $\mathrm{NSPACE}(S(n)) = \mathrm{NSPACE}(cS(n))$, for all $c > 0$.

Linear speedup of time is possible too, but not quite as readily as is linear compression of space. The Linear Speedup Theorem [HS65] asserts that if L is accepted by a k-tape $T(n)$ time-bounded Turing machine, $k \geq 1$, and if

$$\inf_{n \to \infty} T(n)/n = \infty, \tag{1}$$

then for any $c > 0$, L is accepted by a k-tape $cT(n)$ time-bounded Turing machine. Thus, if $\inf_{n \to \infty} T(n)/n = \infty$ and $c > 0$, then

$$\mathrm{DTIME}(T(n)) = \mathrm{DTIME}(cT(n)).$$

Condition 1 stipulates that $T(n)$ grows faster than every linear function. The Linear Speedup Theorem does not apply if $T(n) = cn$, for some constant c. Instead, we have the result that for all $\epsilon \geq 0$,

$$\text{DTIME}(O(n)) = \text{DTIME}((1+\epsilon)n),$$

and the proof actually follows from the proof of the Linear Speedup Theorem. This result cannot be improved, for Rosenberg [Ros67] showed that $\text{DTIME}(n) \neq \text{DTIME}(2n)$.

The two linear speedup theorems also hold for nondeterministic machines. Thus, if $\inf_{n \to \infty} T(n)/n = \infty$ and $c > 0$, then

$$\text{NTIME}(T(n)) = \text{NTIME}(cT(n)).$$

And, for all $\epsilon \geq 0$, $\text{NTIME}(O(n)) = \text{NTIME}((1+\epsilon)n)$.

However, a stronger result is known for nondeterministic linear-time complexity classes. A Turing machine that accepts inputs of length n in time $n+1$ (the time it takes to read the input) is called *realtime*. Nondeterministic Turing machines that accept in time $n+1$ are called *quasi-realtime*. The class of quasi-realtime languages is $\text{NTIME}(n+1)$. Book and Greibach [BG70] showed that $\text{NTIME}(n+1) = \bigcup\{\text{NTIME}(cn) \mid c \geq 1\}$.

Since $\text{DTIME}(n)$ is a proper subset of $\text{DTIME}(2n)$, and

$$\text{DTIME}(2n) \subseteq \text{NTIME}(2n) = \text{NTIME}(n),$$

it follows that $\text{DTIME}(n) \neq \text{NTIME}(n)$. In 1983, Paul, Pippenger, Szemeredi, and Trotter [PPST83] obtained the striking and deep result that

$$\text{DTIME}(O(n)) \neq \text{NTIME}(O(n)).$$

4.2. Inclusion relationships. Now we survey the known inclusion relationships between time-bounded and space-bounded, deterministic and nondeterministic classes.

First of all, it is trivial that for every function f, $\text{DTIME}(f) \subseteq \text{DSPACE}(f)$.

Obviously a Turing machine might enter an infinite loop and still use only bounded space. Nevertheless, if a language L is accepted by an $S(n)$ space-bounded Turing machine, where $S(n) \geq \log(n)$, then L is accepted by an $S(n)$ space-bounded Turing machine that halts on every input. The proof depends on the observation that within space $S(n)$ a Turing machine can enter at most an exponential in $S(n)$ possible distinct configurations. A machine enters an infinite loop by repeating one of these configurations, thus making loop detection possible. Analysis of this result yields the following theorem.

THEOREM 1. $\text{DSPACE}(S(n)) \subseteq \bigcup\{\text{DTIME}(c^{S(n)}) \mid c \geq 1\}$, *for* $S(n) \geq \log(n)$.

THEOREM 2. $\text{NTIME}(T(n)) \subseteq \bigcup\{\text{DTIME}(c^{T(n)}) \mid c \geq 1\}$.

We alluded to this result already. Recall that a nondeterministic $T(n)$ time-bounded Turing machine M and an input x of length n determine a computation tree of depth at most $T(n)$, and that M accepts n only if one of the paths of the tree is an accepting path. Theorem 2 is simply a formal expression of the time it takes to execute a search of this tree.

The theorems just presented are proved by rather straightforward simulations. The next theorems involve deep recursions in their simulations. First we need a couple of technical definitions. The point is that the following results don't seem to hold for all resource bounds, but only for those that are *constructible*. This is not a problem, because it turns out that all the bounds in which we are interested are constructible.

A function $S(n)$ is *space-constructible* if there is some Turing machine M that is $S(n)$ space-bounded, and for each n, there is some input of length n which uses exactly $S(n)$ cells. A function $S(n)$ is *fully space-constructible* if every input of length n uses $S(n)$ cells.

A function $T(n)$ is *time-constructible* if there is some Turing machine M that is $T(n)$ time-bounded, and for each n, there is some input of length n on which M runs for exactly $T(n)$ steps. A function $T(n)$ is *fully time-constructible* if for every input of length n, M runs for exactly $T(n)$ steps.

It is merely an exercise to see that ordinary arithmetic functions such as $\log(n)$, n^c, and c^n, are fully space-constructible and n^c, and c^n when $c \geq 1$, are fully time-constructible.

THEOREM 3 (Savitch [Sav70]). *If S is fully space-constructible and $S(n) \geq \log(n)$, then* $\mathrm{NSPACE}(S(n)) \subseteq \mathrm{DSPACE}(S^2(n))$.

This is a very important result, from which the following corollaries follow. Observe that a standard depth-first search simulation would provide only an exponential upper-bound.

COROLLARY 1.

$$\mathrm{PSPACE} = \bigcup \{\mathrm{DSPACE}(n^c) \mid c \geq 1\} = \bigcup \{\mathrm{NSPACE}(n^c) \mid c \geq 1\}$$

and

$$\mathrm{POLYLOGSPACE} = \bigcup \{\mathrm{DSPACE}(\log(n)^c) \mid c \geq 1\}$$
$$= \bigcup \{\mathrm{NSPACE}(\log(n)^c) \mid c \geq 1\}.$$

For this reason, nondeterministic versions of PSPACE and POLYLOG–SPACE were not defined as standard complexity classes.

COROLLARY 2. $\mathrm{LBA} \subseteq \mathrm{DSPACE}(n^2)$ *and* $\mathrm{NL} \subseteq \mathrm{POLYLOGSPACE}$.

THEOREM 4 ([Coo71b]). *If S is fully space-constructible and $S(n) \geq \log(n)$, then*

$$\mathrm{NSPACE}(S(n)) \subseteq \bigcup \{\mathrm{DTIME}(c^{S(n)}) \mid c > 1\}.$$

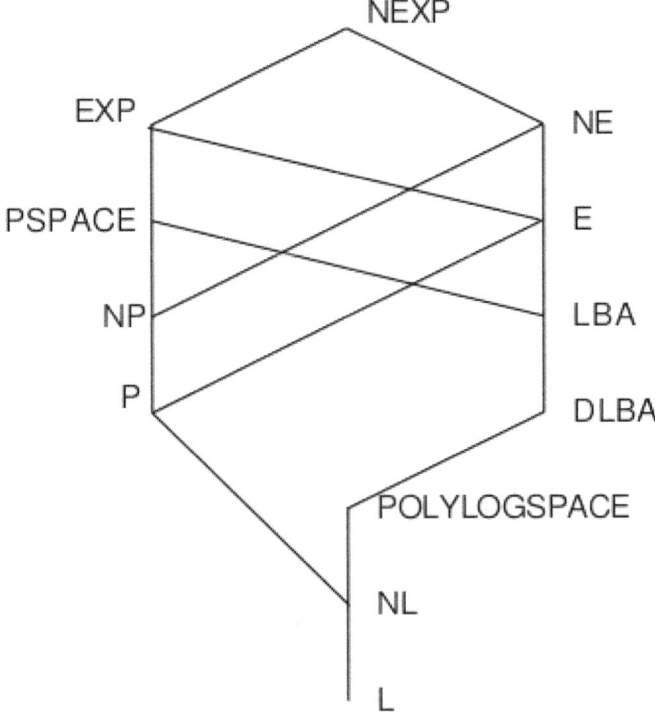

FIGURE 1. Inclusion relations between standard classes.

4.3. Relations between the standard classes. Figure 1 shows the inclusion relations that emerge by application of the results just presented. In this figure, a complexity class C is included in complexity class D if there is a path from C to D reading upward.

From Theorem 1 we learn that $L \subseteq P$, $PSPACE \subseteq EXP$, and $DLBA \subseteq E$. By Theorem 2, we know that $NP \subseteq PSPACE$. By Savitch's Theorem, $LBA \subseteq PSPACE$, and (Corollary 2) $NL \subseteq POLYLOGSPACE$. Theorem 4 is used to conclude that $NL \subseteq P$ and $LBA \subseteq E$. All other inclusions in the figure are straightforward.

4.4. Separation results. Now we consider which of these classes are the same and which are not equal. The following theorems are proved by diagonalization.

The first theorem asserts that if two space bounds differ by even a small amount, then the corresponding complexity classes differ. The second theorem gives a separation result for time, but it requires that the time functions differ by a logarithmic factor. There are technical reasons for this. The intuition that makes results for time harder to obtain than for space is straightforward though. Time marches relentlessly forward, but space can be reused.

THEOREM 5 ([HLS65]). *If S_2 is space-constructible, and*

$$\inf_{n \to \infty} S_1(n)/S_2(n) = 0,$$

then there is a language in $\mathrm{DSPACE}(S_2(n))$ *that is not in* $\mathrm{DSPACE}(S_1(n))$.

(This theorem was originally proved with the additional assumptions that $S_1(n) \geq \log(n)$ and $S_2(n) \geq \log(n)$. Sipser [Sip78] showed that the additional assumptions are unnecessary.) As consequences of this theorem, L $\neq$ POLYLOGSPACE, POLYLOGSPACE $\neq$ DLBA, and DLBA $\neq$ PSPACE. It even follows easily from theorems we described already that LBA is properly included in PSPACE.

THEOREM 6 ([HS65]). *If T_2 is fully time-constructible and*

$$\inf_{n \to \infty} T_1(n) \log(T_1(n))/T_2(n) = 0,$$

then there is a language in $\mathrm{DTIME}(T_2(n))$ *that is not in* $\mathrm{DTIME}(T_1(n))$.

Thus, P $\neq$ EXP. Equality (or inequality) of other inclusion relationships given in the figure are unknown. Amazingly, proper inclusion of each inclusion in the chain

$$L \subseteq NL \subseteq P \subseteq NP \subseteq PSPACE$$

is an open question even though the ends of the chain are distinct (L $\neq$ PSPACE), and it is known that NL $\neq$ PSPACE as well.

Similarly, equality of each inclusion in the chain

$$P \subseteq NP \subseteq PSPACE \subseteq EXP$$

is open, yet P $\neq$ EXP.

Also, it is not known whether any of the inclusions in the chain

$$DLBA \subseteq LBA \subseteq E \subseteq NE$$

are proper.

In addition to the results explicitly mentioned here, there is a nondeterministic space hierarchy theorem due to Ibarra [Iba72]. In 1987 Immerman [Imm88] and Szelepcsényi [Sze88] independently proved that nondeterministic $S(n)$ space-bounded classes are closed under complements for $S(n) \geq \log(n)$. A consequence of this important result is that there is a hierarchy theorem for nondeterministic space that is as strong as that for deterministic space.

The original hierarchy theorem for nondeterministic time is due to Cook [Coo73]. To date, the strongest such theorem known is due to Žák [Zak83]. He proved that if T_1 and T_2 are fully time-constructible functions such that $T_1(n+1) \in o(T_2(n))$, then $\mathrm{NTIME}(T_2(n))$ contains a set that is not in $\mathrm{NTIME}(T_1(n))$. From this hierarchy theorem some interesting separations of complexity classes follow, for example NP ⊂ NE. A new and more elementary proof of Žák's result is contained in Fortnow and Santhanam [FS11].

Book [Boo72, Boo76] has shown that none of the complexity classes POLYLOGSPACE, DLBA, and LBA is equal to either P or NP. He does this by analyzing the closure properties of these classes, and shows that they are different from one another. To this date, it is not known which of these classes contains a language that does not belong to the other.

§5. Nondeterminism and NP-completeness.

Several different additions to the basic deterministic Turing machine model are often considered. These additions add computational power to the model and allow us to classify certain problems more precisely. Often these are important problems with seemingly no efficient solution in the basic model. The question then becomes whether the efficiency provided by the additional power is really due to the new model or whether the added efficiency could have been attained without the additional resources.

The original and most important example of this type of consideration is nondeterminism. For each of the standard nondeterministic complexity classes we have been considering, it is an open question whether the class is distinct from its deterministic counterpart. (The only exception being PSPACE.)

Recall that a nondeterministic Turing machine is one with a multivalued transition function, and recall that such a machine M accepts an input word x if there some computation path of M on x that terminates in an accepting state. Let us assume, for each configuration of M, that there are exactly two possible next configurations, say c_0 and c_1. Then, M can be simulated in the following two-stage manner: (1) Write an arbitrary binary string on a worktape of a deterministic Turing machine M'. (2) Given an input word x, M' deterministically executes the computation path of M on x that is determined by the binary-string written on its work-tape. That is, M' simultaneously simulates M and reads its binary word from left to right; if the symbol currently scanned is 0, then M' continues its simulation in configuration c_0, and if the symbol currently scanned is 1, then M' continues its simulation in configuration c_1. Clearly, there is a binary string r such that M' accepts x when r is written on its work-tape if and only if there is an accepting computation of M on x. Informally, stage (1) comprises a "guess" of a *witness* or *proof* r to the fact that $x \in L(M)$, and stage (2) comprises a deterministic *verification* that r is a correct guess.

This guessing and verifying completely characterizes nondeterminism. Consider the problem (SAT) of determining, given a formula F of propositional logic, whether F is satisfiable. If F has n variables, there are 2^n different truth assignments. All known deterministic algorithms for deciding the satisfiability of F in the worst case are equivalent to a sequential search of each of the assignments to see if one of them leads to satisfaction (i.e., evaluates to the truth value TRUE). Clearly, an exhaustive search algorithm for checking satisfaction takes 2^n steps for a formula of size n, thereby placing SAT into the complexity class E. However, the following nondeterministic algorithm for SAT reduces its complexity to polynomial time: Given an input formula F, (1) guess an assignment to the Boolean variables of F. This takes $O(n)$ steps. (2) Verify that the assignment evaluates to True. This takes $O(n^2)$ steps. Thus, SAT belongs to the class NP.

NP is the most important nondeterministic complexity class. It is the class of languages that have deterministic, polynomial-time verifiers. NP plays a central role in computational complexity as many important problems from computer science and mathematics, which are not known to be solvable deterministically in polynomial time, are in NP. These include SAT, the graph Hamiltonian Circuit problem discussed earlier, various scheduling problems, packing problems, nonlinear programming, and hundreds of others. The most central and well-known open problem in complexity theory is whether P = NP. This problem is one of the several Millenium Prize Problems and the researcher who solves it receives a $1,000,000 prize from the Clay Mathematics Institute [Ins00].

The concept of NP-completeness plays a crucial role here as it gives a method of defining the most difficult NP problems. Intuitively, a problem A in NP is NP-complete if any problem in NP could be efficiently computed using an efficient algorithm for A as a subroutine.

Formally, A is NP-complete if (1) $A \in$ NP, and (2) for every $B \in$ NP, there is a function f that can be computed in polynomial time such that $x \in B$ if and only if $f(x) \in A$. (When property (2) holds for any sets B and A, we say B is *many-one reducible to A in polynomial time*, and we write $B \leq_m^P A$.) This definition captures the intuitive notion: An efficient procedure for determining whether x belongs to B is to compute $f(x)$ and then input $f(x)$ to a simple subroutine for determining membership in A.

Focus on the class NP as well as the discovery that SAT is NP-complete is due to Cook [Coo71a], and to Levin [Lev73], who independently obtained a variant of this result. Karp [Kar72] discovered a wide variety of NP-complete problems, and clearly demonstrated the importance of this new notion. These works opened the floodgates to the discovery of many more NP-complete problems in all areas of computer science and other computational disciplines.

The salient fact about NP-complete problems is that NP = P if and only if P contains an NP-complete problem. Thus, a single problem captures the complexity of the entire class.

§6. Relative computability. In this section we expand more broadly on the idea of using a subroutine for one problem in order to efficiently solve another problem. By doing so, we make precise the notion that the complexity of a problem B is related to the complexity of A—that there is an algorithm to efficiently accept B *relative to* an algorithm to efficiently accept A. Most generally, this should mean that an acceptor for B can be written as a program that contains subroutine calls of the form "$x \in A$," which returns TRUE if the Boolean test is true, and which returns FALSE, otherwise. The algorithm for accepting B is called a *reduction procedure* and the set A is called an *oracle*. The reduction procedure is *polynomial time-bounded* if the algorithm runs in polynomial time when we stipulate that only one unit of time is to be charged for the execution of each subroutine call. Obviously, placing faith in our modified Church–Turing Thesis and in the Cobham–Edmonds Thesis, these ideas are made precise via the oracle Turing machine.

An *oracle* Turing machine[2] is a multitape Turing machine with a distinguished work tape called the oracle tape, and three special states Q, YES, and NO. When the Turing machine enters state Q the next state is YES or NO depending on whether or not the word currently written on the oracle tape belongs to the oracle set. In this way, if the machine is provided with an oracle A, it receives an answer to a Boolean test of the form "$x \in A$" in one move. Let M be an oracle Turing machine, let A be an oracle, and let T be a time complexity function. Oracle Turing machine M with oracle A is $T(n)$ *time-bounded* if for every input of length n, M makes at most $T(n)$ moves before halting. If M is a nondeterministic oracle Turing machine, then every computation of M with A on words of length n must make at most $T(n)$ moves before halting. The language accepted by M with oracle A is denoted $L(M, A)$.

6.1. The polynomial hierarchy. If B is accepted by a polynomial time-bounded oracle Turing machine with oracle A, then we write $B \in \mathrm{P}^A$. P^A is the class of sets acceptable in polynomial time relative to the set A. The class NP^A is defined analogously. P^{NP} is the class of sets accepted in polynomial time using oracles in NP. Similarly, $\mathrm{NP}^{\mathrm{NP}}$ is the class of sets accepted in nondeterministic polynomial time using oracles in NP.

In analogy to the arithmetical hierarchy, there is a hierarchy of complexity classes that lies between P and PSPACE that is generated by using more and more powerful oracles. This is called the *polynomial hierarchy* (cf. [Sto76, Wra76]). The class P forms the bottom of the hierarchy and NP lies on the first level. The second level, denoted Σ_2^{P}, is defined to be $\mathrm{NP}^{\mathrm{NP}}$. The third level, Σ_3^{P}, is $\mathrm{NP}^{\Sigma_2^{\mathrm{P}}}$. In this manner, the classes of the polynomial hierarchy are defined inductively. It is not known whether the polynomial hierarchy is a strict hierarchy, just as it is not known whether $\mathrm{P} \neq \mathrm{NP}$. However, researchers believe the polynomial hierarchy forms an infinite hierarchy of

---

[2]Oracle Turing machines were first discussed by Turing in his 1939 London Mathematical Society paper, "Systems of Logic Based on Ordinals" [Tur39]. They were called o-machines in the paper.

distinct complexity classes, and it serves as a useful classification scheme for combinatorial problems.

If P = PSPACE, then the entire polynomial hierarchy (PH) collapses to P. However, this is not believed to be the case. Researchers believe that the hierarchy is strict, in which case it follows that PH is properly included in PSPACE. And assuming this is so, the structure of PSPACE − PH is very complex [AS89].

6.2. NP-hardness. A set A is NP-*hard* if for all sets $B \in$ NP, $B \in P^A$. Note that NP-hard sets are not required to belong to NP. NP-hard problems are as "hard" as the entire class of problems in NP, for if A is NP-hard and $A \in P$, then P = NP.

Define a set B to be *Turing reducible to A in polynomial time* ($B \leq_T^P A$) if $B \in P^A$. Then, it is interesting to compare $\leq_T^P$ with $\leq_m^P$. (Recall that we needed the latter in order to define NP-complete sets.) The reducibilities "$B \leq_m^P A$" and "$B \leq_T^P A$" are not the same, for they are known to differ on sets within the complexity class E [LLS75]. Define a set A to be $\leq_m^P$-*hard* for NP if for all sets $B \in$ NP, $B \leq_m^P A$. Under the assumption P $\neq$ NP, one can prove the existence of an NP-hard set A that is not $\leq_m^P$-hard for NP [SG77]. Furthermore, NP-hard problems are known that (1) seem not to belong to NP, and (2) no $\leq_m^P$-reduction from problems in NP is known (for example, see [GJ79] page 114, or [HS11], pages 147–148).

6.3. Complete problems for other classes. Completeness is not a phenomenon that applies only to the class NP. For each of the complexity classes in our list of standard classes, it is possible to define appropriate reducibilities and to find complete problems in the class with respect to those reducibilities. That is, there are problems that belong to the class such that every other problem in the class is reducible to them. A problem that is complete for a class belongs to a subclass (that is closed under the reducibility) if and only if the entire classes are the same. Interestingly, each of the sample problems given at the end of Section 3 are complete for the classes to which they belong. So, for example, the graph accessibility problem is complete for NL, and this problem belongs to L if and only if L = NL. The canonical complete problem for PSPACE is formed by fully quantifying formulas of propositional calculus. The problem then is to determine whether the resulting quantified Boolean formula is true.

§7. Nonuniform complexity. Here we briefly digress from our description of computational complexity based on the Turing machine to describe an important finite model of computing, Boolean circuits. Since real computer are built from electronic devices, digital circuits, this is reason enough to consider their complexity. The circuits that we consider are idealizations of digital circuits just as the Turing machine is an idealization of real digital computers. Another reason to study such nonuniform complexity is because interesting connections to uniform complexity are known.

A *Boolean circuit* is a labeled, acyclic, directed graph. Nodes with in-degree 0 are called *input nodes*, and are labeled with a Boolean variable x_i or a constant 0 or 1. Nodes with out-degree 0 are called *output nodes*. *Interior nodes*, nodes other than input nodes represent logical gates: they are labeled with AND, inclusive OR, or NOT. Arbitrary fan-out is allowed.

The constants 1 and 0 are the allowed inputs. If C is a circuit with exactly one output node and n input nodes, then C *realizes* a Boolean function $f: \{0, 1\}^n \mapsto \{0, 1\}$. When the input nodes receive their values, every interior value receives the value 0 or 1 in accordance with the logical gate that the interior node represents. If $x = x_1 x_2 \ldots x_n$ is a string in $\{0, 1\}^n$ and C is a circuit with n input nodes, then we say C *accepts* x if the output node of C has the value 1 when x is the input. Let A be a set of strings of length n over the binary alphabet. Then A is realized by a circuit C if for all strings x of length n, C accepts x if and only if x belongs to A.

As we have described computational complexity thus far, one machine is expected to serve for inputs of all lengths. The Turing machine is "uniform;" it is finitely describable, but might accept an infinite number of strings. A circuit however serves only for inputs of one length. Therefore, it is not a single circuit that corresponds to a machine, or that can accept a language, but a family of circuits, one for each length. A family of circuits is "nonuniform;" it may require an infinite description. We say that a family of circuits $\{C_n\}$ *recognizes* a set A, $A \subseteq \{0, 1\}^*$, if for each n, the circuit C_n, realizes the finite set $A^n = \{x \mid |x| = n \text{ and } x \in A\}$.

It should be obvious that families of circuits are too broad to exactly classify complexity classes, because every tally language[3] is recognized by some family of circuits. So there are families of circuits that recognize undecidable languages. Nevertheless, we proceed.

The *size* of a circuit is the number of nodes and the *depth* of a circuit is the length of the longest path in the circuit. The size of a circuit is a measure of the quantity of the circuit's hardware. We concentrate our focus on polynomial-size families of circuits, because these are the only families that can feasibly be constructed.

Circuit depth corresponds to parallel processing time, but we are not ready to fully justify this claim, which requires understanding the relationship between uniform and nonuniform computation. Basically, every parallel computation (on some parallel processor) can be unraveled to form a family of circuits with constant delay at each gate. So circuit depth is a lower bound on parallel time. If language A is not recognizable by a family of circuits of depth $T(n)$, then no parallel computer can compute A in time $T(n)$.

---

[3] A tally language is a language defined over a unary alphabet.

Savage [Sav72] proved that every set in DTIME($T(n)$) can be recognized by a family of circuits of size $O(T(n))^2$. A tighter result than this is proved by Pippenger and Fischer [PF79]. As an immediate consequence, every language in P has a polynomial-size family of circuits. Hence, if some language in NP does not have a polynomial-sized family of circuits, then P $\neq$ NP.

Largely because of this observation, much effort has gone into obtaining lower-bounds for various classes of circuits. Thus far this effort has met with limited success: Deep results have been obtained for highly restricted classes [FSS84, Ajt83, Yao85]. There is more to this story. Shannon [Sha49], in a nonconstructive proof, showed that almost all languages require exponential size circuits, but little is known about which sets these are. One can construct specific, though quite unnatural languages in ESPACE that require exponential size. Furthermore, Meyer and Stockmeyer [MS72] proved an exponential lower bound on the circuit complexity of deciding the weak monadic second-order theory of one successor (WS1S). From this they are able to conclude that ESPACE contains a binary language of maximum circuit complexity. One of the great challenges is to find lower bounds to the circuit size of families of circuits that recognize other explicit, interesting, combinatorial problems.

Now we must mention the important result of Karp and Lipton [KL80] that states that if NP has a polynomial-size family of circuits, then the polynomial hierarchy collapses to the second level. Here is another important connection between nonuniform and uniform complexity, which supports the conjecture that NP does not have a polynomial-size family of circuits.

§8. **Parallelism.** We consider now the theory of highly parallel computations. VLSI chip technology is making it possible to connect together large numbers of processors to operate together synchronously. The question, of course, is what problems can be implemented more efficiently on such "parallel" machines, and how much can such machines speed up computations? If we assume that a polynomial number of processors is reasonable, and more than a polynomial number of processors unreasonable, then it turns out that parallelism will not make intractable problems tractable. Therefore, if we are to dramatically improve performance it must be by reducing polynomial sequential time to subpolynomial parallel time. We achieve this by trading numbers of processors for speed. We investigate the limits of what can be achieved by parallel machines and whether fewer than a polynomial number of processors can suffice.

Several formal models of parallel computation have been proposed and even though each model can simulate the other without much loss of efficiency, the issue of "correct" model is not quite as settled as it is with sequential computation. In this article we mention two models of parallelism, "alternating Turing machines" and "uniform families of circuits," which play an important role in computational complexity.

The alternating Turing machine, due to Chandra, Kozen, and Stockmeyer [CKS81] is a fascinating extension of the nondeterministic Turing machine. Informally, think of the nondeterministic Turing machine as one consisting of "existential" configurations. That is, when the Turing machine enters an existential configuration, it causes a number of processes to operate in parallel, one process for each nondeterministic choice. If one of these processes accepts, then it reports acceptance back to its parent, and in this manner the computation accepts. With the alternating Turing machine, a process that became active by this action, in turn enters a "universal" configuration that causes another large number of processes to become active. This time, the universal configuration eventually reports acceptance to its parent if and only if all of the processes it spawns accept. Consider such alternating Turing machines that operate in polynomial time, and consider the complexity class AP of languages that are accepted by these devices. It should come as no surprise that alternating Turing machines that perform a constant $k > 0$ number of alternations accept precisely the kth level of the polynomial hierarchy. More surprising, is the theorem that AP = PSPACE. Studies have shown that it is true in general that

parallel time is within a polynomial factor of deterministic space.

The theorem just stated suggests that presumably intractable languages, i.e., those in PSPACE, are capable of enormous speedup by using parallelism. This, however, is an impractical observation. The proof of the theorem requires an exponential number of processes. No one will ever build parallel computers with an exponential number of processes.

Before returning to the question of what are the problems that can take advantage of parallelism, we introduce our second model. Families of circuits would provide a simple and effective model were it not for the fact that small families of circuits can recognize undecidable sets. We repair this flaw by introducing uniformity. Borodin and Cook [Coo79, Bor77] define a family of circuits $\{C_n\}_n$ to be *logspace uniform* if there exists a deterministic Turing machine that on input 1^n, for each $n \geq 1$, computes (an encoding of) the circuit C_n. Then, for example, one interesting result is that P is the class of languages that have logspace uniform, polynomial-size families of circuits.

Now we can return to the question at hand and we do so with a discussion of Greenlaw *et al.* [GHR95]: We have seen already that parallelism will not make intractable problems tractable. Therefore, if we are to dramatically improve performance it must be by reducing polynomial sequential time to subpolynomial parallel time. We achieve this by trading numbers of processors for speed. The goal of practical parallel computation is to develop algorithms that use a reasonable number of processors and are exceedingly fast. What do we mean by that? We assume that a polynomial number of processors is reasonable, and more than that is unreasonable. Can fewer than a polynomial number of processors suffice? To answer this questions observe that

(sequential time)/(number of processors) ≤ (parallel time).

Taking sequential time to be polynomial time, obviously, if parallel time is to be subpolynomial, then a polynomial number of processors must be used. We focus on the class of problems that have uniform families of circuits with polynomial size and polylog, i.e., $(\log n)^{O(1)}$, depth. So highly parallel problems are those for which we can develop algorithms that use a polynomial number of processors to obtain polylog parallel time bounds.

The resulting class of languages is called NC, and is named in honor of Nick Pippenger, who obtained an important characterization [Pip79]. Researchers identify the class NC with the collection of feasibly parallel problems, much as we identify P with the collection of feasibly computable problems.

The models we introduced are quite robust. For example, it is known that NC is identical to the class of languages accepted by alternating Turing machines in simultaneous polylog time and log space.

§9. Probabilistic complexity. In a 1950 paper in the journal *Mind* [Tur50], Alan Turing wrote,

> An interesting variant on the idea of a digital computer is a 'digital computer with a random element'. These have instructions involving the throwing of a die or some equivalent electronic process: one such instruction might for instance be, 'Throw a die and put the resulting number into store 1000.'

Nearly thirty years after this paper appeared the ideas put forth there became a central part of the active area of probabilistic complexity theory. The main computational model used for this study is a probabilistic Turing machine. It has been used to measure and formalize the performance of probabilistic algorithms and to define the main concepts for the theoretical study of probabilistic complexity theory.

The impetus for this work began in the mid-1970s with the work on algorithms for primality testing pioneered by Solovay and Strassen [SS77], and by Miller and Rabin [Mil76], [Rab80]. The two algorithms invented in these papers both have the same character. They both yield methods for testing whether a given input integer is prime that are significantly more efficient than any known deterministic algorithm. Instead they make use of random numbers r provided to the algorithm, and used there to decide primality. While very efficient, the algorithm can, in rare instances of r, make a mistake and wrongly decide that a composite number is actually prime. If the probability of these errors happening is non-negligible, such algorithms have little worth. But in this case it is possible to ensure that the probability of error is extremely small, say smaller than the chances the computer makes a hardware error during the computation, or smaller than 1/(the number of atoms in the known universe). In such cases the algorithms can be thought to be essentially correct and useful for the actual task of finding large prime numbers for various applications, most notably cryptographic applications.

From these ideas arose other examples of strong probabilistic algorithms for important interesting problems. Over the next decade these algorithms gained in impact and theorists took on the task of classifying and examining these algorithms and the strength of the assumptions and method underlying them. The model they used for this study was essentially the model proposed above by Alan Turing in 1950.

These algorithms suggested that we should revise our notion of "efficient computation". Perhaps we should now equate the efficiently computable problems with the class of problems solvable in probabilistic polynomial time. Beginning in the 1970s a new area of complexity theory was developed to help understand the power of probabilistic computation.

Formally, a probabilistic Turing machine is just a nondeterministic Turing machine, but acceptance is defined differently. Each nondeterministic choice is considered as a random experiment in which each outcome has equal probability. We may assume that each nondeterministic branch has exactly two possible outcomes, so that each has possibility $1/2$. A probabilistic Turing machine has three kinds of final states, accepting or a-states, rejecting or r-states, and undetermined or ?-states. The outcome of the machine on an input is now a random variable whose range is $\{a, r, ?\}$. We let $\Pr[M(x) = y]$ denote the probability that machine M on input x halts in a y-state. Note that the probability of a given nondeterministic path is obtained by raising $1/2$ to a power equal to the number of nondeterministic choices along it. The probability that M accepts an input x, $\Pr[M(x) = a]$, is the sum of the probabilities of all accepting paths (that is, paths which end in an a-state).

Using this model we can now define several different useful probabilistic complexity classes. These classes were originally defined by Gill [Gil77] and by Adleman and Manders [AM77]. Each consists of languages accepted by restricting the machines to polynomial time and specifying the probability needed for acceptance. The first class, PP, for probabilistic polynomial time, is the easiest to define and is the most powerful parallel class we consider, but is the least useful. Let χ_A denote the characteristic function of A. PP is the class of all languages for which there is a probabilistic, polynomial time-bounded Turing machine M such that for all x,

$$\Pr[M(x) = \chi_A(x)] > 1/2.$$

That is,

$$x \in A \to \Pr[M(x) = a] > 1/2,$$

and

$$x \notin A \to \Pr[M(x) = r] > 1/2.$$

One would like to increase the reliability of a probabilistic Turing machine by repeating its computations a large number of times and giving as output the majority result. This seems not to be possible with PP, and this is why there is no practical interest in this class; we are at times stuck with a too high probability of error. The following two, quite straightforward facts place PP in the context of the earlier defined complexity classes. While it is not known whether PP is contained in the polynomial hierarchy, it seems unlikely, as in 1989 Toda [Tod89] proved that if PP is so contained then the PH collapses.

In our definition of PP we require that $\Pr[M(x) = \chi_A(x)] > 1/2$ for all x. In fact, we can very slightly weaken this condition to allow that $\Pr[M(x) = r] \geq 1/2$ in the case that $x \notin A$, and still conclude that $A \in$ PP. The proof of this fact is sometimes useful, for example in the next lemma. It can be found in Homer and Selman [HS11], page 227.

LEMMA 1. NP $\subseteq$ PP.

PROOF. Let M be a nondeterministic Turing machine that accepts an NP language L. Now consider the nondeterministic Turing machine M' whose first move is to nondeterministically either simulate M or to go into a special state where the computation, in every succeeding move, just splits into two identical computations until the length of the computation reaches the same length as the computations of M and then halts in an accepting state.

Now if $x \notin L$, then there are exactly as many accepting computations as rejecting computations. Namely, none of the computations whose first step is to simulate M accept while all of the computations whose first step is to go into the special new state accept. These two groups of computations are of the same size.

On the other hand, if $x \in L$, then as above, all computations starting with the special state accept and at least one computation of M accepts and so more computations accept than reject. So the machine M', considered as a probabilistic machine, shows that L is in PP. ⊣

While stated without proof, it is quite straightforward to simulate the answer to PP computations within PSPACE, yielding,

LEMMA 2. PP $\subseteq$ PSPACE.

The error probability is the probability that a probabilistic machine gives the wrong answer on a given input. The next class we define has the error probability bounded away from $1/2$, and that restriction makes it possible to increase reliability. The class BPP, for bounded-error probabilistic polynomial time, is the class of all languages for which there is a probabilistic, polynomial time-bounded Turing machine M and a number ϵ, $0 < \epsilon < 1/2$, such that for

all x,
$$\Pr[M(x) \neq \chi_A(x)] < \epsilon.$$

BPP-type Turing machines are said to be of the "Monte Carlo" type. They model algorithms that are allowed to make mistakes (i.e., terminate in an r-state when $x \in L$, or terminate in an a-state, when $x \notin L$) with some small probability. On the other hand, "Las Vegas" algorithms may terminate with "?" (again with small probability), but they never err. These too can be captured by probabilistic complexity, as we explain below.

The primality algorithms in the examples above were Monte Carlo algorithms. It is not difficult to see that $P \subseteq BPP \subseteq PP$, and BPP is closed under complement. But where does BPP lie with respect to the polynomial hierarchy? Somewhat surprisingly, Sipser, Gacs, and Lautemann proved that $BPP \subseteq \Sigma_2^P$ [Sip83, Lau83].

In fact something stronger can be seen from these algorithms. Namely the probabilistic primality algorithms have the property that when the input to them is prime it is always identified as being prime by the algorithms. It is only when the input in composite that there is some probability of error. To capture this property we define the class, RP, an interesting, one-sided version of BPP.

The class RP, for random polynomial time, is the class of all languages A for which there is a probabilistic, polynomial time-bounded Turing machine M such that for all x,
$$x \in A \rightarrow \Pr[M(x) = a] > 1/2,$$
and
$$x \notin A \rightarrow \Pr[M(x) = a] = 0.$$

Notice that if $M(x)$ is run and outputs a then we know with certainty that $x \in A$. However, if $M(x)$ outputs r or ? then we do not know with certainty whether x is an element of A or not.

In some interesting instances a problem L has been shown to lie in $RP \cap co\text{-}RP$. This implies that there are algorithms for L of the Las Vegas type as described above. To see this consider a probabilistic Turing machine M showing that L is in RP and another probabilistic Turing machine N showing L's complement is in RP. Then if, on an input x, we alternately compute $M(x)$ and $N(x)$ in turn, outputting an answer only when we obtain $M(x) = a$ in which case we know with certainty x is in L, or $N(x)$ outputs a, in which case we know x is not in L. This is a Las Vegas algorithm for L as we never obtain a wrong answer with this method, though with small probability we may compute for a very long time without a definitive answer.

Neither RP nor BPP are known to have complete problems. Several properties of these classes suggest that such problems don't exist. And it is known that oracles exist relative to which neither class has complete problems [Sip82, HH86].

Returning to the idea of primality testing, and the complexity of the set of prime numbers, it is easy to see that primality is in co-NP. An early result of Pratt [Pra75] showed that one could also find a witness for primality which can be checked in polynomial time, putting primality into NP ∩ co-NP.

The probabilistic algorithms of Strassen/Solovay and of Miller/Rabin, discussed earlier, show that primality is in co-RP. Later, research of Adleman and Huang [AH87] improved this to obtain primality in RP ∩ co-RP, implying that primality has a Las Vegas algorithm. Finally, early in this century Agrawal, Kayal and Saxena [AKS04] achieved a breakthrough by finding a deterministic polynomial-time algorithm for primality. If this result had been known in the 70s, perhaps the study of probabilistic algorithms would not have progressed as quickly.

9.1. Derandomization. If you generate a random number on a computer, you do not get a truly random value, but a pseudorandom number computed by some complicated function on some small, hopefully random seed. In practice this usually works well, so perhaps in theory the same might be true. Many of the exciting results in complexity theory in the 1980s and 90s consider this question of derandomization—how to reduce or eliminate the number of truly random bits to simulate probabilistic algorithms.

These derandomization results add to the possibility of eliminating randomness from BPP and RP problems. BPP has many similar properties to P (e.g., they are both closed under complement and $\text{BPP}^{\text{BPP}} = \text{BPP}$) and because there are few natural candidates for problems in BPP − P, many believe that P = BPP. Additionally, it is known that P ≠ BPP implies that E is surprisingly weak, specifically that it implies that every problem in E can be solved by a circuit family of subexponential size [IW97].

One natural and interesting problem which has resisted derandomization is the problem of polynomial identity testing for algebraic circuits. This is one of the few natural problems known to be in BPP and not proven to be in P. The study of this problem dates back to papers of DeMillo and Lipton [DL78], Schwartz [Sch80] and Zippel [Zip79] in the late 1970s. The problem statement is: We are given a field F and an arithmetic circuit with input variables $x_1, x_2, \ldots, x_n$ over F. This circuit defines a polynomial $p(x_1, x_2, \ldots, x_n)$ in the ring $F[x_1, x_2, \ldots, x_n]$. The question is to find an algorithm to decide whether $p(x_1, x_2, \ldots, x_n) = 0$ for all inputs from F^n.

Derandomization considers how to reduce or eliminate randomness from probabilistic algorithms, making them deterministic. For example P = BPP is equivalent to being able to derandomize all BPP algorithms. There has been very interesting progress in this area in the past decade. But it is still unknown whether P = BPP, or even P = RP.

There have been two approaches to eliminating randomness, both of which indicate that strong, general derandomization results may be possible. The first approach arose from cryptography where creating pseudorandomness from cryptographically hard functions was shown by Blum and Micali [BM84]. Subsequently Yao [Yao82] showed how to reduce the number of random bits of any algorithm based on any cryptographically secure one-way permutation. Håstad, Impagliazzo, Levin and Luby [HILL91] building on techniques of Goldreich and Levin [GL89] and Goldreich, Krawczyk and Luby [GKL88] prove that one can get pseudorandomness from any one-way function.

Nisan and Wigderson [NW94] take a different approach. They show how to get pseudorandomness based on a language hard against nonuniform computation. Impagliazzo and Wigderson [IW97] building on this result and upon the work of Babai, Fortnow, Nisan and Wigderson [BFNW93] show that BPP equals P if there exists a language in exponential time that cannot be computed by any subexponential-size circuit. This is a believable hypothesis.

§10. Interactive proof systems. One can think of the class NP as a *proof system*: For example, for the NP problem SAT = {satisfiable Boolean formulas} an arbitrarily powerful prover gives a proof, i.e., a string representing a truth assignment, that shows a formula is satisfiable. This proof can be verified by a polynomial-time verifier, an algorithm that is capable of checking that the string is a short (polynomial-length) proof of satisfiability. We do not put any restriction on how the verifier obtains the proof itself; it is sufficient that the proof exists. For example, the proof might be provided by a powerful prover with unrestricted computational power.

Consider a generalization of these ideas where we allow a protocol or dialogue between the verifier and a prover to decide language membership. As before there is no restriction placed on the computational power of the prover, and the verifier is restricted to be only polynomially powerful. We add one more ingredient to this mix by allowing randomization. That is, we allow the verifier, whose time is polynomially limited, to use a random sequence of bits to aid in the computation, and require only that she be convinced with high likelihood, and not with complete certainty, of the membership claim being made. Specifically the verifier is a probabilistic polynomial-time Turing machine with acceptance probability bounded away from $1/2$, i.e., one implementing a Monte Carlo algorithm.

This model generalizes that of the NP verifier; it is clear that as before NP problems are those for which membership can be proved by a prover giving the verifier a polynomial length "proof" and the verifier validating that proof in deterministic polynomial time. The model also easily captures the class BPP as any BPP problem can be decided by the probabilistic polynomial time verifier without any information from the prover.

The class of probabilistic proofs where the interaction is limited to one "round" of communication, in this case the prover sending a message to the verifier, yields the complexity class of problems called MA. One can also consider extended interaction where the verifier sends messages based on her random bits back to the prover. The (fixed) bounded-round version of this class is denoted AM and the unbounded polynomial-round version is IP. The incredible power of these interactive proof systems has led to several of the most surprising and important recent results in computational complexity theory.

In 1985, Babai [Bab85] defined interactive proof systems, in order to give complexity-theoretic characterizations of some problems concerning matrix groups. An alternative interactive proof system was defined by Goldwasser, Micali and Rackoff [GMR89] as a basis for the cryptographic class zero-knowledge. Zero-knowledge proof systems have themselves played a major role in cryptography.

Goldreich, Micali and Wigderson [GMW91] showed that the set of pairs of nonisomorphic graphs has a bounded-round interactive proof system. Boppana, Håstad and Zachos [BHZ87] showed that if the complement of any NP-complete language has bounded-round interactive proofs then the polynomial-time hierarchy collapses. As a consequence of this result it is known that the graph isomorphism problem is not NP-complete unless the polynomial-time hierarchy collapses.

In 1990, Lund et al. [LFKN90] gave an interactive proof for the permanent problem which implies, as the permanent problem is #P-hard, that every language in the polynomial hierarchy has an interactive proof system. They also showed that the complements of NP-complete languages have unbounded-round interactive proof systems. These techniques were quickly extended by Shamir [Sha90] to show that every language in PSPACE has an interactive proof system. A few years earlier, Feldman [Fel86] had proved that every language with interactive proofs lies in PSPACE. The result that IP = PSPACE is one of the central advances in complexity theory in the last three decades.

It is notable that in general proofs concerning interactive proof systems do not relativize. That is, they are not true relative to every oracle. The classification of interactive proofs turned out not to be the end of the story but only the beginning of a revolution connecting complexity theory with approximation algorithms. For the continuation of this story we turn to probabilistically checkable proofs.

10.1. Probabilistically checkable proofs. Understandably, following on the IP = PSPACE result, there was a flurry of activity in the early 1990s examining and classifying the many variations of proof systems both stronger and weaker than IP. One natural direction of this work was to try to apply these methods to the large, central class of NP problems. After a series of results improving on the understanding of interactive proof systems with multiple provers, Babai et al. [BFLS91] scaled these proof techniques down to develop "holographic" proofs for NP where, with a properly encoded input, the verifier can check the correctness of the proof in a very short amount of time.

Feige et al. [FGL+96] made an amazing connection between probabilistically checkable proofs and the clique problem. By viewing possible proofs as nodes of a graph, they showed that one cannot approximate the size of a clique well without unexpected collapses in complexity classes.

In 1992, Arora et al. [ALM+92] building on work of Arora and Safra [AS98] showed that every language in NP has a probabilistically checkable proof where the verifier uses only a logarithmic number of random coins and a constant number of queries to the proof. Their results have tremendous implications for the class MAXSNP of approximation problems. This class developed by Papadimitriou and Yannakakis [PY91] contains many interesting complete problems such as vertex cover, max-cut, independent set, some variations of the traveling salesman problem and maximizing the number of satisfiable clauses of a formula. Arora et al. shows that, unless P = NP, every MAXSNP-complete set does not have a polynomial-time approximation scheme. Specifically, for each of these problems there is some constant $\delta > 1$ such that they cannot be approximated within a factor of δ unless P = NP.

In the period since these initial results on probabilistically checkable proofs, we have seen a large number of outstanding papers improving the proof systems and getting stronger hardness of approximation results. For a sampling of these results consult Arora [Aro98], Håstad [Has99], [Has01] or Trevisan [Tre04].

§11. **A. M. Turing Award.** The highest honor available to be bestowed on a computer scientist is the ACM's appropriately named A. M. Turing Award. The award is accompanied by a prize of $250,000, with financial support provided by the Intel Corporation and Google Inc. Awardees who we have cited in this paper include Manuel Blum, Stephen Cook, Juris Hartmanis, Richard Karp, Michael Rabin, Dana Scott, Richard Stearns, and Andrew Yao.

§12. **Acknowledgment.** We thank Juris Hartmanis for his invaluable comments on Turing machines and the birth of complexity theory. We thank Lance Fortnow for his permission to use some of the discussion of the history of complexity theory, which first appeared in Fortnow and Homer [FH03]. Figure 1 on page 307 comes from Homer and Selman [HS11] and is used by permission from Springer Science and Business Media.

REFERENCES

[AH87] L. ADLEMAN and M-D. HUANG, *Recognizing primes in random polynomial time*, **Proceedings of the nineteenth annual ACM Symposium on Theory of Computing**, 1987, pp. 462–469.

[AM77] L. ADLEMAN and K. MANDERS, *Reducibility, randomness, and intractability*, **Proceedings of the ninth annual ACM Symposium on Theory of Computing**, 1977, pp. 151–163.

[AKS04] M. AGRAWAL, N. KAYAL, and N. SAXENA, *Primes is in P*, **Annals of Mathematics**, vol. 160 (2004), no. 2, pp. 781–793.

[Ajt83] M. AJTAI, σ_1^1 *formulea on finite structures*, **Journal of Pure and Applied Logic**, vol. 24 (1983), pp. 1–48.

[AS89] K. AMBOS-SPIES, *On the relative complexity of hard problems for complexity classes without complete problems*, **Theoretical Computer Science**, vol. 63 (1989), pp. 43–61.

[Aro98] S. ARORA, *Polynomial time approximation schemes for Euclidean traveling salesman and other geometric problems*, **Journal of the ACM**, vol. 45 (1998), no. 5, pp. 753–782.

[ALM+92] S. ARORA, C. LUND, R. MOTWANI, M. SUDAN, and M. SZEGEDY, *Proof verification and hardness of approximation problems*, **Proceedings of the thirty-third symposium on Foundations of Computer Science**, IEEE Computer Society, 1992, pp. 14–23.

[AS98] S. ARORA and S. SAFRA, *Probabilistic checking of proofs: A new characterization of NP*, **Journal of the ACM**, vol. 45 (1998), no. 1, pp. 70–122.

[Bab85] L. BABAI, *Trading group theory for randomness*, **Proceedings of the 17th annual ACM Symposium on Theory of Computing**, 1985, pp. 421–429.

[BFLS91] L. BABAI, L. FORTNOW, L. LEVIN, and M. SZEGEDI, *Checking computations in polylogarithmic time*, **Proceedings of the 23rd ACM Symposium on the Theory of Computing**, 1991, pp. 21–31.

[BFNW93] L. BABAI, L. FORTNOW, N. NISAN, and A. WIGDERSON, *BPP has subexponential simulations unless EXPTIME has publishable proofs*, **Computational Complexity**, vol. 3 (1993), pp. 307–318.

[BM84] M. BLUM and S. MICALI, *How to generate cryptographically strong sequences of pseudorandom bits*, **SIAM Journal on Computing**, vol. 13 (1984), pp. 850–864.

[Boo72] R. BOOK, *On languages accepted in polynomial time*, **SIAM Journal on Computing**, vol. 1 (1972), no. 4, pp. 281–287.

[Boo76] ———, *Translational lemmas, polynomial time, and* $(\log n)^j$ *-space*, **Theoretical Computer Science**, vol. 1 (1976), pp. 215–226.

[BG70] R. BOOK and S. GREIBACH, *Quasi-realtime languages*, **Mathematical Systems Theory**, vol. 4 (1970), pp. 97–111.

[BHZ87] R. BOPPANA, J. HÅSTAD, and S. ZACHOS, *Does co-NP have short interactive proofs?*, **Information Processing Letters**, vol. 25 (1987), no. 2, pp. 127–132.

[Bor77] A. BORODIN, *On relating time and space to size and depth*, **SIAM Journal on Computing**, vol. 6 (1977), pp. 733–744.

[CKS81] A. CHANDRA, D. KOZEN, and L. STOCKMEYER, *Alternation*, **Journal of the ACM**, vol. 28 (1981), no. 1, pp. 114–133.

[Chu36] A. CHURCH, *An unsolvable problem of elementary number theory*, **American Journal of Mathematics**, vol. 58 (1936), pp. 345–363.

[Cob64] A. COBHAM, *The intrinsic computational difficulty of functions*, **Proceedings of the 1964 international congress for Logic, Methodology, and the Philosophy of Science**, North Holland, 1964, pp. 24–30.

[Coo71a] S. COOK, *Characterizations of pushdown machines in terms of time-bounded computers*, **Journal of the ACM**, vol. 19 (1971), pp. 175–183.

[Coo71b] ———, *The complexity of theorem-proving procedures*, **Proceedings of the third ACM Symposium on Theory of Computing**, 1971, pp. 151–158.

[Coo73] ———, *A hierarchy for nondeterministic time complexity*, **Journal of Computer and System Sciences**, vol. 7 (1973), no. 4, pp. 343–353.

[Coo79] ———, *Deterministic CFL's are accepted simultaneously in polynomial time and log squared space*, **Proceedings of the 11th annual ACM Symposium on Theory of of Computing**, 1979, pp. 338–345.

[Dav06a] MARTIN DAVIS, *The Church–Turing thesis: Consensus and opposition*, **CiE'06**, 2006, pp. 125–132.

[Dav06b] ———, *Why there is no such discipline as hypercomputation*, **Applied Mathematics and Computation**, (2006), pp. 4–7.

[DL78] R. DEMILLO and R. LIPTON, *A probabilstic remark on algebraic program testing*, **Information Processing Letters**, vol. 7 (1978), no. 4, pp. 193–195.

[Edm65] J. EDMONDS, *Paths, trees, and flowers*, **Canadian Journal of Mathematics**, vol. 17 (1965), pp. 449–467.

[ET76] S. EVEN and R. TARJAN, *A combinatorial problem which is complete in polynomial space*, **Journal of the ACM**, vol. 23 (1976), pp. 710–719.

[FGL+96] U. FEIGE, S. GOLDWASSER, L. LOVASZ, S. SAFRA, and M. SZEGEDY, *Interactive proofs and the hardness of approximating cliques*, **Journal of the ACM**, vol. 43 (1996), pp. 268–292.

[Fel86] P. FELDMAN, *The optimum prover lives in PSPACE*, manuscript, 1986.

[FH03] L. FORTNOW and S. HOMER, *A brief history of complexity theory*, **Bulletin of the European Association for Theoretical Computer Science**, vol. 80 (2003), pp. 95–133.

[FS11] L. FORTNOW and R. SANTHANAM, *Robust simulations and significant separations*, **Proceedings of the 38th International Colloquium on Automata, Languages and Programming**, Lecture Notes in Computer Science, vol. 6755, Springer, 2011, pp. 569–580.

[FSS84] M. FURST, J. SAXE, and M. SIPSER, *Parity, circuits, and the polynomial-time hierarchy*, **Mathematical Systems Theory**, vol. 17 (1984), no. 1, pp. 13–28.

[GJ79] M. GAREY and D. JOHNSON, **Computers and intractability: A guide to the theory of NP-completeness**, W. H. Freeman, San Francisco, 1979.

[Gil77] J. GILL, *Computational complexity of probabilistic Turing machines*, **SIAM Journal on Computing**, vol. 6 (1977), no. 4, pp. 675–695.

[GKL88] O. GOLDREICH, H. KRAWCZYK, and M. LUBY, *On the existence of pseudorandom generators*, **Proceedings of the 29th annual IEEE symposium on Foundations of Computer Science**, 1988, pp. 12–24.

[GL89] O. GOLDREICH and L. LEVIN, *Hard-core predicates for any one-way function*, **Proceedings of the 21st annual ACM Symposium on Theory of Computation**, 1989, pp. 25–32.

[GMW91] O. GOLDREICH, S. MICALI, and A. WIGDERSON, *Proofs that yield nothing but their validity or all languages in NP have zero-knowledge proofs*, **Journal of the ACM**, vol. 38 (1991), no. 3, pp. 691–729.

[GMR89] S. GOLDWASSER, S. MICALI, and C. RACKOFF, *The knowledge complexity of interactive proof systems*, **SIAM Journal on Computing**, vol. 18 (1989), no. 1, pp. 186–208.

[GHR95] R. GREENLAW, H. HOOVER, and W. RUZZO, *Limits to parallel computation: P-completeness theory*, Oxford University Press, New York, NY, 1995.

[HH86] J. HARTMANIS and L. HEMACHANDRA, *Complexity classes without machines: on complete languages for UP*, **Proceedings of the 13th International Colloquium on Automata, Languages, and Programming** (Berlin), Lecture Notes in Computer Science, Springer-Verlag, 1986, pp. 123–135.

[HLS65] J. HARTMANIS, P. LEWIS, and R. STEARNS, *Hierarchies of memory limited computations*, **Proceedings of the sixth annual IEEE Symposium on Switching Circuit Theory and Logical Design**, 1965, pp. 179–190.

[HS65] J. HARTMANIS and R. STEARNS, *On the computational complexity of algorithms*, **Transactions of the American Mathematical Society**, vol. 117 (1965), pp. 285–306.

[Has99] J. HÅSTAD, *Clique is hard to approximate within $n^{1-\epsilon}$*, **Acta Mathematica**, vol. 182 (1999), pp. 105–142.

[Has01] ———, *Some optimal inapproximabiity results*, **Journal of the ACM**, vol. 48 (2001), pp. 798–849.

[HILL91] J. HÅSTAD, R. IMPAGLIAZZO, L. LEVIN, and M. LUBY, *Construction of a pseudorandom generator from any one-way function*, **Technical report**, ICSI, Berkely, CA, 1991.

[HS11] S. HOMER and A. SELMAN, *Computability and complexity theory*, 2nd ed., Texts in Computer Science, Springer-Verlag, New York, NY, December 2011.

[Iba72] O. IBARRA, *A note concerning nondeterministic tape complexities*, **Journal of Computer and System Sciences**, vol. 19 (1972), no. 4, pp. 609–612.

[Imm88] N. IMMERMAN, *Nondeterministic space is closed under complementation*, **SIAM Journal on Computing**, vol. 17 (1988), no. 5, pp. 935–938.

[IW97] R. IMPAGLIAZZO and A. WIGDERSON, $P = BPP$ *if E requires exponential circuits: Derandomizing the XOR lemma*, **Proceedings of the 29th ACM Symposium on the Theory of Computing**, ACM, New York, 1997, pp. 220–229.

[Ins00] CLAY MATHEMATICS INSTITUTE, *Millennium prize problems*, http://www.claymath.org/prizeproblems/, 2000.

[Jon73] N. JONES, *Reducibility among combinatorial problems in log n space*, **Proceedings of the seventh annual Princeton Conference on Information Sciences and Systems** (Princeton, NJ), Department of Electrical Engineering, Princeton University, 1973, pp. 547–551.

[Jon75] ———, *Space-bounded reducibility among combinatorial problems*, **Journal of Computer and System Sciences**, vol. 11 (1975), pp. 68–85.

[JS74] N. JONES and A. SELMAN, *Turing machines and the spectra of first-order formulas*, **The Journal of Symbolic Logic**, vol. 29 (1974), pp. 139–150.

[Kar72] R. KARP, *Reducibility among combinatorial problems*, **Complexity of computer computations**, Plenum Press, New York, 1972, pp. 85–104.

[KL80] R. KARP and R. LIPTON, *Some connections between nonuniform and uniform complexity classes*, **Proceedings of the twelfth annual ACM Symposium on the Theory of Computing**, 1980, an extended version has appeared in *L'Enseignement Mathématique*, 2nd series 28, 1982, pp. 191–209, pp. 302–309.

[Kur64] S. KURODA, *Classes of languages and linear bounded automata*, **Information and Control**, vol. 7 (1964), no. 2, pp. 207–223.

[LLS75] R. LADNER, N. LYNCH, and A. SELMAN, *A comparison of polynomial time reducibilities*, **Theoretical Computer Science**, vol. 1 (1975), pp. 103–123.

[Lau83] C. LAUTEMANN, *BPP and the polynomial hierarchy*, **Information Processing Letters**, vol. 17 (1983), pp. 215–217.

[Lev73] L. LEVIN, *Universal sorting problems*, **Problems of Information Transmission**, vol. 9 (1973), pp. 265–266, English translation of original in *Problemy Peredaci Informacii*.

[LFKN90] C. LUND, L. FORTNOW, H. KARLOFF, and N. NISAN, *Algebraic methods for interactive proof systems*, **Proceedings of the 31st IEEE symposium on Foundations of Computer Science**, 1990, pp. 2–10.

[Mag69] G. MAGER, *Writing pushdown acceptors*, **Journal of Computer and System Sciences**, vol. 3 (1969), no. 3, pp. 276–319.

[MS72] A. MEYER and L. STOCKMEYER, *The equivalence problem for regular expressions with squaring requires exponential space*, **Proceedings of the thirteenth IEEE symposium on Switching and Automata Theory**, 1972, pp. 125–129.

[Mil76] G. MILLER, *Reimann's hypothesis and tests for primality*, **Journal of Computer and System Sciences**, vol. 13 (1976), pp. 300–317.

[Myh60] J. MYHILL, *Linear bounded automata*, **WADD 60-165**, Wright Patterson AFB, Ohio, 1960.

[NW94] N. NISAN and A. WIGDERSON, *Hardness vs. randomness*, **Journal of Computer and System Sciences**, vol. 49 (1994), pp. 149–167.

[PY91] C. PAPADIMITRIOU and M. YANNAKAKIS, *Optimization, approximation, and complexity classes*, **Journal of Computer and System Sciences**, vol. 43 (1991), pp. 425–440.

[PPST83] W. PAUL, N. PIPPENGER, E. SZEMERÉDI, and W. TROTTER, *On determinism and nondeterminism and related problems*, **Proceedings of the twenty fourth ACM Symposium on Theory of Computing**, 1983, pp. 429–438.

[Pip79] N. PIPPENGER, *On simultaneous resource bounds*, **Proceedings of the 20th annual IEEE symposium on Foundations of Computer Science**, 1979, pp. 307–311.

[PF79] N. PIPPENGER and M. FISCHER, *Relations among complexity measures*, **Journal of the ACM**, vol. 26 (1979), pp. 361–381.

[Pra75] V. PRATT, *Every prime has a succinct certificate*, **SIAM Journal on Computing**, vol. 4 (1975), pp. 214–220.

[Rab80] M. RABIN, *Probabilistic algorithms for testing primality*, **Journal of Number Theory**, vol. 12 (1980), pp. 128–138.

[RS59] M. RABIN and D. SCOTT, *Finite automata and their decision problems*, **IBM Journal**, (1959), pp. 114–125.

[Ros67] A. ROSENBERG, *Real-time definable languages*, **Journal of the ACM**, vol. 14 (1967), pp. 645–662.

[Sav72] J. SAVAGE, *Computational work and time on finite machines*, **Journal of the ACM**, vol. 19 (1972), pp. 660–674.

[Sav70] W. SAVITCH, *Relationships between nondeterministic and deterministic time complexities*, **Journal of Computer and System Sciences**, vol. 4 (1970), no. 2, pp. 177–192.

[Sch80] J. T. SCHWARTZ, *Fast probabilistic algorithms for verification of polynomial identities*, **Journal of the ACM**, vol. 27 (1980), pp. 701–717.

[Sha90] A. SHAMIR, *IP=PSPACE*, **Proceedings of the 31st IEEE symposium on Foundations of Computer Science**, 1990, pp. 145–152.

[Sha49] C. SHANNON, *The synthesis of two–terminal switching circuits*, **Bell System Technical Journal**, vol. 28 (1949), pp. 59–98.

[SG77] I. SIMON and J. GILL, *Polynomial reducibilities and upward diagonalizations*, **Proceedings of the ninth annual ACM Symposium on Theory of Computing**, 1977, pp. 186–194.

[Sip78] M. SIPSER, *Halting space-bounded computations*, **Proceedings of the 19th annual IEEE symposium on Foundations of Computer Science**, 1978, pp. 73–74.

[Sip82] ———, *On relativization and the existence of complete sets*, **Automata, languages, and programming**, Lecture Notes in Computer Science, vol. 140, Springer-Verlag, 1982.

[Sip83] ———, *A complexity theoretic approach to randomness*, **Proceedings of the fifteenth ACM Symposium on Theory of Computing**, 1983, pp. 330–335.

[SS77] R. SOLOVAY and V. STRASSEN, *A fast Monte-Carlo test for primality*, **SIAM Journal on Computing**, vol. 6 (1977), pp. 84–85.

[Sto76] L. STOCKMEYER, *The polynomial-time hierarchy*, **Theoretical Computer Science**, vol. 3 (1976), pp. 1–22.

[Sze88] R. SZELEPCSÉNYI, *The method of forced enumeration for nondeterministic automata*, **Acta Informatica**, vol. 26 (1988), pp. 279–284.

[Tod89] S. TODA, *On the computational power of PP and $\oplus P$*, **Proceedings of the 30th IEEE symposium on Foundations of Computer Science**, 1989, pp. 514–519.

[Tre04] L. TREVISAN, *Inapproximability of combinatorial optimization problems*, **The Electronic Colloquium in Computational Complexity**, (2004), pp. 1–39, Technical Report 65.

[Tur36] A. M. TURING, *On computable numbers with an application to the Entscheidungsproblem*, **Proceedings of the London Mathematical Society**, vol. 42 (1936), pp. 230–365.

[Tur39] ———, *Systems of logic based on ordinals*, **Proceedings of the London Mathematical Society**, vol. 45 (1939), pp. 161–228.

[Tur50] ———, *Computing machinery and intelligence*, **Mind**, vol. 49 (1950), pp. 433–460.

[Wra76] C. WRATHALL, *Complete sets and the polynomial hierarchy*, **Theoretical Computer Science**, vol. 3 (1976), pp. 23–33.

[Yao82] A. YAO, *Theory and applications of trapdoor functions*, **Proceedings of the 23rd IEEE symposium on Foundations of Computer Science**, 1982, pp. 80–91.

[Yao85] ———, *Separating the polynomial-time hierarchy by oracles*, **Proceedings of the 26th IEEE symposium on Foundations of Computer Science**, 1985, pp. 1–10.

[Zak83] S. ŽÁK, *A Turing machine time hierarchy*, **Theoretical Computer Science**, vol. 26 (1983), pp. 327–333.

[Zip79] R. ZIPPEL, **Probabilistic algorithms for sparse polynomials**, Lecture Notes in Computer Science, vol. 72, Springer, 1979.

DEPARTMENT OF COMPUTER SCIENCE
BOSTON UNIVERSITY
BOSTON, MA 02215, USA

DEPARTMENT OF COMPUTER SCIENCE AND ENGINEERING
STATE UNIVERSITY OF NEW YORK AT BUFFALO
338 DAVIS HALL
BUFFALO, NY 14260, USA

TURING MACHINES TO WORD PROBLEMS

CHARLES F. MILLER III

Abstract. We trace the emergence of unsolvable problems in algebra and topology from the unsolvable halting problem for Turing machines.

§1. Introduction. Mathematicians have always been interested in being able to calculate with or about the things they study. For instance early developers of number theory and the calculus apparently did extensive calculations. By the early 1900s a number of problems were introduced asking for general algorithms to do certain calculations. In particular the tenth problem on Hilbert's influential list asked for an algorithm to determine whether an integer polynomial in several variables has an integer solution.

The introduction by Poincaré of the fundamental group as an invariant of a topological space which can often be finitely described by generators and relations led to Dehn's formulation of the word and isomorphism problem for groups. To make use of such group invariants we naturally want to calculate them and determine their properties. It turns out many of these problems do not have algorithmic solutions and we will trace the history and some of the ideas involved in showing these natural mathematical problems are unsolvable.

In the 1930s several definitions of computable functions emerged together with the formulation of the Church–Turing Thesis that these definitions captured intuitive notions of computability. Church and independently Turing showed that there is no algorithm to determine which formulas of first-order logic are valid, that is, the Entscheidungsproblem is unsolvable.

So it became meaningful to consider whether more standard mathematical questions have algorithmic solutions, and it seems Church encouraged several mathematicians including Emil Post to work on such problems. Indeed Post had also described an equivalent notion of a computing machine, though not in detail, and had anticipated many of these results.

In this article we describe how the unsolvable halting problem for Turing machines can be used to show the unsolvability of several important algorithmic problems in mathematics including the word and isomorphism problems for finitely presented groups and the homeomorphism problem for higher dimensional manifolds. Our intention is to provide some descriptions of what was done following the historical trail and to make some additional comments about related results and about subsequent developments. Usually we give more modern versions of the constructions. We have included somewhat sketchy but essentially complete proofs for the unsolvability of the word and isomorphism problems based ultimately on Turing machines. But for many other results we give only a few comments on how they are proved. Full treatments of these advances are nowadays available in textbooks and we provide references to those sources for further details (see the last section below). Hopefully those readers with less technical interest will gloss over some of the detail and still gain an impression of how these results can be established.

Here is an outline of some milestones on the historical trail to these results.

- 1908 Tietze describes transformations of presentations of groups which realise all isomorphisms and formulates the isomorphism problem.
- 1911 Dehn poses the word, conjugacy and isomorphism problems for finitely presented groups.
- 1914 Thue poses what is essentially the word problem for finitely presented semigroups.
- 1930s Definitions of computable functions emerge and existence of unsolvable problems is established (Gödel, Church, Kleene, Turing, Post and others).
- 1947 Post and Markov independently construct finitely presented semigroups with unsolvable word problem. Post's construction is built on Turing machines while Markov's uses Post's normal systems.
- 1950 Turing shows the word problem for cancellation semigroups is unsolvable by furthering Post's construction.
- 1951 Markov shows the isomorphism problem and recognising most properties of semigroups are unsolvable.
- 1955 Novikov and independently 1957 Boone show the word problem for groups is unsolvable. Novikov uses Turing's cancellation semigroup result while Boone uses Post's semigroup construction. In 1958 Britton gives another proof and later, in 1963, he greatly simplifies Boone's proof.
- 1956 Adian and independently 1958 Rabin show the isomorphism problem and recognising most properties of groups are unsolvable.
- 1958 Markov shows the homeomorphism problem for n-manifolds is unsolvable ($n \geq 4$).
- 1961 Higman characterises the subgroups of finitely presented groups thus connecting recursion theory with group theory in an unexpected way and giving a very different proof of the unsolvability of the word problem.

The mathematical logic "school" in Princeton during this era was actively involved in these developments. Early in his career Post had a post-doctoral fellowship at Princeton and worked on problems in logic. Alonzo Church encouraged Post to work on these problems, and he was the doctoral supervisor of Turing, Boone, Rabin, Scott, and Collins (actually advised by Boone) who all contributed to the developments we describe. Also Gödel gave encouragement to Boone on several occasions.

It sometimes happens in mathematics that the same result is obtained quite independently by different researchers, particularly when answering a widely known problem. Our story has several instances of such independent discoveries. In my view it is desirable to give credit to all involved.

It has been my pleasure to know as friends at least four of the principals in our story: Bill Boone (who was my doctoral supervisor), John Britton, Graham Higman, and Sergei Adian. Likewise I will mention work of several other friends and collaborators. So my account is no doubt influenced by these various relationships.

§2. Presentations of Groups and Semigroups. A familiar example of a group is the dihedral group D_3 of order 6 consisting of the rigid motions of an equilateral triangle onto itself. It can be described by giving two generators which correspond to rotation through 120° and flipping about a central axis. These operations have orders 3 and 2 respectively and the group D_3 is described by the *presentation*

$$D_3 = \langle a, b \mid a^2 = 1, b^3 = 1, a^{-1}ba = b^{-1} \rangle.$$

Here the symbols a and b are called *generators* and the equations they are subjected to are called *defining relations*.

Groups are often described as symmetry groups or permutation groups and defined by actions on some object. A more general and abstract approach emerged describing groups by generators and defining relations thought of as quotients of a free groups. So if $\{a_1, a_2, \dots\}$ is a set of symbols representing generators of a group G, then a *presentation* of G on these generators will have the form

$$G = \langle a_1, a_2, \cdots \mid u_1 = v_1, u_2 = v_2, \dots \rangle.$$

Here the u_i and v_i are words on the generators a_j, that is, they are finite strings of the symbols a_j and a_j^{-1}. Any such piece of notation describes a a group as follows. Let F be the free group with free basis $\{a_1, a_2, \dots\}$ and let N be the smallest normal subgroup of F containing all the words $u_i v_i^{-1}$. The group presented by the above notation is then the quotient group $G = F/N$. Here, as usual, the inverse v^{-1} of the word $v = v_{i_1}^{\epsilon_1} \dots v_{i_k}^{\epsilon_k}$ is the word $v^{-1} = v_{i_k}^{-\epsilon_k} \dots v_{i_1}^{-\epsilon_1}$ where $\epsilon_j = \pm 1$. Note that by letting $r_i = u_i v_i^{-1}$ we can also describe G as

$$G = \langle a_1, a_2, \cdots \mid r_1 = 1, r_2 = 1, \dots \rangle$$

which is clearly equivalent and sometimes more convenient.

A presentation of a group G as above is said to be *finitely generated* if the set of symbols $\{a_1, \ldots, a_n\}$ is finite, *finitely related* if the set of defining relations $\{u_1 = v_1, \ldots, u_m = v_m\}$ is finite and *finitely presented* if both are finite.

Given a finite presentation of the form $G = \langle a_1, \ldots, a_n \mid r_1 = 1, \ldots, r_m = 1 \rangle$ there is a standard recipe for constructing a finite 2-complex Y with fundamental group $G \cong \pi_1(Y, \mathbf{o})$ as follows. Start with a single 0-cell $\mathbf{o}$ and for each of the n generators a_j attach an oriented 1-cell by identifying its endpoints with $\mathbf{o}$. This gives a 1-complex Y^1 with fundamental group which is free on the 1-cells corresponding to the a_i. Now for each relation of the form $r_i = 1$ we attach a 2-cell to Y^1 by subdividing the boundary of the 2-cell into 1-cells labeled according to the word r_i and attaching (respecting orientation) the 2-cell to Y^1. The resulting 2-complex Y is connected, has a single 0-cell, has n 1-cells corresponding to the generators of the presentation for G and has 2-cells corresponding to the defining relations in the presentation for G. Then one shows $G \cong \pi_1(Y, \mathbf{o})$ as desired. Conversely one can show that the fundamental group of any finite cell or simplicial complex is finitely presented. One can extend these observations to show that finitely presented groups are also the fundamental groups of closed, differentiable n-manifolds ($n \geq 4$). So groups with a finite presentation are of interest for geometric reasons and correspond to spaces with a certain geometric finiteness.

In 1908 Heinrich Tietze, in studying fundamental groups of manifolds and their topological invariance, introduced four operations on presentations which are now known as *Tietze transformations*. They can be described as: (1) adding a relation which is a consequence of the others; (2) removing a redundant relation; (3) defining a new symbol as a word in existing symbols; and (4) removing a symbol which can be defined in terms of the remaining symbols. Tietze then showed that if the groups described by two presentations are isomorphic, then there is a sequence of Tietze transformations leading from one of these presentations to the other. Actually this is a result of a very general nature applicable to algebraic systems describable by presentations.

Tietze noted that the problem of finding a method to determine whether two groups given by their presentations are isomorphic was unresolved. Thus he raised the algorithmic question known as the *isomorphism problem for finitely presented groups*.

Soon thereafter, in 1911, Max Dehn highlighted three basic algorithmic problems concerning groups defined by presentations which have played a central role:

word problem: Let G be a group given by a finite presentation. Does there exist an algorithm to determine of an arbitrary word w in the generators of G whether or not $w =_G 1$?

conjugacy problem: Let G be a group given by a finite presentation. Does there exist an algorithm to determine of an arbitrary pair of words u and v in the generators of G whether or not u and v define conjugate elements of G?

isomorphism problem: Does there exist an algorithm to determine of an arbitrary pair of finite presentations whether or not the groups they present are isomorphic?

In geometric terms, the word problem for the fundamental group $\pi_1(Y, \mathbf{o})$ corresponds to the problem of determining whether or not a closed loop at $\mathbf{o}$ in Y is contractible. The conjugacy problem for $\pi_1(Y, \mathbf{o})$ corresponds to the problem of determining whether or not two closed loops are freely homotopic (intuitively whether one can be deformed into the other). Since homeomorphic spaces have isomorphic fundamental groups, the isomorphism problem is related to the *homeomorphism problem*, that is, determining whether given spaces are homeomorphic.

For the fundamental groups of closed 2-manifolds Dehn gave algorithms to solve each of these problem, though in general they turn out to be unsolvable. We will comment further on "Dehn's algorithm" later when we consider the word problem in more detail.

Here are some observations about Dehn's problems. If u, v are any two words in the generators of a group G, then $u =_G v$ if and only if $uv^{-1} =_G 1$ so the problem of deciding equality of words is equivalent to the problem of deciding whether another word is equal to 1. If w is a word of G then $w =_G 1$ if and only if w and 1 are conjugate in G, so the word problem for G is always reducible the conjugacy problem for G.

Recall that if F is a free group with free basis $\{a_1, \ldots, a_n\}$ then every element of F is equal to a unique *freely reduced* word obtained by successively removing (deleting) inverse pairs, that is, removing subwords of the form $a_i a_i^{-1}$ or $a_i^{-1} a_i$. If $G = F/N$ is given by a presentation as above, then the elements of N are those words of F which are freely equal to products of conjugates of the relators r_j and their inverses. In particular one can systematically enumerate all such words, and so $\{w \in F \mid w =_G 1\}$ is a recursively enumerable set of words.

Suppose that $G = \langle a_1, \ldots, a_n \mid r_1 = 1, \ldots, r_m = 1 \rangle$ and $H = \langle b_1, \ldots, b_k \mid q_1 = 1, \ldots, q_p = 1 \rangle$ are two finite presentations, and suppose that $u_1, \ldots, u_k$ are words in the generators a_i of G. Now the function ϕ defined by $b_i \mapsto u_i$ for $i = 1, \ldots, k$ can be formally extended to all words of H of by setting

$$\phi(b_{i_1}^{\epsilon_{i_1}} \ldots b_{i_v}^{\epsilon_{i_v}}) = u_{i_1}^{\epsilon_{i_1}} \ldots u_{i_v}^{\epsilon_{i_v}}.$$

This ϕ defines a homomorphism from H into G if and only if each $\phi(q_j) =_G 1$ for $j = 1, \ldots, p$.

Now further suppose that the map ϕ above defines a monomorphism from H into G so that ϕ embeds H in G. If z is any word in the given generators of H, since we are given the images $\phi(b_i)$ of the generators, we can write down its image $\phi(z)$ as a word in the generators of G. Since we are assuming ϕ is a monomorphism we know $z =_H 1$ if and only if $\phi(z) =_G 1$. Since the words in the generators of G that are equal to 1 are recursively enumerable, it follows that $\{z$ a word on the $b_i \mid z =_H 1\}$ is also recursively enumerable. Moreover, if we have an algorithm to solve the word problem for G we can use it determine whether or not $z =_H 1$. Note that these observations only require that H be finitely generated (not necessarily finitely presented). These observations enable us to conclude the following:

PROPOSITION 1. *Let $G = \langle a_1, \ldots, a_n \mid r_1 = 1, \ldots, r_m = 1 \rangle$ be a finitely presented group given by generators and relations, and suppose $\{u_1, \ldots, u_k\}$ is a finite set of words on the generators of G. Let H denote the subgroup generated by $\{u_1, \ldots, u_k\}$. Then:*

1. *H can be presented on finitely many generators and a recursively enumerable set of defining relations.*
2. *If G has a solvable word problem, then H also has a solvable word problem. More generally the word problem for H is reducible to the word problem for G.*
3. *The solvability of the word problem does not depend on the particular finite set of generators or finite presentation of the group G.* ⊣

Sometimes it is necessary to distinguish carefully between a presentation which is a piece of notation and the group it describes as an algebraic object. So if $\mathcal{P}$ is a presentation, the group it presents is written $gp(\mathcal{P})$. So in the above discussion about the presentation for G we technically have $G = gp(\mathcal{P}) = F/N$. One fact we need to have is the following.

PROPOSITION 2. *The set of pairs of presentations $(\mathcal{P}_1, \mathcal{P}_2)$ of groups, say on some fixed alphabet, such that $gp(\mathcal{P}_1)$ and $gp(\mathcal{P}_2)$ are isomorphic (as abstract groups) is recursively enumerable.* ⊣

Here in outline are two ways to prove this using enumieration. (1) According to Tietze's Theoem if the groups presented are isomorphic there is a sequence to Tietze transformations leading from one presentation to the other. But the Tietze transformations applicable to a presentation are recursively enumerable. Hence by an elaborate enumeration we find a route from one of our presentations to the other. (2) Since the true equations among words in a finite presentation are always recursively enumerable, we can enumerate all pairs of homomorphisms $\phi \colon \mathcal{P}_1 \to \mathcal{P}_2$ and $\psi \colon \mathcal{P}_2 \to \mathcal{P}_1$. As such pairs are enumerated, compute the values of $\phi \circ \psi$ and $\psi \circ \phi$ on each of the respective generating sets. Then search by enumeration for those homomorphism which are inverse pairs and so fix the respective generating sets.

Presentations of semigroups and string rewriting. We recall that a *semigroup with* 1, also called a *monoid*, is a set S with an associative multiplication $\cdot$ and a distinguished element denoted 1 so that $u \cdot 1 = u = 1 \cdot u$ for all $u \in S$. Suppose now that $A = \{a_1, a_2, \ldots\}$ is a set of symbols and let A^* denote the set of all finite strings of symbols from A, also called *words on A*. We allow the empty word which is denoted by 1. We define the product of two words $u, v \in A^*$ by concatenation, so $u \cdot v = uv$, which is clearly an associative multiplication. The resulting system which we denote by $\Phi(A)$ is the *free semigroup with* 1 or *free monoid* with basis A and it enjoys the usual universal mapping properties of free objects.

As in the case of groups one can impose relations on semigroup elements to obtain presentations. Thus a *finitely presented semigroup* S is a semigroup described by a notation such as $S = sgp\langle a_1, \ldots, a_n \mid u_1 = v_1, \ldots, u_m = v_m \rangle$ where we have added the *sgp* to distinguish from the group case. It is important to always keep in mind that there are in general no inverses for elements of a semigroup nor inverse symbols for the generating symbols. The semigroup presented by this notation is the quotient of the free semigroup $\Phi(A)$ by the smallest congruence determined by the defining relations $\{u_1 = v_1, \ldots, u_m = v_m\}$.

A finitely presented semigroup such as S can also be described in terms of string rewriting. A *rewrite rule* is a pair of words $(u, v) \in A^* \times A^*$ which can be applied to strings in A^* of the form $z_1 u z_2$ where $z_1, z_2 \in A^*$. The result of applying the rule (u, v) to this string is $z_1 u z_2$ and we write $z_1 u z_2 \to z_1 v z_2$. The rule (u, v) is usually written as $u \to v$ with the understanding that it enables one to replace any substring u in a word by the substring v. A system consisting of A^* together with a finite set R of rewrite rules on A^* is called a string rewriting system or semi-Thue system after Axel Thue who studied them systematically in 1914 [96]. If the rewrite rules R are also symmetric, that is, if $v \to u$ is a rule whenever $u \to v$ is a rule, the semi-Thue system is sometimes called a Thue system and the rewrite rules are written as $u \leftrightarrow v$. One observes that a Thue system is exactly the same as a finitely presented semigroup as discussed above, but the emphasis is on string substitution rather than abstract algebra. Thue [96] raised the word problem for finitely presented semigroups which in the string rewriting context asks for an algorithm to determine whether, given two words y and z of a semigroup S presented as above, there is a sequence of substring replacements of the form $u_i \leftrightarrow v_i$ leading from y to z.

Below we will need some terminology for the direction of rewrite rules when a semigroup is given by a presentation $S = sgp\langle a_1, \ldots, a_n \mid u_1 = v_1, \ldots, u_m = v_m \rangle$. A string rewrite of the form $z_1 u_i z_s \to z_1 v_i z_2$ is called a *forward* application of the defining relation $u_i = v_i$, so the left-hand side is replaced by the right-hand side of the relation. Similarly a rewrite of the form $z_1 v_i z_s \to z_1 u_i z_2$ is called a *backward* application where the right-hand side is replaced by the left-hand side. A *proof* in S is a sequence of rewrites of the form $w_1 \to w_2 \to \cdots \to w_n$ where each $w_i \to w_{i+1}$ is a single application of a defining relation, so is either a forward or backward application (rewrite). Of course two words are equal in S if and only if there is a proof beginning with one and ending with the other. A proof is said to be *reversal-free* or *without reversals* if it does not contain two successive rewrites of the form $z_1 u_i z_2 \to z_1 v_i z_s \to z_1 u_i z_2$ or $z_1 v_i z_2 \to z_1 u_i z_s \to z_1 v_i z_2$, the second of which undoes the first.

Results analogous to most of those above for groups also apply to the semigroup case. Note that every finitely presented group can be viewed as a finitely presented semigroup by adding new symbols a_i^{-1} and rules $a_i^{-1} a_i = 1 = a_i a_i^{-1}$. So a solution to the word problem for all semigroups would include a solution to the word problem for all groups. But as we shall describe there are semigroups and groups (more difficult to prove) for which such a solution is not possible.

§3. The word problem for semigroups and Post's construction.

In 1947 Emil Post [83] and A. A. Markov [61] independently constructed finitely presented semigroups with unsolvable word problem. This was perhaps the first long standing mathematical algorithmic problem shown to be recursively unsolvable. Post's construction was based rather directly on a Turing machine with unsolvable halting problem while Markov's was based on Post's earlier unsolvability results on "canonical systems".

Markov's father was the mathematician A. A. Markov famous for his work in stochastic processes after whom Markov chains are named. The son, whose name was the same, made a number of important contributions to mathematical logic, in particular "unsolving" the word problem for semigroups, the isomorphism problem for semigroups and the homeomorphism problem for 4-manifolds.

The role of Emil Post in mathematical logic seems underappreciated. For much of his adult life he suffered recurring attacks of manic-depressive disorder (bipolar disorder) and many of his ideas were not published in sufficient detail or a timely way. It seems fair to say that he anticipated many of the important results of Gödel, Church and Turing. The appreciative article [92] on Post's work by Stillwell and Martin Davis's introduction to Post's collected works [40] are recommended for trying to put Post's work into perspective. In particular, in a short 1936 paper [79] Post describes a theory of computation which is easily seen to be equivalent to Turing machines, work done independent of Turing's work. Also his classic 1944 paper [81] is the origin of the structure theory of recursively enumerable sets and degrees of unsolvability.

We are going to describe Post's construction of a finitely presented semigroup with unsolvable word problem in some detail. The starting point is a Turing machine, or rather a slightly adapted form of a Turing machine whose tape is always finite but expandable. The instantaneous descriptions of the Turing machine correspond to certain words of the semigroup and basic moves of the machine to string rewrites in the semigroup. Since we have later applications to the word problem for groups, we also do some tidying up after a successful computation.

Part of our intention here is to make clear the parametric nature of this and some later constructions and that the objects can be written down explicitly in terms of the underlying Turing machine. Much of this is available in textbooks, for instance in various editions of Rotman's text [87].

Turing machine refresher. We now describe Post's construction [83] of a finitely presented semigroup $\gamma(T)$ from a Turing machine T in some detail. The aim is to show that computations in the Turing machine correspond exactly to rewriting sequences for certain special words in the semigroup. The word problem for the semigroup is then reduced to the halting problem for the Turing machine, and so can be unsolvable.

For us a word is a finite string of symbols (all positive in the absence of groups). We want an informal model of a Turing machine computation in which the "tape" is potentially infinite and blank outside some finite segment, yet we want the "blank" to be a symbol in our alphabet. So we think of our tape as finite but expandable in the sense that when we reach either end, we add a blank to that end and proceed as usual. Here is the basic set up (which may be a variation on one the reader is familiar with) and the necessary background on the halting problem we require.

A (deterministic) *Turing machine* T consists of:

1. a finite set of *symbols* $s_0, s_1, \ldots, s_M$, called *the alphabet of* T, including a distinguished symbol s_0 called *blank*;
2. a finite set of *internal states* $q_0, q_1, \ldots, q_N$ with two distinguished states q_1, called the *start state*, and q_0, called the *halting state*; and
3. a non-empty finite set of *quadruples*, which are 4-tuples of one of the following three types: $q_i s_j s_k q_m$ or $q_i s_j R q_m$ or $q_i s_j L q_m$, such that no two quadruples have the same first two letters.

An *instantaneous description* α is a positive word on the letters s_j and exactly one q_i which is not at the right end. For example, $s_2 s_0 q_3 s_5 s_2$ is an instantaneous description which is to be interpreted as "the symbols on the tape are $s_2 s_0 s_5 s_2$ (with blanks elsewhere) and the machine is in state q_3 scanning s_5 (the symbol to the right of q_3).

If T is a Turing machine and α, β are two instantaneous descriptions, then a *basic move* $\alpha \to \beta$ of T is a string replacement of one of the following forms:

1. $\alpha = P q_i s_j Q \to P q_m s_k Q = \beta$ where $q_i s_j s_k q_m \in T$;
2. $\alpha = P q_i s_j s_k Q \to P s_j q_m s_k Q = \beta$ where $q_i s_j R q_m \in T$;
3. $\alpha = P q_i s_j \to P s_j q_m s_0 = \beta$ where $q_i s_j R q_m \in T$;
4. $\alpha = P s_k q_i s_j Q \to P q_m s_k s_j Q = \beta$ where $q_i s_j L q_m \in T$; and
5. $\alpha = q_i s_j Q \to q_m s_0 s_k Q = \beta$ where $q_i s_j L q_m \in T$;

where P and Q are (possibly empty) words on the s_j letters.

The interpretation of the basic move of type (1) is "when in state q_i and scanning s_j, replace s_j with s_k and enter state q_m". The interpretation of type (2) is "when in state q_i and scanning s_j and s_k is the next symbol to the right, move right one space to scan that s_k and enter state q_m". The interpretation of type (3) is "when in state q_i and scanning s_j and there are no further symbols to the right, add an s_0 to the right then move right one space to scan that s_0 and enter state q_m". Similarly for types (4) and (5) with left in place of right.

An instantaneous description is *terminal* if there is no β with $\alpha \to \beta$, that is, no further basic moves can be applied. A *computation* of a Turing machine is a finite sequence of basic moves

$$\alpha_1 \to \alpha_2 \to \cdots \to \alpha_t$$

where α_t is terminal.

Our requirement that T is deterministic, that is, no two quadruples have the same first two letters, implies that at most one basic move is applicable to any instantaneous description. So if $\alpha \to \beta$ and $\alpha \to \delta$ then $\beta = \delta$.

If z is any word on the s_i, we say that $T(z)$ exists if there is a computation of T beginning with $q_1 z$ (recall that q_1 is our machines "start state"). That is, when started in state q_1 with tape containing exactly the word z, the machine eventually *halts* in the sense that the computation reaches a terminal state. The algorithmic problem of determining whether $T(z)$ exists is called the *halting problem* for T.

We will assume henceforth that our Turing machines all have *unique stopping state* q_0, that is, an instantaneous description α is terminal if and only if α involves q_0. There is no loss in this assumption since, by renumbering and adding appropriate quadruples, one can easily modify any machine T to obtain a new machine T' on the same alphabet with unique stopping state q_0 such that $T(z)$ exists if and only if $T'(z)$ exists.

Some fundamental facts about Turing machines are summarised in the following (which may be theorems or definitions depending on your point of view):

THEOREM 3 (Turing, Post). *Let Z be the set of all words on the alphabet $\{s_1, \ldots, s_M\}$. Then:*

1. *A subset $E \subseteq Z$ is recursively enumerable if and only if there is a Turing machine T with alphabet $\{s_0, s_1, \ldots, s_M\}$ such that $E = \{z \in Z \mid T(z) \text{ exists}\}$.*
2. *There is a Turing machine T such that $E = \{z \in Z \mid T(z) \text{ exists}\}$ is not recursive.*

Thus, assuming the Church–Turing thesis, there is a Turing machine T such that there is no algorithm to determine of an arbitrary $z \in Z$ whether or not $T(z)$ exists.

Post's semigroup construction. We now describe Post's construction [83] of a sort of faithful interpretation of the Turing machine T inside a finitely presented semigroup. Some modifications to Post's original construction are included for later application to the word problem for groups.

Let $\gamma(T)$ be the semigroup with presentation

$$\gamma(T) = \langle h, s_0, s_1, \ldots, s_M, q, q_0, q_1, \ldots, q_N \mid R(T) \rangle$$

where the relations $R(T)$ are

$$q_i s_j = q_l s_k \text{ if } q_i s_j s_k q_l \in T,$$

and for all $b = 0, 1, \ldots, M$:

$$\begin{aligned}
q_i s_j s_b &= s_j q_l s_b & \text{if } q_i s_j R q_l \in T, \\
q_i s_j h &= s_j q_l s_0 h & \text{if } q_i s_j R q_l \in T, \\
s_b q_i s_j &= q_l s_b s_j & \text{if } q_i s_j L q_l \in T, \\
h q_i s_j &= h q_l s_0 s_j & \text{if } q_i s_j L q_l \in T,
\end{aligned}$$

$$\begin{aligned}
q_0 s_b &= q_0 \\
s_b q_0 h &= q_0 h \\
h q_0 h &= q.
\end{aligned}$$

Intuitively the symbol h marks the ends of the tape the Turing machine T is reading. A word of $\gamma(T)$ is h-*special* if it has the form $huq_j vh$ where u and v are s-words and $q_j \in \{q, q_0, q_1, \ldots, q_N\}$. The h-special words correspond exactly to the instantaneous descriptions of the Turing machine T with h symbols added at each end.

It is important to observe that the first five types of relations of $\gamma(T)$ that correspond to quadruples of T, when applied in the forward direction, exactly carry out a basic move of the T on that description. So the operation of the Turing machine exactly corresponds to a sequence of forward moves in $\gamma(T)$.

The effect forward applications of the last three types of defining relations in the presentation of $\gamma(T)$ is to take any h-special word $huq_0 vh$, which corresponds to a stopped computation of T and erase symbols as follows: first (1) erase all the s_b symbols to the right of q_0 obtaining an the h-special word $huq_0 h$; then (2) erase all of the s_b symbols to the left of $q_0 h$ to obtain the word $hq_0 h$; and finally (3) enter state q no longer flanked by h symbols. That is, erase the tape and then enter the (new) final state q.

A consequence of these observations is the following.

LEMMA 4. *Suppose z is a word on $\{s_1, \ldots, s_M\}$ such that $T(z)$ exists, that is, the computation of T beginning with $q_1 z$ halts in the stop state q_0. Then there is a proof in $\gamma(T)$ of the form*

$$hq_1 zh \to \cdots \to hq_0 h \to q$$

in which each step is a forward application of a defining relation. Thus in $\gamma(T)$ we have $hq_1zh = q$. ⊣

To show the word problem for $\gamma(T)$ is equivalent to the halting problem for T we need the converse of this lemma, namely if $hq_1zh = q$ then $T(z)$ exists. With a view to subsequent applications we follow the proof in [39], which is a variant of Post's original argument.

Since only the last relation $hq_0h = q$ creates or destroys h, we observe the following:

LEMMA 5. *If w_1 and w_2 are words of $\gamma(T)$ and neither if them is the word q, and if $w_1 \to w_2$ is an application of a relation, then w_1 is h-special if and only if w_2 is h-special.* ⊣

LEMMA 6. *Let $w \equiv huq_jvh$ be an h-special word of $\gamma(T)$. Then at most one of the relations of $\gamma(T)$ has a forward application to w.*

PROOF. Clearly we may assume $q_j \neq q$. If $q_j \neq q_0$ the conclusion is immediate from the deterministic nature of the Turing machine T. So suppose $q_j \equiv q_0$. If u and v are empty then there is a forward application of $hq_0h = q$, but clearly this is the only possibility. If u is non-empty and v is empty, then just one of the relations $s_bq_0h = q_0h$ has a forward application. Finally if v is non-empty, just one of the relations $q_0s_b = q_0$ has a forward application. This completes the proof. ⊣

LEMMA 7. *Let $w \equiv huq_jvh$ be an h-special word of $\gamma(T)$. If*

$$w \equiv w_1 \to w_2 \to \cdots \to w_{n-1} \to q$$

is a proof in $\gamma(T)$, then either this proof contains a reversal or all the applications of relations are forward. In particular, the shortest such proof consists entirely of forward applications of relations.

PROOF. Clearly the application $w_{n-1} \to q$ is forward and $w_{n-1} \equiv hqh$. Assume that not all applications of relations in the given proof are forward. Suppose that $w_{j-1} \to w_j$ is the last backward application of a relation in the proof. Then $w_j \to w_{j+1}$ and all subsequent applications are forward. But $w_{j-1} \to w_j$ backward implies that $w_j \to w_{j-1}$ is a forward application of the same relation.

Now w_{n-1} is h-special and none of $w_j, \ldots, w_{n-1}$ could be q since only a backward application applies to q. So each of $w_j, \ldots, w_{n-1}$ must be h-special. Hence at most one relation has a forward application to w_j by the previous result. Thus $w_{j+1} \equiv w_{j-1}$ and hence $w_{j-1} \to w_j \to w_{j+1}$ is a reversal. This proves the lemma. ⊣

Since forward applications of relations correspond exactly to basic moves of T or to tape erasing moves at the end, we conclude the following.

COROLLARY 8. *Suppose z is a word on $\{s_1, \ldots, s_M\}$. Then $hq_1zh = q$ in the semigroup $\gamma(T)$ of and only if $T(z)$ exists, that is, the Turing machine T halts when started in state q_1 with tape input exactly z.*

By choosing T to be a Turing machine with unsolvable halting problem, we obtain the main result.

THEOREM 9 (Post–Markov). *There is a finitely presented semigroup with unsolvable word problem.*

The proof given above actually shows $\gamma(T)$ has some additional properties which we record for later reference.

COROLLARY 10. *Every non-trivial proof in $\gamma(T)$ of the form*

$$q \to w_2 \to \cdots \to w_{n-1} \to q$$

contains a reversal.

PROOF. Clearly $w_2 \equiv hq_0h$. So $w_2 \to \cdots \to w_{n-1} \to q$ either contains a reversal (as claimed), or consists of all forward applications. Assume all the applications $w_2 \to \cdots \to w_{n-1} \to q$ are forward. But the only forward application of a relation to $w_2 \equiv hq_0h$ is $hq_0h \to q$ so the given proof begins with a reversal. This proves the corollary. ⊣

An intuitive way to express the situation is that if we look the graph with vertices words of $\gamma(T)$ which are equal to q and edges joining words when $w_1 \to w_2$ is a forward application of a relation, the resulting graph is a connected tree. This corresponds to the fact that the instantaneous descriptions (vertices) in terminating computations in our Turing machine with basic moves (edges) are a forest of trees.

For use in a later section, we modify the above by relabelling the alphabet to include h among the s_i and use the term s-word to describe any (positive) word on $\{s_0, \ldots, s_M\}$. Then Post's construction can be recast as yielding a finitely presented semigroup $\Gamma(T)$ of the form

$$\Gamma(T) = \langle q, q_0, \ldots, q_N, s_0, \ldots, s_M \mid F_i q_{i_1} G_i = H_i q_{i_2} K_i, i \in I \rangle$$

where the F_i, G_i, H_i, K_i are (positive) s-words and $q_{i_j} \in \{q, q_0, \ldots, q_N\}$. If T has an unsolvable halting problem, then the problem of deciding for arbitrary (positive) s-words X, Y whether or not $Xq_1Y = q$ in $\Gamma(T)$ is recursively unsolvable.

§4. **Turing on the word problem for cancellation semigroups.** Following on from Post's construction, Turing [99] constructed a finitely presented cancellation semigroup with unsolvable word problem published in 1950. It is reported that he initially thought he had proved the desired result for groups but before presenting his work had realized that it only gave the cancellation semigroup case.

We will not look at Turing's construction in any detail, partly because modern approaches achieve the desired unsolvable word problem for groups more simply and directly. But while going from semigroups to cancellation semigroups may seem but a step on the road to the goal of the result for groups, Turing's paper played a very important role in developments. Novikov's 1955 paper [76] containing the first published proof of the unsolvability of the word problem for groups is based on Turing's result for cancellation semigroups. Moreover Boone's independent 1957 proof of the result for groups, while based only on Post's construction, used a new "phase change" idea which was suggested by Turing's work.

Turing's paper [99] suffers from a number of unfortunate misprints and other technical difficulties and is apparently difficult to follow. In his 1952 review of the paper for the Journal of Symbolic Logic [16] Boone outlines the proof and concludes:

> The paper is rendered difficult to read by many minor errors of detail and arrangement, corrections of which must be omitted here for lack of space. In spite of these it is to be emphasized that Turing's proof is in principle correct, as the reviewer has fully verified.

Some years later in 1958 Boone published [18] his explication and analysis of Turing's paper. While Boone's 1952 review vouches for the correctness of Turing's result, Novikov's reliance on it was a potential concern. Then in 1958 Novikov and Adian published [77] a proof, based directly on the Post–Markov result for semigroups, that the word problem for one-sided cancellation semigroups is unsolvable. They then show how to modify Novikov's proof for groups so that one-sided cancellation is all that is required.

§5. Further unsolvable problems for semigroups.
In this section we describe a few further developments concerning decision problems for semigroups which have grown out of the above considerations.

Markov's isomorphism and recognition results. We will call a property $\mathcal{P}$ of finitely presented semigroups *abstract* if it does not depend on the particular presentation chosen.

DEFINITION 11. *An abstract property $\mathcal{P}$ of finitely presented semigroups is a* Markov *property if there are finitely presented semigroups S_+ and S_- such that*:

1. $S_+ \in \mathcal{P}$; *and*
2. *if S_- is embedded in a finitely presented semigroup T then $T \notin \mathcal{P}$.*

S_+ *is called the* positive witness *and S_- is called the* negative witness *for $\mathcal{P}$.*

For example the properties "being trivial" (that is, having only one element) is a Markov property and "being finite" is a Markov property of semigroups. There is of course an analogous definition of a "Markov property of groups" which will be considered below. Notice that "being a group" is a Markov property of semigroups since there are semigroups not embeddable in a group. More examples will be given below for the case of groups. In 1951 Markov [62] proved the following:

THEOREM 12 (Markov). *Let $\mathcal{P}$ be a Markov property of semigroups. There is no algorithm to determine of an arbitrary finite presentation Π of a semigroup whether or not the semigroup presented by Π has property $\mathcal{P}$, that is symbolically, whether or not $sgp(\Pi) \in \mathcal{P}$.*

An immediate consequence of this theorem is that there is no algorithm to recognize the trivial semigroup and so the isomorphism problem is unsolvable.

COROLLARY 13 (Markov). *There is no algorithm to determine of two finite presentations of semigroups whether or not the semigroups they present are isomorphic.* ⊣

PROOF. We sketch a proof of Markov's theorem. We may suppose that we have three finite presentations of semigroups on disjoint sets of generators as follows:

1. $S_0 = sgp\langle A_0 \mid R_0\rangle$ which has unsolvable word problem;
2. $S_+ = sgp\langle A_+ \mid R_+\rangle$ which is a positive witness for property $\mathcal{P}$; and
3. $S_- = sgp\langle A_- \mid R_-\rangle$ which is a negative witness for $\mathcal{P}$.

For each pair of words u, v in the generators A_0 for S_0 we consider the presentation $S_{u,v}$ having generators $A_0 \cup A_- \cup \{c, d\}$ where c, d are new symbols, and defining relations $R_0 \cup R_-$ together with the relations $cud = 1$ and the relations $cvd = cvd\beta$ where β ranges over all the generators of $S_{u,v}$.

Clearly if $u = v$ in S_0 then in $S_{u,v}$ we have $cud = cvd = 1$; so by the last collection of relations all the generators of $S_{u,v}$ are equal to 1 and $S_{u,v}$ is trivial.

Suppose now that $u \neq v$ in S_0. We claim that S_- is embedded in $S_{u,v}$, that is, if y, z are two words on the generators of S_- then $y = z$ in S_- if and only if $y = z$ in $S_{u,v}$. The necessity is clear. For sufficiency we argue by induction on the number n of applications of the relation $1 \to cud$ in the indicated direction thought of as a rewrite rule. In case $n = 0$, no c or d can appear in the proof and so the same proof is valid in the (ordinary) free product $S_0 * S_-$ and so must give a proof in S_- which is embedded. So suppose $n > 0$ and consider the last application of $1 \to cud$. The only way to get rid of that c and d is by applying the rule $1 \leftarrow cud$ since $u \neq v$ in S_0. So this last application must be reversed and can be omitted to give a shorter proof. This proves the claim.

It follow that the free product $S_+ * S_{u,v} \in \mathcal{P}$ if and only if $u = v$ in S_0. Since the word problem for S_0 is unsolvable, this proves the theorem. ⊣

Further remarks on the word problem for semigroups. It turns out that groups defined by a single defining relation have a recursively solvable word problem. This was proved by Magnus using induction on the length of the relation (see below). But the word problem for semigroups with a single defining relation remains a seemingly difficult open problem.

Mathematicians seem to have some desire or impulse to write down quite explicitly a finite presentation of semigroup (or group) with an unsolvable word problem, that is, not depending on a parameter such as a Turing machine. The simplest examples (with respect to the total length of the presentation) of semigroups with unsolvable word problem were constructed independently in 1956 by G. S. Tseitin [98] and D. Scott [88]. Both presentations contain seven defining relations. A proof can be found in [5] for Tseitin's example which we record and follows:

THEOREM 14. *The semigroup TS presented by the five generators a, b, c, d, e and the seven defining relations*

$$ac = ca; \quad ad = da; \quad bc = cb; \quad bd = db; \quad ce = eca; \quad de = edb; \quad cca = ccae$$

has a recursively unsolvable word problem. ⊣

The idea here is that the semigroup TS is in a sense universal for all semigroups G that have a presentation of a given special form. So deductions of equations in any such G have parallel deductions of appropriately encoded equations in this one fixed semigroup TS. See [5] for a full account.

Matiyasevich [67] constructed a semigroup with 3 defining relations and an unsolvable word problem. Two of his relations are short but the third, much longer relation identifies two words with several hundred symbols each.

See the survey [5] by Adian and Durnev for more information on these and many other aspects of the word problem for semigroups.

§6. The word problem for groups. As we previously stated, Novikov in 1955 [76] and independently Boone in 1957 [15] showed the word problem for finitely presented groups is recursively unsolvable. Even in the many modern improved versions, this is still a hard result requiring about ten pages of technical argument and a substantial knowledge of combinatorial group theory.

Actually it was not certain for groups the word problem would turn out to be unsolvable. Indeed there are many large classes of groups for which the word problem can be solved. So before discussing how the finitely presented groups with unsolvable word problem were constructed, we indicate some solutions to the word problem for special classes including several known by the mid 1950s. We also discuss an interesting example (Higman's non-hopfian group) that appears as a subgroup in both the Novikov and Boone groups and recall some relevant tools from combinatorial group theory.

6.1. Many, even most, groups have solvable word problem.

Dehn's Algorithm: When Dehn introduce the word, conjugacy and isomorphism problems he provided algorithms for their solution for closed 2-manifolds. In fact the algorithms he described for the word and conjugacy problems work for a very large class of groups. In the case of the word problem *Dehn's algorithm* for deciding whether a word w is equal to 1 works by successively trying to apply the defining relations to find a way to shorten w.

DEFINITION 15. *A finite presentation $G = gp\langle A \mid R\rangle$ is called a* Dehn *presentation if $R = \{u_1v_1^{-1}, \ldots, u_nv_n^{-1}\}$ where the lengths of the words satisfy $|v_i| < |u_i|$ and the following holds*: *if w is a freely reduced word of G with $w =_G 1$ then at least one of the u_i is a subword of w.*

Thus, if $w =_G 1$ and is freely reduced then for some i, we have w is $w_1 u_i w_2$ and $w =_G w_1 v_i w_2$ where the right hand side is shorter than the left. So after at most $|w|$ steps of this kind or free reductions we must reach the identity 1. This is known as *Dehn's algorithm* for solving the word problem.

Dehn showed that the fundamental group of a closed orientable surface of genus $g \geq 2$ has a Dehn presentation. It is easy to see any finite group has a Dehn presentation by considering a multiplication table. However the free abelian group $\mathbb{Z}^n$ of rank $n \geq 2$ does not have a Dehn presentation (though it is easy to solve the word problem for free abelian groups). One can show that having a Dehn presentation does not depend on the chosen set of generators for the group.

If a group has a Dehn presentation, it is usually convenient to close the relations under inverses and cyclic permutations obtaining a *symmetrized set* of defining relations. Tartakovski [93], [94], [95] considered symmetrized presentations which have the properly that any two non-inverse relations $r_i, r_j \in R$ cancel less than one-sixth their length in freely reducing their product $r_i r_j$ (as words in the free group on the generators). He showed groups satisfying such a *cancellation condition* have a Dehn presentation and so Dehn's algorithm solves their word problem. Moreover, they have solvable conjugacy problem. These and related *small cancellation* groups were later studied geometrically by Lyndon and Schupp and a readable account of their theory can be found in the last chapter of [56].

Almost all finitely presented groups have solvable word problem: In the 1980s Gromov [46] introduced the study hyperbolic groups which have various definitions based on metric properties of their Cayley graphs. Among other characterisations of these groups, Gromov showed that a finitely presented group is hyperbolic if and only if it has a Dehn presentation. Intuitively the relators in a "random" finitely presentation of a group should not have long subwords in common. So they might be expected to satisfy a small cancellation condition and hence be hyperbolic. While this intuition is not exactly correct, it turns out that almost all groups are hyperbolic in a probabilistic sense made precise in Gromov [46] and in Olshanskii [78]. We record this informally as follows:

THEOREM 16 (Gromov–Olshanskii). *In a certain probabilistic sense, almost all finitely presented groups are hyperbolic and so have Dehn presentations and solvable word and conjugacy problems.* ⊣

The geometry and algorithmic aspects of hyperbolic groups and more generally groups of non-positive curvature are topics of on going interest. A useful reference is the monograph [25] by Bridson and Haefliger.

Generic-case complexity and the word problem: Another interesting probabilistic aspect of the word problem arises from the observation that in many groups it is easy see that certain plentiful types of words are not equal to 1. For example, suppose the finitely presented group $G = \langle a_1, \ldots, a_n \mid r_1 = 1, \ldots, r_m = 1 \rangle$ has an infinite cyclic quotient group, or equivalently that its abelianization $G/[G, G]$ is infinite. (The various groups with unsolvable word problem constructed by Novikov and by Boone all have this property.) Then we can easily adjust the presentation by making Tietze transformations so that at least one of the generators, say a_1, appears with exponent sum 0 in each of the relations. So if a word w in the generators of G has $w =_G 1$ then w has exponent sum 0 on a_1, that is, a_1 and a_1^{-1} each occur the same number of times in w. Otherwise the image of w in $G/[G, G]$ would be different from 1.

Now a "random" word of G almost never has exponent sum 0 on a_1. The set of words with non-zero exponent sum is *generic* in the sense that, probabilistically, almost all words have the property. Even though there may be no general algorithm for deciding whether or not $w =_G 1$, there is a partial algorithm which correctly answers "No" on a generic set of words, namely those with non-zero exponent sum on a_1.

Such observations have been made precise by Kapovich, Myasnikov, Schupp and Shpilrain in [54] where they introduce a notion of *generic-case complexity* $\mathcal{C}$ where $\mathcal{C}$ is a complexity class such as linear time or polynomial time. Here the complexity concerns the behavior of the partial algorithm only on the generic set, completely ignoring the behavior elsewhere. The following is one of their results.

THEOREM 17 ([54]). *Let G be a finitely generated group. Suppose that H is a finite index subgroup of G that possesses an infinite quotient group $\overline{H}$ for which the word problem is solvable in the complexity class $\mathcal{C}$. Then the word problem for G has generic-case complexity in the class $\mathcal{C}$.*

The full worst-case word problem in groups covered by this theorem can be unsolvable, for instance in the groups of Novikov and of Boone. But the types of words which witness this unsolvability are of a very restricted form and lie outside a generic set.

Similar results are also obtained in [54] for the conjugacy problem and for the problem of deciding membership in a finitely generated subgroup. It does not seem to be known whether there is a finitely presented group whose word problem is not generically computable. But there are finitely presented groups which do not satisfy the hypothesis of the above theorem. By using a pair of effectively inseparable sets, the author has constructed (see [72]) a finitely presented group G such that every non-trivial quotient group of G has unsolvable word problem.

Many interesting classes of groups have solvable word problem: In different direction, Wilhelm Magnus (who was a student of Dehn) had earlier studied groups given by a single defining relation and proved an important result called the Freiheitssatz [57]: a subset of the generators which omits a symbol in the relation freely generates a free subgroup. Magnus then used his methods to solve the word problem for one-relator groups [58].

As further large classes of groups having solvable word problem we mention finitely generated linear groups and finitely presented residually finite groups. A group is *linear of degree n* if it has a faithful representation as a group of $n \times n$-matrices over a field, that is, it is isomorphic to a group of matrices. In an important paper in 1940, A. I. Malcev [60] used the compactness theorem from mathematical logic to prove that groups which are locally linear of degree n, that is, for which every finitely generated subgroup is linear of degree n, are (globally) linear of degree n.

We recall some definitions. A group G is said to be *residually finite* if the intersection of the subgroups of finite index in G is the trivial subgroup. Equivalently, G is residually finite if for every non-trivial element $1 \neq g \in G$ there is a homomorphism $\theta \colon G \to H$ from G to a finite group such that $\theta(g) \neq_H 1$ (where θ and H depend on g). This the fact that $1 \neq g \in G$ is witnessed in the finite quotient H of G. We further recall that H. Hopf (1932) had asked for for finitely generated groups G whether $G \cong G/N$ implies $N = 1$. Nowadays we call a group G *hopfian* if it satisfies the equivalent condition that every surjective map $\phi \colon G \to G$ is an isomorphism. In this 1940 paper [60] on linear groups, Malcev also proved the following results related to our story.

THEOREM 18 (Malcev).

1. *Finitely generated linear groups of degree n are residually finite.*
2. *Finitely generated residually finite groups are hopfian.*
3. *Hence, finitely generated linear groups are hopfian.* ⊣

There are several ways these results are connected with the word problem. In 1943 J. C. C. McKinsey [68] gave a solution to the decision problem for "open sentences" in certain classes of algebras which he termed "finitely reducible". In the case of groups this is essentially the finitely presented residually finite groups (as later observed by Verena Huber-Dyson [42]), though he only mentions finitely generated abelian groups explicitly.

THEOREM 19 (McKinsey). *Finitely presented, residually finite groups have a solvable word problem.*

PROOF. Let G be finitely presented and residually finite. Since G is finitely presented the N set of words equal to 1 in G is recursively enumerable. To decide whether an arbitrary word $w \in G$ is equal to 1, we start enumerating N. At the same time we begin enumerating all homomorphisms from $\phi \colon G \to H$ where H is a finite group, say given by a finite multiplication table. We can check whether a map on generators defines a homomorphism by checking whether the images of the relators are equal to 1 in H using its multiplication table. As such homomorphisms are enumerated, evaluate them on w and check to see whether $\phi(w) =_H 1$. If $w \neq_G 1$, eventually one of its images will be different from 1, so we have an answer. But if $w =_G 1$ then w will appear in the list of N. Thus we have an algorithm to solve the word problem for G. ⊣

While this style of argument is familiar in logic and recursion theory, many mathematicians find the argument disconcerting since there is no easy estimate in advance of how long one might have to wait for an answer.

Concerning finitely generated linear groups G of degree n, they may not be finitely presented so McKinsey's result may not apply. Nevertheless such a linear group over a given field is also linear over a larger field and so should be linear over a field in which one can actually do the field arithmetic computationally. Hence by multiplying the matrices corresponding to the group generators, one should be able to determine whether a product is equal to the identity matrix. This has all been made precise, in particular by Rabin [85] (see also Lipton and Zalcstein [55]).

THEOREM 20 (Rabin). *A finitely generated linear group over a field has solvable word problem.* ⊣

Linear groups are quite a large class of groups and many of the infinite groups mathematicians have studied historically are linear or nearly so, often for geometric reasons. So not only do "most" finitely presented groups have solvable word problem, so also do those arising in many traditionally studied contexts.

For a survey of results on decision problems for various classes of groups as of about 1990 see [72].

6.2. Some constructions in group theory and Higman's non-hopfian group. In this subsection we discuss some examples relevant to our story and review some standard constructions from combinatorial group theory which will be used to "unsolve" the word problem below.

The group $BS(1, 2)$. Consider the group $G = \langle x, s \mid xs = sx^2 \rangle$, which is often referred to as $BS(1, 2)$ for reasons we indicate below. This interesting group appears as a subgroup in important ways in the (very different) groups with unsolvable word problem constructed by Novikov and by Boone, and we want to examine it more closely. In fact this group $G = BS(1, 2)$ is a linear group and one can show the map

$$x \mapsto \begin{bmatrix} 1 & 1 \\ 0 & 1 \end{bmatrix}, \quad s \mapsto \begin{bmatrix} \frac{1}{2} & 0 \\ 0 & 1 \end{bmatrix}$$

is a homomorphism which embeds G as a subgroup of $GL(2, \mathbb{Q})$. Hence G is residually finite and hopfian and has solvable word problem.

But to better understand G we describe an elementary solution to the word problem for G. The defining relation can be also written in a number of equivalent forms: $s^{-1}xs = x^2$ or $x = sx^2s^{-1}$ or $s^{-1}x = x^2s^{-1}$ or $x^{-1}s = sx^{-2}$ or $s^{-1}x^{-1} = x^{-2}s^{-1}$. We observe that, if one starts with any word w on the generators x and s, the relation can be applied to move the letter s from right to left over $x^{\pm 1}$ symbols creating additional x's. Similarly an s^{-1} can be moved from left to right over $x^{\pm 1}$ symbols. So $w =_G s^i x^j s^{-k}$ where $i \geq 0$ and $k \geq 0$ and $j \in \mathbb{Z}$. In case $j = 2m$ is even and both $i > 0$ and $k > 0$ we can apply the relation $x = sx^2s^{-1}$ to deduce that $w =_G s^{i-1}x^m s^{-(k-1)}$ which has fewer s-symbols. This process is called *pinching* a pair of s-symbols, or an *s-pinch*. Continuing one obtains $w =_G s^i x^j s^{-k}$ where either j is odd or at least one of i or k is 0—in either case no further pinches are possible. If the right hand side of this equation is not the trivial word, then one can show $w \neq_G 1$. So the method described solves the word problem for G.

Notice that the algorithm removes inverse pairs of s-symbols during the pinching operation, but inverse pairs of s-symbols were never inserted. Also observe that for the word sxs^{-1} an s-pinch is not possible and this word is not equal in G to any word with fewer s-symbols.

Amalgamated free products. We briefly recall the amalgamated free product construction. Suppose H and K are two groups with respective subgroups A and B which are isomorphic via an isomorphism $\phi \colon A \to B$. We can assume we have presentations $H = \langle X_H \mid R_H \rangle$ and $K = \langle X_K \mid R_K \rangle$ on disjoint alphabets. Then the *free product of H and K with amalgamated subgroup $A = B$* is the group with presentation

$$H \underset{A=B}{\star} K = \langle X_H, X_K \mid R_H, R_K, a = \phi(a) \text{ for } a \in A \rangle.$$

There are two familiar (but not obvious) facts we need to know about this construction: (1) the inclusion maps on generators embed both H and K in $H \underset{A=B}{\star} K$; and (2) if $w = h_1 k_i h_2 k_2 \ldots h_m k_m$ is an alternating product of elements of H and K (where possibly h_1 or k_m does not appear) and if $w = 1$ in $H \underset{A=B}{\star} K$, then for some $i \in \{1, \ldots, m\}$ either $h_i \in A$ or $k_i \in B$ and so relations of the form $a = \phi(a)$ can be used to reduce the number of alternations.

HNN extensions. The group $BS(1, 2)$ above is an example of a very general construction, usually called an *HNN extension*, which is closely related to free products with amalgamated subgroups. It was introduced by Higman, Neumann and Neumann in their famous 1948 paper [50] and used to prove that any countable group C can be embedded in a two generator group H_C spending on C. As a consequence they showed that there are continuously many two generator groups.

The amalgamated free product construction identifies isomorphic subgroups of two different groups. The *HNN extension* or *HNN construction* takes two isomorphic subgroups of the same group and realises this isomorphism as a conjugation in a universal way. So suppose that the group $H = \langle X_H \mid R_H \rangle$ has a pair of isomorphic subgroups via the isomorphism $\phi \colon A \to B$. Then the *HNN extension of H with associated subgroups A and B and stable letter s* is the group with presentation

$$H \underset{\phi}{\star} = \langle X_H, s \mid R_H, s^{-1}as = \phi(a) \text{ for } a \in A \rangle$$

where s is a new symbol not in X_H called the *stable letter*. There are two important facts we need to know about this HNN construction (they are similar to the above statements about amalgamated free products):

LEMMA 21. *Let $H \underset{\phi}{\star}$ be an HNN extension as above. Then*

1. *(Higman–Neumann–Neumann) the inclusion map on generators embeds H into $H \underset{\phi}{\star}$; and*

2. *(Britton) if w is a word of $H \underset{\phi}{\star}$ in which s or s^{-1} appears and if $w = 1$ in $H \underset{\phi}{\star}$, then w contains either a subword of the form $s^{-1}as$ or of the form $s\phi(a)s^{-1}$ and so the relations of the form $s^{-1}as = \phi(a)$ can be used to perform an s-pinch and reduce the number of s-symbols in w.* ⊣

The first assertion of the lemma is a theorem Higman, Neumann and Neumann from [50]. The second assertion is usually known as *Britton's Lemma* whose role in our story will be explained below. There are very similar statements in case one has several pairs of isomorphic subgroups and isomorphisms $\phi_i \colon A_i \cong B_i$ and corresponding new stable letters s_i, but for notational simplicity we here concentrate on the one stable letter case.

The amalgamated free product and HNN extension constructions are closely related and basic facts about one often can be deduced from the analogous facts about the other (see for instance [56]). They also play an important role in geometric topology, particularly the study of 3-manifolds. Suppose one has a nicely embedded two sided surface Y in a 3-manifold M (possibly with boundary) such that the inclusion embeds the fundamental group of Y in the fundamental group of M. If one cuts M along Y, the resulting manifold (with boundary) either (1) has 2 components so Y separates and the fundamental group of M is the corresponding amalgamated free product; or (2) has one component so Y does not separate and the fundamental group of M is the corresponding HNN extension. (The associated isomorphic subgroups come from the two copies of Y in each case.)

As an example of an HNN extension we note that the group $G = \langle x, s \mid xs = sx^2 \rangle = \langle x, s \mid s^{-1}xs = x^2 \rangle$ introduced above is an HNN extension of the infinite cyclic group generated by x with associated subgroups the cyclic groups $\langle x \rangle$ and $\langle x^2 \rangle$ with the isomorphism ϕ defined by $\phi(x) = x^2$. (Notice that we only need to include the relations $a = \phi(a)$ for a ranging over a generating set for the subgroup A.) One consequence of Britton's Lemma for this group is that in the word sxs^{-1} the subword x is not of the form $\phi(x^i)$ and so no s-pinch is possible and sxs^{-1} is not equal to a word with fewer s-symbols.

Using Britton's Lemma. In our later discussion we will assume some familiarity with various ways of applying Britton's Lemma, so we pause here to introduce some terminology and make a few observations.

Continuing with the above notation, suppose $H \underset{\phi}{\star}$ is the HNN extension of H with stable letter s and associated subgroups A and B which are isomorphic via the map $\phi \colon A \to B$. If w is any word of $H \underset{\phi}{\star}$ (*not necessarily* $= 1$), then an *s-pinch* in w is a subword of the form $s^{-1}us$ where $u \in A$ or sus^{-1} where $u \in B$ (here we are assuming in either case that u does not contain $s^{\pm 1}$). A word w is said to be *s-reduced* it does not contain any s-pinch.

Changing our point of view slightly, suppose that H is given by a presentation $H = \langle X_H \mid R_H \rangle$ and that the associated subgroups are given by sets of generators, say $A = gp\langle a_1, a_2, \dots \rangle$ and $B = gp\langle b_1, b_2, \dots \rangle$ so that $\phi(a_i) = b_i$ where the a_i and b_i are certain words in the generators X_H for H. Then the HNN extension can be presented as

$$H \underset{\phi}{\star} = \langle X_H, s \mid R_H, \; s^{-1}a_1 s = b_1, \; s^{-1}a_2 s = b_2, \dots \rangle$$
$$\cong \langle X_H, s \mid R_H, \; a_1 s = s b_1, \; a_2 s = s b_2, \dots \rangle.$$

Suppose the word w contains and s-pinch. For instance suppose it contains a subword of the form $s^{-1}us$ with $u \in A$. This means there is a word $a_{i_1}^{\epsilon_1} \dots a_{i_k}^{\epsilon_k}$ on the generators of A so that $u =_H a_{i_1}^{\epsilon_1} \dots a_{i_k}^{\epsilon_k}$ is a consequence of the defining relations R_H of H. Then, applying the relations of $H \underset{\phi}{\star}$ involving s in the form $a_i s = s b_i$ repeatedly moving s from right to left, we have the following calculation:

$$w \equiv w_1 s^{-1} u s w_2 = w_1 s^{-1} a_{i_1}^{\epsilon_1} \dots a_{i_k}^{\epsilon_k} s w_2$$
$$= w_1 s^{-1} s b_{i_1}^{\epsilon_1} \dots b_{i_k}^{\epsilon_k} w_2$$
$$= w_1 b_{i_1}^{\epsilon_1} \dots b_{i_k}^{\epsilon_k} w_2$$

in which the result has fewer s-symbols. The analogous process transforms an s-pinch of the form sus^{-1} with $u \in B$: first use the relations of H to express u as a word in the b_i and then move s from left to right over b_i's to give an equation

$$w \equiv w_1 s u s^{-1} w_2 = w_1 s b_{i_1}^{\epsilon_1} \dots b_{i_k}^{\epsilon_k} s^{-1} w_2$$
$$= w_1 a_{i_1}^{\epsilon_1} \dots a_{i_k}^{\epsilon_k} w_2$$

with fewer s-symbols. Either process is called *pinching out an s* or *removing an s-pinch*.

There is subtle but important difference here in the form we choose to express the defining relations. From a group theoretic point of view, the form $s^{-1}a_i s = b_i$ emphasises that the subgroups A and B are conjugate realising the isomorphism $\phi(a_i) = b_i$. But thinking in terms of rewrite rules transforming words, to show say $s^{-1}a_1 a_2 s = b_1 b_2$ we need to insert an inverse pair of s-symbols. Thus the calculation in detail would be $s^{-1}a_1 a_2 s = s^{-1}a_1 s s^{-1} a_2 s = s^{-1}a_1 s b_2 = b_1 b_2$. By constrast, using the relations in the form $a_i s = s b_i$ as we have above, we have the detailed calculation $s^{-1}a_1 a_2 s = s^{-1}a_1 s b_2 s = s^{-1}s b_1 b_2 = b_1 b_2$ which does not involve the insertion of inverse pairs of s-symbols.

Clearly, by successively pinching out s-symbols we eventually arrive at an s-reduced word $\overline{w}$ such that $w =_{H *_\phi} \overline{w}$. It is fairly easy to show that *a word w of $H *_\phi$ is s-reduced if and only if it is not equal to any word with fewer s-symbols.* For if a word w is not equal to a word with fewer s-symbols it must clearly be s-reduced by the above. But if w and w' are two s-reduced words, then $1 = w^{-1} w'$ must reduce to an s-free word (= word without s-symbols) and in the reduction process each pinch must involve an $s^{\pm 1}$ from each. So w and w' are s-parallel in the sense that they have the same signed sequence of s symbols.

Finally we note that (using the $a_i s = s b_i$ form of the defining relations) the process of removing s-pinches never introduces an inverse pair of s symbols. So if w is a word of $H *_\phi$ which is equal to an s-free word w' it can be transformed into w' without introducing any inverse pairs of s-symbols. This fact is due to Novikov (his Principal Lemma). Novikov's Principal Lemma is essentially equivalent to Britton's Lemma as will be discussed below.

Higman's non-hopfian group. In 1951 Graham Higman [48] gave the first example of a finitely presented non-hopfian group. Higman's non-hopfian group can be presented as $H = \langle x, s_1, s_2 \mid s_1^{-1} x s_1 = x^2, s_2^{-1} x s_2 = x^2 \rangle$. Now the map $\theta \colon H \to H$ defined by $s_i \mapsto s_i$ and $x \mapsto x^2$ is a surjective homomorphism from H onto itself (easy check). Observe that the element $s_1 x s_1^{-1} s_2 x^{-1} s_2^{-1} \neq_H 1$ since, if we view it as an HNN extension with two stable letters s_1 and s_2, there is no possible s_i-pinch. But

$$\theta(s_1 x s_1^{-1} s_2 x^{-1} s_2^{-1}) = (s_1 x^2 s_1^{-1})(s_2 x^{-2} s_2^{-1}) =_H x x^{-1} = 1$$

and so θ is not injective and H is non-hopfian. Hence H is not residually finite.

There are several alternative ways to view Higman's group H. For instance it is the amalgamated free product of two copies $G_i = \langle x, s_i \mid s_i^{-1} x s_i = x^2 \rangle$

amalgamated along the subgroup $\langle x \rangle$, that is, $H = G_1 \underset{\langle x \rangle = \langle x \rangle}{\star} G_2$. Introducing a new symbol t a setting $t = s_1 s_2^{-1}$ one can apply a few Tietze transformations to obtain $H \cong \langle x, t, s_1 \mid xt = tx, s_1^{-1} x s_1 = x^2 \rangle$ which is an HNN extension of the free abelian group on x, t with stable letter s_1.

It happens that Higman's non-hopfian group also appears in important ways as subgroups of the (quite different) groups constructed by Novikov and by Boone. So they visibly contain a non-hopfian group and can not be residually finite (for reasons other than having an unsolvable word problem). We will see in Boone's groups, for instance, the subgroup

$$\mathcal{B}_1 = \langle x, s_0, \ldots, s_M \mid s_b^{-1} x s_b = x^2 \text{ for } b = 0, \ldots, M \rangle$$

which is an enlarged version of Higman's group with more stable letters. A key technical property, proved in Lemma 29 below, is that certain words contain only positive occurrences of s_b symbols.

As further examples we mention the *Baumslag–Solitar groups* commonly denoted

$$BS(n, m) = \langle x, s \mid s^{-1} x^n s = x^m \rangle$$

studied by Baumslag and Solitar in [8]. Among other things they showed that the one-relator group $BS(2, 3)$ is non-hopfian. This can be easily deduced from Britton's Lemma. The map $x \to x^2$ and $s \to s$ defines a homomorphism ϕ from $BS(2, 3)$ onto itself. Now the commutator $[x, s^{-1} x s] = x^{-1} s^{-1} x^{-1} s x s^{-1} x s \neq 1$ by Britton's Lemma since no s-pinch is possible because $x^{\pm 1} \notin \langle x^2 \rangle$ and $x^{\pm 1} \notin \langle x^3 \rangle$. But $\phi([x.s^{-1} x s]) = [x^2, s^{-1} x^2 s] = [x^2, x^3] = 1$ and so $BS(2, 3)$ is non-hopfian.

6.3. Novikov's groups and principal lemma. In 1952 [74] Novikov announced the construction of a finitely presented group with unsolvable word problem, providing a sketch of the construction but no proofs. According to reviews by Hirsch [51] and by Boone [17] the construction is based on an unsolvable problem of Post concerning canonical systems and contains a number of misprints which make the exact meaning unclear.

In 1954 [75] Novikov published a construction of a finitely presented group with unsolvable conjugacy problem. While this follows trivially from the unsolvability of the word problem, Novikov provided a separate proof for the conjugacy problem which is considerably simpler and has many fewer generators and relations. Again this result is based on "Post systems". See the reviews by Hirsch [52] and Britton [28]

Then in 1955 [76] Novikov published a rather lengthy paper proving the unsolvability of the word problem based on Turing's cancellation semigroup with unsolvable word problem. The review by A. A. Markov [63] of this 1955 proof concludes as follows:

> The exposition is rather inadequate, which makes for very difficult reading. There are numerous inaccuracies and some untrue assertions. In addition there are many misprints, of which only a part appear on the list of errata appended to the monograph. All these defects, however, do not affect the essence of the matter. After a careful study of the monograph the reviewer has come to the conclusion that the small errors in it can all be easily corrected, and hence that the result of this remarkable work is valid.

The review [27] for the Journal of Symbolic Logic by Britton (based on Hirsch's translation of the monograph) notes that subsequent proofs have been given by Boone and by Britton himself and includes the comment

> With a paper of this size and complexity, involving such profusion of detail, it is difficult to feel convinced that one has overlooked nothing. However, the reviewer now firmly believes that this proof is correct and contains no serious gaps.

The group constructed in this monograph differs significantly from the one outlined in the 1952 announcement and it is not so clear whether the earlier claimed proof held up. But it is clear that Novikov gave the first published proof of the unsolvability of the word problem for finitely presented groups. For some related historical comments, see the obituary of Boone [37].

We will not look at Novikov's construction in any detail since the Post–Boone–Britton sequence of constructions provides a more straightforward approach than the Post–Turing–Novikov sequence of constructions. But we want to record an important tool in Novikov's proof which he calls the "Principal Lemma".

Novikov's general approach to the word problem was combinatorial and more in the spirit of mathematical logic than group theory. As above for semigroups, relations are used to transform or rewrite words. But in addition to the defining relations specified in the presentation one has the *inverse pair* relations $xx^{-1} = 1$ and $x^{-1}x = 1$ for each generating symbol x. So two words are equal in the group defined by a presentation if and only if one can be obtained from the other by a sequence of *elementary transformations* of the form: (1) $z_1 u_i z_2 \to z_1 v_i z_2$ or $z_1 v_i z_2 \to z_1 u_i z_2$ where $u_i = v_i$ is a defining relation; or (2) inserting or deleting inverse pairs, as in $z_1 z_2 \to z_1 x x^{-1} z_2$ or $z_1 x x^{-1} z_2 \to z_1 z_2$.

Suppose the group $H = \langle X_H \mid R_H \rangle$ has a pair of isomorphic subgroups via the isomorphism $\phi \colon A \to B$. Let

$$H \underset{\phi}{\star} = \langle X_H, s \mid R_H,\ as = s\phi(a) \text{ for } a \in A \rangle$$

be the corresponding HNN extension with associated subgroups A and B and stable letter s. Note that if A is generated by $\{a_1, a_2, \ldots, a_m\}$ we can use

instead the more efficient presentation
$$H\underset{\phi}{\star} = \langle X_H, s \mid R_H, a_i s = s\phi(a_i) \text{ for } i = 1, \ldots, m \rangle.$$

In particular if H is finitely presented and A is finitely generated then $H\underset{\phi}{\star}$ is finitely presented.

THEOREM 22 (Novikov's Principal Lemma). *Let $H\underset{\phi}{\star}$ be an HNN extension as above. Suppose that* (1) *w, z are two words of such that $w = z$ in $H\underset{\phi}{\star}$ and* (2) *the symbols $s^{\pm 1}$ do not appear in z. Then there is an elementary sequence of transformations*
$$w \equiv w_1 \to w_2 \to \cdots \to w_n \equiv z$$
that contains no insertions of the letter s.

We first observe that in the case w also does not contain the symbols $s^{\pm 1}$, then the Principal Lemma implies $w =_H z$ and so H is embedded in $H\underset{\phi}{\star}$. This fact was originally proved by Higman, Neumann and Neumann their 1948 paper [50]. We further note that, in the case w does contain the symbols $s^{\pm 1}$, the assertion of the Principal Lemma is easily seen to be equivalent to Britton's Lemma. However, as we shall see below, the particular form of Britton's Lemma makes it very useful for group-theoretic style arguments.

As we have already noted, following Novikov's 1955 proof of the unsolvability of the word problem, Boone gave a rather different proof in 1957 based directly on Post's semigroup. In 1958, Britton [29] gave a somewhat different proof based on Turing's cancellation semigroups result and some of his own earlier work on solutions to the word problem. In this paper a version of what is called Britton's Lemma appears but under the unnecessarily restrictive assumption that the associated subgroups intersect trivially. As his review above indicates, Britton would have been familiar with Novikov's work and Principal Lemma, but he establishes it using the earlier paper of Higman, Neumann and Neuman [50].

In 1959 Boone [19] published a simplified version of his construction. Then in 1963 Britton [30] further simplified Boone's construction and gave a proof in a much more group-theoretic style using Britton's Lemma which first appears there in full generality.

In later years Boone, who seemingly never tried to read Novikov's work in detail, asked the writer if there might be a more combinatorial style proof of Britton's Lemma. He seemed surprised when I explained to him that Novikov had proved (in a combinatorial manner) his Principal Lemma which is easily shown to be equivalent to Britton's Lemma.

6.4. Boone's groups. Boone's initial construction of a group with unsolvable word problem was developed in a six part series of papers over 5 years

[15]. In the first four articles it is shown the unsolvability of the "quasi-magnus problem", that is, whether or not an arbitrary word is expressible as a product of positive powers of a certain proper subset of the generators. The unsolvability of the word problem itself occupies the last two parts which are in principle independent of the first four, but the exposition relies on the earlier parts.

We now describe a variant of Boone's construction of a finitely presented group with unsolvable word problem. The construction was modified a number of times: first by Boone himself [19] and then by Britton [30] who gave a much more group theoretic proof. Further variations and simplifications were then contributed by Boone, Collins and myself, and were presented in various seminars. Eventually these were included in later editions of Rotman's group theory text [87]. We also borrow from the discussion in [39] which provides for additional applications to asphericity. The theme is still the same so we continue to refer to them as Boone's groups. An elegant further simplification of Boone's approach was introduced by Borisov [24] and is incorporated in the exposition in [5], but we have chosen to follow the earlier treatments since the involvement of Turing machines is more apparent.

A very attractive aspect of Boone's group is that it encodes Post's semigroup construction in a very direct way so that the underlying Turing machine and its computations are still visible in the group. Actually, as Boone liked to describe the situation, the words for which the word problem is shown to be difficult encode four copies of the Turing machine operating simultaneously. In addition it is exhibited as a tower of successive HNN extensions which enables one to use techniques of combinatorial group theory.

So the construction of Boone's group begins with a Turing machine T having an unsolvable halting problem. The semigroup construction of Post is then applied to obtain a finitely presented semigroup $\Gamma(T)$ of the form

$$\Gamma(T) = \langle q, q_0, \ldots, q_N, s_0, \ldots, s_M \mid F_i q_{i_1} G_i = H_i q_{i_2} K_i \ (i \in I) \rangle$$

where the F_i, G_i, H_i, K_i are positive s-words and $q_{i_j} \in \{q, q_0, \ldots, q_N\}$. In the present context, an s-word is a word on the symbols $s_0, \ldots, s_M$ and their inverses. An s-word is *positive* if it contains no s_i^{-1} symbols. By Post's results:

LEMMA 23. *The problem of deciding for an arbitrary pair of positive s-words X, Y whether or not $Xq_i Y = q$ in $\Gamma(T)$ is recursively unsolvable.* ⊣

As before we use $X \equiv Y$ to mean the words X and Y are identical (letter by letter). If $X \equiv s_{b_1}^{e_1} \ldots s_{b_m}^{e_m}$ is an s-word, we define $X^\# \equiv s_{b_1}^{-e_1} \ldots s_{b_m}^{-e_m}$. Note that $X^\#$ is not the same as X^{-1}. Also, if X and Y are s-words, then $(X^\#)^\# \equiv X$ and $(XY)^\# \equiv X^\# Y^\#$.

Boone's group $\mathcal{B} = \mathcal{B}(T)$ is then the finitely presented group depending on $\Gamma(T)$ described as follows:

generators: $q, q_0, \ldots, q_N, s_0, \ldots, s_M, r_i \ (i \in I), x, t, k$;

relations: for all $i \in I$ and all $b = 0, \ldots, M$,

$$\left.\begin{array}{l} xs_b = s_b x^2 \quad \;\;\;\;\;\;\;\;\;\;\;\;\;\;\; \left.\right] \Delta_1 \\ r_i s_b = s_b x r_i x \\ r_i^{-1} F_i^{\#} q_{i_1} G_i r_i = H_i^{\#} q_{i_2} K_i \\ tr_i = r_i t \\ tx = xt \\ kr_i = r_i k \\ kx = xk \\ k(q^{-1}tq) = (q^{-1}tq)k. \end{array}\right\} \begin{array}{l} \\ \\ \Delta_2 \\ \\ \\ \\ \end{array} \right\} \Delta_3$$

The subsets $\Delta_1 \subset \Delta_2 \subset \Delta_3$ of the relations each define a presentation of a group $\mathcal{B}_i$ generated by the symbols appearing in the Δ_i.

DEFINITION 24. *A word Σ is special if $\Sigma \equiv X^{\#} q_j Y$ where X and Y are positive s-words and $q_j \in \{q, q_0, \ldots, q_N\}$.*

The main technical result linking the word problem in $\mathcal{B}(T)$ as presented above to the word problem in Post's semigroup $\Gamma(T)$ is the following:

LEMMA 25 (Boone's Lemma). *If $\Sigma \equiv X^{\#} q_j Y$ is a special word in $\mathcal{B}(T)$, then*

$$k(\Sigma^{-1} t \Sigma) = (\Sigma^{-1} t \Sigma) k \text{ in } \mathcal{B}(T)$$

if and only if $X q_j Y = q$ in $\Gamma(T)$.

Because of the properties (Lemma 23) of Post's semigroup, Boone's Lemma immediately implies the finitely presented group $\mathcal{B}(T)$ has an unsolvable word problem.

THEOREM 26 (Novikov–Boone). *There exists a finitely presented group $\mathcal{B}$ with recursively unsolvable word problem.* ⊣

We here give a sketchy but essentially complete proof of Boone's Lemma and make some other observations so as to give some feeling for why the construction works. For additional details the reader is referred to the text book by Rotman [87].

From the very form of the various defining relations for $\mathcal{B} = \mathcal{B}(T)$ above, we suspect that the group $\mathcal{B}$ is built up by successively forming HNN extensions. To make this precise we let $\mathcal{B}_0$ denote the infinite cyclic group generated by x, and let Q denote the free group with basis $\{q, q_0, \ldots, q_N\}$. Using some standard facts from combinatorial group theory, one can easily show the following (this is Lemma 12.11 in [87]).

LEMMA 27. *In the chain*

$$\mathcal{B}_0 \leq \mathcal{B}_1 \leq \mathcal{B}_1 * Q \leq \mathcal{B}_2 \leq \mathcal{B}_3 \leq \mathcal{B}$$

each group is an HNN-extension of its predecessor; moreover, the free product $\mathcal{B}_1 * Q$ is an HNN-extension of $\mathcal{B}_0$. ⊣

One consequence of this is that copies of the group $BS(1,2) = \langle x, s \mid xs = sx^2 \rangle$ and also Higman's non-hopf group are embedded as subgroups of $\mathcal{B}$. Indeed we observe that $\mathcal{B}_1$ is an enlarged version of Higman's group obtained by adding even more stable letters s_b, and it is non-hopfian by the same argument. As we have noted before these groups also appear as subgroups of Novikov's groups and play an important role there as well.

We previously observed in our discussion of $BS(1,2)$ that we can move s_b symbols from right to left over $x^{\pm 1}$. So if R_1 is any word on the r_i and x, then $R_1 s_b = s_b R_2$ where $R_2 = s_b^{-1} R_1 s_b$ is another word on r_i and x.

Similarly, if L_1 is a word on r_i and x, then s_b^{-1} can be moved from left to right over L_1. So $s_b^{-1} L_1 = L_2 s_b^{-1}$ where $L_2 = s_b^{-1} L_1 s_b$ is again another word on r_i and x.

More generally these considerations apply to moving any positive s-word V from right to left and $V^\#$ from left to right across a word in r_i and x. Also observe that only the relations of $\mathcal{B}_2$ have been used in these calculations. We state this as follows:

LEMMA 28. *Suppose that V is a positive s-word and that R_1 and L_1 are words on r_i and x. Then there are words R_2 and L_2 on r_i and x so that $R_1 V =_{\mathcal{B}_2} V R_2$ and $V^\# L_1 =_{\mathcal{B}_2} L_2 V^\#$.* ⊣

Next consider one of the defining relations of $\mathcal{B}_2$ of the form $r_i^{-1} F_i^\# q_{i_1} G_i r_i = H_i^\# q_{i_2} K_i$. Recall that this relation comes from Post's construction so it corresponds a basic step in a Turing machine. Suppose that U, V are two positive s-words. Then, by the relation and the previous Lemma,

$$U^\#(H_i^\# q_{i_2} K_i) V = U^\#(r_i^{-1} F_i^\# q_{i_1} G_i r_i) V = L U^\# F_i^\# q_{i_1} G_i V R$$

for certain words L, R on r_i and X. Similarly

$$U^\#(r_i H_i^\# q_{i_2} K_i r_i^{-1}) V = U^\#(F_i^\# q_{i_1} G_i) V = L_1 U^\# F_i^\# q_{i_1} G_i V R_1$$

for certain words L_1, R_1 on r_i and X. We use these observations in the following:

PROOF OF SUFFICIENCY IN BOONE'S LEMMA. Suppose X, Y are two positive s-words such that $Xq_j Y = q$ in $\Gamma(T)$. Then there is a sequence of applications of the defining relations of $\Gamma(T)$, say

$$Xq_j Y \equiv w_1 \to w_2 \to \cdots \to w_n \equiv q$$

so that for each i, one of the words w_i and w_{i+1} has the form $U^\# F_i^\# q_{i_1} G_i V$ and the other has the form $U^\# H_i^\# q_{i_2} K_i V$.

Now if we start with a special word $\Sigma \equiv X^\# q_j Y$ and apply the corresponding relations in $\mathcal{B}(T)$ (in the same direction as the above sequence $w_i \to w_{i+1}$) and move the s-words towards the q_i as in the above discussion we get a corresponding sequence of equalities in $\mathcal{B}(T)$ of the form $w_i^* = L_i w_{i+1}^* R_i$ where w_i^* is the special word of $\mathcal{B}$ corresponding to w_i and L_i, R_i are words on the r_j and x. Setting $L = L_1 \cdots L_{n-1}$ and $R = R_{n-1} \cdots R_1$, it follows that in $\mathcal{B}(T)$ we have $\Sigma = LqR$ where L, R are words in r_j and x. Note that in arriving at this equation only the relations of $\mathcal{B}_2$ are used.

But the generators k and t commute with the r_j and with x so they commute with L and R. Therefore

$$k(\Sigma^{-1}t\Sigma) = k(R^{-1}q^{-1}L^{-1}tLqR) = k(R^{-1}q^{-1}tqR) = R^{-1}k(q^{-1}tq)R$$
$$= R^{-1}(q^{-1}tq)kR \quad \text{(since } k \text{ commutes with } q^{-1}tq\text{)}$$
$$= (R^{-1}q^{-1}tqR)k = (R^{-1}q^{-1}L^{-1}tLqR)k = (\Sigma^{-1}t\Sigma)k$$

as claimed. This proves sufficiency in Boone's Lemma. ⊣

DISCUSSION. Conceptually, the proof of sufficiency establishes two connected implications between equations in various systems which we can write as

$$Xq_jY \underset{\Gamma(T)}{=} q \implies \Sigma \equiv X^{\#}q_jY \underset{\mathcal{B}_2(T)}{=} LqR \implies k(\Sigma^{-1}t\Sigma) \underset{\mathcal{B}(T)}{=} (\Sigma^{-1}t\Sigma)k$$

where X, Y are positive words on the s_b and L, R are words on the r_i and x. In the proof of necessity, the converse of each of these implications is established. The converse of the second implication is fairly straight forward using Britton's Lemma. The converse of the first implication is more delicate because one needs to show certain words on the s_b are positive and hence the semigroup rewrites can be carried out.

PROOF OF NECESSITY IN BOONE'S LEMMA. Suppose $\Sigma \equiv X^{\#}q_jY$ is a special word in $\mathcal{B} = \mathcal{B}(T)$ and that $k(\Sigma^{-1}t\Sigma) =_\mathcal{B} (\Sigma^{-1}t\Sigma)k$. Then

$$k(\Sigma^{-1}t\Sigma)k^{-1}(\Sigma^{-1}t^{-1}\Sigma) =_\mathcal{B} 1$$

and since $\mathcal{B}$ is an HNN extension of $\mathcal{B}_3$ with stable letter k, there must be a k-pinch. There are exactly two $k^{\pm 1}$ which must pinch over the word between them and hence $\Sigma^{-1}t\Sigma$ must be equal in $\mathcal{B}_3$ to a word W on $\{r_i(i \in I), x, q^{-1}tq\}$. Now $\mathcal{B}_3$ is an HNN extension of $\mathcal{B}_2$ with stable letter t which commutes with $\{r_i(i \in I), x\}$. So pinching as many t-symbols from W as possible, we obtain a t-reduced word $W_1 =_{\mathcal{B}_3} W$ and still on $\{r_i(i \in I), x, q^{-1}tq\}$. Thus $\Sigma^{-1}t\Sigma =_{\mathcal{B}_3} W_1$ and both sides are t-reduced and hence W_1 has a single $q^{-1}tq$. Thus we may suppose $W_1 \equiv R^{-1}q^{-1}tqR_0$ where R, R_0 are words on $\{r_i(i \in I), x\}$.

Thus $\Sigma^{-1}t^{-1}\Sigma R^{-1}q^{-1}tqR_0 =_{\mathcal{B}_3} 1$ and by Britton's Lemma there must be a t-pinch and

$$\Sigma R^{-1}q^{-1} \equiv X^{\#}qYR^{-1}q^{-1} =_{\mathcal{B}_2} L$$

where L is a word on $\{r_i(i \in I), x\}$. Thus in $\mathcal{B}_2$ we have the equation

$$\Sigma \equiv X^{\#}qY =_{\mathcal{B}_2} LqR$$

where L, R are words on $\{r_i \, (i \in I), x\}$.

REMARK. The argument so far establishes the converse of the second implication in the "Discussion" above. So in $\mathcal{B}$ we know that $k(\Sigma^{-1}t\Sigma) = (\Sigma^{-1}t\Sigma)k$ if and only if $\Sigma \equiv X^{\#}q_j Y = LqR$. The remainder of the proof establishes the converse of the first implication, namely that if $\Sigma \equiv X^{\#}q_j Y = LqR$ in $\mathcal{B} = \mathcal{B}(T)$, then $Xq_j T = q$ in $\Gamma(T)$.

Now $\mathcal{B}_2$ is an HNN extension of $\mathcal{B}_1 * Q$ with stable letters the $\{r_i (i \in I)\}$ and it is easy to show that the elements $\{r_i (i \in I), x\}$ freely generate a free subgroup of $\mathcal{B}_2$ (see Lemma 31 below). So we may assume that L and R are freely reduced and hence r_i-reduced for $i \in I$. But by Britton's Lemma we must be able to successively transform LqR by carrying out r_i-pinches across what is between them to obtain a word which no longer involves r_i and is in fact equal in $\mathcal{B}_1 * Q$ to Σ. In particular L^{-1} and R must be r_i-parallel in the sense that the sequences of r_i that appear must be the same.

Rewriting the equation $\Sigma = LqR$ instead as $L^{-1}\Sigma R^{-1} = q$, the sequence of successive r_i-pinches must act on $\Sigma \equiv X^{\#}q_j Y$ eventually leading to the halting state q. Intuitively this sequence of r_i-pinches just encodes a proof in Post's semigroup $\Gamma(T)$ that $Xq_j Y =_{\Gamma(T)} q$. That is exactly what the remainder of our proof is devoted to showing.

So let us examine the first such r_i-pinch. We can write $L^{-1} \equiv L_1 r_i^{-\epsilon} x^{\alpha}$ and $R^{-1} \equiv x^{\beta} r_i^{\epsilon} R_1$ where ϵ is ± 1 and α, β are integers and L_1, R_1 are the remainder of L^{-1} and R^{-1}. We know here that $r_i^{-\epsilon} x^{\alpha} X^{\#} q_j Y x^{\beta} r_i^{\epsilon}$ is a pinch and thus $x^{\alpha} X^{\#} q_j Y x^{\beta}$ is equal in $\mathcal{B}_1 * Q$ to a word in the appropriate subgroup associated to r_i depending on ϵ.

Consider the case $\epsilon = +1$. We then have

$$x^{\alpha} X^{\#} q_j Y x^{\beta} = (x^{\alpha} X^{\#} F_i^{\#-1}) F_i^{\#} q_{i_1} G_i (G_i^{-1} Y x^{\beta})$$

where q_{i_1} is q_j and $(G_i^{-1} Y x^{\beta})$ and $(x^{\alpha} X^{\#} F_i^{\#-1})$ are equal in $\mathcal{B}_1$ to words in the elements $s_b x$ corresponding to the (free) generators of the left hand associated subgroup in the relations $r_i^{-1} s_b x r_i = s_b x^{-1}$.

We now prove an elementary but important lemma about certain words in $\mathcal{B}_1$ which explains why the group $\mathcal{B}_1$ plays a crucial role in the construction.

LEMMA 29. *Suppose that U and V are positive word in the s_b-symbols and that $U^{-1}V$ is freely reduced as written, that is, the last symbol of U^{-1} is not the inverse of the first symbol of V. If the word $U^{-1}Vx^{\beta}$ is equal in $\mathcal{B}_1$ to a word in the elements $s_b x$, then U must be empty. Similarly, if $U^{-1}Vx^{\beta}$ is equal to a word in the elements $s_b x^{-1}$, then U must be empty.*

PROOF. Note that the s_b-symbols freely generate a free subgroup which is a retract of $\mathcal{B}_1$. Suppose that U is not the empty word. If we write out $U^{-1}Vx^{\beta}$ in detail it has the form

$$U^{-1}Vx^{\beta} \equiv s_{b_1}^{-1} \cdots s_{b_\lambda}^{-1} s_{c_1} \cdots s_{c_\rho} x^{\beta}.$$

Assume this is equal in $\mathcal{B}_1$ to a word in the $s_b x$ which must have the same retraction onto the free group on the s_b. So we must have

$$U^{-1} V x^\beta \equiv s_{b_1}^{-1} \cdots s_{b_\lambda}^{-1} s_{c_1} \cdots s_{c_p} x^\beta =_{\mathcal{B}_1} x^{-1} s_{b_1}^{-1} \cdots x^{-1} s_{b_\lambda}^{-1} s_{c_1} x \cdots s_{c_p} x.$$

Equivalently this can be expressed as

$$x^{-\beta} s_{c_p}^{-1} \cdots s_{c_1}^{-1} s_{b_\lambda} \cdots s_{b_1} x^{-1} s_{b_1}^{-1} \cdots x^{-1} s_{b_\lambda}^{-1} s_{c_1} x \cdots s_{c_p} x =_{\mathcal{B}_1} 1.$$

Now $\mathcal{B}_1$ is an HNN extension with stable letters the s_b-symbols so by Britton's Lemma there must be an s_b-pinch. But by the assumptions on free reductions, the only place such a pinch could occur is at $s_{b_1} x^{-1} s_{b_1}^{-1}$. But this is not a pinch since the relevant relation is $x = s_{b_1} x^2 s_{b_1}^{-1}$ and x^{-1} does not lie in the subgroup generated by x^2. So we have a contradiction, proving the claim. The proof for equality to words in $s_b x^{-1}$ is very similar with x^{-1} in place of x. ⊣

Returning now to the proof of necessity in Boone's Lemma, we have shown that $(G_i^{-1} Y x^\beta)$ is equal in $\mathcal{B}_1$ to a word in the elements $s_b x$. So by Lemma 29 if we freely reduce $G_i^{-1} Y$ there are no negative symbols remaining. Hence Y begins with the word G_i and $Y \equiv G_i Y_1$ where Y_1 is the remainder of Y and is a positive s-word. Similar considerations apply to $(x^\alpha X^\# F_i^{\#-1})^{-1} = F_i^\# (X^\#)^{-1} x^{-\alpha}$ and we conclude that $X^\# \equiv X_1^\# F_i^\#$ where X_1 is a positive s-word. That is, the r_i-pinch we are examining sends

$$x^\alpha X^\# q_j Y x^\beta \equiv x^\alpha X_1^\# F_i^\# q_{i_1} G_i Y_1 x^\beta$$

to

$$x^{-\alpha} X_1^\# H_i^\# q_{i_2} K_i Y_1 x^{-\beta}$$

which corresponds exactly to the rewrite $X_1 F_i q_{i_1} G_i Y_1 \to X_1 H_i q_{i_2} K_i Y_1$ in the semigroup $\Gamma(T)$.

The case $\epsilon = -1$ is entirely analogous except that one uses Lemma 29 for words in $s_b x^{-1}$ because the associated subgroup relations are $s_b x = r_i s_b x^{-1} r_i^{-1}$.

It now follows inductively that the sequence of pinches is exactly parallel to the corresponding sequence of rewrites (including direction) in $\Gamma(T)$ and so in particular $X q_j Y =_{\Gamma(T)} q$ as claimed. This completes the proof of necessity in Boone's Lemma. ⊣

Given the recursive unsolvability of the halting problem for a suitable Turing machine T we have now constructed explicitly in terms of T a finitely presented semigroup $\Gamma(T)$ with unsolvable word problem (following Post) and incorporated that into an explicit construction of a finitely presented group $\mathcal{B}(T)$ with unsolvable word problem (following Boone et al.).

It is sometimes useful that we know quite a lot about the structure of $\mathcal{B}(T)$ via the combinatorial group theory of HNN extensions. For instance it is quite easy to prove the $\mathcal{B}(T)$ is torsion free. We pause here to record another property of $\mathcal{B}(T)$ observed by Collins and Miller [39] which has interesting applications for decision problems in topology.

We first observe there is a somewhat different way to view $\mathcal{B}_3$ as a tower of HNN extensions. Let $\mathcal{A}$ be the group with presentation

$$\mathcal{A} = \langle x, s_0, \ldots, s_M, q, q_0, \ldots, q_N, t \mid t^{-1}xt = x, s_b^{-1}xs_b = x^2, b = 0, \ldots, M \rangle.$$

Then $\mathcal{A}$ is an HNN-extension of $\mathcal{B}_0 = \langle x \mid \rangle$ with all the listed generators other than x as stable letters. In particular, the associated subgroups are either cyclic or trivial and hence are finitely generated free groups. We now have the following easy fact.

LEMMA 30. *In the chain*

$$\mathcal{B}_0 \leq \mathcal{A} \leq \mathcal{B}_3$$

each group is an HNN-extension of its predecessor; moreover, the associated subgroups are finitely generated free groups.

PROOF. Let $\mathcal{F}$ denote the free group on the stable letters of $\mathcal{A}$ and $\phi\colon \mathcal{A} \to \mathcal{F}$ the retraction sending stable letters to themselves and x to 1. One of the associated subgroups for r_i is generated by the $M+3$ elements $\{F_i^\# q_{i_1} G_i, t, s_0 x, \ldots, s_M x\}$. The image of this subgroup under ϕ is easily seen to be the (free) subgroup of $\mathcal{F}$ generated by $\{q_{i_1}, t, s_0, \ldots, s_M\}$ which has rank $M+3$. Hence the associated subgroup is free on the given generators. Similar considerations show the other associated subgroup for r_i is free. This completes the proof. ⊣

We have previously made use of the following easy result.

LEMMA 31. *The elements $\{x, r_i, i \in I\}$ freely generate a free subgroup of $\mathcal{B}_3$.*

PROOF. Consider $\mathcal{B}_3$ as an HNN-extension of $\mathcal{A}$ and adopt the notation in the previous proof. If w is a (non-empty) freely reduced word in x and the r_i and if $w = 1$ in $\mathcal{B}_3$ then Britton's Lemma implies w contains a subword of the form $r_i^{-e} x^n r_i^e$ and x^n belongs to one of the associated subgroups of r_i depending on the sign of e. But $x \in \ker \phi$ while $\ker \phi$ intersects the associated subgroups in the identity. Thus $n = 0$ and w is not freely reduced which is a contradiction. ⊣

The result we want to state is the following improvement of the previous lemma.

LEMMA 32 (Collins–Miller). *The elements $\{q^{-1}tq, x, r_i, i \in I\}$ freely generate a free subgroup of $\mathcal{B}_3$. Hence in the chain*

$$\mathcal{B}_0 \leq \mathcal{A} \leq \mathcal{B}_3 \leq \mathcal{B}$$

each group is an HNN-extension of its predecessor; moreover, the associated subgroups are finitely generated free groups. ⊣

The proof of this is more difficult and uses some details of our arguments above as well as the determinism of Turing machines via our earlier Corollary 10. Intuitively, just as halting computations form a tree, so do semigroup proofs of $Xq_j Y = q$ and group theory proofs of $\Sigma = LqR$. This tree-like behaviour implies a uniqueness showing the subgroup is free. The reader is referred to [39] for more details. We record one consequence of this lemma as follows.

COROLLARY 33 (Collins–Miller). *The finitely presented group* $\mathcal{B} = \mathcal{B}(T)$ *having unsolvable word problem is of cohomological dimension* 2. ⊣

Later we explain how Lemma 32 combines with other results to show one cannot recognise asphericity for finite 2-complexes.

In the 1960s, L. A. Bokut' introduced a more systematic study of towers of HNN extensions of the type found in the constructions of Boone and Novikov. He described certain normal forms in such groups, calling them *groups with standard normal form*. His methods are very useful in studying the word and conjugacy problems in such groups, and in obtaining recursively enumerable degree results. See the article by Bokut' [13] and the monograph by Bokut' and Kukin [14] for expositions of these methods and for versions of the Novikov and Boone constructions.

6.5. An explicit presentation for a group with unsolvable word problem. As for the case of semigroups, sometimes there is interest in writing down explicitly a finite presentation of a group having unsolvable word problem. Here we describe a reasonably small presentation provided by Collins [38] based on results of Tseitin and Borisov. The starting point is another small presentation of a semigroup with unsolvable word problem due to Tseitin [98] that is a variant of the one described above in Theorem 14. We call this presentation $\mathcal{C}$ and it is given by:

Generators:
$$a, b, c, d, e.$$

Relations:
$$ac = ca, \ ad = da, \ bc = cb, \ bd = db,$$
$$ce = eca, \ de = edb, \ cdca = cdcae,$$
$$caaa = aaa, \ daaa = aaa.$$

Tseitin proved that the problem of deciding if $w =_\mathcal{C} aaa$, for arbitrary w, is unsolvable. The transition from this particular word problem for a semigroup presentation to the word problem for a group presentation is based on Boone's construction described above but incorporating an elegant simplification of Boone's approach by Borisov [24].

Applied to Tseitin's presentation $\mathcal{C}$, Borisov's method yields the following presentation $\mathcal{B}$ of a group:

Generators:
$$a, b, c, d, e, p, q, r, t, k.$$

Relations:

$$p^{10}a = ap, \ p^{10}b = bp, \ p^{10}c = cp, \ p^{10}d = dp, \ p^{10}e = ep,$$
$$qa = aq^{10}, \ qb = bq^{10}, \ qc = cq^{10}, \ qd = dq^{10}, \ qe = eq^{10},$$
$$ra = at, \ rb = br, \ rc = cr, \ rd = dr, \ re = er,$$
$$pacqr = rpcaq, \ p^2adq^2r = rp^2daq^2,$$
$$p^3bcq^3r = rp^3cbq^3, \ p^4bdq^4r = rp^4dbq^4,$$
$$p^5ceq^5r = rp^5ecaq^5, \ p^6deq^6r = rp^6edbq^6, \ p^7cdcq^7r = p^7cdceq^7,$$
$$p^8caaaq^8r = rp^8aaaq^8, \ p^9daaaq^9r = rp^9aaaq^9,$$
$$p = tp, \ qt = tq,$$
$$k(aaa)^{-1}t(aaa) = (aaa)^{-1}t(aaa)k.$$

This presentation $\mathcal{B}$ has 27 relations among 10 generators which require 421 occurrences of a generator. Borisov proved that for any semigroup word $w = w(a, b, c, d, e)$

$$k(w^{-1}tw) =_{\mathcal{B}} (w^{-1}tw)k \text{ if and only if } w =_{\mathcal{C}} aaa$$

and the unsolvability of the word problem for $\mathcal{B}$ follows from Tseitin's result for the semigroup $\mathcal{C}$

§7. The isomorphism problem for groups and recognizing properties.
We now apply the existence of a finitely presented group with unsolvable word problem to show that a wide variety of group theoretic questions cannot be answered algorithmically. First we prove the Theorem of Adian [2], [4], [3] and Rabin [84] that Markov properties of groups are not recursively recognisable. In particular, the problem of deciding of a finite presentation of a group whether or not the group is trivial is recursively unsolvable. Hence, the isomorphism problem for groups is recursively unsolvable.

Using a finitely presented group with unsolvable word problem, one can produce a finitely generated subgroup H of the direct product $F \times F$ of two non-abelian free groups such that the problem of deciding membership in H is recursively unsolvable. Combining this with the Adian–Rabin Theorem, we show the problem of whether a finite subset of $F \times F$ generates the whole group is recursively unsolvable. We also show that recognising certain standard properties of elements, such as being a commutator $[x, y]$, can be unsolvable in certain finitely presented groups.

7.1. The Adian–Rabin Theorem. By analogy to Markov's results for semigroups we look more generally at abstract properties not depending on the presentation. The notion of a Markov property for groups is entirely analogous to the semigroup case.

DEFINITION 34. *An abstract property $\mathcal{P}$ of finitely presented groups is said to be a* Markov property *if there are two finitely presented groups G_+ and G_- such that*:

1. G_+ has the property $\mathcal{P}$; and
2. if G_- is embedded in a finitely presented group H then H does not have property $\mathcal{P}$.

These groups G_+ and G_- will be called the positive *and* negative witnesses *for the Markov property $\mathcal{P}$ respectively.*

It should be emphasized that if $\mathcal{P}$ is a Markov property then the negative witness does not have the property $\mathcal{P}$, nor is it embedded in any finitely presented group with property $\mathcal{P}$.

For example the property of "being finite" is a Markov property. For G_+ one can take $\langle a \mid a^2 = 1 \rangle$ which is a finite group. For G_- one can take the group $\langle b, c \mid b^{-1}cb = c^2 \rangle$ which is an infinite group and therefore not embedded in any finite group. The property "being infinite" is not a Markov property, but it is a *co-Markov property* in the sense that its complement is Markov.

An example of a property which is neither a Markov property nor a co-Markov property is the property of being perfect, that is $G/[G, G] \cong 1$.

An abstract property $\mathcal{P}$ of finitely presented groups is *hereditary* if H is finitely presented and embedded in G and $G \in \mathcal{P}$ imply that $H \in \mathcal{P}$, that is, the property $\mathcal{P}$ is inherited by finitely presented subgroups. A property of finitely presented groups P is *non-trivial* if it is neither the empty property nor is it enjoyed by all finitely presented groups. Suppose $\mathcal{P}$ is a non-trivial, hereditary property of finitely presented groups. Then, since $\mathcal{P}$ is non-trivial, there are groups $G_+ \in \mathcal{P}$ and $G_- \notin \mathcal{P}$. But if G_- is embedded in a finitely presented group H, then $H \notin \mathcal{P}$ because $\mathcal{P}$ is hereditary. Thus $\mathcal{P}$ is a Markov property with witnesses G_+ and G_-. This shows the following:

LEMMA 35. *If $\mathcal{P}$ is a non-trivial hereditary property of finitely presented groups, then $\mathcal{P}$ is a Markov property.*

As discussed below, Graham Higman has constructed a universal finitely presented group, say U. If $\mathcal{P}$ is a Markov property with positive and negative witnesses G_+ and G_-, then G_- is embedded in U so $U \notin \mathcal{P}$. Moreover, if U is embedded in a finitely presented group H then so is G_- and hence $H \notin \mathcal{P}$. Thus $\mathcal{P}$ is a Markov property with positive and negative witnesses G_+ and U. Hence U is a negative witness for every Markov property.

An abstract property $\mathcal{P}$ of finitely presented groups is *recursively recognizable* if there is an recursive method which when applied to an arbitrary finite presentation π determines whether or not $gp(\pi)$ has the property $\mathcal{P}$. That is, if $\{\pi \mid gp(\pi) \in \mathcal{P}\}$ is a recursive set of finite presentations (where all presentations are assumed to be on some fixed alphabet).

The main unsolvability result concerning the recognition of properties of finitely presented groups is the following:

THEOREM 36 (Adian–Rabin). *If $\mathcal{P}$ is a Markov property of finitely presented groups, then $\mathcal{P}$ is not recursively recognizable.*

Before indicating a proof of this result, we note the following easy corollaries:

COROLLARY 37. *The following properties of finitely presented groups are not recursively recognizable*:

1. *being the trivial group*;
2. *being finite*;
3. *being abelian*;
4. *being nilpotent*;
5. *being solvable*;
6. *being free*;
7. *being torsion-free*;
8. *being residually finite*;
9. *having a solvable word problem*;
10. *being simple*; *and*
11. *being automatic.*

Each of (1) through (9) is a non-trivial, hereditary property and hence is a Markov property. For (10), it is known that finitely presented, simple groups have solvable word problem and hence a group with unsolvable word problem is a negative witness. Similarly for (11), automatic groups have solvable word problem and so being automatic is a Markov property.

COROLLARY 38. *The isomorphism problem for finitely presented groups is recursively unsolvable.*

For by (1) in the previous corollary there is no algorithm to determine of an arbitrary presentation π whether or not $gp(\pi) \cong 1$.

PROOF OF THE ADIAN–RABIN THEOREM. We are going to give a simple proof of the Adian–Rabin Theorem which is our modification [72] of one given by Gordon. The construction is quite straightforward and variations on the details can be applied to obtain further results. So suppose that $\mathcal{P}$ is a Markov property and that G_+ and G_- are witnesses for $\mathcal{P}$. We also have available a finitely presented group U having unsolvable word problem.

Using these three items of initial data, we construct a recursive family of finite presentations $\{\pi_w \mid w \in U\}$ indexed by the words of U so that if $w =_U 1$ then $gp(\pi_w) \cong G_+$ while if $w \neq_U 1$ then G_- is embedded in U. Thus $gp(\pi_w) \in \mathcal{P}$ if and only if $w =_U 1$. Since U has unsolvable word problem, it follows that $\mathcal{P}$ is not recursively recognizable.

The family $\{\pi_w \mid w \in U\}$ is rather like a collection of buildings constructed from playing cards standing on edge. Such a building can be rather unstable so that if an essential card is removed (corresponding to $w =_U 1$) then the entire structure will collapse. The technical result we need provides such a construction and is somewhat independent of the present context.

LEMMA 39 (Main Technical Lemma). *Let K be a group given by a presentation on a finite or countably infinite set of generators, say*

$$K = \langle x_1, x_2, \cdots \mid r_1 = 1, r_2 = 1, \ldots \rangle.$$

For any word w in the given generators of K, let L_w be the group with presentation obtained from the given one for K by adding three new generators a, b, c together with defining relations

$$a^{-1}ba = c^{-1}b^{-1}cbc, \qquad (1)$$
$$a^{-2}b^{-1}aba^2 = c^{-2}b^{-1}cbc^2, \qquad (2)$$
$$a^{-3}[w,b]a^3 = c^{-3}bc^3, \qquad (3)$$
$$a^{-(3+i)}x_i ba^{(3+i)} = c^{-(3+i)}bc^{(3+i)}, \quad i = 1, 2, \ldots \qquad (4)$$

where $[w, b]$ is the commutator of w and b. Then:
1. *if $w \neq_K 1$ then K is embedded in L_w by the inclusion map on generators;*
2. *the normal closure of w in L_w is all of L_w; in particular, if $w =_K 1$ then $L_w \cong 1$, the trivial group; and*
3. *L_w is generated by the two elements b and ca^{-1}.*

If the given presentation of K is finite, then the specified presentation of L_w is also finite.

PROOF. Suppose first that $w \neq_K 1$. In the free group $\langle b, c \mid \rangle$ on generators b and c consider the subgroup C generated by b together with the right hand sides of the equations (1) through (4). It is easy to check that the indicated elements are a set of free generators for C since in forming the product of two powers of these elements or their inverses some of the conjugating symbols will remain uncancelled and the middle portions will be unaffected.

Similarly, in the ordinary free product $K * \langle a, b \mid \rangle$ of K with the free group on generators a and b consider the subgroup A generated by b together with the left hand sides of the equations (1) through (4). Using the assumption that $w \neq_K 1$ it is again easy to check that the indicated elements are a set of free generators for A.

Thus assuming $w \neq_K 1$, the indicated presentation for L_w together with the equation identifying the symbol b in each the two factors is the natural presentation for the free product with amalgamation

$$(K * \langle a, b \mid \rangle) * \langle b, c \mid \rangle.$$
$$A = C.$$

So if $w \neq_K 1$, then K is embedded in L_w establishing the first claim.

Now let N_w denote the normal closure of w in L_w. Clearly $[w, b] \in N_w$ so by equation (3), $b \in N_w$. But equations (1) and (2) ensure that a, b, c are all conjugate and so a, b, c all belong to N_w. Finally, since each of the system of equations (4) can be solved to express x_i in terms of a, b, c, it follows that $x_i \in N_w$ for $i = 1, 2, \ldots$. Thus each of the generators of L_w belongs to N_w and so $L_w = N_w$. This verifies the second assertion.

Finally, let M be the subgroup of L_w generated by b and ca^{-1}. Equation (1) can be rewritten as $b(ca^{-1})b(ca^{-1})^{-1}b^{-1} = c$ so that $c \in M$. But then from $ca^{-1} \in M$ it follows that $a \in M$. Finally from the system of equations (4) which can be solved for the x_i in terms of a, b, c it follows that $x_i \in M$ for $i = 1, 2, \ldots$ and so $M = L_w$. (For later use we note that neither equation (2) nor equation (3) was used in the proof of the final assertion). This completes the proof of the lemma. ⊣

Using this technical lemma it is easy to complete the proof of the Adian–Rabin Theorem. We are given the three finitely presented groups U, G_+ and G_- which can be assumed presented on disjoint alphabets as follows:

$$U = \langle y_1, \ldots, y_k \mid r_1 = 1, \ldots, r_\rho = 1 \rangle,$$
$$G_- = \langle s_1, \ldots, s_m \mid u_1 = 1, \ldots, u_\sigma = 1 \rangle, \text{ and}$$
$$G_+ = \langle t_1, \ldots, t_n \mid v_1 = 1, \ldots, v_\tau = 1 \rangle.$$

Let $K = U * G_-$ the ordinary free product of U and G_- presented as the union of the presentations of its factors. Since U has unsolvable word problem, K also has unsolvable word problem. Also both U and G_- are embedded in K by the inclusion map on generators. For any word w in the generators of U (these are also generators of K) form the presentation L_w as in the Main Technical Lemma. Finally we form the ordinary free product $L_w * G_+$.

A presentation π_w for these groups $L_w * G_+$ can be obtained by simply writing down all of the above generators together with all of the above defining equations. Such a presentation is defined for any word w in U whether or not $w \neq_U 1$. But it follows from the lemma that if $w \neq_U 1$ then the group G_- is embedded in $gp(\pi_w) = L_w * G_+$ and so $gp(\pi_w) \notin \mathcal{P}$ by the definition of a Markov property. On the other hand, if $w =_U 1$ then by the lemma $L_w \cong 1$ and so $gp(\pi_w) \cong G_+$ and hence $gp(\pi_w) \in \mathcal{P}$.

Thus we have shown that the recursive collection of presentations

$$\{\pi_w \mid w \text{ a word in } U\}$$

has the property that $gp(\pi_w) \in \mathcal{P}$ if and only if $w =_U 1$. Since U has unsolvable word problem, it follows that $\mathcal{P}$ is not recursively recognizable. This completes the proof of the Adian–Rabin Theorem. ⊣

We remark that among other consequences of the above technical lemma (Lemma 39) is the theorem of Higman, Neumann and Neumann [50] that any countable group K can be embedded in a two generator group.

7.2. Membership and generation problems.
Suppose G is a finitely presented group and H is a finitely generated subgroup given by a finite set of generating words. The *membership problem* or *generalised word problem* for H in G is to determine of an arbitrary word w in the generators of G whether or not w represents an element of H, that is, whether or not w is equal to some word in the given generators of the subgroup H. Notice that the word problem is just the membership problem for the trivial subgroup generated by $1 \in G$.

Using a group with unsolvable word problem, Mikhailova [69] the membership problem is unsolvable in what at first seem rather elementary groups.

THEOREM 40 (Mikhailova). *Let F be a non-abelian free group of finite rank. Then the direct product $F \times F$ has a finitely generated subgroup H for which membership is unsolvable.*

PROOF. Let $F = \langle a_1, \ldots, a_n \mid \; \rangle$ be a non-abelian free group and let $U = F = \langle a_1, \ldots, a_n \mid r_1 = 1, \ldots, r_m = 1 \rangle$ be a finite presentation of a group with unsolvable word problem (on the same set of generating symbols). We let $\phi \colon F \to U$ denote the quotient homomorphism from F onto U. Now form the direct product of two copies of $F \times F$ for which one can easily write down a finite presentation.

Let $H \subseteq F \times F$ be the pullback or fibre product of two copies of ϕ. Then

$$H_U = \{(x, y) \in F \times F \mid \phi(x) = \phi(y)\} = \{(x, y) \in F \times F \mid x =_U y\}.$$

It is a fairly straightforward exercise to see that a set of generators for H_U is

$$S = \{(a_1, a_1), \ldots, (a_n, a_n), (1, r_1), \ldots, (1, r_m)\}.$$

These pairs clearly belong to H. If K is the subgroup they generate then the intersection of the normal closure in K of the $(1, r_i)$ with the second factor is just $\ker \phi$. And since the diagonal belongs to K the intersection with the first factor is again $\ker \phi$. So $K = H_U$. Now $(x, y) \in H_U$ if and only if $x =_U y$ and so the membership problem for H_U is unsolvable. ⊣

In [70] (see also [72]) the author combined this with the Adian–Rabin construction to show the following result on generation of groups.

THEOREM 41 (Miller). *Let F be a non-abelian free group of finite rank. Then there is no algorithm to determine of an arbitrary finite set of elements in $F \times F$ whether or not they generate $F \times F$.*

PROOF. Let $F = \langle a_1, \ldots, a_n \mid \; \rangle$ be a non-abelian free group. Suppose the finitely presented group U has unsolvable word problem. Let L_w for $w \in U$ be as in the proof if Adian–Rabin for the Markov property "being trivial". So $L_w \cong 1$ if and only if $w =_U 1$. Now each L_w can be written as a presentation on two generators a_1, a_2 and we add more generators $a_3 = 1, \ldots, a_n = 1$ if $n > 2$ so that $\phi_w \colon F \to L_w$ is a surjective homomorphism As in the previous proof, let S_w be the specified finite generating set for the pullback H_{L_w} of two copies of ϕ_w. Then S_w generates $F \times F$ if and only if $H_{L_w} = F \times F$ if and only if $w =_U 1$. But this last is a recursively unsolvable problem. ⊣

We remark that in case $w \neq_U 1$ the subgroup H_{L_w} is finitely generated but not finitely presented [47], [7] and has an unsolvable conjugacy problem [70]. In particular it cannot be isomorphic to $F \times F$ even as an abstract group. From this we conclude the following:

COROLLARY 42. *There is no algorithm to determine of two finite sets of elements in $F \times F$ whether or not the subgroups they generate are isomorphic.*

Recall that a free group F has a faithful representation as a group of two-by-two integer matrices. So $F \times F$ is a subgroup of $SL(4, \mathbb{Z})$. Hence these results carry over to analogous statements for certain matrix groups and subgroups generated by finite set of matrices.

7.3. Some other decision problems about elements. We next briefly consider a few other local decision problems concerning elements in a group. The observations here are from the paper [6] by Baumslag, Boone and Neumann.

The structure of finitely generated abelian groups can be completely determined from a finite presentation of such a group, and in particular one can solve the word problem for such groups. Consequently, if G is an arbitrary finitely presented group one can effectively determine the structure of its abelianization $G/[G, G]$. So for instance, there is an algorithm to decide whether G is perfect, that is $G = [G, G]$. Moreover, since one can solve the word problem for $G/[G, G]$ it follows that one can decide of a arbitrary word w of G whether or not $w \in [G, G]$.

However, it seems that almost any property of elements of a finitely presented group which is not determined by the abelianization $G/[G, G]$ will be recursively unrecognizable. The following result shows a few common properties of elements are not recognizable.

THEOREM 43 (Baumslag, Boone and Neumann). *There is a finitely presented group G such that there is no algorithm to determine whether or not a word in the given generators represents*:

1. *an element of the center of G;*
2. *an element which commutes with a given element of G;*
3. *an n-th power, where $n > 1$ is a fixed integer;*
4. *an element whose class of conjugates is finite;*
5. *a commutator; and*
6. *an element of finite order > 1.*

PROOF. Fix a finitely presented group U having unsolvable word problem. Define G to be the ordinary free product of U with a cyclic group of order 3 and an infinite cyclic group, that is,

$$G = U * \langle s \mid \rangle * \langle t \mid t^3 = 1 \rangle.$$

We use the commutator notation $[x, y] = x^{-1}y^{-1}xy$. In the following, w is a variable for an arbitrary word in the generators of U.

The center of G is trivial so w lies in the center of G if and only if $w =_U 1$. So there is no algorithm to determine whether an arbitrary word of G lies in the center. This gives the first assertion. Similarly, w commutes with s if and only if $w =_U 1$ which establishes the second assertion. The element $s^n[t, w]$ is an n-th power if and only if $w =_U 1$ establishing the third assertion. The conjugacy class of w is finite if and only if $w =_U 1$ since if $w \neq_U 1$ the conjugates $s^{-i}ws^i$ would all be distinct. This gives the fourth assertion. For the fifth assertion, note that $[s, t]w$ is a commutator if and only if $w =_U 1$. Finally for the sixth assertion, observe that tw has infinite order if and only if $w \neq_U 1$, while if $w =_U 1$ then tw has order 3. This completes the proof. ⊣

§8. The Higman Embedding Theorem.

Recall that a finitely generated, recursively presented group is one defined by a presentation $H = \langle a_1, \ldots, a_n \mid R \rangle$ where R is a recursively enumerable set of defining relations. In contrast to the difficulties encountered for finitely presented groups, it is easy to give examples of finitely generated, recursively presented groups with unsolvable word problem. For example, let $S \subset \mathbb{N}$ be a recursively enumerable set of natural numbers which is not recursive. Define the recursively presented group

$$H_S = \langle a, b, t \mid t^{-1}a^{-i}ba^i t = a^{-i}ba^i \; \forall i \in S \rangle.$$

Now H_S can be described as the HNN extension of the free group $\langle a, b \mid \rangle$ with associated subgroup (freely) generated by the $a^{-i}ba^i, i \in S$. It follows from Britton's lemma that the commutator $[t, a^{-i}ba^i] =_{H_S} 1$ if and only if $i \in S$. Thus the word problem for H_S is recursively unsolvable.

Graham Higman observed that if one could embed this group H_S in a suitable finitely presented group, say G_S then, by Proposition 1, the group G_S would also have an unsolvable word problem. So finding such and embedding would give an alternative proof of the Novikov–Boone Theorem. In [49] Higman succeeded in showing such an embedding exists. Moreover, Higman gave a complete characterisation of the finitely generated subgroups of finitely presented groups—namely they are the recursively presented groups.

THEOREM 44 (Higman Embedding Theorem). *A finitely generated group H can be embedded in a finitely presented group if and only if H is recursively presented.*

This is an important and, at the time, totally unanticipated result connecting group theory and recursive function theory.

That finitely generated subgroups of finitely presented groups are recursively presented is contained in our Proposition 1 above. The difficult part of this theorem is to show that a recursively presented group can be embedded in a finitely presented group.

As already pointed out, the Novikov–Boone Theorem is an easy corollary of Higman's Embedding Theorem. Another consequence is the existence of universal finitely presented groups.

COROLLARY 45 (Higman). *There exists a universal finitely presented group; that is, there exists a finitely presented group G which contains an isomorphic copy of every finitely presented group.*

To prove this one systematically enumerates all finite presentations on a fixed countable alphabet. Using the theorem of Higman, Neumann and Neumann, the free product of all of these can be embedded in a two generator group which will be recursively presented. This group can then be embedded in a finitely presented group which is the desired universal group.

Higman's original proof of the Embedding Theorem has three stages. First he recasts the notion of an recursively enumerable set in more algebraic terms. He then introduces the following definition:

DEFINITION 46. *A subgroup H of a finitely generated group G is called* benign *in G if the group $G \underset{H}{\star} = \langle G, t \mid t^{-1}ht = h, h \in H \rangle$ is embeddable in a finitely presented group.*

Of course $G \underset{H}{\star}$ is the HNN extension in which the stable letter t just commutes with the subgroup. Notice that if $H \subseteq G \subseteq K$ and if H is benign in K, then H is also benign in G.

Observe that *if H is benign in G, then the set of words in the generators of G and their inverses that represent elements of H is recursively enumerable.* For if H is benign in G then $G \underset{H}{\star}$ is embeddable in a finitely presented group and so we can enumerate the set of words u in the generators of G such that $[t, u] = 1$ which are just the elements of H. The main task in Higman's proof is showing the following converse for the case of free groups.

LEMMA 47 (Higman). *A subgroup of a finitely generated free group is benign if and only if it is recursively enumerable.*

The hard work then is to show that if H is a recursively enumerable subgroup of a free group F, then the HNN extension $F \underset{H}{\star}$ can be embedded in a finitely presented group. As previously noted this already implies the unsolvability of the word problem (Novikov–Boone Theorem) which was Higman's starting point.

But having proved this lemma, he then gives a very brief but beautiful argument—dubbed the "Higman Rope Trick" in the monograph [56] by Lyndon and Schupp—which completes the full embedding result. I once asked Higman why he had placed this at the end of the paper since it seemed to explain why he did certain earlier steps. He responded that the paper was in the order he found the results, so he had proved the unsolvability of the word problem before he realised this beautiful additional argument would prove the embedding and characterisation results. We include this argument here.

LEMMA 48 (The Higman Rope Trick). *If R is a benign normal subgroup of the finitely generated free group F, then F/R is embeddable in a finitely presented group.*

PROOF. By hypothesis the group $F \underset{R}{\star} = \langle F, t \mid t^{-1}rt = r, r \in R \rangle$ is embeddable in a finitely presented group H. Suppose the given generators of F are $x_1, \ldots, x_n$. Applying Tietze transformations if necessary we can assume the x_i are among the generating symbols of the given finite presentation of H. Let $\overline{F}$ be an isomorphic copy of F with generators $\overline{x}_1, \ldots, \overline{x}_n$. If w is a word in the x_i, we denote by $\overline{w}$ the corresponding word in the $\overline{x}_i$.

In $F \underset{R}{\star}$ the subgroup L generated by F and $t^{-1}Ft$ is isomorphic to their amalgamated free product with the subgroup $R = t^{-1}Rt$ amalgamated. Define a map $\phi \colon L \to \overline{F}/\overline{R}$ by $\phi(w) = \overline{w}$ and $\phi(t^{-1}wt) = 1$. Since the definitions of ϕ agree on the amalgamated subgroup, ϕ defines a homomorphism.

Consider the direct product $H \times \overline{F}/\overline{R}$ for which we will use ordered pair notation. Viewing L as a subgroup of H, the we can define a map $\psi \colon L \to L \times \overline{F}/\overline{R}$ by $\psi(z) = (z, \phi(z))$ for $z \in L$. Clearly ψ is an injective homomorphism. Hence we can form the HNN extension

$$K = \langle H \times \overline{F}/\overline{R}, s \mid s^{-1}(z, 1)s = (z, \phi(z)) \text{ for } z \in L \rangle.$$

This is the required finitely presented group containing $\overline{F}/\overline{R}$. It is visibly finitely generated. As defining relations we may take the relations of H, the relations of $\overline{F}/\overline{R}$ (that is, $\overline{r} = 1$ for $\overline{r} \in \overline{R}$), the relations which say the generators of H commute with the generators of $\overline{F}/\overline{R}$, and the relations $s^{-1}(z, 1)s = (z, \phi(z))$ for z in a set of generators of L. Each of these collections of relations is finite except for the relations of $\overline{F}/\overline{R}$, so that it is sufficient to show these are redundant. For any word w on the generators of F, by the relations for conjugation by s and commutativity of H and $\overline{F}/\overline{R}$ we have

$$s^{-1}(w, 1)s = (w, \overline{w}) \text{ and } s^{-1}(t^{-1}wt, 1)s = (t^{-1}wt, 1).$$

But if $r \in R$, the relations of H imply that $r = t^{-1}rt$ and so applying the previous equations we have

$$(r, \overline{r}) = s^{-1}(r, 1)s = s^{-1}(t^{-1}rt, 1)s = (t^{-1}rt, 1) = (r, 1).$$

Hence $\overline{r} = 1$. That is the relations of $\overline{F}/\overline{R}$ are consequences of the remaining relations and so are redundant, as claimed. This completes the proof. ⊣

In 1970 Aanderaa [1] found another proof of Higman's Embedding Theorem based upon using suitable HNN extensions of Boone's groups and their built-in Turing machines. Aanderaa's proof can also be found in Rotman's textbook [87]. Other proofs are to be found in Lyndon–Schupp [56] and the appendix to Shoenfield's logic text [90].

§9. Decision problems in geometric topology.

In 1958 Markov [66], [64] announced the unsolvability of the homeomorphism problem for n-manifolds for $n \geq 4$ and provided a brief sketch.

In the classic textbook by Seifert and Threlfall [89] there is a construction from an arbitrary finite presentation $G = gp\langle X \mid R \rangle$ of a n-manifold M^n for $n \geq 4$ with $\pi_1(M^n) \cong G$ roughly as follows. First construct in the manner we described before a finite 2-complex Y using the presentation with $\langle X \mid R \rangle$ with a single 0-cell, 1-cells corresponding to the generators and 2-cells corresponding to the relators so that fundamental group of Y is G. Then Y can be embedded in Euclidean space $\mathbb{R}^{n+1}$ and we take as $M^n(X; R)$ the boundary of a suitable neighbourhood of the embedded image of Y. Then $M^n(X; R)$ is a closed (i.e., compact, no boundary) n-manifold and one can check that the fundamental group of $M^n(X; R)$ is isomorphic to G.

By the proof of the Adian–Rabin Theorem given above we know there is a recursive collection of finite presentations $\langle X_i \mid R_i \rangle$ all having the same set of generators and the same number of defining relations—indeed all except one defining relation are the same for all the presentations since they have the form

$$P(w) = \langle x_1, \ldots, x_n \mid r_1(w) = 1, r_2 = 1, \ldots, r_m = 1 \rangle$$

where w is a word which is the only parameter. (What is important here is that the number of generators and the number of relations are constant.) Recall that there is an auxiliary group U having unsolvable word problem and w is also word of U and $gp(P_w) \cong 1$ if and only if $w =_U 1$. So the triviality problem for $P(w)$ is unsolvable. So "being simply connected", that is, having trivial fundamental group, is not recursively recognisable for 2-complexes or n-manifolds.

THEOREM 49. *The problem of deciding whether or not a finite 2-complex is simply connected is recursively unsolvable. Similarly the problem of deciding whether or not a closed n-manifold $n \geq 4$ is simply connected is recursively unsolvable.*

Markov modifies the above construction by adding a fixed number t of copies of the trivial relation $1 = 1$ to each of the presentations so the presentations become

$$P_{n,m,t}(w) = \langle x_1, \ldots, x_n \mid r_1(w) = 1, r_2 = 1, \ldots, r_m = 1, 1 = 1 \ (t \text{ times}) \rangle.$$

The point adding these trivial relations is that if $P_{u,t}$ and $P_{w,t}$ are isomorphic then there is a sequence of "restricted" Tietze transformations on the presentations which never changes the quantity $m + t - n$. Here the Tietze transformations are restricted to certain operations which (1) operate on relations by multiplying one by another, taking inverses, taking cyclic permutations and inserting or deleting inverse pairs but do not change the number of relations; and (2) add a new generator and a new relator setting the new generator equal to a word in the existing generators, or the inverse of this operation. The copies of the trivial relation provide room for the necessary adding of consequences or relations to transform $P_{n,m,t}(w)$ to another presentation $P'_{n',m',t'}(w')$ of the same group provided $m' + t' - n' = m + t - n$ and $t \geq m + n'$ and $t' \geq m' + n$ (see [21] for more details).

Then Markov shows that for the manifolds $M^n(P_{n,m,t}(w))$ constructed from the presentations $P_{n,m,t}(w)$, if the groups are isomorphic, then one manifold can be obtained from the other by a corresponding sequence of homeomorphisms involving sliding of handles. Of course if the fundamental groups are not isomorphic, the manifolds could not be homeomorphic. In light of the Adian–Rabin Theorem this proves the following:

THEOREM 50 (Markov). *The homeomorphism problem for closed n-manifolds with $n \geq 4$ is recursively unsolvable.*

The thoughtful reader may be wondering whether this makes sense since there are certainly continuously many closed n-manifolds and in what manner are the ones we are interested in described. Also what is the situation for diffeomorphism as opposed to homeomorphism? Markov's two announcements [66], [64] on the result provide a reasonable outline of his proofs. A third announcement [65] describes the analogous result for combinatorial equivalence.

In a 1968 paper [21], Boone, Haken and Poénaru dealt with these questions in considerable detail. They clarified what might be meant by presentation of a manifold and a differentiable structure. They then showed that Markov's result holds for several different kinds of equivalence. We state their result as follows.

THEOREM 51 (Markov–Boone–Haken–Poénaru). *For each dimension $n \geq 4$, there is a recursive class $\mathcal{C}$ of finite presentations of n-manifolds, endowed with a differentiable and a compatible combinatorial structure so that the problem of deciding for an arbitrary pair $M_1, M_2 \in \mathcal{C}$ whether or not $M_1 \sim M_2$ is recursively unsolvable where $\sim$ is any of diffeomorphic, homeomorphic, combinatorially equivalent, or homotopy equivalent.*

Their paper [21] is mainly concerned with recursively enumerable degree analogs of Markov's unsolvability results, but it also provides details and foundational material on these questions.

One question left unanswered by Markov's Theorem and construction is whether one can determine whether or not a closed n-manifold is homeomorphic to the n-sphere S^n. The manifolds given by the construction outlined above are never spheres since they have the wrong homology groups because the number of relations is much larger than the number of generators. Using universal central extensions and additional topological techniques S. P. Novikov (who is the son of P. S. Novikov) in 1962 proved the sphere recognition problem is unsolvable.

THEOREM 52 (S. P. Novikov). *The problem of determining whether or not a closed n-manifold with $n \geq 5$ is homeomorphic to the n-sphere S^n is recursively unsolvable.*

A more recent exposition of the topological unsolvability results of Markov and S. P. Novikov can be found in the paper by Chernavsky and Leksine [31].

Recall that a finite complex Y is aspherical if its higher dimensional homotopy groups vanish, that is, $\pi_n(Y) = 0$ for $n \geq 2$, or equivalently if its universal covering space $\tilde{Y}$ is contractible. As we mentioned earlier, by combining the stronger properties of Boone's groups and the above version of the Adian–Rabin construction together with results of Chiswell, Collins and Huebschmann [32], one can show [39] that asphericity of 2-complexes is not recognisable.

THEOREM 53 (Collins–Miller). *There is a recursive class of finite 2-complexes $\mathcal{C}$ such that the problem of determining whether or not a complex $Y \in \mathcal{C}$ is aspherical is recursively unsolvable.*

This result follows from the constructions we have described and the fact that asphericity of presentations is preserved under the amalgamated free product construction and HNN extensions provided the associated subgroups are finitely generated free groups.

§10. **Some connections with recursive function theory.** As the Higman Embedding Theorem suggests, there are numerous connections between the word problem for groups and the theory of recursively enumerable sets. In his classic 1944 paper [81] Post introduced several notions of reducibility and in particular *Turing reducibility*: if A and B are two sets of objects, then $A \leq_T B$ means that an (hypothetical) algorithm to answer questions about membership in B would yield and an algorithm to answer questions about membership in A. This can be made precise for subsets of the natural numbers $\mathbb{N}$ in the following manner. The recursive functions can be defined by declaring that a certain collection of basic functions (addition, multiplication, projection) are recursive and then closing under the usual operations of composition, mineralization and recursion. A function is said to be *B-recursive* if it is among the functions obtained from the base functions together with the characteristic function of B by closing under the usual operations. Then $A \leq_T B$ or A is Turing reducible to B is defined to mean the the characteristic function of A is B-recursive. Of course if B is already recursive (that is, membership in B is decidable) and $A \leq_T B$, then A is also recursive.

Two sets (of natural numbers) are *Turing equivalent* $A \equiv_T B$ if both $A \leq_T B$ and $B \leq_T A$. One can check that $\equiv_T$ is an equivalence relation on the natural numbers $\mathbb{N}$ and hence it defines a partition of the power set of $\mathbb{N}$ into disjoint collections of subsets called (*Turing*) *degrees of unsolvability*. Those degrees of unsolvability which contain a recursively enumerable set are called *r.e. degrees of unsolvability* and are of particular interest. The r.e. degrees are partially ordered by $\leq_T$. There is a smallest r.e. degree denoted by **0** which consists of the recursive sets and a largest r.e. degree denoted **0′** which is the degree of the general halting problem for all Turing machines. In his 1944 paper [81] Post asked asked whether there are any more r.e. degrees, which is known as *Post's problem*. In the late 1950s this problem was solved in the affirmative independently by Friedberg and Muchnik (see the textbooks [86] or [91]) who introduced the priority method to show there is a pair of mutually incomparable r.e. degrees strictly between **0** and **0′**. It is now known the r.e. degrees have a rich structure; for example, they are dense with respect to the partial order $\leq_T$.

These concepts can, as usual, be carried over to the realm of finitely presented groups by using a suitable coding device or Gödel numbering. From our previous discussion we know that if $\mathcal{P}$ is a finite presentation of a group, then the set of words on the generators which are equal to 1 is recursively enumerable. Quite generally one can show that if $\mathcal{P}_1$ and $\mathcal{P}_2$ are two presentations on finite generating sets for the same group G, then set of words equal to 1 in $\mathcal{P}_1$ is Turing equivalent to the set of words equal to 1 in $\mathcal{P}_2$. So if we use the notation $WP(G)$ for the word problem for G, that is, the algorithmic problem of deciding which words are equal to 1, then the degree of unsolvability of the word problem for G does not depend on the presentation. Similar considerations apply to the degree of the conjugacy problem for G, thought of as a set of pairs of conjugate elements in G, which we denote by $CP(G)$.

We observe that if G is finitely presented then both $WP(G)$ and $CP(G)$ are r.e. degrees. From our previous discussion we know $WP(G) \leq_T CP(G)$. Moreover, if H is a finitely generated subgroup of G, then $WP(H) \leq_T WP(G)$.

The solution of Post's problem occurred around the time that Novikov and Boone proved there are finitely presented groups with unsolvable word problem. So it was natural to ask whether the word problem for groups could have any of the many intermediate r.e. degrees between **0** and **0′**. Not so surprisingly the answer is affirmative, but there are considerable technical difficulties to overcome since one needs to understand the word problem in the groups in question rather thoroughly. There were a number of independent proofs of the result, all at about the same time in the mid 1960s.

THEOREM 54 (Bokut' [9, 10], Boone [20], Clapham [33], Fridman [44, 45]). *Let* **D** *be an r.e. degree of unsolvability. Then there is a finitely presented group G with word problem of degree* **D**. *In more detail, there is an explicit construction which, when applied to a Turing machine T with halting problem of degree* **D**, *yields a finitely presented group G(T) such that the WP(G(T)) is Turing equivalent to the halting problem for T*.

The proofs of Bokut', Boone and Fridman are all based on variants of the Post–Boone construction that we have discussed above. Clapham's proof is based on the proof of Higman's Embedding Theorem, and in a subsequent paper he shows the embedding can be done in a "degree preserving" way.

THEOREM 55 (Clapham [34]). *If H is a finitely generated, recursively presented group, then H can be embedded in a finitely presented group G such that $WP(H) \equiv_T WP(G)$.*

Another natural question is whether the word and conjugacy problems must have the same degree, or more generally what r.e. degrees of unsolvability can they be? As we mentioned earlier, in [75] Novikov gave a construction of a finitely presented group with an unsolvable conjugacy problem which was different from and considerably simpler than his construction [76] for the word problem. In 1960 Fridman [43] showed that these groups of Novikov with unsolvable conjugacy problem have solvable word problem. So indeed the two problems can be different. Then in the mid 1960s Bokut' and Collins independently showed there are finitely presented groups with solvable word problem but conjugacy problem of arbitrary degree.

THEOREM 56 (Bokut' [11, 12], Collins [35], Fridman [43]). *Let $\mathbf{D}$ be an r.e. degree of unsolvability. Then there is a finitely presented group G with solvable word problem but conjugacy problem $\mathbf{D}$.*

Finally Collins showed that, in a suitable variant, the Post–Boone–Britton construction gives groups whose word and conjugacy problem have the same degree. So taking a free product with the groups of the previous theorem, one gets a complete picture.

THEOREM 57 (Collins [35]). *Let $\mathbf{D}_1$ and $\mathbf{D}_2$ be two r.e. degrees of unsolvability and assume $\mathbf{D}_1 \leq_T \mathbf{D}_2$. Then there is a finitely presented group G with word problem of degree $\mathbf{D}_1$ and conjugacy problem of degree $\mathbf{D}_2$.*

The explicit nature of the construction of groups with word problem equivalent to the halting problem for a Turing machine in Theorem 54 has useful consequences. Since "having a solvable word problem" is a hereditary property, we know from the Adian–Rabin Theorem that it is not recursively recognisable. As we indicated earlier, there are many interesting classes of groups for which the word problem can be solved. So one might hope there is some sort of unified approach to solving the word problem. But Boone and Rogers [23] combined a bit of recursive function theory with the explicit nature of Theorem 54 to observe the following.

THEOREM 58 (Boone–Rogers). *The set of all (Gödel numbers of) finitely presented groups with solvable word problem is Σ_3^0-complete in the arithmetic hierarchy.*

As a consequence of this one can easily show the following statements which (negatively) answer questions raised by Whitehead, Church and Higman respectively.

COROLLARY 59.   1. *There is no recursive enumeration of all finite presentations of groups having solvable word problem.*
2. *There is no single partial algorithm which solves the word problem for all finite presentations of groups having solvable word problem.*
3. *There is no universal solvable word problem group, that is, there is no group with solvable word problem which contains an isomorphic copy of every finitely presented group with solvable word problem.*

Despite all these negative results about the word problem, Boone and Higman managed to give an algebraic characterisation of groups with solvable word problem. We have previously mentioned that finitely presented simple groups have solvable word problem. In fact somewhat more is true as we now sketch. Suppose that we have three groups $H \subseteq S \subseteq G$ where G is finitely presented, S is simple and H is finitely generated. Notice there are no finiteness or recursion assumptions about S. We claim that H has a solvable word problem. Fix a non-trivial element $1 \neq s_o \in S$. For any word w in the generators of H we let G_w be the presentation obtained from the given presentation of G by adding the relation $w = 1$. We now enumerate the list L of words equal to 1 in G and the list L_w of all words equal to one in G_w. Now if $w =_H 1$, then w will appear in L. But if $w \neq_H 1$, then $s_0 \in L_w$ since $1 \neq w \in S$ and S is simple. Exactly one of these must eventually occur enabling us to decide whether or not $w =_H 1$. This proves the claim which is half of the following Boone–Higman result.

THEOREM 60 (Boone–Higman [22]). *A finitely generated group has a solvable word problem if and only if it can be embedded in a simple subgroup of a finitely presented group.*

Sometimes technical results and methods from recursive function theory find application to decision problems about groups. As one example, the author obtained the following which has a number of applications, for instance to the study of algebraically closed groups.

THEOREM 61 (Miller [71, 72]). *There is a finitely presented group G with unsolvable word problem such that every non-trivial quotient group of G has unsolvable word problem.*

The proof uses a pair of disjoint recursively enumerable sets A and B which are recursively inseparable (from recursive function theory). The construction applies the Higman Embedding Theorem together with techniques from the proof of the Adian–Rabin Theorem.

A remarkable instance of unsolvability has recently been obtained by Myasnikov and Osin who were partly motivated by questions about the existence of groups with "generically hard" word problem. They define a finitely generated group G to be *algorithmically finite* if there is no algorithm which enumerates an infinite set of pairwise distinct elements of G. An algorithmically finite group G which is infinite must have an unsolvable word problem (though it may not even be recursively presented).

Recall that a *section* of a group G is a quotient group of a subgroup of G. It is easy to see that a finitely generated section of an algorithmically finite group is either finite or has an unsolvable word problem. So such a group must be a torsion group. (Of course the powers of an element of infinite order would be a recursively enumerable set of distinct elements.)

It is not known whether infinite, finitely presented torsion groups exist, so it is also unknown whether infinite, finitely presented algebraically finite groups exist. But Myasnikov and Osin show the following striking result.

THEOREM 62 (Mysanikov–Osin [73]). *There is an infinite, finitely generated recursively presented group G which is algorithmically finite.*

They call such a group a "Dehn monster". The proof of the theorem uses the construction of Golod and Shafarevich of an infinite torsion group combined with an analog of Post's [81] simple set construction.

Finally we mention that generic case complexity, which we informally described in an earlier section, has been widely studied in connection with decision problems for groups. A partial motivation for these investigations has been the potential for applications to cryptography. But quite recently these notions have been studied in the context of recursive function theory by Jockusch, Schupp and Downey (see in [53]). This is an area of ongoing research.

§11. Hints about further directions and references. In this section we make some comments and suggestions for further reading and references concerning the material covered above. A particularly relevant group theory reference is the textbook by Rotman [87] which includes a lot of the material with more details and proofs in its final chapter.

A standard reference for combinatorial group theory is the monograph [59] by Magnus, Karrass and Solitar but it doesn't deal with unsolvability results. Of more direct relevance for the present article is the classic monograph [56] by Lyndon and Schupp which contains a great deal of useful material as well as treatments of many of the results described above, but usually from a different point of view. Combinatorial group theory has in many ways evolved towards geometric group theory which was given great impetus by Gromov's work [46]. The monograph [25] by Bridson and Haefliger has a wealth of material concerning geometric group theory up to about 1999. Algorithmic questions play a substantial role in geometric group theory which continues to be a very active area.

In about 1990, I wrote the survey article [72] which tries to describe the status of decision problems in group theory at the time. While it has some unfortunate omissions (such as neglecting Bokut's work), it hopefully gives a fairly broad picture of developments.

Of more specific relevance to the present article is the survey [5] by Adian and Durnev which contains a great deal of material concerning decision problems for both groups and semigroups, including a lot of proofs. It is recommended for further mathematical details about various topics, particularly concerning semigroups. Another useful work is the monograph [14] by Bokut' and Kukin which discusses algorithmic questions in a wider variety systems including associative algebra, lie algebras and varieties. They also discuss the Novikov and Boone constructions and subsequent related results including Bokut's standard normal form approach.

Finally I would like to thank Gilbert Baumslag, Don Collins, Rod Downey, Paul Schupp and Hamish Short for reading various drafts of this article and providing many useful comments, suggestions and corrections.

REFERENCES

[1] S. AANDERAA, *A proof of Higman's embedding theorem using Britton extensions of groups*, **Word problems**, Studies in Logic and the Foundations of Mathematics, vol. 71, 1973, pp. 1–18.

[2] S. I. ADIAN, *Algorithmic unsolvability of problems of recognition of certain properties of groups*, **Doklady Akademii Nauk SSSR. New Series**, vol. 103 (1955), pp. 533–535, (Russian).

[3] ———, *Finitely presented groups and algorithms*, **Doklady Akademii Nauk SSSR. New Series**, vol. 117 (1957), pp. 9–12, (Russian).

[4] ———, *Unsolvability of some algorithmic problems in the theory of groups*, **Trudy Moskovskogo Matematicheskogo Obshchestva**, vol. 6 (1957), pp. 231–298, (Russian).

[5] S. I. ADIAN and V. G. DURNEV, *Decision problems for groups and semigroups*, **Uspekhi Matematicheskikh Nauk**, vol. 55 (2000), pp. 3–94, translated in **Russian Mathematical Surveys**, vol. 55 (2000), pp. 207–296.

[6] G. BAUMSLAG, W. W. BOONE, and B. H. NEUMANN, *Some unsolvable problems about elements and subgroups of groups*, **Mathematica Scandinavica**, vol. 7 (1959), pp. 191–201.

[7] G. BAUMSLAG and J. E. ROSEBLADE, *Subgroups of direct products of free groups*, **Journal of the London Mathematical Society. Second Series**, vol. 30 (1984), pp. 44–52.

[8] G. BAUMSLAG and D. SOLITAR, *Some two-generator one-relator non-hopfian groups*, **Bulletin of the American Mathematical Society**, vol. 68 (1962), pp. 199–201.

[9] L. A. BOKUT', *On a property of the Boone groups*, **Algebra i Logika**, vol. 5 (1966), pp. 5–23.

[10] ———, *On a property of the Boone groups II*, **Algebra i Logika**, vol. 6 (1967), pp. 15–24.

[11] ———, *On the Novikov groups*, **Algebra i Logika**, vol. 6 (1967), pp. 25–38.

[12] ———, *Degrees of unsolvability of the conjugacy problem for finitely presented groups*, **Algebra i Logika**, vol. 7 (1968), pp. 4–70.

[13] ———, *Mal'cev's problem and groups with a normal form*, **Word problems II (Conference on Decision Problems in Algebra, Oxford, 1976)**, Studies in Logic and the Foundations of Mathematics, vol. 95, North-Holland, Amsterdam-New York, 1980, with the collaboration of D. J. Collins, pp. 29–53.

[14] L. A. BOKUT' and G. P. KUKIN, *Algorithmic and combinatorial algebra*, Mathematics and Its Applications, vol. 255, Kluwer Academic Publishers, Dordrecht-Boston-London, 1994.

[15] W. W. BOONE, *Certain simple unsolvable problems in group theory, I, II, III, IV, V, VI*, **Koninklijke Nederlandse Akademie van Wetenschappen. Indagationes Mathematicae. Series A**,

vol. 57 (1954) (= *Indagationes Mathematicae*, vol. 16) pp. 231–237 and 492–497; vol. 58 (1955) (= *Indagationes Mathematicae*, vol. 17) pp. 252–256 and 571–577; vol. 60 (1957) (= *Indagationes Mathematicae*, vol. 19) pp. 22–27 and 227–232.

[16] ——— , *Review of* [99], **The Journal of Symbolic Logic**, vol. 17 (1952), pp. 207–265.

[17] ——— , *Review of* [74], **The Journal of Symbolic Logic**, vol. 19 (1954), pp. 58–60.

[18] ——— , *An analysis of Turing's "The Word Problem in Semi-Groups with Cancellation"*, **Annals of Mathematics**, vol. 67 (1958), pp. 195–202.

[19] ——— , *The word problem*, **Annals of Mathematics**, vol. 70 (1959), pp. 207–265.

[20] ——— , *Word problems and recursively enumerable degrees of unsolvability. A sequel on finitely presented groups*, **Annals of Mathematics**, vol. 84 (1966), pp. 49–84.

[21] W. W. BOONE, W. HAKEN, and V. POÉNARU, *On recursively unsolvable problems in topology and their classification*, **Contributions to mathematical logic** (K. Schütte, editor), North-Holland, Amsterdam, 1968, pp. 13–74.

[22] W. W. BOONE and G. HIGMAN, *An algebraic characterization of the solvability of the word problem*, **Journal of the Australian Mathematical Society**, vol. 18 (1974), pp. 41–53.

[23] W. W. BOONE and H. ROGERS, JR., *On a problem of J. H. C. Whitehead and a problem of Alonzo Church*, **Mathematica Scandinavica**, vol. 19 (1966), pp. 185–192.

[24] V. V. BORISOV, *Simple examples of groups with unsolvable word problem*, **Matematicheskie Zametki**, vol. 6 (1969), pp. 521–532, translated in **Mathematical Notes**, vol. 6, pp. 768–775.

[25] M. R. BRIDSON and A. HAEFLIGER, *Metric spaces of non-positive curvature*, Grundlehren der Mathematischen Wissenschaften, vol. 319, Springer-Verlag, Heidelberg-Berlin, 1999.

[26] J. L. BRITTON, *Solution to the word problem for certain types of groups, I, II*, **Proceedings of the Glasgow Mathematical Association**, vol. 3 (1956), pp. 45–54, vol. 3 (1957), pp. 68–90.

[27] ——— , *Review of* [76], **The Journal of Symbolic Logic**, vol. 23 (1958), pp. 50–52.

[28] ——— , *Review of* [75], **The Journal of Symbolic Logic**, vol. 23 (1958), pp. 52–54.

[29] ——— , *The word problem for groups*, **Proceedings of the London Mathematical Society**, vol. 8 (1958), pp. 493–506.

[30] ——— , *The word problem*, **Annals of Mathematics**, vol. 77 (1963), pp. 16–32.

[31] A. V. CHERNAVSKY and V. P. LEKSINE, *Unrecognizability of manifolds*, **Annals of Pure and Applied Logic**, vol. 141 (2006), pp. 325–335.

[32] I. M. CHISWELL, D. J. COLLINS, and J. HUEBSCHMANN, *Aspherical group presentations*, **Mathematische Zeitschrift**, vol. 178 (1981), pp. 1–36.

[33] C. R. J. CLAPHAM, *Finitely presented groups with word problems of arbitrary degrees of insolubility*, **Proceedings of the London Mathematical Society. Third Series**, vol. 14 (1964), pp. 633–676.

[34] ——— , *An embedding theorem for finitely generated groups*, **Proceedings of the London Mathematical Society. Third Series**, vol. 17 (1967), pp. 419–430.

[35] D. J. COLLINS, *Recursively enumerable degrees and the conjugacy problem*, **Acta Mathematica**, vol. 122 (1969), pp. 115–160.

[36] ——— , *Representation of Turing reducibility by word and conjugacy problems in finitely presented groups*, **Acta Mathematica**, vol. 128 (1972), pp. 73–90.

[37] ——— , *Obituary of William Werner Boone*, **Bulletin of the London Mathematical Society**, vol. 17 (1985), pp. 168–174.

[38] ——— , *A simple presentation of a group with unsolvable word problem*, **Illinois Journal of Mathematics**, vol. 30 (1986), pp. 230–234.

[39] D. J. COLLINS and C. F. MILLER, III, *The word problem in groups of cohomological dimension 2*, **Group St. Andrews 1997 in Bath, I** (C. M. Campbell, E. F. Robertson, N. Ruskuc, and G. C. Smith, editors), London Mathematical Society Lecture Notes, vol. 260, Cambridge University Press, 1999, pp. 211–218.

[40] M. Davis (editor), **Solvability, provability, definability: The collected works of Emil L. Post**, Birkhauser, Boston, 1994.

[41] M. DEHN, *Über unendliche diskontinuerliche Gruppen*, **Mathematische Annalen**, vol. 69 (1911), pp. 116–144.

[42] V. H. DYSON, *The word problem and residually finite groups*, **Notices of the American Mathematical Society**, vol. 11 (1964), p. 734.

[43] A. A. FRIDMAN, *On the relation between the word problem and the conjugacy problem in finitely defined groups*, **Trudy Moskovskogo Matematicheskogo Obshchestva**, vol. 9 (1960), pp. 329–356, (Russian).

[44] ———, *Degrees of unsolvability of the word problem for finitely presented groups*, **Doklady Akademii Nauk SSSR**, vol. 147 (1962), pp. 805–808, (Russian).

[45] ———, *Degrees of unsolvability of the word problem for finitely defined groups*, Izdatel'stvo "Nauka", Moscow, 1967.

[46] M. GROMOV, *Hyperbolic groups*, **Essays on group theory** (S. Gersten, editor), Mathematical Sciences Research Institute series, vol. 8, Springer-Verlag, 1987, pp. 75–263.

[47] F. GRUNEWALD, *On some groups which cannot be finitely presented*, **Journal of the London Mathematical Society. Second Series**, vol. 17 (1978), pp. 427–436.

[48] G. HIGMAN, *A finitely related group with an isomorphic proper factor*, **Journal of the London Mathematical Society**, vol. 26 (1951), pp. 59–61.

[49] ———, *Subgroups of finitely presented groups*, **Proceedings of the Royal Society of London. Series A**, vol. 262 (1961), pp. 455–475.

[50] G. HIGMAN, B. H. NEUMANN, and H. NEUMANN, *Embedding theorems for groups*, **Journal of the London Mathematical Society**, vol. 24 (1949), pp. 247–254.

[51] K. A. HIRSCH, *Review of [74]*, **Mathematical Reviews**, (1953), MR0052436 (14,618h) 20.0X.

[52] ———, *Review of [75]*, **Mathematical Reviews**, (1956), MR0075196 (17,706a) 20.0X.

[53] C. G. JOCKUSCH, JR. and P. E. SCHUPP, *Generic computability, turing degrees, and asymptotic density*, **Journal of the London Mathematical Society. Second Series**, vol. 85 (2012), pp. 472–490.

[54] I. KAPOVICH, A. MYASNIKOV, P. SCHUPP, and V. SHPILRAIN, *Generic-case complexity, decision problems in group theory, and random walks*, **Journal of Algebra**, vol. 264 (2003), pp. 665–694.

[55] R. J. LIPTON and Y. ZALCSTEIN, *Word problems solvable in logspace*, **Journal of the Association for Computing Machinery**, vol. 24 (1977), pp. 522–526.

[56] R. C. LYNDON and P. E. SCHUPP, **Combinatorial group theory**, Ergebnisse der Mathematik und ihrer Grenzgebiete, vol. 89, Springer, Berlin-Heidelberg-New York, 1977.

[57] W. MAGNUS, *Über diskontinuierliche Gruppen mit einer definierenden Relation (Der Freiheitssatz)*, **Journal für die Reine und Angewandte Mathematik**, vol. 163 (1930), pp. 141–165.

[58] ———, *Das Identitätsproblem für Gruppen mit einer definierenden Relation*, **Mathematische Annalen**, vol. 106 (1932), pp. 295–307.

[59] W. MAGNUS, A. KARRASS, and D. SOLITAR, **Combinatorial group theory**, Wiley, New York, 1966, (also corrected Dover reprint 1976).

[60] A. I. MALCEV, *On isomorphic matrix representations of infinite groups*, **Matematicheskiĭ Sbornik**, vol. 8 (1940), pp. 405–422, translated as *On the faithful representation of infinite groups by matrices*, **Translations of the American Mathematical Society. Second Series**, vol. 45 (1965), pp. 1–18.

[61] A. A. MARKOV, *On the impossibility of certain algorithms in the theory of associative systems*, **Doklady Akademii Nauk SSSR. New Series**, vol. 55 (1947), pp. 587–590.

[62] ———, *The impossibility of certain algorithms in the theory of associative systems*, **Doklady Akademii Nauk SSSR. New Series**, vol. 77 (1951), pp. 19–20.

[63] ———, *Review of [76]*, **Mathematical Reviews**, (1956), MR0075197 (17,706b) 20.0X.

[64] ———, *The insolubility of the problem of homeomorphy*, **Doklady Akademii Nauk SSSR. New Series**, vol. 121 (1958), pp. 218–220.

[65] ———, *Unsolvability of certain problems in topology*, **Doklady Akademii Nauk SSSR. New Series**, vol. 123 (1958), pp. 978–980.

[66] ———, *Insolubility of the problem of homeomorphy*, **Proceedings of the International Congress of Mathematicians, Cambridge 1958**, Cambridge University Press, Cambridge, 1960, pp. 300–306.

[67] Yu. V. Matiyasevich, *Simple examples of undecidable associative calculi*, **Doklady Akademii Nauk SSSR. New Series**, vol. 173 (1967), pp. 555–557.

[68] J. C. C. McKinsey, *The decision problem for some classes of sentences without quantifiers*, **The Journal of Symbolic Logic**, vol. 8 (1943), pp. 61–76.

[69] K. A. Mikhailova, *The occurrence problem for direct products of groups*, **Doklady Akademii Nauk SSSR**, vol. 119 (1958), pp. 1103–1105.

[70] C. F. Miller, III, *On group-theoretic decision problems and their classification*, Annals of Mathematics Studies, vol. 68, Princeton University Press, 1971.

[71] ———, *The word problem in quotients of a group*, **Aspects of effective algebra** (J. N. Crossley, editor), Upside Down A Book Company, Steel's Creek, 1981, Proceedings of a conference at Monash University August 1979, pp. 246–250.

[72] ———, *Decision problems for groups—survey and reflections*, **Algorithms and classification in combinatorial group theory** (G. Baumslag and C. F. Miller, III, editors), MSRI Publications, vol. 23, Springer-Verlag, 1992, pp. 1–59.

[73] A. Myasnikov and D. Osin, *Algorithmically finite groups*, **Journal of Pure and Applied Algebra**, vol. 215 (2011), pp. 2789–2796.

[74] P. S. Novikov, *On the algorithmic unsolvability of the problem of identity*, **Doklady Akademii Nauk SSSR**, vol. 85 (1952), pp. 709–712.

[75] ———, *Unsolvability of the conjugacy problem in the theory of groups*, **Izvestiya Akademii Nauk SSSR Seriya Matematicheskaya**, vol. 18 (1954), pp. 485–524.

[76] ———, *On the algorithmic unsolvability of the word problem in group theory*, **Trudy Matematischeskogo Instituta Imeni V. A. Steklov**, vol. 44 (1955), pp. 1–143, translated as *American Mathematical Society Translations*. Second Series, vol. 9 (1958), pp. 1–122.

[77] P. S. Novikov and S. I. Adyan, *Das Wortproblem für Halbgruppen mit einseitiger Kürzungsregel*, **Zeitschrift für Mathematische Logik und Grundlagen der Mathematik**, vol. 4 (1958), pp. 66–88, (Russian. German summary).

[78] A. Yu. Ol'shanskii, *Almost every group is hyperbolic*, **International Journal of Algebra and Computation**, vol. 2 (1992), pp. 1–17.

[79] E. L. Post, *Finite combinatory processes — Formulation 1*, **The Journal of Symbolic Logic**, vol. 1 (1936), pp. 103–105.

[80] ———, *Formal reductions of the general combinatorial decision problem*, **American Journal of Mathematics**, vol. 65 (1943), pp. 197–215.

[81] ———, *Recursively enumerable sets of positive integers and their decision problems*, **Bulletin of the American Mathematical Society**, vol. 50 (1944), pp. 281–316.

[82] ———, *A variant of a recursively unsolvable problem*, **Bulletin of the American Mathematical Society**, vol. 54 (1946), pp. 264–268.

[83] ———, *Recursive unsolvability of a problem of Thue*, **The Journal of Symbolic Logic**, vol. 12 (1947), pp. 1–11.

[84] M. O. Rabin, *Recursive unsolvability of group theoretic problems*, **Annals of Mathematics. Second Series**, vol. 67 (1958), pp. 172–194.

[85] M. O. Rabin, *Computable algebra, general theory and theory of computable fields*, **Transactions of the American Mathematical Society**, vol. 95 (1960), pp. 341–360.

[86] H. Rogers, Jr., **Theory of recursive functions and effective computability**, McGraw-Hill, 1967.

[87] J. J. Rotman, **An introduction to the theory of groups**, fourth ed., Graduate Texts in Mathematics, vol. 148, Springer-Verlag, Berlin-Heidelberg-New York, 1995.

[88] D. SCOTT, *A short recursively unsolvable problem (abstract)*, **The Journal of Symbolic Logic**, vol. 21 (1956), pp. 111–112.

[89] H. SEIFERT and W. THRELFALL, **Lehrbuch der Topologie**, B. G. Teubner, Leipzig and Berlin, 1934, (Since WWII this book has been reprinted by Chelsea Publishing Company and is now in the AMS–Chelsea book series published by the American Mathematical Society.).

[90] J. R. SHOENFIELD, **Mathematical logic**, Addison-Wesley, Reading, MA, 1967.

[91] R. I. SOARE, **Recursively enumerable sets and degrees**, Perspectives in Mathematical Logic, Springer-Verlag, 1987.

[92] J. STILLWELL, *Emil Post and his anticipation of Gödel and Turing*, **Mathematics Magazine**, vol. 77 (2004), pp. 3–14.

[93] V. A. TARTAKOVSKII, *The sieve method in group theory*, **Matematicheskiĭ Sbornik**, vol. 25 (1949), pp. 3–50.

[94] ———, *Application of the sieve method to the solution of the word problem for certain types of groups*, **Matematicheskiĭ Sbornik**, vol. 25 (1949), pp. 251–274.

[95] ———, *Solution of the word problem for groups with a k-reduced basis for $k > 6$*, **Izvestiya Akademii Nauk SSSR Seriya Matematicheskaya**, vol. 13 (1949), pp. 483–494.

[96] A. THUE, **Probleme üher Veränderungen von Zeichenreihen nach gegeben Regeln**, Skrifter utgit av Videnskapsselskapet i Kristiania, I. Mathematisk-naturvidenskabelig klasse 1914, no. 10, 1914.

[97] H. TIETZE, *Üher die topologischen Invarienten mehrdimensionalen Mannigfaltigkeiten*, **Monatshefte für Mathematik und Physik**, vol. 19 (1908), pp. 1–118.

[98] G. S. TSEITIN, *Associative calculus with insoluble equivalence problem*, **Doklady Akademii Nauk SSSR**, vol. 107 (1956), pp. 370–371.

[99] A. M. TURING, *The word problem in semi-groups with cancellation*, **Annals of Mathematics**, vol. 52 (1950), pp. 491–505.

DEPARTMENT OF MATHEMATICS AND STATISTICS
UNIVERSITY OF MELBOURNE
MELBOURNE 3010, AUSTRALIA
E-mail: c.miller@ms.unimelb.edu.au

MUSINGS ON TURING'S THESIS

ANIL NERODE

I take Turing's thesis (equivalently, Church's thesis) to assert that those functions on the integers which can be computed by a human being following any fixed algorithm on pencil and paper can also be computed by a Turing machine algorithm (or alternately by a lambda calculus algorithm). This thesis can be formulated using any of the many definitions of algorithm developed in the past eighty years which compute the same functions of integers. This has often been implicitly replaced by what I would call the physical version of Turing's thesis. This asserts that those functions on the integers which can be computed on any physical machine can be computed by a Turing algorithm. If the brain is regarded as a physical machine, this version subsumes the first version. But not everyone regards the brain as entirely physical ("Mathematics is a free creation of the human mind"—Brouwer). So we separate these formulations.

The meaning of Turing's thesis depends on determining what algorithms are possible, deciding whether algorithms should be defined to allow unbounded search using potentially infinite time and space, and what algorithms the brain can execute. The meaning of the physical Turing thesis depends in addition on determining what can be manufactured in the physical world. Neither the capabilities of the brain nor the capabilities of physical materials have been or are likely to be characterized by science. These questions have an intuitive, informal, and inexhaustibly open character. This is why Turing's thesis is just a thesis and the physical Turing's thesis is just a thesis. It is instructive to trace the evolution of these intuitions into well defined mathematical concepts. My perspective is conditioned by my work in and out of logic.

Early Impact of Turing on Mathematicians

I learned computability theory as an entering graduate student in mathematics at the University of Chicago in 1950. Over the subsequent 63 years I have known many of the founders, including Gödel, Church, Curry, Kleene, and Rosser. I did not meet Turing or Post. Both died while I was in graduate school. I and my fellow students Raymond Smullyan, Stanley Tennenbaum, and Bill Howard met while attending Rudolf Carnap's logic seminars in the Chicago philosophy department. No logic courses existed in that eminent mathematics department. The ringleader of the Bourbaki, the most irascible and most famous faculty member, Andre Weil, was so antithetical to logic as a branch of mathematics that he not only verbally attacked students who expressed interest in logic but also visiting logicians such as John Myhill. I think the mathematics community was not willing to accept computability as a mathematical subject at that time.

As we absorbed the work of Gödel, Church, Kleene, Curry, Rosser and Post, Saunders MacLane took a mild interest. He was an exception because his thesis at Göttingen was in logic and signed by Bernays in lieu of Hilbert. MacLane was a friend of both Kleene and Rosser, knew Church and Gödel, and read Turing's paper with us in a seminar at my request. (I sometimes think the invention of categories and functors was a consequence of MacLane having studied with Bernays, the main developer of Von Neumann's class theory.) It was only Gödel in logic who was famous in the world of mathematics. He was recognized as the one who had torpedoed Hilbert's program for establishing the consistency of mathematics, though very few had read or understood his work. At that time Turing was virtually unknown outside the few mathematical logicians. His cryptoananalytic genius and his computers were totally invisible outside the secret cryptoanalytic coummunty. It was only in the early 1960s that I heard about his work on building computers from his stellar cryptoanalytic World War II colleagues: I. Jack Good; and Bill Tutte, and that was not in a public arena.

Later I spent 1957–1958 with Gödel at IAS. There was still the same general lack of interest in computability both at IAS and at Princeton outside the logicians. Here I count Von Neumann as a logician. Von Neumann's dissertation invented class theory and the modern definition of ordinal number. He published papers in the 1920s contributing to Hilbert's program. He had understood what Gödel's incompleteness theorem meant as soon as Gödel gave a lecture on it. He was instrumental in bringing Gödel to IAS. At IAS he had designed and built one of the earliest postwar computers. He was familiar with the work of Church, Kleene, Rosser, and Turing. In his last years, dying of cancer, he developed the universal blueprint machine which, fed its own blueprint and an inventory of parts, reproduces itself. This was a mechanical form of reproduction mimicking the construction of Turing's universal machine, and a prescient precursor of the 3-D printers we can buy for a few thousand dollars, some of which can reproduce themselves from their own blueprints.

The visibility of Turing in the scientific community outside logic only rose when more and more powerful computers, supporting higher level languages, emerged.

Primitive Recursion and Unbounded Search:
Dedekind, Hilbert, Skolem, Ackermann, Gödel, Kleene, and Turing

Recall that the Hilbert program stemmed from discovery of the Russell and Burali Forti paradoxes. His program for proving the consistency of a branch of mathematics was to reformulate a mathematical theory as a formal system dealing with manipulation of strings of letters from finite alphabets without reference to what we now call a semantical interpretation. He defined logical formulas and axioms as sets of strings, proof rules as rules for inferring strings from strings, and proofs as finite sequences of strings such that each string is either an axiom or derived from previous formulas of the proof by one of the inference rules. As early as 1904–1905, in an article reproduced in van Heijenoort, Hilbert said that proving consistency of a formal system containing arithmetic is the problem of finding a property possessed by the axioms, preserved under the rules of inference, and not possessed by $0 = 1$. He and his school observed that all syntactical notions such as "x is a formula" or "x is a proof of y" are defined by nests of simple inductive definitions with all variables except the induction variable as fixed parameters. These we now call primitive recursive definitions. In Dedekind's essay on the theory of numbers (1870s) he gave the first statement of, and complete set-theoretic justification for inductive definitions of functions of integers such as addition and multiplication and exponentiation. He is generously credited nowadays with having invented the class of primitive recursive functions, even though he never gave such a definition. One can view the Hilbert program as one attempting to give a finitary justification for more and more elaborate inductive definitions of primitive recursive functions of higher type. We do not define this notion here.

The goal of a Hilbert consistency proof for a formal system was to show by a transparent "finitary" proof that the primitive recursions that produce proofs do not ever produce the theorem $0 = 1$. So giving a Hilbert consistency proof reduces to a question about finitary properties of particular inductive definitions. Primitive recursive function arithmetic and primitive recursive functions were well understood. Skolem's bounded quantifier arithmetic of 1923 is built up from atomic formulas denoting primitive recursive functions and sets, propositional connectives, and quantifiers with primitive recursive bounds. Skolem proved that in this theory every first order formula or function denotes a primitive recursive set or function. This development is reproduced and used in Kleene's "Introduction to Metamathematics" of 1952. Ackermann's 1924 thesis under Hilbert writes out what is a doubly recursive enumeration of the primitive recursive functions as a primitive recursion of higher type. Hilbert even thought in 1926 that he could prove the continuum hypothesis using primitive recursive functionals of transfinite type. Ackermann's 1924 dissertation and 1928 paper exhibit the fundamental role primitive recursion played in Hilbert's program. Why mention here the history of a subject no longer pursued?

Gödel's 1931 paper was entitled "On the Incompleteness of Russell's Principia Mathematica". (I had the misfortune to read the latter work of Russell in its entirety before knowing better.) Gödel realized that his incompleteness proof worked for any formal system in which certain computation procedures could be represented. He asked in a footnote whether there was an absolute definition of computability. Kleene recognized early on after the Church–Turing thesis was formuated that the lambda calcus formulation was not likely to be read, and recognized that the primitive recursive mechanisms in Gödel's 1931 paper, from the primitive recursive function theory of the Hilbert school plus unbounded search, gave a definition of computably enumerable set as the projection of a primitive recursive relation which avoids the language of logic and is more palatable to mathematicians.

Gödel's technology for expressing the syntactical predicates such as "x is a formula" and "x is a proof of y" consists of formalizing definitions of primitive recursive functions such as "the nth term of the code m of a finite sequence of numbers is primitive recursive." He formalizes "x is a proof of y" as a primitive recursive predicate. Let us depart from Gödel's notation and use modern notation from computability theory. Let $\phi_m(v)$ be the mth formula in one free variable v. Define ω_m as the set of integers n such that $\phi_m(\underline{n})$ is provable, where $\underline{n}$ is the numeral denoting n. This is an effective enumeration ω_m of all computably enumerable sets computably isomorphic, by an early theorem of Hartley Rogers to those we use nowadays, but at the cost of detouring through the primitive recursive proof predicate of a logical system rather than through a direct definition. Kleene's theorem that computably enumerable sets are exactly the ranges of primitive recursive functions follows immediately. To me this indicates the historical inevitability of the definition of computably enumerable sets as an outgrowth of the Hilbert school's introduction of primitive recursive functions in its consistency proofs.

Here is a generalized description of Kleene's formulation of computability. In executing any algorithm in any language, computation states evolve in time. It is straightforward to code computation states as strings of symbols containing all the information needed to determine the next state so that the following are primitive recursive on strings: coding an input into an initial computation state, transitioning from a computation state to the next computation state, recognizing that a computation state contains a coded value of a desired answer, and extracting that coded value from a computation state. Given a decent definition of the string coding a state, any student can use Skolem's 1923 theorem to write out the simple primitive recursive definitions. Simulating execution of an algorithm primitive recursively comes down to encoding the input primitive recursively, evolving the state primitive recursively, checking whether the state encodes an answer primitive recursively; if it does, stopping the computation and reading off the answer primitive recursively. This is the analysis underlying Kleene's characterization of computability. The extension from Hilbert school primitive recursive functions to Turing computable functions is solely unbounded search. This provides a fairly transparent transition from the Hilbert school's technology of primitive recursive functions for consistency proofs to Gödel's paper, where the set of theorems is a projection of the primitive recursive proof predicate, to Kleene's computably enumerable sets as projections of a primitive recursive relation.

Inductive definitions are natural for the mathematical mind. It is easy to take any algorithm given in any form, define a notion of computational state for that algorithm, then to write out the primitive recursions which define: evolution of state, coding inputs into a state, recognizing that a state contains a desired output, reading out that output. That is, simulation of any algorithm by a primitive recursive process with unbounded search is easy. Post said, in his famous 1945 paper, that reasoning about computability is so intuitive that it can be carried out without reference to any exact definition. Over the years I have read a myriad of papers following that advice, and observed that they can, without alteration in form or order, be routinely transcribed as a series of primitive recursive definitions and unbounded searches. Occasionally they give intuitions using machine models such as pinball machines, but the proofs are based on primitive recursive definitions and unbounded searches. Gödel's 1931 paper contained implictly the computational state decribed by the primitive recursive proof predicate, but he apparently did not think in computational terms about looking for a proof of a statement as an unbounded primitive recursive search. That Gödel later rejected Kleene's definition of computability by primitive recursion plus unbounded search based on a notion of primitive recursive computational state and accepted the Turing machine model with a Turing computational state which also evolves primitive recursively and computes its answers by unbounded search seems odd to me nowadays. The advance in defining computability over the primitive recursions of the Hilbert school was solely unbounded search on state evolution as Kleene formulated it. Perhaps it was the explicit introduction of computational state in Turing's construction of a universal machine that was so convincing, since the concept of computational state is hidden in Kleene's and Church's constructions of the universal algorithms which can simulate any algorithm. Channeling Kleene, I have always found the concept of computational state and of universal algorithm quite clear in Gödel's 1931 primitive recursive proof predicate. It does implement a universal Turing algorithm.

Turing's thesis

Definitions equivalent to computability by Turing machines have been discovered all over mathematics, such as in number theory (in Hilbert's tenth problem) and group theory (in the word problem for finitely presented groups). This was evidence that Turing computability is natural in mathematics and has deep properties. But I do not regard these successes as evidence at all for either version of Turing's thesis.

Both are based on the belief that what we see as possible for human beings now and what we see as possible for machines now is all we will ever see as possible. There is an element of hubris here. Some people lack this hubris. In 1963, when I was again at IAS, I asked Gödel if he has spent any time in later years trying to discover a more general notion than Turing Computability. He paused for awhile, and then said yes, but that nothing had come of it. I also asked Church the same question. He said he had tried quite hard to define the most general "honest" potentially infinite environment generalizing the Turing tape. I believe that "honest" meant that the combinatorial structure of the environment does not code a lot of hidden information. I recall that he published some work in this direction.

In 1953 I had the opportunity of spending some hours with Brouwer on his visit to the University of Chicago. His time was free because I was the only one there interested in constructive mathematics, and the only one who knew that there were classical computable analysis counterparts to constructive analysis. I acquired from Brouwer's papers the conviction that fundamental mathematical intuitions have so much power, and fundamental physical phenomena have so much complexity, that assuming that known mathematical constructions or known physical models capture all of an intuition or all of a physical phenomenon is almost always a mistake. Permitting the possibility of as yet unknown extensions of our knowledge is the basis of Brouwer's and Heyting's interpretation of logical connectives.

—Gödel's interpretation of classical mathematics within intuitionistic logic demonstrates that classical proofs and theorems survive in a constructive context but with a reinterpretation, that is, a different semantics.

—The theorems of Lobachevski and Bolyai geometries survive in the usual models within Euclidean space with a reinterpretation of points, lines, and distances, that is, a different semantics.

—Quantum mechanics refined classical mechanics when physicists had to account for indeterminacy, bosons vs fermions, entanglement, etc. Nineteenth century Hamiltonian–Jacobi theory for mechanical dynamical systems was replaced by a quantized version. Real valued measurements were replaced by Hermitian operator valued measurements. Commutative geometry over the reals was replaced by non-commutative geometry over Hermitian operators.

The history of science is replete with contractions, extensions, revisions of powerful intuitions stemming from new discoveries.

What functions of integers *can* the brain or a physical machine compute using unbounded search? It is perfectly possible that a later generation may discover that human beings can compute functions of integers not Turing computable using an as-yet-unimagined or un-utilized non-Turing atomic act that cannot be reduced to reading, writing, changing discrete state, and moving a finite set of symbols from one place to another in a simple discrete environment. This might be interesting mathematics if it is not just a relativised version of current computability.

Unbounded Search

We usually think of Turing machines as discrete, deterministic, and capable of runs of any length. But realizing unbounded search in the physical world contradicts basic tenets of contemporary physics. Unbounded search is a mathematical construction in a mathematical universe, not realizable in the world in which we live. A physically realized Turing machine doing an unbounded search with a potentially infinite tape must retain all information on the tape till a computation is finished. The second law of thermodynamics implies that entropy is constantly increasing; the tape will decay to unreadability in a finite time unless you have an agent constantly counteracting the decay of each and every tape square. A long enough search will eventually require for repair more energy than the total energy of the universe. So the machine will inevitably behave incorrectly on very long tapes. You cannot get around this by developing a potentially infinite tape in a finite region of space-time using smaller and smaller tape squares. At the quantum level, as the tape squares get smaller, quantum probabilistic effects turn reading and writing into probabilistic phenomena and unreliable. Not only that but occasional quantum entanglement can alter the state of any tape square. This applies to all digital computers. Physics does nor support the possibility of unbounded search, reliable or not.

For any physical digital computer, based on knowing its finite resources, in principle we can estimate how long a run we can rely on with confidence. Also since any physical computer has bounded resources, it is a finite automaton which can only simulate a program until its memory resources are used up, after which it is useless.

Suppose we eliminate unbounded searches and limit ourselves to runs of bounded size. Then we can try to compare digital and continuous computations executed on radically different machines. They need not be based on the Turing paradigm. We can compare efficiency within a run length, for instance, of digital simulations of continuous phenomena and of continuous simulations of digital phenomena.

I am of the opinion that analog computing is as fundamental as digital computing, and often provides the most efficient tool for solving many problems. This is because a very significant part of the time during a run on a digital microprocessor is devoted to waiting for an electron to settle down at a new energy level before using its energy level to determine the next computation step. This digital delay is absent in continuous simulations.

I have had a special interest in continuous simulations since I worked on U. S. Air Force simulations on Reeves differential analyzers and also on the RCA prototype called "Junior" in 1954 and 1955. I and Wolf Kohn have a 1990s patent for an (unbuilt) continuous wave quantum processor for discrete algorithms. Discrete algorithms are coded into "continualized" algorithms which are input to the quantum wave processor, which operates without discrete transitions. Answers are read as non-demolition readings on the wave. Finally, if the qubit modelers of quantum computation solve the decoherence problem, we will have another physical realization of a non-Turing model for computation. One should keep an open mind.

Computation and Control

My views are conditioned by having worked in computability since 1950 and also worked in hybrid control with Wolf Kohn since 1991. We introduced the systematic study of hybrid control, that is, study of feedback systems of digital controllers controlling continuous "plants". For those in complete ignorance of control, a feedback control system consists of a controller and a controlled plant. The controller senses the state of the plant at discrete or continuous or mixed times, and in response changes its internal state and sends a signal to the plant which changes plant state. The state of the completed feedback controlled system is the simultaneous state of the controller and the plant and the signals passing between them. Any physical computer is a feedback control system. In the case of the Turing machine model, the controller state is the internal state of the (read-write-move) tapehead. A plant state is a printed tape and the square the tapehead is reading. The controller senses (reads) a symbol, sends a control signal to simultaneously alter the plant state (writes, moves), and alters its internal state. This describes the evolution of computation state of any computer. In the past sixty years digital computation based on control has spawned giant industries which have revolutionized modern society. We want to emphasize that advances in computing depend on advances in control, and vice versa.

Historically advances in computers have fed on advances in feedback control as they develop on new physical substrates using new engineering technologies. In sequence: Jacquard cards were programs for weaving cloth patterns on looms, powered by hand, steam or water. Babbage's unfinished machine was constructed of wood and brass to compute mathematical tables of values of functions by finite differences. Hollerith cards for sorting census data were programs using electrically powered controllers. DC generator power was used for programs (circuit diagrams) executed by electromagnetic relay circuits in telephone switching and subway traffic control. Armstrong's' triode tube (1917) led to the high gain amplifiers of Bell Labs and were used for vacuum tube computers. Silicon semiconductor solid state physics led to transistor based computers, integrated circuits, and microchips. In every case physical feedback control is what implements iteration and looping for changes of state.

Vice versa, advances in feedback control have fed on advances in computers in such areas as automated factories, space vehicles, automobiles, and household appliances. In sequence: From about 1900 to 1960, controllers were linear controllers for linear plants. The methods used stemmed from the design methods of Henrik Bode's group in the 1920s at Bell Labs for high gain amplifiers. The mathematical algorithms used for designing such linear controllers for linear plants were transfer functions and frequency response curves from linear algebra, linear differential equations, the Laplace transform, and complex integration. The most useful controllers were linear controllers. These methods were sharpened during World War II at the MIT radiation laboratories by first rate mathematicians such as Norbert Wiener. They used them to design controllers for RADAR tracking of airborne targets. At the end of the war, the MIT radiation laboratories published a red series of books on control. In 1954, during the Korean war, I worked on control systems for air to air missiles based on these books, books which I read from cover to cover. In the real world most systems (called plants in control) are not linear. Desired control cannot be effected by a linear controller. The ad hoc solution of control engineers was to break up the operating regime of the plant into regions so small that piecewise linear controls could be designed for each region and switched on or off as one entered or left a region. This switching can be implemented by a digital controller, leading to a hybrid system consisting of a digital controller and a continuous plant. This is now often called embedded digital control. In the 1990s Wolf Kohn and I used measure valued calculus of variations which proves the existence of measure valued optimal controls (which are not physically realizable) for a wide class of non-linear systems, and gave algorithms to extract close to optimal (physically realizable) piecewise linear controls. This gave a theoretical base and algorithms for what had been an ad hoc extension of linear control to non-linear control. Embedded digital control of non-linear plants is an example in which advances in digital computers depend on advances in control.

There is an old credo, "computation = logic plus control." The Turing thesis and its physical version are in that sense not models for computation because they do not model control.

Physical Realizability

There is no definition of "physically realizable machine." In the past 200 years we have had in succession computers based on mechanics (Babbage), electromagnetic relay switching (telephone exchanges, subway traffic controllers), electronic tube amplifiers, transistors, and integrated circuits, with no end in technological sight. All of these support executing Turing algorithms. There are even proposals for subatomic computing within the orbits of a single atom, where already millions of bytes of information have been stored and retrieved. Computation and control are being developed in all sorts of physical media, liquid, solid, gaseous, DNA, in nano superparticles, with states evolving according to macroscopic or quantum systems of pdes.

Who knows what physical models for computation will arise? Analog computing of solutions of ordinary and partial differential equations is as fundamental as digital computing and is slowly merging with digital computing in hybrid systems, and often provides the most efficient tools for solving many practical problems. In engineering we distinguish between the mathematical design for a machine which meets a performance specification (a mathematical model), and the question whether such a design can be physically built. This is usually referred to as physical realizability. Control engineering is centered around this topic. Patent applications are often rejected because the machines described are pretty mathematics but can't be built. Sometimes this is because they violate the second law of thermodynamics and are perpetual motion machines.

Conclusion

There is a direct line from Hilbert's failed attempts to establish the consistency of mathematics, through Gödel's proof, to Kleene's characterization of computably enumerable sets, and to Turing's introduction of computation state as a tool for constructing his universal machine. Turing's thesis in its physical realizability version leads to analyzing, for both discrete and continuous problems, relative efficiency of programs computing solutions in bounded space and time using all possible physical computing machines, discrete or continuous. The secondary literature has become vast. Here are a few sources, mostly primary.

REFERENCES

[1] W. ACKERMANN, *Zum Hilbertschen Aufbau der reellen Zahlen*, **Mathematische Annalen**, vol. 99 (1923), pp. 118–133.

[2] P. J. ANTSAKLIS and A. NERODE, *Hybrid control systems*, **IEEE Transactions on Automatic Control**, vol. 43 (1998), no. 4.

[3] M. DAVIS, ***The undecidable: Basic papers on undecidable propositions, unsolvable problems and computable functions***, Raven Press, New York, 1965.

[4] J. M. DAVOREN and A. NERODE, *Logics for hybrid systems*, **Proceedings of the IEEE**, vol. 88 (2000), no. 7, pp. 985–1010.

[5] Solomon Feferman et al.(editors), ***K. Gödel: Collected works***, vol. 3, Oxford University Press, USA, 1986–.

[6] Jens Erik Fenstad (editor), ***T. Skolem: Selected works in logic***, Universitetsforlaget, Oslo, 1970.

[7] D. HILBERT, ***Gesammelte Abhandlungen***, Chelsea Publishing Company, 1981.

[8] S. C. KLEENE, ***Introduction to metamathematics***, Van Nostrand, New York, 1952.

[9] W. KOHN, V. BRAYMAN, and A. NERODE, *Control synthesis in hybrid systems with Finsler dynamics*, **Houston Journal of Mathematics**, vol. 28 (2002), no. 2, pp. 353–375, (issue dedicated to S. S. Chern).

[10] W. KOHN, A. NERODE, and J.B. REMMEL, *Hybrid systems as Finsler manifolds: Finite state control as approximation to connections*, **Hybrid systems II**, Springer, 1995.

[11] A. H. Taub (editor), ***J. Von Neumann: Collected works***, 1961–3.

[12] A. M. TURING, *Pure Mathematics: with a Section on Turing's Statistical Work by I. J. Good*, (J.L. Britton, editor), North-Holland, Amsterdam, 1992.

[13] J. VAN HEIJENOORT, ***From Frege to Gödel***, Harvard University Press, 1967.

[14] R. ZACH, *The practice of finitism: Epsilon calculus and consistency proofs in Hilbert's Program*, **Synthese**, vol. 137 (2003), pp. 211–259.

MATHEMATICS DEPARTMENT
CORNELL UNIVERSITY
ITHACA, NY 14853, USA
E-mail: anil@math.cornell.edu

HIGHER GENERALIZATIONS OF THE TURING MODEL

DAG NORMANN

Abstract. The "Turing Model", in the form of "Classical Computability Theory", was generalized in various ways. This paper deals with generalizations where computations may be infinite. We discuss the original motivations for generalizing computability theory and three directions such generalizations took. One direction is the computability theory of ordinals and of admissible structures. We discuss why Post's problem was considered the test case for generalizations of this kind and briefly how the problem was approached. This direction started with metarecursion theory, and so did the computability theory of normal functionals. We survey the key results of the computability theory of normal functionals of higher types, and how, and why, this theory led to the discovery and development of set recursion. The third direction we survey is the computability theory of partial functionals of higher types, and we discuss how the contributions by Platek on the one hand and Kleene on the other led to typed algorithms of interest in Theoretical Computer Science. Finally, we will discuss possible ways to axiomatize parts of higher computability theory.

Throughout, we will discuss to what extent concepts like "finite" and "computably enumerable" may be generalized in more than one way for some higher models of computability.

§1. Introduction. In this paper we will survey what we may call *higher analogues* of the Turing model. The Turing model, in the most restricted interpretation of the term, consists of the Turing machines as a basis for defining computable functions, decidable languages, semi-decidable languages and so forth. By a higher generalization of the model, using this restricted interpretation, we would mean something resembling a Turing machine, but operating in infinite time, having infinite sets of states, handling infinite words over infinite languages or in other ways being generalized Turing machines. Over the past decade, there has been an interesting development of higher generalizations of the Turing model in this restricted sense. This will be treated in another paper in this volume, see Welch [74].

The importance of the Turing model in this restricted sense is indisputable, it provides us with a conceptually sound mathematical model of what we may mean by *computable*. Turing machines are also used to define the important concepts of *complexity theory*, like space and time complexity classes. However, when the basic properties of computable sets of integers, of computably enumerable sets and of relativized computations are proved in a beginners' course in computability theory, one soon leaves the Turing machine approach, for instance using primitive recursion and the μ-operator as the basis instead. The early generalizations of computability theory reflect this, and so will this survey.

Now, what will we mean by a higher generalization of the Turing model? As we will see from this paper, our interpretation is actually very wide. First of all, we are talking about *generalizations*. This means that we are investigating mathematical structures where the objects of interest share some of the properties of computations, computable functions or other items studied in classical computability theory. With *higher* generalizations we will mean such generalizations where computations terminating in the sense of the generalization may be infinite, and where algorithms may accept infinite entities of some sort as input data and as output data. This normally means that the rôle of *finiteness* is replaced by levels of *definability*. We will explain this in detail for some of the generalizations we consider.

Kreisel [33] pointed out that one of the motivations for generalizing computability theory is to obtain a better understanding of the classical proofs, for instance of which properties of the finite sets that are used in proofs and constructions. This led to a line of research where one tried to extend the classical solution to Post's problem and other results from degree theory to a wide range of mathematical structures.

Another important motivation is that of applicability. Concepts and methods from classical computability are tools that can be used in other settings. For instance, there are close links between higher computability theory and descriptive set theory. We will not discuss the motivations for generalizing computability theory in any depth, but refer the interested reader to Kreisel [33] for a discussion.

One important line of generalizations was to replace the natural numbers with an ordinal. In Section 2 we will survey the development of this line from metarecursion theory, i.e., computability over the first non-computable ordinal, through α-recursion theory and branching to β-recursion theory and to computability over admissible structures.

Another line is computations relative to the so called *normal functionals* of arbitrarily high finite types, a line that led to set recursion. This will be surveyed in Section 3, where we also briefly discuss computability relative to non-normal functionals.

In Section 4 we will discuss two alternative ways to define computability relative to functionals, this time relative to partial ones.

An unforeseen bonus of the attempts to make a decent theory for computations relative to functionals of higher, finite types was the modern theory of denotational semantics for programs, and we will briefly describe this important path from classical computability theory, via higher computability theory, to theoretical computer science.

As the number of examples of generalizations of classical computability theory grew, the efforts to axiomatize computability theory came as a natural consequence. In Section 5 we will discuss the nature of such generalizations.

In Section 6 we will briefly mention some directions that may be considered to be a part of higher computability theory, but that we could not cover in this survey paper.

One of the basic concepts in classical computability theory is that of a *computably enumerable set*. Originally, the importance of this concept was due to the fact that the set of Gödel numbers of theorems in an axiomatizable theory is computably enumerable, and more generally, that a formal grammar will generate a computably enumerable set of words. Computably enumerable sets also turn up when classical computability theory is applied to algebra, analysis, and other parts of mathematics. What is of interest to us is that the various characterizations of the computably enumerable sets may generalize in different ways in higher computability theory. By definition, a set is *computably enumerable* if it is empty or the range of a total, computable function. A set is *semicomputable* if it is the domain of a partial computable function and it is Σ_1^0 if it is the projection of a primitive recursive set. These three concepts are equivalent in the classical case, but, as we shall see, not necessarily so in all generalized settings.

Higher computability theory is the product of the 1960s and 1970s with some development stretching into the 1980s, and this survey will cover, only in parts, what went on in those years. We will occasionally use the jargon of the time, talking about "finite sets" that actually are only generalized finite according to a given generalization of computability theory. We will use "α-recursion theory", "β-recursion theory" and "set recursion" as that was the established terminology for specific generalizations. However, we will also use the terminology of today, and talk about "generalized computability theory", "computably enumerable set", "semicomputable set" and so forth. We do not expect any confusion arising from this.

Prerequisites. We will assume that the reader is familiar with the basic concepts of classical computability theory like *primitive recursion*, the *Kleene T-predicate*, the *recursion theorem* and *Turing degrees*, but not that the reader is a trained computability theorist. It will be an advantage to know the basic ideas behind priority arguments and the classical solution to Post's problem.

We will also assume that the reader knows basic set theory, including ordinal arithmetic and the definition of Gödel's L. This includes knowledge of the terminology related to formal languages for number theory, analysis and set theory, using the terminology Π_n^0 and Σ_n^0 in the standard way for statements in first order number theory and Π_n^1 and Σ_n^1 for statements in second order number theory. We use Π_n and Σ_n for statements in the language of set theory. Sets are classified by their defining formulas, and if a set or a class is both in a Π-class and the corresponding Σ-class, we say that the set is in the corresponding Δ-class in the usual way.

Formulas in the language of set theory will, by default, be interpreted in the structure under discussion. When we consider fragments of set theory we will only consider transitive models, and then a Δ_0 set will be a set defined by a formula using bounded quantifiers only.

The paper will be self contained for any reader who has passed basic courses in computability theory, logic and set theory. By being self contained, we mean that the intended reader should be able to follow the exposition. As we are covering a part of computability theory that would actually take several textbooks to introduce, we will skip almost all proofs and even skip some technical definitions. We will, however, try to give the intuition behind technical concepts. Our ambition is to give the reader an idea of what the different directions in higher computability theory are, an idea of the diversity and importance of higher computability theory and partly to tell where to find further literature on the subject.

Acknowledgements. During the preparation of this paper, the author benefitted from discussions with Johan Moldestad, Viggo Stoltenberg-Hansen and Philip Welch.

An anonymous referee suggested several linguistic improvements and supplementary references, in addition to pointing out one mathematical error in the first version.

§2. Extending classical computability theory to ordinals. In this section we will consider generalizations of μ-recursion where the set $\mathbb{N}$ or ω of natural numbers is replaced by another structure, primarily replacing the classical μ-operator with search over an ordinal. This story started with the exploration of the *hyperarithmetical* sets, an application of computability theory to definability theory and descriptive set theory. Then one gradually observed that the class of hyperarithmetical sets, and related set theoretical structures, may themselves serve as domains of generalizations of classical computability theory. The so called *companion* of the hyperarithmetical sets is the least admissible structure, i.e., model of Kripke–Platek set theory, containing ω as an element. See Subsection 2.2 for an introduction to KP-theory. The companion will be the class of sets that can be coded by a hyperarithmetical set in a natural way. The companion of the hyperarithmetical sets turned out to be the fragment of Gödel's L up through the computable ordinals. The computability theory of the hyperarithmetical sets was called *metarecursion theory*.

The next step was to generalize computability to allow search over admissible ordinals in general, α-recursion theory. Much of the inspiration came from the desire to generalize Turing degrees, and since being an admissible ordinal does not relativize to predicates, it was natural to develop tools where the ordinal is not necessarily admissible. This led to β-recursion theory.

Another further generalization of α-recursion theory is the extension of search computability to other admissible structures. In this section, we will survey some of the mathematics developed in the attempts to understand what the consequences are of replacing search over $\mathbb{N}$ with search over higher structures and then generalize the theory of Turing degrees.

We suggest Sacks [56] for further reading. We also suggest the survey papers Chong and Friedman [3] and Hinman [18] in Griffor [13].

2.1. The hyperarithmetical sets. One of the consequences of the existence of a universal Turing machine is that there are universal semicomputable sets U_n in all dimensions, i.e., there are semicomputable sets $U_n \subset \mathbb{N}^{n+1}$ such that all semicomputable sets $V \subset \mathbb{N}^n$ will be of the form

$$V = \{(a_1, \ldots, a_n) \in \mathbb{N}^n \mid (e, a_1, \ldots, a_n) \in U_n\}$$

for some $e \in \mathbb{N}$.

On the other hand, a simple diagonalization argument shows that there cannot be a universal computable predicate, i.e., a computable predicate such that all other computable predicates are slices of the given one as described above. In the early days of computability theory there was already another example of a similar phenomenon, there is no universal Borel subset of the topological space $(\mathbb{N}^\mathbb{N})^2$, but there is a projection of a Borel set that is universal for all other projections of Borel sets. Projections of Borel sets are called *analytic sets*. The analogy goes a bit deeper: It is well known that there are disjoint semicomputable sets that cannot be separated by a computable set, while any two disjoint complements of semicomputable sets can be separated. One interesting analogy with the Borel hierarchy is that two disjoint analytic sets can be separated by a Borel set, while there are Borel-inseparable disjoint pairs of complements of analytic sets. This analogy suggests that while *Borel* corresponds to *computable*, *co-analytic* will correspond to *semicomputable*. Considering the classical proof of the separation theorem for disjoint analytic sets, we will see that it is based on the fact that there is an $\aleph_1$-approximation from within of any co-analytic set, via co-analytic pre-wellorderings where all proper initial segments are Borel. A selection based on search along such approximations to the complements of two disjoint analytic sets is used to separate them. Thus the methods of proof have some analogies as well. The consequences are also analogues, a set is Borel if the set and its complement are co-analytic, and a set is computable if the set and its complement are semicomputable. Since it is outside the scope of this paper, we do not offer detailed references to the classical results from descriptive set theory, see Kechris [21] or Moschovakis [41] for more details.

Computability theory offers an effective analogue to this early descriptive set theory, the *hyperarithmetical sets*, the Δ_1^1-sets and the Π_1^1 and Σ_1^1-sets. If we let ϕ_e be the partial computable function indexed by e (e.g., using a Gödel enumeration of the Turing machines), we may define the *indexed set* of hyperarithmetical sets H_e as follows:

1. $H_{\langle 0,a\rangle} = \{a\}$.
2. If H_e is defined, then $H_{\langle 1,e\rangle} = \mathbb{N} \setminus H_e$.
3. If ϕ_e is total and $H_{\phi_e(a)}$ is defined for all $a \in \mathbb{N}$, then
$$H_{\langle 2,e\rangle} = \cup_{a\in\mathbb{N}} H_{\phi_e(a)}.$$
4. H_e is undefined unless it is defined according to 1.–3.

Here $\langle\ \rangle$ will be some standard sequence number constructor.

The analogy to the Borel hierarchy is evident, singletons play the rôle of basic open sets, and we close under complements and effective countable unions. The indexing is needed in order to make the latter precise. If we use open intervals in $\mathbb{R}$ with rational endpoints instead of singleton subsets of $\mathbb{N}$ at the base of this definition, we get the hyperarithmetical subsets of $\mathbb{R}$, and this corresponds exactly to being *computably Borel*. In this paper we will only be concerned with hyperarithmetical subsets of $\mathbb{N}$.

The original definition goes back to Kleene [23]. For a detailed introduction, see Rogers [50] or Sacks [56].

We see that the definition above actually is a positive, arithmetical inductive definition of the relation $a \in H_e$, and then, that relation will be Π_1^1. Each hyperarithmetical set H_a will be Δ_1^1, as both it and its complement will be Π_1^1 by the observation above. What is more important, if we let α be a computable ordinal, in the sense that there is a computable well ordering of $\mathbb{N}$ of the same order type as α, then iterating the inductive definition only as far as α will give us a hyperarithmetical set that is universal for all hyperarithmetical sets definable before α (see Kleene [24], Spector [68]). Moreover, the set of indices for computable well orderings of $\mathbb{N}$ of length $< \alpha$ will be hyperarithmetical. As consequences we may obtain:

> There are universal Π_1^1-sets and the set of indices for computable well orderings of $\mathbb{N}$ is complete Π_1^1 (Kleene [23]).
>
> Every Δ_1^1-set is hyperarithmetical (Kleene [24]).
>
> If an ordinal is hyperarithmetical, then it is computable (Kleene [24]).

If we replace *computable set* with *hyperarithmetical set* and *semicomputable set* with Π_1^1-*set* we get an analogue to classical computability theory. We may even replace *partial computable function* with *function having a Π_1^1-graph*, and thereby we have analogues of the main concepts. There are, however, a few observations that make these analogues a bit superficial:

> The range of a total, hyperarithmetical function is always hyperarithmetical. Thus all computably enumerable sets are computable in this

setting, and the equivalence between semicomputability and computable enumerability fails. Instead, in this generalization, we will have that a set is computable if and only if it is computably enumerable.

Any Π_1^1-set A is either complete Π_1^1 (in the sense that all other Π_1^1-sets are hyperarithmetical in A) or hyperarithmetical, so there is no interesting degree theory of the semicomputable sets for this generalization.

There is seemingly no concept that is similar to the Σ_1^0-characterization of semicomputability, but as we will see, only seemingly.

In spite of the poor analogy, hyperarithmetical theory was in many respects the start of higher computability theory. The main conceptual step was to realize that the reason for the poor analogy was that one had chosen the wrong analogues. In Kreisel and Sacks [34], the authors observed that the right analogues were:

The class Hyp of hyperarithmetical subsets of $\mathbb{N}$ generalize the class of finite sets.

Subsets of Hyp that are Δ_1^1 over Hyp generalize computable sets.

Π_1^1-classes restricted to Hyp generalize semicomputable sets.

Partial computable functions will be generalized by partial functions

$$f \colon Hyp \to Hyp$$

with a Π_1^1-graph.

In fact, Π_1^1 restricted to Hyp coincides with Σ_1^1 interpreted over Hyp, (Spector [69] and Gandy [9]) so the equivalence between being semicomputable, computably enumerable and Σ_1 over the domain is restored.

Cutting the history very short and skipping intermediate stages and explicit references, the analogies can be based on the following observations. We let ω_1^{CK} be the first non-computable ordinal, where CK honors Church and Kleene.

The hyperarithmetical sets are exactly the subsets of $\mathbb{N}$ that are in $L(\omega_1^{CK})$.

For each set x, the structure $\langle trcl(x), x, \in \rangle$ is isomorphic to a hyperarithmetical structure $\langle Y, X, E \rangle$ with $Y \subseteq \mathbb{N}$, $X \subseteq Y$, and $E \subseteq \mathbb{N}^2$ if and only if $x \in L(\omega_1^{CK})$. Here $trcl(x)$ denotes the transitive closure of x.

Every Π_1^1 subset of $\mathbb{N}$ is Σ_1-definable over $L(\omega_1^{CK})$ and vice versa.

The true analogies are that $L(\omega_1^{CK})$ replaces the *hereditarily finite sets*, ω_1^{CK} replaces $\mathbb{N} = \omega$, the Δ_1-definable subsets of ω_1^{CK} replace the computable subsets of $\mathbb{N}$ and the Σ_1-definable subsets of ω_1^{CK} replace the semicomputable subsets of $\mathbb{N}$.

We have generalized the basic concepts of classical computability theory using definability characterizations. We will obtain an equivalent generalization if we define the class of partial computable functions using the characteristic functions of $<$ and $=$ over ω_1^{CK}, primitive recursion and the μ-operator over

ω_1^{CK}, so the generalization is indeed that of replacing $\mathbb{N}$ with ω_1^{CK}. Note that all Δ_0 predicates on ω_1^{CK} will be computable.

The analogies are nevertheless not complete, in this model there will be semicomputable subsets of "finite" sets that are not "finite". Any complete Π_1^1-subset of $\mathbb{N}$ will be like this. The problem with this is that if we try to directly translate a construction from classical computability theory to this generalized setting, the construction may not work. We will discuss one example, the solution of Post's problem.

A set $A \subset \omega_1^{CK}$ is *regular* if $A \cap \beta$ is in $L(\omega_1^{CK})$ for each $\beta < \omega_1^{CK}$. If A and B are subsets of ω_1^{CK}, we say that A is *reducible* to B if there is a set R, that is Σ_1 over $L(\omega_1^{CK})$, of quadruples $\langle x, y, u, v \rangle$ of subsets of ω_1^{CK} in $L(\omega_1^{CK})$ such that for all x and y

$$x \subset A \wedge A \cap y = \emptyset \Leftrightarrow \exists u \, \exists v \, (u \subset B \wedge v \cap B = \emptyset \wedge \langle x, y, u, v \rangle \in R),$$

where all quantifiers are over $L(\omega_1^{CK})$. This express that we can decide a finite amount of membership in A from a finite amount of membership in B, generalizing Turing reducibility to this setting. When R works this way, we call R a *reduction procedure*.

If A is Σ_1 and not regular, then there is a Π_1^1-subset B of ω reducible to A, where B is not hyperarithmetical, and ω_1^{CK} is computable in B. The set A is not, however, necessarily complete, due to the fact that reductions must take place within $L(\omega_1^{CK})$ (Shore [60]).

There is a regular, complete Σ_1-set. If $f \colon \omega_1^{CK} \to \mathbb{N}$ is Δ_1 and one-to-one, then the *deficiency set*

$$D_f = \{\gamma < \omega_1^{CK} \mid \exists \delta > \gamma \, (f(\delta) < f(\gamma))\}$$

will be both complete and regular.

In the classical solution of Post's problem we enumerate all possible reduction procedures, and produce two computably enumerable sets A and B satisfying that for each number e, A is not reducible to B via procedure number e, and vice versa. We solve the conflicts between trying to satisfy one of these requirements and protecting an earlier attempt to satisfy another requirement by giving the requirements a priority order.

The generalization of the classical solution to Post's problem will be to construct two regular Σ_1 sets that are not reducible to each other, and a direct translation of the classical construction would be to enumerate all reductions R over ω_1^{CK} and carry out the priority construction. The problem with this naive approach is that we cannot prove the priority lemma. Since it is undecidable, in fact not even semidecidable, when a requirement is finally met, we may use the full time of ω_1^{CK} while satisfying a bounded set of requirements. $L(\omega)$ satisfies full replacement, while $L(\omega_1^{CK})$ only satisfies Σ_1-replacement, and this is not sufficient to prove the naive version of the priority lemma.

The solution, in this first generalization to higher computability theory, is that we in a way enumerate the requirements using ω, and then use this enumeration as a basis for assigning priority. This is possible since there is a total, computable and one-to-one map $\rho \colon \omega_1^{CK} \to \omega$. We give the requirement R_e a priority according to the value of $\rho(e)$. Then our construction will be a genuine, and not generalized, finite injury argument. Thus the classical solution to Post's problem carries over to this first instance of higher computability theory. In fact, much of the classical degree theory can be translated using similar compressions of the relevant set of requirements.

It is worth noticing that the set of requirements of higher priority than a given one is not uniformly computable, even if it is finite. This is not a genuine obstacle, since given a requirement and a stage in the construction, the set of requirements that are both of higher priority than the given one, and has been handled at an earlier stage, is computable uniformly from the data.

The Turing model has been generalized to other ordinals as well, and much of the degree theory for ω_1^{CK} can be seen as a special case of a further generalization. Thus we move directly to the next step in the sequence of generalizations.

2.2. Computing over an admissible ordinal. Computational models, where the domain is an admissible ordinal, is historically known as α-*recursion theory*. In this section we will give a brief introduction to the most important aspects of α-recursion theory. Further references are Sacks [56] and Chong and Friedman [3].

We let *an admissible structure* be a transitive model of *Kripke–Platek set theory*. This is the fragment of Zermelo–Fraenkel set theory without choice where we omit the power set axiom and restrict the comprehension axioms and replacement axioms to Δ_0-formulas. The following are axiom schemes in this theory:

$$\forall x \, \exists y \, \forall z \, (z \in y \leftrightarrow z \in x \wedge \Phi(z, x, \vec{x})),$$

and

$$\forall x \in u \, \exists y \, \Phi(x, y, \vec{x}) \to \exists v \, \forall x \in u \, \exists y \in v \, \Phi(x, y, \vec{x}),$$

where Φ will be a Δ_0 statement, u is a free variable and $\vec{x}$ is any list of free variables.

It is easy to see, using the Δ_0-definability of pairing, that the replacement scheme holds for Σ_1-statements Φ in general, and that we have comprehension for Δ_1-predicates.

An ordinal α is admissible if $L(\alpha)$ is an admissible structure. For some years, there was a rapid development of α-recursion theory, with the Sacks school at MIT and Harvard as the driving force. The focus was on generalizing classical degree theory to α-recursion theory.

First of all, we let a set $A \subset \alpha$ be α-*computable* if A is Δ_1 definable over $L(\alpha)$ with parameters from $L(\alpha)$, and we let A be α-*semicomputable* if it is Σ_1-definable. These definitions are direct translations of those we used for $\alpha = \omega_1^{CK}$, and they are equally natural. Working over $L(\alpha)$ has the effect that we can skip some coding. The well ordering of $L(\alpha)$ has order type α and is α-computable. Our definition of computable and semicomputable sets may alternatively be based on a generalized computability theory involving primitive recursion over α, the μ-operator over α and elementary operations. This theory will satisfy the S_n^m-theorem, the recursion theorem, a variant of Kleene's T-predicate and other basic ingredients of classical computability theory.

We generalize the definition of regular sets directly, and let a set $A \subset \alpha$ be *regular* if $A \cap \beta \in L(\alpha)$ for all $\beta < \alpha$. Our definition of reducibility for subsets of ω_1^{CK} translates directly from ω_1^{CK} to α. Then, for every admissible α, we have the following result due to Sacks [51]:

THEOREM 1. *Every Σ_1-set A is of the same degree as a regular Σ_1-set B.*

The key to the proof is to let B be the deficiency set of a one-to-one enumeration of A.

If A and B are subsets of α, we let $A \leq_\alpha B$ if A is reducible to B in the above sense. Note that an alternative definition is also that the sets $\mathcal{P}(A) \cap L(\alpha)$ and $\mathcal{P}(\overline{A}) \cap L(\alpha)$ are both Σ_1 in B. Here $\mathcal{P}$ is the power set operator and $\overline{A}$ is the complement of A. This is stronger than just requiring that A and $\overline{A}$ are Σ_1 in B. If only the latter is the case, we say that A is *weakly α-recursive* in B, and write $\leq_{w\alpha}$ for this relation.

$\leq_{w\alpha}$ is in general not transitive, while $\leq_\alpha$ is transitive. A strong solution to Post's problem will be a pair of Σ_1-subsets of α that are not weakly reducible to each other.

As a consequence of Theorem 1 we see that since there is a universal Σ_1-set, there will be a complete, regular Σ_1-set. Another consequence is that we may restrict the α-degree theory of semicomputable sets to regular sets.

Several of the prominent results of classical degree theory can be generalized to this setting, e.g., the solution to Post's problem (Sacks–Simpson [57] and Lerman [35]), the *splitting theorem* (Shore [61]) and the *density theorem* (Shore [62]). The proofs are nontrivial, and often combine insight from the fine structure theory for L and computability theory. One of the useful concepts inherited from fine structure theory is that of a *projectum*. We will define the simplest projectum and explain intuitively why this projectum is an important tool.

We let α^* be the least ordinal $\leq \alpha$ such that there is a 1-1 α-computable map ρ of α into α^*.

α^* is called the Σ_1-*projectum* of α. There will be other projecta used in solving harder problems than Post's, but we will not need them in this paper.

One important property of α^* is that any bounded Σ_1 subset of α^* will be in $L(\alpha)$. One might hope that this is sufficient in order to set up at least the simplest priority constructions. Any decent set of requirements can be enumerated as $\{R_\gamma \mid \gamma < \alpha\}$, and we may give R_γ a priority according to the value of $\rho(\gamma)$. Then the set of requirements of higher priority than a given one will always be an element of $L(\alpha)$, even though we cannot compute it in a uniform way. The problem comes, even for generalizations of *finite protection, finite injury*-arguments, when we want to prove that we at some stage stop injuring a given requirement. If $L(\alpha)$ satisfies replacement for sets bounded below α^*, there is no problem. Then we can show that if $\beta < \alpha^*$ and all actions for any requirement R_γ with $\rho(\gamma) < \beta$ will stop at some level before α, then there is a common level where all actions related to all requirements R_γ for $\rho(\gamma) < \beta$ has stopped. This is indeed the case when $\alpha = \omega_1^{\mathrm{CK}}$. Then $\alpha^* = \omega$ and β is truly finite when $\beta < \alpha^*$. In other cases, we need more sophisticated arguments.

There are two papers from 1972 where Post's problem for α-recursion theory is solved, Lerman [35] and Sacks–Simpson [57]. We cannot give any of the constructions in full in this paper, but we will discuss one of the cases in the Sacks–Simpson argument.

If α is admissible, an α-cardinal will be an ordinal $\kappa < \alpha$ that cannot be put in a 1-1-correspondence with a smaller ordinal via a function in $L(\alpha)$. Note that if $\alpha^* < \alpha$, then α^* is an α-cardinal, and the largest one. We will only consider the case when $\alpha^* < \alpha$ and α^* is a regular α-cardinal, meaning that

$$L(\alpha) \models cf(\alpha) = \alpha.$$

As above, let $\rho \colon \alpha \to \alpha^*$ be computable and 1–1. Let $\{R_e\}_{e \in \alpha}$ be an enumeration of all the standard requirements we want to satisfy in order to solve Post's problem for α-recursion theory. We let R_e have higher priority than R_d if $\rho(e) < \rho(d)$. Thus, every requirement has a priority rank below α^*, the set of requirements of higher priority than R_e will be α-finite, and we have to prove the induction step: If each requirement of higher priority than R_e is injured only $< \alpha^*$ many times, then R_e is injured $< \alpha^*$ many times.

The order type of the number of injuries a given requirement must suffer is computably enumerable, uniformly in the index of the requirement. Sacks and Simpson [57] showed a general lemma that we for our purposes simplify to:

If $A \subseteq (\alpha^*)^2$ is semicomputable such that
$$\{\beta < \alpha \mid A_\beta \neq \emptyset\} \text{ is bounded below } \alpha^*,$$
and
$$\text{each } A_\beta \text{ is bounded below } \alpha^* \text{ for } \beta < \alpha^*,$$
where $A_\beta = \{\gamma < \alpha^* \mid (\beta, \gamma) \in A\}$,
then $A \in L(\alpha)$.

By this lemma, and using induction on the priority ordering, we can prove that every requirement is satisfied.

Focussing on this case can be seen as making α-recursion theory an injustice, since we have avoided the main problem. We normally need a more condensed ordering of the requirements than what we have used here. Even for the case that we discussed we do not have an α-computable control of which requirements will eventually have higher priority than the one we are currently dealing with. The set of requirements of higher priority than R_e is computably enumerable in e, but not computable. Using more condensed priorities, the set of requirements of higher priority than a given one will not even be semicomputable.

It is the different ways one handles priority constructions, where on the one hand we need a condensed ordering of the priorities and on the other hand produce semicomputable sets, or c.e. degrees, that gives the flavor to α-recursion theory. In the beginning one would use fine structure arguments from the theory of L, but later also more dynamic constructions less dependent on Löwenheim–Skolem arguments came into use.

There is another property of interest in α-recursion theory, and that is *hyper-regularity*. A set $A \subset \alpha$ is *hyper-regular* if the structure $\langle L(\alpha), A \rangle$ satisfies Δ_0-comprehension and Δ_0-replacement. The comprehension requirement is satisfied if and only if A is regular, so this is a strengthening of regularity. The advantage of having a hyper-regular set is that constructions depending on admissibility can be relativized to hyper-regular sets, and thus there is a challenge to find c.e. degrees with hyper-regular elements whenever possible. We will not discuss how this is achieved in this survey.

For further reading, we also recommend Simpson [64].

2.3. Computing over admissible structures in general. Even though an ordinal α is a natural generalization of $\mathbb{N}$, and $L(\alpha)$ then is the natural generalization of the hereditarily finite sets, the successful generalization of classical computability to α begged the question if a computable well ordering of the domain is necessary for generalizing e.g., priority constructions. Moschovakis [40] suggested an axiomatic approach and defined what he called *Friedberg theories*, and he conjectured that Post's problem could be solved in the traditional way for all Friedberg theories. We will leave the discussion of attempts to axiomatize generalized computability theory to Section 5, and restrict ourselves to models of Kripke–Platek set theory here.

We will consider transitive sets M with urelements, where M satisfies the axioms of admissibility. *Urelements* are objects that are not themselves sets, so M will not satisfy extensionality. If there is a Σ_1-definable well ordering of M we will be close enough to α-recursion theory to have no serious problems generalizing the simplest priority constructions to M. However, Simpson [63] proved, under the assumption of determinacy, that the admissible closure $\mathbb{R}^+$

of the reals $\mathbb{R}$ offers a negative solution to Post's problem. In this example, there is a pre-wellordering of the model such that all initial segments are elements of the model. Thus admissibility together with the existence of a nice pre-wellordering do not suffice to generalize priority constructions.

Some research clarifying what was needed for generalizing priority constructions beyond what can be well ordered was carried out, e.g., in the thesis [71] by Stoltenberg-Hansen, but this line of research was never brought far.

2.4. Computing over an inadmissible ordinal. As we have mentioned, the fine structure analysis of L, as developed by Jensen [20], was an important tool in α-recursion theory. Admissibility is not an important property in the fine structure theory, and one motivation for extending the computability theory on ordinals to inadmissible fragments of L was to exploit fine structure theory in a more substantial way. It is also the case that even in α-recursion theory, we will be interested in computability relative to some predicate $A \subseteq L(\alpha)$ where $\langle L(\alpha), A \rangle$ is not necessarily admissible.

In β-recursion theory we work with *rudimentary closed* $L(\beta)$. $L(\beta)$ is rudimentary closed when it is closed under pairing and union, and when $L(\beta)$ satisfies Δ_0-comprehension. For a detailed definition of the rudimentary functions, see the section on set recursion.

The basic definitions of being computable, semicomputable and so forth are direct translations of the corresponding concepts in α-recursion theory. The lack of admissibility is an extra challenge if one tries to adjust arguments from α-recursion theory to β-recursion theory, and not all constructions can be generalized. There is a distinction between *weakly admissible ordinals* and *strongly inadmissible ordinals*. The ordinal β is weakly admissible if the Σ_1 cofinality of β is not smaller than the Σ_1-projectum. For an admissible α, the Σ_1-cofinality of α is α itself, but it turns out that many classical arguments depends on weak admissibility and not on full admissibility.

β-Recursion theory has received new interest in the context of hypercomputability, see [74]. Readers desiring a detailed introduction may consult the papers by Friedman [8] and Maass [36], and for a less detailed introduction, consult Chong and Friedman [3].

§3. Total functionals of finite types. It is well know that computations can be performed relative to an oracle in the form of a subset of $\mathbb{N}$ or in the form of a function $f: \mathbb{N} \to \mathbb{N}$. The rich theory of Turing-degrees of total functions is based on this. Mathematically, there is nothing stopping us from considering an oracle as an *input* to the algorithm. We do so when we use Turing machines or equivalent formulations to define what we mean by a partial computable function $F: A \to \mathbb{N}$ where A is a subset of $\mathbb{N}^{\mathbb{N}}$. The domain of F will be an open set and F will be continuous on the domain in the sense of the standard topologies on $\mathbb{N}$ and $\mathbb{N}^{\mathbb{N}}$.

Since all computations, also relativized ones, will be finite, it is also a basic fact that when F is computable, $F(f)$ makes sense for some partial inputs f as well, and that if F terminates on f, then F terminates on a finite, partial subfunction of f.

Computing with oracles does not belong to higher computability theory, but if we ask what it means to compute relative to a functional, to use a functional as an oracle, the matter changes. If we want to compute, using a functional F of type 2 (or even more complex) as an oracle, we have to face the problem that the oracle must be fed with infinite information before we can expect an answer. As we will see, there are various ways to make sense of such oracle calls, either by restricting our attention to functionals of a special kind or by strongly generalizing our idea of what a computation might be. In this section we will survey various directions where we only consider typed structures of functionals that are total. In the next section we will consider various directions where we compute relative to partial functionals. Some of these directions actually have important applications in theoretical computer science.

As was the case when computations were generalized to ordinals, the original motivation when computations were relativized to functionals was to have a tool for working with hyperarithmetical sets. However, while we in Section 2 considered direct extensions of the concepts from μ-recursion to ordinals, and it was easy to catch the intuition on the basis of our prerequisites, it is less obvious how to relativize computations to functionals. Thus we have to give a more detailed introduction.

3.1. Functional arguments. Let us consider the functional $^2E : \mathbb{N}^\mathbb{N} \to \mathbb{N}$ defined by

$$^2E(f) = \begin{cases} 1 & \text{if } \exists n\,(f(n) > 0), \\ 0 & \text{if } \forall n\,(f(n) = 0). \end{cases}$$

The functional ^2E is not continuous, so there is no way we can compute ^2E, not even from a function oracle. This means that if we want to ask ^2E a question, i.e., we want to use ^2E as an oracle, we somehow have to present a function f to ^2E and then ask for the value of $^2E(f)$. By presenting f to ^2E we will mean that we have an algorithm for computing f. In this algorithm, we may use ^2E as an oracle.

There are two reasons why it is natural to require that this f is total. If f is a partial subfunction of the constant zero function, we simply do not have the information required for ^2E to know what the value should be. This is one simple reason. The other reason is more technical. In order to have a precise definition of a computation as a well founded tree of local calculations, we need to know exactly which part of f that is used in establishing a value for

$^2E(f)$, and then the simplest alternative is to let $^2E(f)$ be undefined unless f is total.

Later, we will give a precise definition of what we mean by computing relative to ^2E. Let us prepare for this definition by showing how the hyperarithmetical set H_e, defined in Section 2.1, can be computed from e and ^2E. We compute the characteristic function K_e of H_e by the following "algorithm":

If $e = \langle 0, a \rangle$ then $K_e(a) = 1$ and $K_e(x) = 0$ when $x \neq a$.
If $e = \langle 1, d \rangle$ then $K_e(x) = 1 - K_d(x)$.
If $e = \langle 2, d \rangle$ and $x \in \mathbb{N}$, let $f_{e,x}(a) = K_{\phi_d(a)}(x)$, then
$$K_e(x) = {}^2E(f_{e,x}).$$

We let $K_e(x)$ be undefined otherwise.

In the last clause, if $\phi_d(a)$ is undefined, our algorithm for computing $f_{e,x}(a)$ will fail to terminate for all x, and $^2E(f_{e,x})$ will be undefined. Thus, if e is not an index for a hyperarithmetical set, we may prove that $K_e(x)$ is undefined for all x. If e is an index for a hyperarithmetical set we will be able to prove, having given the formal definitions, that K_e is computable in ^2E and is the characteristic function of H_e. The proof requires transfinite induction, and what we will consider to be terminating computations will actually be infinite well founded trees that code the evaluation tree for the characteristic function of H_e uniformly in e.

Kleene [26] extended these ideas to functionals of arbitrary high pure type, where we consider a functional as an argument by accepting oracle calls to that functional. We will give the precise definition, discuss the original motivation and survey some of the directions the investigation of Kleene's concept has taken. In contrast to α- and β-recursion theory, Kleene's concept cannot be seen as a direct generalization of μ-recursion, and the basic definition of a computable functional and semicomputable set is much more involved.

Also, the main motivation was not to see what is needed in order to carry out degree-theoretical arguments in a generalized setting, but to apply computability theory to definability theory.

3.2. Kleene's model. By recursion, we let

$$\text{Tp}(0) = \mathbb{N}$$

and

$\text{Tp}(k + 1)$ be the set of all functions $F\colon \text{Tp}(k) \to \mathbb{N}$.

We call $\{Tp(k) \mid k \in \mathbb{N}\}$ the *hierarchy of total functionals of pure types*. Kleene [26] defined a global relation

$$\{e\}(\Phi_1^{k_1}, \ldots, \Phi_n^{k_n}) \simeq a$$

where e, n, k_i vary over $\mathbb{N}$ and $\Phi_i^{k_i}$ varies over $\text{Tp}(k_i)$. Here, a may either be in $\mathbb{N}$ or *undefined*, and $\simeq$ means that either both sides are undefined or both sides are defined and equal. The definition is by a positive induction, using nine clauses known as the *schemes* S1–S9.

In the definition, we will let x and y be natural numbers. We will ignore the superscripts k_i, only indicating the *arity* of the $\{e\}$, but of course, $\{e\}$ will have different interpretations for different arities. In some expositions, the arity is coded into the *index* e.

REMARK 2. Unfortunately, the word "index" is traditionally not only used for the Gödel numbers used to code algorithms, but also for the upper and lower indices we always use in mathematics. In the definition below, an upper index will mark the type of an argument.

S1 If $e = \langle 1 \rangle$ then $\{e\}(x, \Phi_2, \ldots, \Phi_n) = x + 1$.
S2 If $e = \langle 2, q \rangle$ then $\{e\}(\Phi_1, \ldots, \Phi_n) = q$.
S3 If $e = \langle 3 \rangle$ then $\{e\}(x, \Phi_2, \ldots, \Phi_n) = x$.
S4 If $e = \langle 4, e_1, e_2 \rangle$ then
$$\{e\}(\Phi_1, \ldots, \Phi_n) \simeq \{e_1\}(\{e_2\}(\Phi_1, \ldots, \Phi_n), \Phi_1, \ldots, \Phi_n).$$
S5 If $e = \langle 5, e_1, e_2 \rangle$ then
$$\{e\}(0, \Phi_2, \ldots, \Phi_n) \simeq \{e_1\}(\Phi_2, \ldots, \Phi_n)$$
$$\{e\}(x+1, \Phi_2, \ldots, \Phi_n) \simeq \{e_2\}(x, \{e\}(x, \Phi_2, \ldots, \Phi_n), \Phi_2, \ldots, \Phi_n).$$
S6 If $e = \langle 6, d, \tau(1), \ldots, \tau(n) \rangle$ where τ is a permutation of $\{1, \ldots, n\}$, then
$$\{e\}(\Phi_1, \ldots, \Phi_n) \simeq \{d\}(\Phi_{\tau(1)}, \ldots, \Phi_{\tau(n)}).$$
S7 If $e = \langle 7 \rangle$ then $\{e\}(x, f^1, \Phi_3, \ldots, \Phi_n) = f(x)$
S8 If $e = \langle 8, e_1 \rangle$ then
$$\{e\}(\Phi_1^{k+2}, \Phi_2, \ldots, \Phi_n) \simeq \Phi_1(\lambda \xi^k \{e_1\}(\xi, \Phi_1, \ldots, \Phi_n)).$$
S9 If $e = \langle 9, m \rangle$ where $2 \leq m \leq n$, then
$$\{e\}(d, \Phi_2, \ldots, \Phi_n) \simeq \{d\}(\Phi_2, \ldots, \Phi_m).$$
– $\{e\}(\Phi_1, \ldots, \Phi_n)$ is undefined otherwise.

In S8 we used the standard notation for function abstraction, where $\lambda \xi - \xi -$ denotes the function that to an argument ξ gives the interpretation of $-\xi-$ as its value.

S1–S9 can be viewed as the clauses in a strictly positive inductive definition, and as usual with such definitions, there will be a unique well founded tree verifying that
$$\{e\}(\Phi_1, \ldots, \Phi_n) = a$$
whenever this is the case. Applications of S1, S2, S3 and S7 give us the leaf nodes, and we call them *initial computations*.

A terminating computation $\{e\}(\Phi_1, \ldots, \Phi_n) = a$, where e is an index as constructed in S4, S5, S6 or S9, will have one or two immediate subcomputations. In the case of S8, however, there will be one subcomputation

$\{e_1\}(\xi, \Phi_1, \ldots, \Phi_n)$ for each ξ of type k, and thus there will be an infinite branching of the computation tree. This means that computing relative to functionals of type 2 and above belongs to higher computability theory, at least in the sense that we generalize to the infinite (but well founded) what it means to be a terminating computation. Note that the nodes of a computation tree will be labelled with *computation tuples*

$$\langle e, \Phi_1, \ldots, \Phi_n, a \rangle,$$

i.e., the labels will consist of the index, the input and the value or output.

Kleene's definition of computing relative to functionals is not unproblematic from a conceptual point of view. S9 axiomatizes that there is a universal algorithm, something that in other connections are seen as theorems. This is the main conceptual problem. Another problem is the understanding of S8, oracle calls for functionals. Since we restrict ourselves to the hierarchy of *total* functionals, the scheme requires that $\{e_1\}(\xi, \Phi_2, \ldots, \Phi_n)$ terminate for all functionals ξ of type k. If Φ_1 is given by some algorithm, it may be that if we apply this algorithm to $\lambda \xi \{e_1\}(\xi, \Phi_1, \ldots, \Phi_n)$, we may get termination also in some cases where the input functional is not total. Kleene's notion of computability is not closed under composition, information is lost by requiring that everything is total. This is of course problematic if the aim is to extend the concept of computability to functionals in "the correct way".

On the positive side we can observe that the S_n^m-theorem will hold, with a simple proof, and the recursion theorem follows. Experience has shown that there is no reasonable Church–Turing thesis for computing with functionals, and Kleene's concept must be judged from what it can be used for. Kleene's ambition was partly to find higher versions of the hyperarithmetical hierarchy, i.e., to investigate natural transfinite extensions of the analytical hierarchy, the Π_n^1- and Σ_n^1-hierarchy, and even higher analogues.

In the first years, much of the research around S1–S9 concentrated on *normal functionals*, and this will be our theme for the next section. If Ψ^{k+1} and Φ are functionals, we say that Ψ is *Kleene computable* in Φ if there is an index e such that for all ξ of type k:

$$\Psi(\xi) = \{e\}(\Phi, \xi).$$

The 1-*section* of a functional Φ is the set of functions f that are Kleene computable in Φ. The 1-section is of course closed under Turing computability. The 1-section tells us something about the computational power of a functional, but not everything. The 1-section of ^2E is the set of hyperarithmetical functions. Grilliot [15] showed that a functional F of type 2 either resembles a continuous functional or ^2E is computable in F. This was later improved into the dichotomy [45]:

THEOREM 3. *Let Φ be a total functional of finite type. Then one of the two statements will hold*:
1. *There is an f computable in Φ such that the 1-section of Φ is generated by its c.e.(f) elements.*
2. *^2E is Kleene computable in Φ.*

As a corollary, if the 1-section of Φ is closed under jump, then ^2E is computable in Φ.

Bergstra [2] and Wainer [73] developed the methods used in [15] further. The main construction in [45] combines constructions due to Grilliot, Berger and Wainer with the help of the recursion theorem.

3.3. Normal functionals. We have seen that every hyperarithmetical set is computable in ^2E, and that the 1-section of ^2E is the set of hyperarithmetical functions on $\mathbb{N}$. This suggests that if we consider computations

$$\{e\}(^2E, a_1, \ldots, a_n)$$

we get an alternative concept of computing within the realm of hyperarithmetical sets than what we obtained from μ-recursion over ω_1^{CK}. It turns out that the two approaches are equivalent. Gandy [10] showed that there is a partial selection operator over sets semicomputable relative to ^2E, and this can be used to reduce the μ-operator over ω_1^{CK} to Kleene computability relative to ^2E. The point behind the proof is that every terminating computation in the sense of Kleene is equipped with a *well founded computation tree*, and comparison of the heights of two computation trees will be semicomputable. This, combined with the recursion theorem, is used in Gandy selection.

A functional F of type 2 is called *normal* if ^2E is Kleene computable relative to F. The computability theory of a normal F shares many of the same properties as that of ^2E, and we will not go into any detail here. If it is just the 1-sections we are interested in, the following theorem due to Sacks [52] tells us that using normal functionals of type 2 has a high level of generality:

THEOREM 4. *Let M be a countable, transitive model of Kripke–Platek set theory. Then there is a normal functional F of type 2 such that $M \cap \mathbb{N}^{\mathbb{N}}$ is the 1-section of F.*

The proof is by a forcing construction, where F will be generic over M. The construction is actually a case of *class forcing* over a countable, admissible set.

The definition of being normal extends naturally to functionals of types higher than 2.

DEFINITION 5. Let $k \geq 2$ and define ^kE as the functional of type k where:
$^kE(F) = 0$ if $F(f) = 0$ for all f of type $k - 2$.
$^kE(F) = 1$ if $F(f) > 0$ for some f of type $k - 2$.

Here F ranges over all total functionals of type $k - 1$.

We say that Φ of type k is *normal* if ^kE is Kleene computable in Φ.

In this survey, we will restrict ourselves mostly to $k = 3$ and to the computability theory of ^3E, though most of what we say will be valid for larger k and for normal functionals in general. There are some interesting new aspects turning up at type 4, but the search for new phenomena at type 5 has been in vain. We discuss the new aspects of type 4 at the end of this section.

First of all, the proof of Gandy selection works as long as we search for integers. Moschovakis [39] and Platek [48] established this. We can use this fact to show that the 1-section of any normal Φ satisfies the requirement of Sacks' forcing argument, and thus that there will be a normal functional of type 2 with the same 1-section. This is a special case of Sacks' *plus one theorem* [53], a similar statement for higher level sections.

The type is better reflected in the properties of the semicomputable sets. Ideally a set is semicomputable if there is an algorithm such that the corresponding *computation tree* is well founded exactly when the argument is in the set. The problem with this is that the natural computation trees connected to Kleene computations are defined bottom-up, following the inductive definition of the relation $\simeq$, and then there is no tree at all for non-terminating computations. Moschovakis [39] showed how we may give a top-down definition of a computation tree such that the computation tree of $\{e\}(^3E, f_1, \ldots, f_n)$ is semicomputable in ^3E. Here $f_i \in \mathbb{N}^\mathbb{N}$ or $f_i \in \mathbb{N}$.

In contrast to the computation trees we discussed above, nodes in a Moschovakis-style computation tree will only be labeled with sequences of the form

$$\langle e, \Phi_1, \ldots, \Phi_n \rangle,$$

i.e., we do not require any value of a computation. Thus we may give a meaning to computation trees of non-terminating computations, and generate the computation tree of any expression

$$\{e\}(\Phi_1, \ldots, \Phi_n)$$

independent of termination.

In all cases except S4 and S5 we know what the arguments of all the immediate subcomputations will be in case the computation terminate. In the cases of S4 and S5 we will have a binary branching of the computation tree, where we have to compute the result of one branch in order to know the input of the other branch. In the computation tree of Moschovakis, there will only be one immediate subcomputation if the first subtree is not well founded, and two if it is. The tree constructed by this top-down procedure will be well founded if and only if the computation terminates. The computation tree is uniformly semicomputable. Moreover, the set of infinite branches in the tree will be semicomputable. The key observation is that an infinite descending sequence in a computation tree now will be of the same type as the arguments, and that *not being well founded* can be expressed as a Σ_1 statement where a semicomputable set is used in a positive way. The consequence, due to Moschovakis, is that the class of subsets of $\mathbb{N}^\mathbb{N}$ semicomputable in ^3E is not closed under existential quantifiers. The corresponding classes of sets fail to satisfy the Kripke–Platek axioms, and at this level some new and interesting phenomena occur. One of these phenomena is *reflection*. In order to explain this, we need to define the *companion* of ^3E in precise terms.

An ordinal α is computable in ^3E and f if there is a pre-wellordering of $\mathbb{N}^\mathbb{N}$ computable in f and ^3E of hight α. We let κ_f be the supremum of the ordinals computable in ^3E and f, and we let κ be the supremum of all κ_f. Let $L[\mathbb{N}^\mathbb{N}](\alpha)$ be the analogue of the Gödel hierarchy, where $L[\mathbb{N}^\mathbb{N}](0) = \mathbb{N}^\mathbb{N}$. Then the companion is $M = L[\mathbb{N}^\mathbb{N}](\kappa)$.

The elements of the companion will be coded by subsets of $\mathbb{N}^\mathbb{N}$ computable in ^3E and some f, and each f defines a non-transitive substructure M_f of those sets with codes computable in ^3E and f. Each M_f satisfies Δ_0-comprehension, while the family as a whole satisfies a modified collection principle as a consequence of Gandy selection (MacQueen [37]).

Let ϕ be a Δ_0-statement with parameters from M_f. Then

$$\forall g \in \mathbb{N}^\mathbb{N} \exists y \in M_{\langle f,g \rangle}\, \phi(g,y) \Rightarrow \exists v \in M_f \forall g \in \mathbb{N}^\mathbb{N} \exists y \in v\, \phi(g,y).$$

Now let us return to the reflection property:
Let O denote the constant zero function, and let

$$O'(e) = \begin{cases} 0 & \text{if } \{e\}(^3E, e)\downarrow, \\ 1 & \text{if } \{e\}(^3E, e)\uparrow, \end{cases}$$

where $\downarrow$ means *is defined* and $\uparrow$ means *is undefined*.

The reflection principle (Harrington [17]) is as follows:

> If ϕ is a Σ_1-statement with parameters from M_O true in $M_{O'}$, then ϕ is true in M_O.

O' is the jump of O, and of course the jump operator can be generalized.

We will give a brief outline of the proof of the reflection property:

Relative to ^3E, any Π_1^1 predicate is computable, and thus in particular the set of codes for countable well orderings will be computable in ^3E. If f codes a well ordering of length α, and ϕ is true in $M_{O'}$, then either ϕ is true in M_f or there will be a set A in M_f of terminating computations with integer inputs only (together with ^3E of course) such that the order type of the set of ordinal ranks of the computations in A is at least α. Using the modified collection principle, we can find a level in M_O where for each code f of a well ordering, one of the two will be satisfied. Since we cannot satisfy the second part at a computable level for each f, the first part must be satisfied.

While the urge to see how far priority arguments can be pushed was the key motivation behind α- and β-recursion theory, this was not so for the theory of computations relative to normal functionals. Nevertheless, attempts to solve Post's problem and to prove other results of a degree-theoretical nature were made by MacQueen [37], by Normann [46, 43] and later by Sacks and his, at the time, students Griffor [12] and Slaman [65, 67, 66] in the setting of set recursion. Normann solved Post's problem, proved the splitting theorem and constructed a minimal pair of c.e. degrees for recursion relative to ^3E and the analogues of even higher types under essentially the assumption that V=L. Through the contributions of Sacks [54] and Slaman [67], reflection phenomena turned out to be useful in the investigation of the finer properties of recursion in ^3E. For instance, a finite injury argument can be replaced by a "wait and see"-argument. Requirements will be met by adding objects to the sets under construction and by setting up protections. Thanks to the reflection principle there will be no need to injure these protections, so there is no need to give the requirements a priority ordering. The reader should consult Sacks [56] in order to see the details of how the reflection principle is used, but simplifying the matter we can say that since a requirement can be met at the level of the jump, it can be met at a computable level by reflection.

The proof of the reflection theorem can be relativized, so M_f will be a Σ_1-elementary substructure of M_g, where g is obtained from f by iterated jump. Recall that κ_f denotes the supremum of f-computable ordinals (relative to ^3E). We let κ_f^r be the largest ordinal such that $L[\mathbb{N}^\mathbb{N}](\kappa_f)$ is a Σ_1-elementary substructure of $L[\mathbb{N}^\mathbb{N}](\kappa_f^r)$, where we only accept parameters from M_f. κ_f^r will be the level where we can verify that there are Moschovakis witnesses, and in general, not before.

We will not enter deeper into the computability theory of ^3E, but just mention two facts:

> If $A \subset \mathbb{N}^\mathbb{N}$ is semicomputable such that the complement is nonempty, then there is some $f \notin A$ such that $\kappa_f < \kappa_O^r$.
> There is a semicomputable, nonempty set $A \subset \mathbb{N}^\mathbb{N}$ with no nonempty, computable subset.

The first result is known as the *Kechris Basis Theorem*. The second result shows that semicomputable sets are not computably enumerable in any decent sense for this kind of higher computability. The set

$$A = \{f \mid \kappa_f > \kappa_0^r\}$$

will satisfy the second claim.

In addition to the plus 1 theorem, there is also a *pluss* 2-theorem, due to Harrington [17]. If Φ is a normal functional of type $\geq k + 2$ we define the *k-envelope* of Φ as the collection of subsets of $\mathrm{Tp}(k-1)$ semicomputable in Φ. Harrington showed that the k-envelope cannot characterize the type of a normal functional, by proving

THEOREM 6. *If Φ is a normal functional of type $\geq k+2$ there will be a normal functional Ψ of type $k+2$ with the same k-envelope.*

Now one may ask if there is anything exiting going on if one moves from type 3 to higher types. The proof of Gandy selection works for all normal functionals of type ≥ 2. At type 3, there is a nonempty semicomputable subset of $\mathbb{N}^\mathbb{N}$ with no nonempty computable subset, so there is no way to select nonempty computable subsets of semicomputable sets. However, if Φ is a normal functional of type 4, Grilliot [14] showed that we may uniformly select a nonempty subset computable in Φ of any nonempty subset of $\mathbb{N}^\mathbb{N}$ semicomputable in Φ. Grilliot selection is a new phenomenon, in the sense that at type 3 it coincides with Gandy selection.

Reflection phenomena and their consequences are valid for computations relative to normal functionals of any type ≥ 3.

For a survey of the computability theory of normal functionals of higher types, see Moldestad [38].

3.4. The Superjump. Even though most of the early research on the Kleene model for computing with functionals of higher types were concerned with computations relative to normal functionals, there was one non-normal functional that called for some attention, *the Superjump*.

The Superjump S is a functional of type 3 defined as the jump operator for type two functionals:

$$S(F, e) = \begin{cases} 0 & \text{if } \{e\}(F)\downarrow, \\ 1 & \text{if } \{e\}(F)\uparrow. \end{cases}$$

It is easy to see that ^2E is computable in S: There is an index e for the algorithm that given f searches for the least n such that $f(n) > 0$. If this algorithm does not terminate, $^2E(f) = 0$ and if it terminates, $^2E(f) = 1$. Since $\mathbb{N}^\mathbb{N}$ is a computable retract of $\mathrm{Tp}(2)$ it follows that ^2E is computable in S.

The superjump S is of special interest because of the links to the first recursively Mahlo ordinal ρ. Harrington [16] showed that the 1-section of S is exactly $\mathbb{N}^\mathbb{N} \cap L(\rho)$. In his proof he essentially replaced S with a functional $\hat{S}$ also operating on partial F. For $\hat{S}(F)$ to be defined, F has to be total on its own 1-section only. Apart from this, the definition is the same as that of S. The Mahlo-property has caught the interest of people working in proof theory and in constructive type theory as one possible limit of closure under predicative constructions.

With the resent development of hypercomputability theory, see [74] of this volume, one natural problem suggests itself:

> Is there a non-normal functional T of type 3 such that the computability theory of T is equivalent to that of hyper Turing machines?

We deliberately are not precise about what we mean by being equivalent here, but at least we must ask for the semicomputable subsets of $\mathbb{N}$ to be the same in the two models for the problem to be of interest.

3.5. Accepting sets as arguments of computations. When we work with computations relative to a normal functional of type $k+2$ we have essentially given ourselves the power to check quantifiers over total predicates of type k objects. So in a sense, we consider $\mathrm{Tp}(k)$ to be a finite set. The philosophy behind *set recursion* is that as soon as we accept an arbitrary set x as an input to an algorithm, we consider x to be generalized finite.

The original motivation for defining set recursion was to extend computations relative to functionals to computations operating on their companions. The final definition was also inspired by the lectures on set theory given by R.B. Jensen in Oslo at the time. If we extend the construction of the rudimentary functions with a Kleene enumeration and a Kleene-scheme for a universal algorithm, we get a unified theory that generalizes computations relative to any normal functional. The precise definition, as given by the author in [44], is:

DEFINITION 7. Let R be a class. We define the class relation
$$\{e\}^R(x_1,\ldots,x_n) \simeq y$$
by the following positive inductive definition:
1. If $e = \langle 1, n, i\rangle$ then $\{e\}^R(x_1,\ldots,x_n) \simeq x_i$.
2. If $e = \langle 2, n, i, j\rangle$ then $\{e\}^R(x_1,\ldots,x_n) \simeq x_i \setminus x_j$.
3. If $e = \langle 3, n, i, j\rangle$ then $\{e\}^R(x_1,\ldots,x_n) \simeq \{x_i, x_j\}$.
4. If $e = \langle 4, n, d\rangle$ then $\{e\}^R(x_1,\ldots,x_n) \simeq \cup_{y \in x_1}\{d\}^R(y, x_1,\ldots,x_n)$.
5. If $e = \langle 5, n, d, e_1,\ldots,e_m\rangle$ then
$$\{e\}^R(x_1,\ldots,x_n) \simeq \{d\}^R(\{e_1\}^R(x_1,\ldots,x_n),\ldots,\{e_m\}^R(x_1,\ldots,x_n)).$$
6. If $e = \langle 6, n, i\rangle$ then $\{e\}^R(x_1,\ldots,x_n) \simeq R \cap x_i$.
7. If $e = \langle 7, n, m\rangle$ then $\{e\}^R(d, x_1,\ldots,x_n, y_1,\ldots,y_m) \simeq \{d\}^R(x_1,\ldots,x_n)$.

If we restrict this definition to 1.–5. we define the *rudimentary functions*. They serve as the basis for defining Jensen's [20] J-hierarchy used in the fine structure analysis of L.

If α is an admissible ordinal, then $L(\alpha)$ will be closed under the schemes of set recursion. The corresponding computability theory will not necessarily correspond to α-recursion, since set recursion will not automatically provide us with search over α.

Even though set recursion was constructed as a tool for translating arguments from α- and β-recursion theory to the setting of normal functionals, it proved to be of independent interest. Set recursion lives in the world of well founded sets. Since the successor operator $x \mapsto x \cup \{x\}$ is rudimentary, it is also set recursive. Combining this with the recursion theorem we see that the rank function will be set recursive and the transitive closure operator will be set recursive. Thus, accepting a set x as an input, set recursion will consider the transitive closure of x as finite.

The *set-recursive closure* of x will be the least set $E(x)$ closed under the schemes of set recursion, containing the transitive closure of x as a subset and x as an element. Sacks [55] proved that in general, $E(x)$ is not semicomputable in x in the sense of set recursion, in contrast with what is the case with computability theories based on search over the admissible closure of a set.

Set recursion is also known as E-recursion theory, see e.g., Sacks [56]. This reflects that it was intended as a generalization of computing relative to the functionals $^k E$. In contrast to α-recursion theory, the power of set recursion depends only on the input sets, and not on any surrounding environment. We are not given any external power of search, and selection theorems will often require complicated proofs. Given a terminating computation

$$\{e\}(x_1, \ldots, x_n) \simeq y,$$

there will be a well founded computation tree witnessing this fact, and this tree is uniformly computable from $e, x_1, \ldots, x_n$ (and y). Given two well founded trees, it will be within the power of set recursion to compare the ordinal ranks of the trees. As a consequence, the proof of the Gandy selection theorem can be directly applied to the case of set recursion. As a further consequence, a transitive set closed under set recursive functions will satisfy the modified collection principle discussed in connection with $^3 E$ and normal functionals.

The most advanced selection theorems are generalizations and improvements of Grilliot selection. The typical format of a selection theorem for $x \in M$, where M is closed under set recursion, is that if x is small in the sense that

$$2^x \cap M \in M$$

then we can uniformly find a nonempty computable subset B of any nonempty semicomputable subset A of x. We have simplified the actual statements here.

Associated reflection theorems are important tools in set recursion. We will not go into any details of the more intricate aspects of set recursion, but mention one result first proved by van de Wiele [72] using Π_2^1-logic, but later proved by more elementary means:

THEOREM 8. *Let F be a Σ_1 class function in the universe of well founded sets that is total and uniformly Σ_1-definable in all transitive, admissible structures. Then F is set recursive.*

3.6. Continuous functionals. The same year that Kleene published his celebrated schemes S1–S9 he also published a paper [25] where he isolated a sub-hierarchy of the hierarchy of all total functionals, a sub-hierarchy consisting of what he called *countable functionals*. They are related to the hereditarily total and continuous functionals that originally were constructed in Kreisel [32], though they are not the same. These days, Kreisel's continuous functionals are normally introduced via Scott–Ershov domains, following a characterization due to Ershov [4], and they are often called *Kleene–Kreisel continuous functionals*. Let us look at the original definition by Kleene:

We simultaneously define what we mean by an associate for a functional and by a countable functional:

Each $n \in \mathbb{N}$ is countable and is its own associate.
Each $f \in \mathbb{N}^\mathbb{N}$ is countable and is its own associate.
If $k \geq 1$, $F\colon \mathrm{Tp}(k) \to \mathbb{N}$ and $\alpha \in \mathbb{N}^\mathbb{N}$ we say that α is *an associate* for F if for all $f \in \mathrm{Tp}(k)$ with an associate β there is an $n \in \mathbb{N}$ such that
$$\alpha(\bar{\beta}(n)) = F(f) + 1$$
$$\alpha(\bar{\beta}(m)) = 0 \text{ for } m < n.$$

Here $\bar{\alpha}(n) = \langle \alpha(0), \ldots, \alpha(n-1) \rangle$ as usual.

$F \in \mathrm{Tp}(k+1)$ is *countable* if F has an associate.

Kleene proved that if Φ is computable in $\Psi_1, \ldots, \Psi_n$ and each of $\Psi_1, \ldots, \Psi_n$ are countable, then Φ is countable.

It can be questioned whether the computability theory of countable functionals belongs to the realm of higher computability theory. We include it here, since the computation trees are infinite, and since testing if a certain index with a certain input terminates is in general not arithmetical.

One problem with this definition is that there will be many functionals sharing the same associate, and there will be many associates for the same functional. Fortunately, if two functionals share one associate, they will have the same set of associates.

Kleene's requirement that oracle calls can only be made via total inputs, will have as a result that even the set of semicomputable subsets of $\mathbb{N}$ relative to a countable functional is not determined by the associate alone.

As a consequence of this, it is now more common to consider the *hereditarily total continuous functionals*. The definition of an associate remains the same, but we define $\mathrm{Ct}(k+1)$ as the set of functions $F\colon \mathrm{Ct}(k) \to \mathbb{N}$ having an associate. This version can be viewed as the extensional collapse of Kleene's original construction. Following Kreisel [32] we call these objects *continuous*. There is no problem in interpreting S1–S9 over the continuous functionals instead of over the countable ones. The definitions will not be equivalent though, since the requirements for termination in S8 are weaker in the continuous case than in the countable case, we only require subcomputations to terminate for continuous arguments and not for all total arguments. Oddly enough, this weakening of requirements for termination implies an increase in the complexity of basic concepts, like semicomputability. If ^4O denotes the constant zero functional of type 4, all Π_2^1-sets in $\mathbb{N}$ will be semicomputable in ^4O over the continuous functionals while all subsets of $\mathbb{N}$ semicomputable in ^4O will be Π_1^1 if we use the original definition of Kleene. The reason for the latter simplicity is that we may use a Löwenheim–Skolem argument and show that a computation terminates in the full type structure if and only if it does so in all countable type structures closed under S1–S9, and this is a Π_1^1-characterization. The reason of the high complexity in the first case is that the set of total continuous functionals of type 2 is a complete Π_1^1-set, and thus being total on this set has complexity Π_2^1. When the level of the type increases, the complexity of semicomputability relative to computable functionals does not increase in the original Kleene model, while it does so, from Π_n^1 to Π_{n+1}^1 at each step, in the hereditarily continuous model.

The computability theory of the hereditarily continuous total continuous functionals is well treated in the literature, and since it is on the borderline of what might be described as *higher computability theory*, we leave these functionals here.

§4. Partial functionals of finite types.

4.1. Platek's model. We may see S1–S9 as a contribution to *definability theory*, and for that purpose the requirement that an object can only be used in an oracle call when it is fully defined makes sense. Then we can also make use of the fact that whenever

$$\{e\}(\Phi_1, \ldots, \Phi_n) \simeq a$$

there is a unique *computation tree* representing the verification of this statement. We might obtain computation trees from objects accepting partial inputs as well, but in order to have uniqueness we must at least demand that whenever $\Phi(F)$ is defined, there is a least subfunction $G \subseteq F$ for which $\Phi(G)$ is defined. This is, however, not a requirement that is satisfied in natural models.

Platek [48] came up with an alternative definition of computing with functionals as inputs. His approach is based on two principles. One is that *recursion* means *finding least fixed points* of monotone operators. The other is that relativized computations represent ways of transforming bits of information to bits of information, and they must be monotone in the sense that the more information we have about the input, the more information we obtain about the output. We will now define the *hereditarily consistent* functionals of finite types as considered by Platek. We define the finite types by induction as:

0 is a type.
If τ and δ are types, then $(\tau \to \delta)$ is a type.

The types are formally words in the alphabet $\{0, (,), \to\}$, but we will interpret each type σ as a partially ordered set

$$(\mathrm{Pf}(\sigma), \sqsubseteq_\sigma)$$

as follows:

$\text{Pf}(0) = \mathbb{N} \cup \{\bot\}$ where $\sqsubseteq_0$ is the least reflexive relation such that $\bot \sqsubseteq_0 n$ for all $n \in \mathbb{N}$. We use the word "bottom" for $\bot$.

If $\sigma = (\tau \to \delta)$, we let $\text{Pf}(\sigma)$ consist of all functions $F\colon \text{Pf}(\tau) \to \text{Pf}(\delta)$ such that for all $f, g \in \text{Pf}(\tau)$:

$$f \sqsubseteq_\tau g \Rightarrow F(f) \sqsubseteq_\delta F(g).$$

We order $\text{Pf}(\tau \to \delta)$ by the pointwise ordering.

There will be no problem introducing product types $\sigma \times \tau$ as well, and it is a matter of taste if one will do so. Normally, the exposition is more elegant if product types are left out, while the types are more convenient for practical purposes if we have product types.

In the literature, the base type 0 is often denoted by the Greek letter ι. Each type σ will have a least element $\bot_\sigma$, essentially the constant $\bot$ functional. Moreover, each type will be chain complete in the sense that if $\{F_i \mid i \in I\}$ is a chain in $\text{Pf}(\sigma)$ then the least upper bound will also be in $\text{Pf}(\sigma)$. As a consequence, each $F \in \text{Pf}(\sigma \to \sigma)$ will have a least fixed point in $\text{Pf}(\sigma)$, a fixed point we reach by iterating Φ on $\bot_\sigma$ over the ordinals, using least upper bounds at limit stages. Platek considered these typed *least fixed point operators* as basic computable operators.

There are two approaches to untyped λ-calculus, on the one hand the standard one with variables, λ-abstraction, application and conversion rules for the terms, and on the other hand the variable free use of combinators. The expressive power is the same. The use of combinators simplifies the language at the cost of the readability of expressions. Platek actually used typed combinators to define the class of computable functionals, while we choose an equivalent definition in the form of a typed λ-calculus:

- For each natural number n, k_n is a constant of type 0.
- S is a constant of type $0 \to 0$, and the rules are $Sk_n \to k_{n+1}$ for each $n \in \mathbb{N}$.
- P is a constant of type $0 \to 0$ and the rules are $Pk_{n+1} \to k_n$ for each $n \in \mathbb{N}$.
- $\supset$ is a constant of type $0, 0, 0 \to 0$ and the rules are
  - $\supset k_0 MN \to M$
  - $\supset k_{n+1} MN \to N$ for each $n \in \mathbb{N}$.
- Y_σ is a constant of type $\sigma \to \sigma$ for each type σ and the rule is

$$Y_\sigma M \to M(Y_\sigma M)$$

for terms M of type $\sigma \to \sigma$.

Here we have followed the convention used in type theory where product types are left out, we write $0, 0, 0 \to 0$ instead of $0 \to (0 \to (0 \to 0))$ and we think of elements of this type as functions of three variables. Our definition is very close to that of PCF, see Section 4.3.

There is a natural interpretation of the terms in this language, and the value under this interpretation will be unchanged when we use the conversion rules. The domain of the interpretation will be the typed hierarchy of hereditarily consistent functionals:

Y_σ will be interpreted as the least fixed point operator in $\text{Pf}(\sigma \to \sigma)$. S is interpreted as the successor function on $\mathbb{N}$, with $S(\bot) = \bot$. P is interpreted as the predecessor function on $\mathbb{N}$ where we also have that $P(0) = \bot$. The interpretation of $\supset (x, y, z)$ is that the value is y if $x = 0$ and the value is z if $x > 0$, but in $\mathbb{N}$. Then all computable partial functionals in the sense of Platek can be defined using a term in this λ-calculus.

The schemes S1–S9 also make sense for the hereditarily consistent functionals of pure types, and Platek actually showed that the Kleene approach interpreted on the hereditarily consistent functionals of pure types will be equivalent to his approach that can be seen as a typed λ-calculus. The proof of the equivalence is worked out in detail in Moldestad [38]. The fact that $\mathbb{N}_\bot = \mathbb{N} \cup \{\bot\}$ is not computably isomorphic to $(\mathbb{N}_\bot)^2$ is a technical obstacle in the proof.

There is a link between the hierarchy of *total* functionals and Platek's hereditarily consistent ones, and the link is that the total functionals can be identified with equivalence classes of partial equivalence relations on $\text{Pf}(\sigma)$ for each type σ.

If we let
- $x \sim_0 y \Leftrightarrow x = y \in \mathbb{N}$ on $\text{Pf}(0)$,
- if $\sigma = \delta \to \tau$, $f, g \in \text{Pf}(\sigma)$ and $x, y \in \text{Pf}(\delta)$ then
$$f \sim_\sigma g \Leftrightarrow (f(x) \sim_\tau g(y) \text{ whenever } x \sim_\delta y),$$

it is easy to see that each relation $\sim_\sigma$ is reflexive and transitive, and that there is a 1-1 correspondence between equivalence classes in $\text{Pf}(\sigma)$ and the elements of $\text{Tp}(\sigma)$, a correspondence commuting with application. Termination of Kleene computations will be preserved when total functionals are replaced by hereditarily consistent ones, but not the other way.

Historically, Platek's work had an important impact on the theory of functional programming. In Section 4.3, we will explain this.

4.2. Kleene revisited. Kleene expressed dissatisfaction with his own definition of relative computability via S1–S9. For instance, he saw the lack of closure under composition as a problem. In a series of "revisited"-papers, [27, 28, 29, 30, 31], he introduced an alternative definition. His idea was that functional application represents some kind of communication between the two objects, a dialogue. The linguistic complexity of this explanation reflects the mathematical complexity of Kleene's new model. It would be too space consuming if we went into the details of Kleene's alternative definitions, so let us give just the intuitive idea. A computation will be a transfinite, sequential dialogue between the inputs, where they ask each other questions in the form of envelopes of dialogue descriptions of other functionals, and it is up to the object receiving the envelope to answer without opening the envelope, refuse to answer independent of the content of the envelope or to engage in a conversation with the object described by the content of the envelope in order to decide if it wants to terminate with a value for the input represented by the envelope. This sub-dialogue is then a part of the full dialogue. The new, and conceptually important, idea was that a computation can be seen as a sequential process, and that one might isolate the partial functionals of higher type that in some sense are sequential. We may say that the objects considered by Kleene are extensional, but that they will have intentional realizers, i.e., instructions for how to engage in a dialogue with a realizer of an input functional in order to "calculate", in a sequential transfinite way, its value on that input.

The hierarchy of functionals definable from dialogues is not entirely compatible with the hereditarily consistent ones. There is one trivial functional of one function variable that cannot be computed by any sequential procedure:

$$F(f) = \begin{cases} 0 & \text{if } f(0) = 1 \vee f(1) = 1, \\ 1 & \text{if } f(0) = f(1) = 0, \\ \bot & \text{otherwise.} \end{cases}$$

This is a variant of the *nondeterministic* **or**. The argument is that a sequential process computing $F(f)$ will have to ask for the value of $f(0)$ or of $f(1)$ before asking for the other one. Then, if the response to this first enquiry is $\bot$, the computation comes to a halt with undefined value, even if $F(f)$ is defined. Nondeterministic **or** is an element of Platek's model, but is not sequential. For a detailed prof, see e.g., Plotkin [49] where the same statement is proved for the language PCF (see Section 4.3).

Thus, at type 2, Kleene's revisited approach and Platek's approach are separated. Since discussing the relationship further would require a space consuming introduction of details, we stop here.

Kleene's revision of higher type computability was developed further by Kierstead, [22] who provided a model for the calculus. The influence on higher computability theory was nevertheless limited. However, in the same way as Platek's computability theory had some influence in theoretical computer science, Kleene's revised approach has influenced the understanding of the fully abstract models of functional programming. Functional programming and the semantics of such programs do not really belong to higher computability theory. We still include a brief survey of this direction, since it was strongly influenced by concepts developed for the analysis of computations relative to functionals in general.

4.3. Continuous and sequential models. The work of Kleene, Platek, and then Kleene again as described above, belong to higher computability theory. Computations will either be infinite well founded trees or manipulations/dialogues of ordinal length. The more recent development, starting as early as with Scott [59] in 1969, has left traditional higher computability theory and may now be seen as a different kind of generalization of classical computability theory, where one does not seek to generalize concepts like "finite set".

We have already given a brief introduction to the countable functionals, where we combine totality with a restriction to a kind of continuity. We also mentioned that an alternative definition of the total, continuous functionals is via the partial continuous functionals. Scott [59] was inspired by Platek's work when he defined a logic for computable functionals, LCF, together with a model based on both partiality and consistency as with Platek, and on continuity as with Kleene and Kreisel. Later, Plotkin [49] reformulated LCF to PCF. Our definition of Platek's computations is based on PCF. Now the interest in the computability theory of functionals moved from mathematics to theoretical computer science. This development takes us out of the scope of this survey, but there is one more observation to make: As a model for LCF and PCF, the set of partial continuous functionals is too rich. The nondeterministic **or** that we considered above is an example of a simple, continuous functional that cannot be programmed in PCF. For some time, efforts have been made to characterize a natural class of *sequential functionals* serving as a model without 'junk'. Important contributions were made by Sazonov [58], Abramsky, Jagadeesan and Malacaria [1] and by Hyland and Ong [19]. Though the influence of Kleene on each of these papers is disputable, we find the same philosophy behind these approaches as we found in the revisited-papers by Kleene.

For a more detailed survey of these aspects of computability in higher types, see Normann [47].

§5. Axiomatizing generalized computability theory. At the end of the 1960s there were several important examples of generalizations of classical computability theory. Thus there was a natural urge to define an abstract concept of a *computation theory*. This both involved deciding what the important concepts to be axiomatized are, and what the axioms should be.

An early, and important, contribution was given by Moschovakis [40]. Later Fenstad [5] developed quite a lot of generalized computability theory based on a modification of what was originally suggested by Moscovakis. Our discussion is based on [40] and [5], but is not fully faithful to either of them.

First, we restrict our attention to *computation domains* $\mathcal{U} = (U, N, S, P, L, R)$ where U is the global set, $N \subset U$, $S: N \to N$, $P: U^2 \to U$ is total and injective and L and R are single valued functions from U to U. The point is that (N, S) is isomorphic to $(\mathbb{N}, +1)$, and L is a left inverse and R is a right inverse of the *pairing* function P.

All examples we have considered up to now can be squeezed into the format of Kleene computations

$$\{e\}(\vec{x}) \simeq y$$

where the index e is chosen from a subset of the domain U, $\vec{x}$ is a sequence of inputs from the domain, and y is a possible output. In this paper, we have only considered examples where the value is unique, but we have seen in the examples of selection theorems that we sometimes prove that a nonempty semicomputable set contains a nonempty computable set. In the abstract approach, such operators may be considered to be multivalued *selection operators*, so the abstract approach suggests for considering multivalued functions, functions that actually are total functions from U^n to the powerset of U for some n. We write $f(\vec{x}) \to y$ instead of $y \in f(\vec{x})$, and we write $f(\vec{x}) = y$ if $f(\vec{x}) = \{y\}$.

A *computation set* Θ over a computation domain $\mathcal{U}$ will be a set of ordered finite sequences

$$(e, \vec{x}, y)$$

where e, $\vec{x}$ and y will be from U and $\vec{x}$ is a sequence of length ≥ 0. A partial, multivalued function $f \colon U^n \to U$ will then be considered to be computable with respect to Θ if there is an element $e \in U$ such that for all $\vec{x}$ of length n and all y we have that

$$(e, \vec{x}, y) \in \Theta \Leftrightarrow f(\vec{x}) \to y.$$

We write $f = \{e\}_\Theta$ when this is the case, omitting the index Θ if we do not need to know which computation theory we work with.

We consider 0 and 1 as elements of N, and then we assume that the characteristic function of N and the functions S, P, L and R are Θ-computable.

Given a computation domain and a computation set, we have enough language to be able to define what we mean by a computable functional:

If ϕ is a partial multivalued function taking a partial multivalued function f on U and a sequence $\vec{x}$ from U as arguments, we say that ϕ is Θ-computable if ϕ is monotone in the function variable, and there is an index e such that

$$\{e\}(d, \vec{x}) \simeq \phi(\{d\}, \vec{x})$$

for all d and $\vec{x}$. This is trivially extended to more than one function argument.

For the weakest form of a computation theory, we axiomatize closure under composition, primitive recursion, the existence of the S_m^n-theorem and so forth. It can be proved that given a computation domain and a finite set $\vec{f}$ of partial multivalued functions on U, we can define a minimal extension

$\Theta[\vec{f}]$ of Θ, in a uniform way, such that each f_i in the sequence will be $\Theta[\vec{f}]$-computable (Moschovakis [40]).

This enables us define a more restricted concept of a Θ-computable functional ϕ, we require an index e such that

$$\phi(f, \vec{x}) \simeq \{e\}_{\Theta[f]}(\vec{x})$$

for all f and $\vec{x}$. We then say that ϕ is *strongly* Θ-computable.

We now have the concepts we need in order to define, in abstract terms, forms of Θ-finite sets. We say that a Θ-computable set $A \subset U$ is *finite* if quantification over predicates on A is Θ-computable, while A is *strongly finite* if quantification over partial predicates on A also is Θ-computable. In the case of α-recursion theory, α is not finite, any ordinal $\beta < \alpha$ is finite and any ordinal $\beta < \alpha^*$ is strongly finite. In the case of recursion in ^3E, $\mathbb{N}^{\mathbb{N}}$ is finite and $\mathbb{N}$ is strongly finite, and, by Grilliot selection, $\mathbb{N}^{\mathbb{N}}$ is strongly finite for recursion in ^4E.

Another important aspect of our examples is the *length of a computation*, or the *ordinal rank* of a computation. In the axiomatization of a *computation theory* from [40] there is an ordinal $|(e, \vec{x}, y)|$ attached to each computation tuple $(e, \vec{x}, y) \in \Theta$, the rank of the computation. The key assumption is that the set of computation tuples bounded in rank by that of a fixed one is uniformly finite in the fixed one. It is also assumed that application of computable functionals, the S_m^n-theorem and e.g., composition will increase the ordinal rank of a computation. This will enable us to prove Gandy selection at this level of generality. Friedberg theories [40] are examples of computation theories that satisfy some further axioms, an attempt to capture what is needed for the Friedberg–Muchnic solution to Post's problem.

Fenstad [5] suggested refining the measurement of a length of computation to an identification of a well founded *computation tree*. In the view of the 70s, in a well behaved computability theory Θ, all computation trees must be Θ-finite. This captured many of the important examples of the time, and in particular those for which Post's problem was of interest. It is a matter of taste if it is the concept of computation tree and the corresponding concept of a subcomputation that is the natural primitive, or if it is simply the concept of the rank of a computation. In any case, the axiomatic approach gives a unified terminology, and it is possible to classify examples of generalizations of the Turing model according to which axioms are satisfied. For example, Platek's model for computability in higher types satisfies the basic axioms, but not those involving lengths of computations or computation trees. Kleene's revised model somehow brings back generalized computations, but only relative to sequential representations of the input. In Kleene's original model, computation trees played a prominent rôle, and theorems are proved by induction on the hight of a computation tree, even in cases where this tree is not finite from the view of the generalized theory.

The axiomatization also makes it easier to analyze exactly what is needed in order to state and/or solve problems like Post's in a generalized setting, although the interest in questions like this seems to have withered over the last two or three decades.

For further reading on axiomatized computability theory we recommend Moschovakis [40] and Fenstad [5].

§6. **Concluding remarks.** In a survey paper like this, one has to make a selection of topics, and most readers with an interest in the history of generalized computability will miss the treatment of important directions in higher computability theory. We have for instance not mentioned the theory of inductive definitions at all, even though the research on monotone and non-monotone inductive definitions brought both deep and applicable results. We also neglected the important work by Spector on the structure of predicate classes and on the structural analysis of the hyperarithmetical hierarchy [68]. Inductive definitions and the theory of *Spector classes* will to some extent be treated in Welch [74].

We have also left out Spector's *bar recursion* and *bar induction* [70]. This will at the same time be higher computability theory in the sense that it captures transfinite recursion and enter as a principle in constructive mathematics used e.g., in *proof mining*. There are now various formulations of bar recursion around, each with its own proof theoretic strength.

We have indicated the early starts of higher computability in the form of hyperarithmetical theory. After a while, there was an awareness of *Generalized Recursion Theory*, we will now say *Generalized Computability Theory*, as a subject on its own, with interesting and related sub-subjects in α–recursion theory, recursion in higher types, inductive definability etc. With this awareness, some conferences and workshops would accidentally or deliberately focus on generalized recursion theory. The proceedings [11] from Logic Colloquium '69 in Manchester is one of the milestones in the development of higher computability theory, and [7], the proceedings of a meeting in Oslo in 1972, concentrating on Generalized Recursion Theory, contains several breakthrough papers in the area. With the following-up-meeting in Oslo in 1977 the area had matured and the first excitements had cooled off, but also this proceedings, [6], contains papers that had impact on the future research in the area. In 1982, the summer symposium of the American Mathematical Society was in Recursion Theory, and the proceedings [42] edited by Nerode and Shore offered further advances in generalized computability theory.

For some time it seemed that the research activity in higher computability theories and the axiomatizations relevant for them was modest. The reason is probably that many of the basic questions were answered to a sufficient degree of satisfaction. A revival may depend on prospects for applications, but the recent interest in higher automata theory may also lead to a revival of the classical aspects of higher computability theory. There are both possibilities of direct applications and analogies to be explored. We believe that combining the methods of fine structure theory, the theory of inductive definitions and the classical theory of computations over ordinals and relative to normal functionals will lead to new developments in higher computability theory.

REFERENCES

[1] S. ABRAMSKY, R. JAGADEESAN, and P. MALACARIA, *Full abstraction for PCF* (*extended abstract*), **Theoretical aspects of computer software** (M. Hagiya and J. C. Mitchell, editors), Springer-Verlag, 1994, Full paper: **Information and Computation**, vol. 163 (2000), pp. 409–470, pp. 1–15.

[2] J. A. BERGSTRA, **Computability and continuity in finite types**, Thesis, University of Utrecht, 1976.

[3] C. T. CHONG and S. D. FRIEDMAN, **Ordinal recursion theory**, In Griffor [13], 1999, pp. 277–299.

[4] YU. L. ERSHOV, *Computable functionals of finite type*, **Algebra and Logic**, vol. 11 (1972), pp. 203–277.

[5] J. E. FENSTAD, **General recursion theory**, Springer-Verlag, 1980.

[6] J. E. Fenstad, R. O. Gandy, and G. E. Sacks (editors), **Generalized recursion theory II**, North-Holland, 1978.

[7] J. E. Fenstad and P. G. Hinman (editors), **Generalized recursion theory**, North-Holland, 1974.

[8] S. D. FRIEDMAN, *An introduction to β-recursion theory*, In Fenstad et al. [6], 1978, pp. 111–126.

[9] R. O. GANDY, *Proof of Mostowski's conjecture*, **Bulletin d l'Académie Polonaise des Sciences Mathématiques, Astronomiques et Physiques**, vol. 8 (1960), pp. 571–575.

[10] ———, *General recursive functionals of finite type and hierarchies of functionals*, **Annales de la Faculté des Sciences de l'Université de Clermont-Ferrand**, vol. 35 (1967), pp. 202–242.

[11] R. O. Gandy and C. E. M. Yates (editors), **Logic colloquium '69**, North-Holland, 1971.

[12] E. R. GRIFFOR, *E-recursively enuemrable degrees*, Ph.D. thesis, MIT, 1980.

[13] E. R. Griffor (editor), **Handbook of computability theory**, Elsevier, 1999.

[14] T. J. GRILLIOT, *Selection functions for recursive functionals*, **Notre Dame Journal of Formal Logic**, vol. 10 (1969), pp. 333–346.

[15] ———, *On effectively discontinuous type-2 objects*, **The Journal of Symbolic Logic**, vol. 36 (1971), pp. 245–248.

[16] L. A. HARRINGTON, *The superjump and the first recursively Mahlo ordinal*, In Gandy and Yates [11], 1971, pp. 43–52.

[17] ———, *Contributions to recursion theory on higher types*, Thesis, MIT, 1973.

[18] P. G. HINMAN, *Recursion on abstract structures*, In Griffor [13], 1999, pp. 315–359.

[19] J. M. E. HYLAND and C.-H. L. ONG, *On full abstraction for PCF: I. Models, observables and the full abstraction problem, II. Dialogue games and innocent strategies, III. A fully abstract and universal game model*, **Information and Computation**, vol. 163 (2000), pp. 285–408.

[20] R. B. JENSEN, *The fine structure of the constructible universe*, **Annals of Mathematical Logic**, vol. 4 (1972), pp. 229–308.

[21] A. S. KECHRIS, **Classical descriptive set theory**, Springer-Verlag, 1995.

[22] D. P. KIERSTEAD, *A semantics for Kleene's j-expressions*, **The Kleene Symposium** (J. Barwise, H. J. Keisler, and K. Kunen, editors), North-Holland, 1980, pp. 353–366.

[23] S. C. KLEENE, *Arithmetical predicates and function quantifiers*, **Transactions of the American Mathematical Society**, vol. 79 (1955), pp. 312–340.

[24] ——, *Hierarchies of number-theoretic predicates*, **American Mathematical Society. Bulletin**, vol. 61 (1955), pp. 193–213.

[25] ——, *Countable functionals*, **Constructivity in mathematics** (A. Heyting, editor), North-Holland, 1959, pp. 81–100.

[26] ——, *Recursive functionals and quantifiers of finite types I*, **American Mathematical Society. Bulletin**, vol. 91 (1959), pp. 1–52.

[27] ——, *Recursive functionals and quantifiers of finite types revisited I*, In Fenstad et al. [6], 1978, pp. 185–222.

[28] ——, *Recursive functionals and quantifiers of finite types revisited II*, **The Kleene Symposium** (J-Barwise, H. J. Keisler, and K. Kunen, editors), North-Holland, 1980, pp. 1–29.

[29] ——, *Recursive functionals and quantifiers of finite types revisited III*, **Patras logic symposion** (G. Metakides, editor), North-Holland, 1982, pp. 1–40.

[30] ——, *Unimonotone functions of finite types* (*Recursive functionals and quantifiers of finite types revisited IV*), In Nerode and Shore [42], 1985, pp. 119–138.

[31] ——, *Recursive functionals and quantifiers of finite types revisited V*, **Transactions of the American Mathematical Society**, vol. 325 (1991), pp. 593–630.

[32] G. KREISEL, *Interpretation of analysis by means of functionals of finite type*, **Constructivity in mathematics** (A. Heyting, editor), North-Holland, 1959, pp. 101–128.

[33] ——, *Some reasons for generalizing recursion theory*, In Gandy and Yates [11], 1971, pp. 139–198.

[34] G. KREISEL and G. E. SACKS, *Metarecursive sets*, **The Journal of Symbolic Logic**, vol. 30 (1965), pp. 318–338.

[35] M. LERMAN, *On suborderings of the α-recursively enumerable degrees*, **Annals of Mathematical Logic**, vol. 4 (1972), pp. 369–392.

[36] W. MAASS, *Inadmissibility, tame r.e. sets and the admissible collapse*, **Annals of Mathematical Logic**, vol. 13 (1978), pp. 149–170.

[37] D. B. MACQUEEN, **Recursion in finite types**, Ph.D. thesis, MIT, 1972.

[38] J. MOLDESTAD, **Computations in higher types**, Springer-Verlag, 1977.

[39] Y. N. MOSCHOVAKIS, *Hyperanalytic predicates*, **Transactions of the American Mathematical Society**, vol. 129 (1967), pp. 249–282.

[40] ——, *Axioms for computation theories—first draft*, In Gandy and Yates [11], 1971, pp. 199–255.

[41] ——, **Descriptive set theory**, second ed., American Mathematical Society, 2009.

[42] A. Nerode and R. A. Shore (editors), **Proceedings of the AMS–ASL summer institute on Recursion Theory**, Proceedings of Symposia in Pure Mathematics, vol. 42, American Mathematical Society, 1985.

[43] D. NORMANN, *Recursion in ^3E and a splitting theorem*, **Essays on mathematical and philisophical logic** (J. Hintikka, I. Niinilouto, and E. Saarinen, editors), D. Reidel Publishing Company, 1978, pp. 275–285.

[44] ——, *Set recursion*, In Fenstad et al. [6], 1978, pp. 303–320.

[45] ——, *A classification of higher type functionals*, **Proceedings from 5th Scandinavian Logic Symposium** (F. V. Jensen, B. H. Mayoh, and K. Møller, editors), Aalborg University Press, 1979, pp. 301–308.

[46] ——, *Degrees of functionals*, **Annals of Mathematical Logic**, vol. 16 (1979), pp. 269–304.

[47] ——, *Computing with functionals—computability theory or computer science?*, **The Bulletin of Symbolic Logic**, vol. 12 (2006), pp. 43–59.

[48] R. A. PLATEK, **Foundations of recursion theory**, Ph.D. thesis and supplement, Stanford, 1966.

[49] G. PLOTKIN, *LCF considered as a programming language*, **Theoretical Computer Science**, vol. 5 (1977), pp. 223–255.

[50] H. ROGERS JR., **Theory of recursive functions and effective computability**, McGraw-Hill, 1967.

[51] G. E. SACKS, *Post's problem, admissible ordinals, and regularity*, **Transactions of the American Mathematical Society**, vol. 124 (1966), pp. 1–23.

[52] ———, *The 1-section of a type n object*, In Fenstad and Hinman [7], 1974, pp. 81–93.

[53] ———, *The k-section of a type n-object*, **American Journal of Mathematics**, vol. 99 (1977), pp. 901–917.

[54] ———, *Post's problem in E-recursion theory*, In Nerode and Shore [42], 1985, pp. 177–193.

[55] ———, *On the limits of E-recursive enumerability*, **Annals of Pure and Applied Logic**, vol. 31 (1986), pp. 87–120.

[56] ———, **Higher recursion theory**, Springer-Verlag, 1990.

[57] G. E. SACKS and S. G. SIMPSON, *The α-finite injury method*, **Annals of Mathematical Logic**, vol. 4 (1972), pp. 343–367.

[58] V. YU. SAZONOV, *On semantics of the applicative algorithmic languages*, Thesis (in Russian), Novosibirsk, 1976.

[59] D. SCOTT, *A type-theoretical alternative to ISWIM, CUCH, OWHY*, unpublished manuscript, 1969, Oxford; later published with comments in **Theoretical Computer Science**, vol. 121 (1993), pp. 411–440.

[60] R. A. SHORE, *The irregular and non-hyperregular α-r.e. degrees*, **Israel Journal of Mathematics**, vol. 22 (1975), pp. 123–155.

[61] ———, *Splitting an α-recursively enumerable set*, **Transactions of the American Mathematical Society**, vol. 204 (1975), pp. 65–78.

[62] ———, *The recursively enumerable α-degrees are dense*, **Annals of Mathematical Logic**, vol. 9 (1976), pp. 123–155.

[63] S. G. SIMPSON, *Post's problem for admissible sets*, In Fenstad and Hinman [7], 1974, pp. 437–441.

[64] ———, *Short course on admissible recursion theory*, In Fenstad et al. [6], 1978, pp. 355–390.

[65] T. A. SLAMAN, *Aspects of E-recursion*, Ph.D. thesis, Harvard, 1981.

[66] ———, *The E-recursively enumerable degrees are dense*, In Nerode and Shore [42], 1985, 195–213.

[67] ———, *Reflection and forcing in E-recursion theory*, **Annals of Pure and Applied Logic**, vol. 29 (1985), pp. 79–106.

[68] C. SPECTOR, *Recursive well orderings*, **The Journal of Symbolic Logic**, vol. 20 (1955), pp. 151–163.

[69] ———, *Hyperarithmetical quantifiers*, **Fundamenta Mathematicae**, vol. 48 (1959), pp. 313–320.

[70] ———, *Provably recursive functionals of analysis: a consistency proof of analysis by an extension of principles formulated in current intuitionistic mathematics*, **Recursive function theory** (J. C. E. Dekker, editor), Proceedings of Symposia in Pure Mathematics, vol. V, American Mathematical Society, 1962, pp. 1–27.

[71] V. STOLTENBERG-HANSEN, **On priority arguments in Friedberg theories**, Thesis, University of Toronto, 1973.

[72] J. VAN DE WIELE, *Recursive dilators and generalized recursion*, **Proceedings of the Herbrand Symposium** (J. Stern, editor), North-Holland, 1982, pp. 325–332.

[73] S. S. WAINER, **The 1-section of a non-normal type-2 object**, In Fenstad et al. [6], 1978, pp. 407–417.

[74] P. D. WELCH, *Transfinite machine models*, this volume.

DEPARTMENT OF MATHEMATICS
THE UNIVERSITY OF OSLO
P.O. BOX 1053
BLINDERN N-0316 OSLO, NORWAY
E-mail: dnormann@math.uio.no

STEP BY RECURSIVE STEP: CHURCH'S ANALYSIS OF EFFECTIVE CALCULABILITY

WILFRIED SIEG

> In fact, the only evidence for the freedom from contradiction of *Principia Mathematica* is the empirical evidence arising from the fact that the system has been in use for some time, many of its consequences have been drawn, and no one has found a contradiction. (Church in a letter to Gödel, July 27, 1932)

Abstract. Alonzo Church's mathematical work on computability and undecidability is well known indeed, and we seem to have an excellent understanding of the context in which it arose. The approach Church took to the underlying conceptual issues, by contrast, is less well understood. Why, for example, was "Church's Thesis" put forward publicly only in April 1935, when it had been formulated already in February/March 1934? Why did Church choose to formulate it then in terms of Gödel's general recursiveness, not his own λ-definability as he had done in 1934? A number of letters were exchanged between Church and Paul Bernays during the period from December 1934 to August 1937; they throw light on critical developments in Princeton during that period and reveal novel aspects of Church's distinctive contribution to the analysis of the informal notion of *effective calculability*. In particular, they allow me to give informed, though still tentative answers to the questions I raised; the character of my answers is reflected by an alternative title for this paper, *Why Church needed Gödel's Recursiveness for his Thesis*. In section 5, I contrast Church's analysis with that of Alan Turing and explore, in the very last section, an analogy with Dedekind's investigation of continuity.

§0. Proem on Church & Gödel. Church's mathematical work on computability and undecidability is well known, and its development is described, for example, in informative essays by his students Kleene and Rosser. The study of the Church *Nachlass* may provide facts for a fuller grasp of this evolution, but it seems that we have an excellent understanding of the context in which the work arose.[1] By contrast, Church's approach to the underlying

---

This paper is dedicated to the memory of Alonzo Church.—A number of colleagues and friends helped me to improve it significantly: J. Avigad, A. Blass, J. Byrnes, M. Davis, S. Feferman, W. W. Tait, and G. Tamburrini. I am very grateful to J. Dawson for providing me with a most informative and reflective letter from Church that is reproduced in Appendix D.

[1] For additional background, cf. Appendix 2 in (Sieg 1994) and Church's letter in Appendix D. It would be of great interest to know more about the earlier interaction with leading logicians and mathematicians: as reported in (Enderton 1995), Church spent part of his time as a National Research Fellow from 1927 to 1929 at Harvard, Göttingen, and Amsterdam.

conceptual issues is less well understood, even though a careful examination of the published work is already quite revealing. Important material, relevant to both historical and conceptual issues, is contained in the Bernays *Nachlass* at the Eidgenössische Technische Hochschule in Zürich. A number of letters were exchanged between Church and Bernays during the period from December 1934 to August 1937; they throw light on critical developments in Princeton[2] and reveal novel aspects of Church's contribution to the analysis of the informal notion of *effective calculability*. That contribution has been recognized: the identification of effective calculability with Gödel's general recursiveness, or equivalent notions, is called *Church's Thesis*.

Church proposed the *definitional* identification publicly for the first time in a talk to the American Mathematical Society on April 19, 1935; the abstract of the talk had been received by the Society already on March 22. Some of the events leading to this proposal (and to the first undecidability results) are depicted in Martin Davis' fascinating paper *Why Gödel did not have Church's Thesis*: Church formulated a version of his thesis via λ-definability in conversations during the period from late 1933 to early 1934.[3] At that time, the main reason for proposing the identification was the quasi-empirical fact he expressed also strongly in a letter to Bernays dated January 23, 1935:

> The most important results of Kleene's thesis concern the problem of finding a formula to represent a given intuitively defined function of positive integers (it is required that the formula shall contain no other symbol than λ, variables, and parentheses). The results of Kleene are so general and the possibilities of extending them apparently so unlimited that one is led to the conjecture that a formula can be found to represent any particular constructively defined function of positive integers whatever.

How strongly such quasi-empirical evidence impressed Church is illustrated by the quotation from his letter to Gödel in the motto of my paper; that was written in response to Gödel's question concerning Church's 1932 paper: "If the system is consistent, won't it then be possible to interpret the basic notions in a system of type theory or in the axiom system of set theory, and is there,

---

[2] This correspondence shows also how closely Bernays followed and interacted with the work of the Princeton group; this is in striking contrast to the view presented in (Gandy 1988). It should also be noted that Bernays was in Princeton during the academic year 1935–1936, resulting in (Bernays 1935–1936). I assume that Bernays spent only the Fall term in Princeton, roughly from late September 1935 to around February 1936. In any event, he made the transatlantic voyage, starting from Le Havre on September 20, in the company of Kurt Gödel and Wolfgang Pauli. Due to health reasons, Gödel left Princeton at the very end of November; cf. (Dawson 1997), pp. 109–110.

[3] Cf. section 1 below and, in particular, Rosser's remarks quoted there.

apart from such an interpretation, any other way at all to make plausible the consistency?"[4]

In the 1935 abstract the thesis was formulated, however, in terms of general recursiveness, and the sole stated reason for the identification is that "other plausible definitions of effective calculability turn out to yield notions that are either equivalent to or weaker than recursiveness". For Davis, this wording "leaves the impression that in the early spring of 1935 Church was not yet certain that λ-definability and Herbrand–Gödel recursiveness were equivalent."[5] Davis' account continues as follows, specifying a particular order in which central results were obtained:

> Meanwhile, Church and Kleene each proved that all λ-definable functions are recursive. Church submitted an abstract of his work on [sic] March 1935, basing himself on recursiveness rather than λ-definability. By the end of June 1935, Kleene had shown that every recursive function is λ-definable, after which Church (1936) was able to put his famous work into its final form. Thus while Gödel hung back because of his reluctance to accept the evidence for Church's thesis available in 1935 as decisive, Church (who after all was right) was willing to go ahead, and thereby launch the field of recursive function theory. (p. 12)

The accounts in (Kleene 1981) and (Rosser 1984), together with the information provided by Church in his letters to Bernays, make it perfectly clear that the λ-definability of the general recursive functions was known at the very beginning of 1935; it had been established by Rosser and Kleene. The converse was not known when Church wrote his letter of January 23, 1935, but had definitely been established by July. Church wrote on July 15, 1935 his next letter to Bernays and pointed to "a number of developments" that had taken place "in the meantime". These developments had led to a(n impressive) list of papers, including his own (1935a) and (1936), Kleene's (1936 and 1936a), Rosser's (1935), and the joint papers with Kleene, respectively Rosser. Contrary to Davis' "impression", the equivalence was known already in March of 1935 when the abstract was submitted: if the inclusion of λ-definability in recursiveness had not also been known by then, the thesis could not have been formulated coherently in terms of recursiveness.

The actual sequence of events thus differs in striking ways from Davis' account (based on more limited historical information); most importantly, the order in which the inclusions between λ-definability and general recursiveness

---

[4] The German original: Falls das System widerspruchsfrei ist, wird es dann nicht möglich sein, die Grundbegriffe in einem System mit Typentheorie bzw. im Axiomensystem der Mengenlehre zu interpretieren, und kann man überhaupt auf einem anderen Wege als durch eine solche Interpretation die Widerspruchsfreiheit plausibel machen?

[5] l.c., p. 10.

were established is reversed. This is not just of historical interest, but important for an evaluation of the broader conceptual issues. I claim, and will support through the subsequent considerations, that Church was reluctant to put forward the thesis in writing—until the equivalence of λ-definability and general recursiveness had been established. The fact that the thesis was formulated in terms of recursiveness indicates also that λ-definability was at first, even by Church, not viewed as one among "equally natural definitions of effective calculability": the notion just did not arise from an analysis of the intuitive understanding of effective calculability. I conclude that Church was cautious in a similar way as Gödel. Davis sees stark contrasts between the two: in the above quotation, for example, he sees Gödel as "hanging back" and Church as "willing to go ahead"; Gödel is described as reluctant to accept the "evidence for Church's Thesis available in 1935 as decisive". The conversations on which the comparison between Church and Gödel are based took place, however, already in early 1934.[6] Referring to these same conversations, Davis writes (and these remarks immediately precede the above quotation):

> The question of the equivalence of the class of these general recursive functions with the effectively calculable functions was ... explicitly raised by Gödel in conversation with Church. Nevertheless, Gödel was not convinced by the available evidence, and remained unwilling to endorse the equivalence of effective calculability, either with recursiveness or with λ-definability. He insisted ... that it was 'thoroughly unsatisfactory' to *define* the effectively calculable functions to be some particular class without first showing that 'the generally accepted properties' of the notion of effective calculability necessarily lead to this class.

Again, the evidence for the thesis provided by the equivalence, if it is to be taken as such, was not yet available in 1934. Church's and Gödel's developed views actually turn out to be much closer than this early opposition might lead one to suspect. That will be clear, I hope, from the detailed further discussion. The next section reviews the steps towards Church's "conjecture"; then we will look at the equivalence proof and its impact. Church used implicitly and Gödel formulated explicitly an "absoluteness" property for the rigorous concept that is based on an explication of the informal notion as "computation in some logical calculus". Sections 3 and 4 discuss that explication and absoluteness notion. In the next to last section I contrast their explication with the analysis of Alan Turing and explore, in the final section, an

---

[6]In footnote 18 of his (1936) Church remarks: "The question of the relationship between effective calculability and recursiveness (which it is here proposed to answer by identifying the two notions) was raised by Gödel in conversation with the author. The corresponding question of the relationship between effective calculability and λ-definability had previously been proposed by the author independently."

analogy between Turing's analysis and Dedekind's investigation of continuity. Let me mention already here that both Church and Gödel recognized and emphasized the special character of Turing's analysis: Church pointed out that Turing's notion has the advantage (over recursiveness and λ-definability) "of making the identification with effectiveness in the ordinary (not explicitly defined) sense evident immediately"; Gödel asserted that Turing's work gives an analysis of the concept of mechanical procedure and that this concept is *shown to be equivalent* with that of a 'Turing machine'.[7]

§1. Effective calculability: a conjecture. The first letter of the extant correspondence between Bernays and Church was mentioned above; it was written by Church on January 23, 1935 and responds to a letter by Bernays from December 24, 1934. Bernays' letter is not preserved in the Zürich *Nachlass*, but it is clear from Church's response, which issues had been raised in it: one issue concerned the applicability of Gödel's Incompleteness Theorems to Church's systems in the papers (Church 1932 and 1933), another the broader research program pursued by Church with Kleene and Rosser. Church describes in his letter two "important developments" with respect to the research program. The first development contains Kleene and Rosser's proof, published in their (1935), that the set of postulates in (Church 1932 and 1933) is contradictory. For the second development Church refers to the proof by Rosser and himself that a certain subsystem is free from contradiction. (Cf. Church's letter in Appendix D for a description of the broader context.) The second development is for our purposes particularly significant and includes Kleene's dissertation work. Church asserts that the latter provides support for the conjecture that "a formula can be found to represent any particular constructively defined function of positive integers whatever". He continues:

> It is difficult to prove this conjecture, however, or even to state it accurately, because of the difficulty in saying precisely what is meant by "constructively defined". A vague description can be given by saying that a function is constructively defined if a method is given by which its value could be actually calculated for any particular positive integer whatever. Every recursive definition, of no matter how high an order, is constructive, and as far as I know, every constructive definition is recursive.[8]

---

[7]Gödel made this brief and enigmatic remark (as to a proof of the equivalence) in his 1964 Postscriptum to (Gödel 1934); it is elucidated in (Sieg & Byrnes 1996 and 1997).

[8]The quotation continues directly the above quotation from this letter.—Church's paper (1934) was given on December 30, 1933 to a meeting of the Mathematical Association; incidentally, Gödel presented his (1933o) in the very same session of that meeting. Cf. also (Gödel 1936b), reviewing (Church 1935).

The last remark is actually reminiscent of part of the discussion in (Church 1934), where Church claims that " ... it appears to be possible that there should be a system of symbolic logic containing a formula to stand for every definable function of positive integers, and I fully believe that such systems exist". (p. 358) From the context it is clear that "constructive definability" is intended, and the latter means minimally that the values of the function can be calculated for any argument. It is equally clear that the whole point of the paper is to propose plausible formal systems that, somehow, don't fall prey to Gödel's Incompleteness Theorems.

> A system of this sort [with levels of different notions of implications, WS] not only escapes our unpleasant theorem that it must be either insufficient or oversufficient, but I believe that it escapes the equally unpleasant theorem of Kurt Gödel to the effect that, in the case of any system of symbolic logic which has a claim to adequacy, it is impossible to prove its freedom from contradiction in the way projected in the Hilbert program. This theorem of Gödel is, in fact, more closely related to the foregoing considerations than appears from what has been said. (p. 360)

Then Church refers to a system of postulates whose consistency can be proved and which probably is adequate for elementary number theory; it seems to be inconceivable to Church that all formal theories should fail to allow the "representation" of the constructively definable functions. Indeed, for the λ-calculus, the positive conjecture had been made by Church in conversation with Rosser tentatively late in 1933, with greater conviction in early 1934. Rosser describes matters in his (1984) as follows:

> One time, in late 1933, I was telling him [Church, WS] about my latest function in the LC [Lambda Calculus, WS]. He remarked that perhaps every effectively calculable function from positive integers to positive integers is definable in LC. He did not say it with any firm conviction. Indeed, I had the impression that it had just come into his mind from hearing about my latest function. With the results of Kleene's thesis and the investigations I had been making that fall, I did not see how Church's suggestion could possibly fail to be true. ... After Kleene returned to Princeton on February 7, 1934, Church looked more closely at the relation between λ-definability and effective calculability. Soon he decided they were equivalent, ... (p. 345)

Kleene put all of these events, except for Church's very first speculations, after his "return to Princeton on February 7, 1934, and before something like the end of March 1934"; see (Davis 1982), p. 8. Church discussed these issues also with Gödel who was at that time, early 1934, not convinced by the proposal to identify effective calculability with λ-definability: he called the proposal "thoroughly unsatisfactory".[9] This must have been discouraging to Church, in particular, as Gödel suggested a different direction for supporting such a claim and made later in his lectures a different proposal for a broader notion; Church reports in a letter to Kleene of November 29, 1935:

> His [Gödel's, WS] only idea at the time was that it might be possible, in terms of effective calculability as an undefined notion, to state a set of axioms which would embody the generally accepted properties of this notion, and to do something on that basis. Evidently it occurred to him later that Herbrand's definition of recursiveness, which has no regard to effective calculability, could be modified in the direction of effective calculability, and he made this proposal in his lectures. At that time he did specifically raise the question of the connection between recursiveness in this new sense and effective calculability, but said he did not think that the two ideas could be satisfactorily identified "except heuristically".[10]

This was indeed Gödel's view and was expressed in Note 3 of his 1934 Princeton lectures. The note is attached to the remark that primitive recursive functions have the *important property* that their unique value *can be computed by a finite procedure*—for each set of arguments.

> The converse seems to be true if, besides recursions according to the schema (2) [of primitive recursion; WS], recursions of other forms (e.g., with respect to two variables simultaneously) are admitted. This cannot be proved, since the notion of finite computation is not defined, but it serves as a heuristic principle.

To some it seemed that the note expressed a form of Church's Thesis. However, in a letter of February 15, 1965 to Martin Davis, Gödel emphasized that no formulation of Church's Thesis is implicit in the conjectured equivalence; he explained:

> ... it is *not true* that footnote 3 is a statement of Church's Thesis. The conjecture stated there only refers to the equivalence of "finite (computation) procedure" and "recursive procedure". However, I was, at the time of these lectures, not at all convinced that my concept of recursion comprises all possible recursions; and in fact the equivalence between my definition and Kleene's ... is not quite trivial.[11]

---

[9] Church in a letter to Kleene, dated November 29, 1935, and quoted in (Davis 1982), p. 9. The conversation took place, according to Davis, "presumably early in 1934"; that is confirmed by Rosser's account on p. 345 of (Rosser 1984).

[10] This is quoted in (Davis 1982), p. 9, and is clearly in harmony with Gödel's remark quoted below. As to the relation to Herbrand's concept, see the critical discussion in (Sieg 1994), pp. 83–85.

[11] (Davis 1982), p. 8.

In the *Postscriptum* to his (1934) Gödel asserts that the question raised in footnote 3 can now, in 1965, be "answered affirmatively" for his recursiveness "which is equivalent with general recursiveness as defined today", i.e., with Kleene's μ-recursiveness. I do not understand, how *that* definition could have convinced Gödel that it captures "all possible recursions", unless its use in proofs of Kleene's normal form theorem is also considered. The ease with which "the" normal form theorem allows to establish equivalences between different formulations makes it plausible that *some* stable notion has been isolated; however, the question, whether that notion corresponds to effective calculability has to be answered independently. — The very next section is focused on the equivalence between general recursiveness and λ-definability, but also the dialectical role this mathematical result played for the first published formulation of Church's Thesis.

§2. Two notions: an equivalence proof. In his first letter to Bernays, Church mentions in the discussion of his *conjecture* two precise mathematical results: all primitive recursive, respectively general recursive functions in Gödel's sense can be represented, and that means that they are λ-definable. The first result is attributed to Kleene[12] and the second to Rosser. The letter's remaining three and a half pages (out of a total of six pages) are concerned with an extension of the pure λ-calculus for the development of elementary number theory, consonant with the considerations of (Church 1934) described above. The crucial point to note is that the converse of the mathematical result concerning general recursive functions and, thus, the equivalence between λ-definability and general recursiveness is *not* formulated.

Bernays had evidently remarked in his letter of December 24, 1934, that some statements in (Church 1933) about the relation of Gödel's theorems to Church's formal systems were not accurate, namely, that the theorems might not be applicable because some very special features of the system of *Principia Mathematica* seemed to be needed in Gödel's proof.[13] Church responds that Bernays' remarks are "just" and then describes Gödel's response to the very same issue:

> Gödel has since shown me, that his argument can be modified in such a way as to make the use of this special property of the system of Principia unnecessary. In a series of lectures here at Princeton last spring he presented this generalized form of his argument, and

---

[12] This fact is formulated also in (Kleene 1935, Part II) on p. 223.

[13] Gödel and Church had a brief exchange on this issue already in June and July of 1932. In his letter of July 27, 1932, Church remarks that von Neumann had drawn his attention "last fall" to Gödel's paper (1931) and continues: "I have been unable to see, however, that your conclusions in §4 of this paper apply to my system. Possibly your argument can be modified so as to make it apply to my system, but I have not been able to find a modification of your argument." Cf. Appendix D.

was able to set down a very general set of conditions such that his theorem would hold of any system of logic which satisfied them.

The conditions Church alludes to are found in section 6 of Gödel's lectures; they include one precise condition that, according to Gödel, *in practice suffices as a substitute for the unprecise* requirement that the class of axioms and the relation of immediate consequence be constructive. The unprecise requirement is formulated at the beginning of Gödel's lectures to characterize crucial normative features for a *formal mathematical system*:

> We require that the rules of inference, and the definitions of meaningful formulas and axioms, be constructive; that is, for each rule of inference there shall be a finite procedure for determining whether a given formula B is an immediate consequence (by that rule) of given formulas $A_1, \ldots, A_n$, and there shall be a finite procedure for determining whether a given formula A is a meaningful formula or an axiom. (p. 346)

The precise condition replaces "constructive" by "primitive recursive".[14] Not every constructive function is primitive recursive, however: Gödel gives in section 9 a function of the Ackermann type, asks what one might mean "by every recursive function", and defines in response the class of *general recursive functions* via his equational calculus.

Clearly, it is of interest to understand, why Church *publicly* announced the *thesis* only in his talk of April 19, 1935, and why he formulated it then in terms of general recursiveness, not λ-definability. Here is the full abstract of Church's talk:

> Following a suggestion of Herbrand, but modifying it in an important respect, Gödel has proposed (in a set of lectures at Princeton, N.J., 1934) a definition of the term *recursive function*, in a very general sense. In this paper a definition of *recursive function of positive integers* which is essentially Gödel's is adopted. And it is maintained that the notion of an effectively calculable function of positive integers should be identified with that of a recursive function, since other plausible definitions of effective calculability turn out to yield notions that are either equivalent to or weaker than recursiveness. There are many problems of elementary number theory in which it is required to find an effectively calculable function of positive integers satisfying certain conditions, as well as a large number of problems in other fields which are known to be reducible to problems in number theory of this type. A problem of this class is the problem to find a complete set of invariants of

---

[14] Here and below I use "primitive recursive" where Gödel just says "recursive" to make explicit the terminological shift that has taken place since (Gödel 1934).

formulas under the operation of conversion (see abstract 41.5.204). It is proved that this problem is unsolvable, in the sense that there is no complete set of effectively calculable invariants.[15]

Church's letter of July 15, 1935, to Bernays explicitly refers to this abstract and mentions the paper (Church 1936) as "in the process of being typewritten"; indeed, Church continues " ... I will mail you a copy within a week or two. All these papers will eventually be published, but it may be a year or more before they appear." His mailing included a copy of a joint paper with Rosser, presumably their (1936), and an abstract of a joint paper with Kleene, presumably their (1935). Of historical interest is furthermore that Kleene's papers *General recursive functions of natural numbers* and *λ-definability and recursiveness* are characterized as "forthcoming", i.e., they had been completed already at this time.

The precise connection between recursiveness and λ-definability or, as Church puts it in his abstract, "other plausible definitions of effective calculability" had been discovered in 1935, between the writing of the letters of January 23 and July 15. From the accounts in (Kleene 1981) and (Rosser 1984) it is quite clear that Church, Kleene, and Rosser contributed to the proof of the equivalence of these notions. Notes 3, 16, and 17 in (Church 1936) add detail: consistently with the report in the letter to Bernays, the result that all general recursive functions are λ-definable was first found by Rosser and then by Kleene (for a slightly modified definition of λ-definability); the converse claim was established "independently" by Church and Kleene "at about the same time". However, neither from Kleene's or Rosser's historical accounts nor from Church's remarks is it clear, *when* the equivalence was actually established. In view of the letter to Bernays and the submission date for the abstract, March 22, 1935, the proof of the converse must have been found after January 23, 1935, but before March 22, 1935. So one can assume with good reason that this result provided to Church the additional bit of evidence for actually publishing the thesis.[16]

That the thesis was formulated for general recursiveness is not surprising when Rosser's remark in his (1984) about this period is seriously taken into account: "Church, Kleene, and I each thought that general recursivity seemed to embody the idea of effective calculability, and so each wished to show it equivalent to λ-definability". (p. 345) There was no independent motivation for λ-definability to serve as a concept to capture effective calculability, as the historical record seems to show: consider the surprise that the predecessor function is actually λ-definable and the continued work in 1933/4 by Kleene

---

[15](Church 1935). In the next to last sentence "abstract 41.5.204" refers to (Church and Rosser 1935).

[16]This account should be compared with the more speculative one given in (Davis 1982), for example in the summary on p. 13.

and Rosser to establish the λ-definability of more and more constructive functions. In addition, Church argued for the correctness of the thesis when completing the 1936 paper (before July 15, 1935); his argument took the form of an *explication* of effective calculability with a central appeal to "recursivity". Kleene referred to Church's analysis, when presenting his (1936b) to the American Mathematical Society on January 1, 1936, and made these introductory remarks (on p. 544): "The notion of a recursive function, which is familiar in the special cases associated with primitive recursions, Ackermann–Péter multiple recursions, and others, has received a general formulation from Herbrand and Gödel. The resulting notion is of especial interest, since the intuitive notion of a 'constructive' or 'effectively calculable' function of natural numbers can be identified with it very satisfactorily." λ-definability was not even mentioned.

§3. Reckonable functions: an explication. The paper *An unsolvable problem of elementary number theory* was published, as Church had expected, in (early) 1936. Church restates in it his proposal for identifying the class of effectively calculable functions with a precisely defined class, so that he can give a rigorous mathematical definition of the class of number theoretic problems of the form: "Find an effectively calculable function that is the characteristic function of a number theoretic property or relation." Church describes this and an additional crucial point as follows:

> .... The purpose of the present paper is to propose a definition of effective calculability which is thought to correspond satisfactorily to the somewhat vague intuitive notion in terms of which problems of this class are often stated, and to show, by means of an example, that not every problem of this class is solvable.[17]

In section 7 of his paper, Church presents arguments in support of the proposal to use general recursiveness[18] as the precise notion; indeed, the arguments are to justify the identification "so far as positive justification can ever be obtained for the selection of a formal definition to correspond to an intuitive notion".[19] Two methods to characterize effective calculability of number-theoretic functions suggest themselves. The first of these methods makes use of the notion of *algorithm*, and the second employs the notion of *calculability in a logic*. Church argues that neither method leads to a definition more general than recursiveness. The two arguments have a very

---

[17] (Church 1936) in (Davis 1965) pp. 89 and 90.

[18] The fact that λ-definability is an equivalent concept adds for Church " ... to the strength of the reasons adduced below for believing that they [these precise concepts] constitute as general a characterization of this notion (i.e., effective calculability) as is consistent with the usual intuitive understanding of it". (Church 1936), footnote 3, p. 90 in (Davis 1965).

[19] (Church 1936), in (Davis 1965), p. 100.

similar structure, and I will discuss only the one pertaining to the second method.[20] Church considers a logic **L**, whose language contains the equality symbol $=$, a symbol $\{\ \}(\ )$ for the application of a unary function symbol to an argument, and numerals for the positive integers. For unary functions F he defines: F is *effectively calculable* if and only if there is an expression f in the logic **L** such that: $\{f\}(\mu) = \nu$ is a theorem of **L** iff $F(m) = n$; here, μ and ν are expressions that stand for the positive integers m and n.[21] Such functions F are recursive, if it is *assumed* that **L** satisfies conditions that make **L**'s theorem predicate recursively enumerable. To argue then for the recursive enumerability of the theorem predicate, Church formulates conditions *any* system of logic has to satisfy if it is "to serve at all the purposes for which a system of symbolic logic is usually intended".[22] These conditions, Church remarks in footnote 21, are "substantially" those from Gödel's Princeton lectures for a formal mathematical system: (**i**) each rule must be an effectively calculable operation, and (**ii**) the set of rules and axioms (if infinite) must be effectively enumerable. Church supposes that these conditions can be *interpreted* to mean that, via a suitable Gödel numbering for the expressions of the logic, ($\mathbf{i}_C$) each rule must be a recursive operation, ($\mathbf{ii}_C$) the set of rules and axioms (if infinite) must be recursively enumerable, and ($\mathbf{iii}_C$) the relation between a positive integer and the expression which stands for it must be recursive. The theorem predicate is thus indeed recursively enumerable, but the *crucial interpretative step* is not argued for at all.

Church's argument in support of the recursiveness of effectively calculable functions may appear to be viciously circular. However, our understanding of the general concept of calculability is explicated in terms of derivability in a logic, and the conditions ($\mathbf{i}_C$)–($\mathbf{iii}_C$) sharpen the idea that within such a logical formalism one operates with an effective notion of immediate consequence.[23] The "thesis" is appealed to in a special and narrower context, and it is precisely here that we encounter the real stumbling block for Church's analysis. Given

---

[20] An argument following quite closely Church's considerations pertaining to the first method is given in (Shoenfield 1967), p. 120.—For the second argument, Church uses the fact that Gödel's class of general recursive functions is closed under the μ-operator, then still called Kleene's p-function. That result is not needed for the first argument on account of the determinacy of algorithms.

[21] This concept is an extremely natural and fruitful one and is directly related to "Entscheidungsdefinitheit" for relations and classes introduced by Gödel in his 1931 paper and to representability of functions used in his 1934 Princeton lectures. As to the former, compare **Collected Works I**, pp. 170 and 176 (Gödel 1986); as to the latter, see p. 58 in (Davis 1965).

[22] (Church 1936) in (Davis 1965), p. 101. As to what is intended, namely for **L** to satisfy epistemologically motivated restrictions, see (Church 1956), section 07, in particular pp. 52–53.

[23] Compare footnote 20 on p. 101 in (Davis 1965) where Church remarks: "In any case where the relation of immediate consequence is recursive it is possible to find a set of rules of procedure, equivalent to the original ones, such that each rule is a (one-valued) recursive operation, and the complete set of rules is recursively enumerable."

the crucial role this observation plays, it is appropriate to formulate it as a normative requirement:

CHURCH'S CENTRAL THESIS. The steps of any effective procedure (governing proofs in a system of symbolic logic) must be recursive.

If the central thesis is accepted, the earlier considerations indeed prove that all effectively calculable functions are recursive. Robin Gandy called this Church's "step-by-step argument".[24]

The idea that computations are carried out in a logic or simply in a deductive formalism is also the starting point of the considerations in a supplement to Hilbert and Bernays' book *Grundlagen der Mathematik II*. Indeed, Bernays' letter of December 24, 1938 begins with an apology for not having written to Church in a long time:

> I was again so much occupied by the working at the "Grundlagenbuch". In particular the "Supplemente" that I found desirable to add have become much more extended than expected. By the way: one of them is on the precising [sic!] of the concept of computability. There I had the opportunity of exposing some of the reasonings of yours and Kleene on general recursive functions and the unsolvability of the Entscheidungsproblem.

Bernays refers to the book's Supplement II, entitled "Eine Präzisierung des Begriffs der berechenbaren Funktion und der Satz von Church über das Entscheidungsproblem". A translation of the title, not quite adequate to capture "Präzisierung", is "A precise explication of the concept of calculable function and Church's Theorem on the decision problem".

In this supplement Hilbert and Bernays make the core notion of *calculability in a logic* directly explicit and define a number theoretic function to be *reckonable* (in German, *regelrecht auswertbar*) when it is computable in some deductive formalism satisfying three recursiveness conditions. The crucial condition is an analogue of Church's Central Thesis and requires that the theorems of the formalism can be enumerated by a primitive recursive function or, equivalently, that the proof predicate is primitive recursive. Then it is shown (1) that a special, very restricted number theoretic formalism suffices to compute the reckonable functions, and (2) that the functions computable

---

[24] It is most natural and general to take the underlying generating procedures directly as finitary inductive definitions. That is Post's approach via his production systems; using Church's central thesis to fix the restricted character of the generating steps guarantees the recursive enumerability of the generated set. Cf. Kleene's discussion of Church's argument in IM, pp. 322–323. Here it might also be good to recall remarks of C. I. Lewis on "inference" as reported in (Davis 1995) on page 273: "The main thing to be noted about this operation is that it is not so much a piece of reasoning as a mechanical, or strictly mathematical, operation for which a rule has been given. No "mental" operation is involved except that required to recognize a previous proposition followed by the main implication sign, and to set off what follows that sign as a new assertion."

in this formalism are exactly the general recursive ones. The analysis provides, in my view, a natural and most satisfactory capping of the development from *Entscheidungsdefinitheit* of relations in (Gödel 1931) to an "absolute" notion of computability for functions, because it captures directly the informal notion of rule-governed evaluation of effectively calculable functions and isolates appropriate restrictive conditions.

§4. **Absoluteness and formalizability.** A technical result of the sort we just discussed was for Gödel in 1935 the first hint that there might be a precise notion capturing the informal concept of effective calculability.[25] Gödel defined an absoluteness notion for the specific formal systems of his paper (1936a). A number theoretic function $\phi(x)$ is said to be *computable in S* just in case for each numeral m there exists a numeral n such that $\phi(m) = n$ is provable in S. Clearly, all primitive recursively defined functions, for example, are already computable in the system S_1 of classical arithmetic, where S_i is number theory of order i, for i finite or transfinite. In the Postscriptum to the paper Gödel observed:

> It can, moreover, be shown that a function computable in one of the systems S_i, or even in a system of transfinite order, is computable already in S_1. Thus the notion 'computable' is in a certain sense 'absolute', while almost all metamathematical notions otherwise known (for example, provable, definable, and so on) quite essentially depend upon the system adopted. (p. 399)

A broader notion of *absoluteness* was used in Gödel's contribution to the Princeton bicentennial conference, i.e., in (Gödel 1946). Gödel starts out with the following remark:

> Tarski has stressed in his lecture (and I think justly) the great importance of the concept of general recursiveness (or Turing's computability). It seems to me that this importance is largely due to the fact that with this concept one has for the first time succeeded in giving an absolute definition of an interesting epistemological notion, i.e., one not depending on the formalism chosen. (p. 150)

For the publication of the paper in (Davis 1965) Gödel added a footnote to the last sentence:

> To be more precise: a function of integers is computable in any formal system containing arithmetic if and only if it is computable

---

[25]Cf. (Sieg 1994), p. 88 and note 52. The latter asserts that the content of (Gödel 1936a) was presented in a talk in Vienna on June 19, 1935. An interesting question is, certainly, how much Gödel knew then about the ongoing work in Princeton reported in Church's 1935 letters to Bernays. I could not find any evidence that Gödel communicated with Bernays, Church, or Kleene on these issues at that time.

in arithmetic, where a function f is called computable in S if there is in S a computable term representing f. (p. 150)

Both in 1936 and in 1946, Gödel took for granted the formal character of the systems and, thus, the elementary character of their inference or calculation steps. Gödel's claim that "an absolute definition of an interesting epistemological notion" has been given, i.e., a definition that does not depend on the formalism chosen, is only partially correct: the definition does not depend on the details of the formalism, but depends crucially on the fact that we are dealing with a "formalism" in the first place. In that sense absoluteness has been achieved only relative to an un-explicated notion of an elementary formalism. It is in this conceptual context that Church's letter from June 8, 1937 to the Polish logician Józef Pepis should be seen.[26] Church brings out this "relativism" very clearly in an indirect way of defending his thesis; as far as I know, this broadened perspective, though clearly related to his earlier explication, has not been presented in any of Church's writings on the subject.

Pepis had described to Church his project of constructing a number theoretic function that is effectively calculable, but not general recursive. In his response Church explains, why he is "extremely skeptical". There is a minimal condition for a function f to be effectively calculable, and "if we are not agreed on this then our ideas of effective calculability are so different as to leave no common ground for discussion": for every positive integer a there must exist a positive integer b such that the proposition $f(a) = b$ has a "valid proof" in mathematics. But as all of extant mathematics is formalizable in *Principia Mathematica* or in one of its known extensions, there actually must be a *formal proof* of a suitably chosen formal proposition. However, if f is not general recursive then, by the considerations of (Church 1936), for every definition of f within the language of *Principia Mathematica* there exists a positive integer a such that for no b the proposition $f(a) = b$ is provable in *Principia Mathematica*; that holds again for all known extensions. Indeed, Church claims this holds for "any system of symbolic logic whatsoever which to my knowledge has ever been proposed". Thus, to satisfy the above minimal condition and to respect the quasi-empirical fact that all of mathematics is formalizable, one would have to find "an utterly new principle of logic, not only never before formulated, but also never before actually used in a mathematical proof".

Moreover, and here is the indirect appeal to the recursivity of steps, the new principle "must be of so strange, and presumably complicated, a kind that its metamathematical expression as a rule of inference was not general recursive", and one would have to scrutinize the "alleged effective applicability of the principle with considerable care". The dispute concerning a proposed

---

[26]The letter to Pepis was partially reproduced and analyzed in (Sieg 1994); it is reprinted in Appendix A.

effectively calculable, non-recursive function would thus for Church center around the required new principle and its effective applicability as a rule of inference, i.e., what I called Church's Central Thesis. If the latter is taken for granted (implicitly, for example, in Gödel's absoluteness considerations), then the above minimal understanding of effective calculability and the quasi-empirical fact of formalizability block the construction of such a function. This is not a completely convincing argument; Church is extremely skeptical of Pepis' project, but mentions that "this [skeptical] attitude is of course subject to the reservation that I may be induced to change my opinion after seeing your work".

On April 22, 1937, Bernays wrote a letter to Church and remarked that Turing had just sent him the paper (Turing 1936); there is a detailed discussion of some points concerned with Turing's proof of the undecidability of the *Entscheidungsproblem*. As to the general impact of Turing's paper Bernays writes:

> He [Turing] seems to be very talented. His concept of computability is very suggestive and his proof of the equivalence of this notion with your λ-definability gives a stronger conviction of the adequacy of these concepts for expressing the popular meaning of "effective calculability".

Bernays does not give in this letter (or in subsequent letters to Church and to Turing) a reason, why he finds Turing's concept "suggestive"; strangely enough, in Supplement II of ***Grundlagen der Mathematik II***, Turing's work is not even mentioned. It is to that work that I shall turn now to indicate in what way it overcomes the limitations of the earlier analyses (all centering around the concept of "computability in a formal logic").

§5. Computors, boundedness, and locality.

The earlier detailed reconstruction of Church's justification for the "selection of a formal definition to correspond to an intuitive notion" and the pinpointing of the crucial difficulty show, first of all, the sophistication of Church's methodological attitude and, secondly, that at this point in 1935 there is no major opposition to Gödel's cautious attitude. These points are supported by the directness with which Church recognized in 1937, when writing a review of (Turing 1936) for ***The Journal of Symbolic Logic***, the importance of Turing's work as making the identification of effectiveness and (Turing) computability "immediately evident". That review is quoted now in full:

> The author proposes as criterion that an infinite sequence of digits 0 and 1 be "computable" that it is possible to devise a computing machine, occupying a finite space and with working parts of finite size, which will write down a sequence to any desired number of terms if allowed to run for a sufficiently long time. As a matter of

convenience, certain further restrictions are imposed on the character of the machine, but these are of such a nature as obviously to cause no loss of generality—in particular, a human calculator, provided with pencil and paper and explicit instructions, can be regarded as a kind of Turing machine. It is thus immediately clear that computability, so defined, can be identified with (especially, is no less general than) the notion of effectiveness as it appears in certain mathematical problems (various forms of the Entscheidungsproblem, various problems to find complete sets of invariants in topology, group theory, etc., and in general any problem which concerns the discovery of an algorithm).

The principal result is that there exist sequences (well-defined on classical grounds) which are not computable. In particular the *deducibility problem* of the functional calculus of first order (Hilbert and Ackermann's engere Funktionenkalkül) is unsolvable in the sense that, if the formulas of this calculus are enumerated in a straightforward manner, the sequence whose nth term is 0 or 1, according as the nth formula in the enumeration is or is not deducible, is not computable. (The proof here requires some correction in matters of detail.)

In an appendix the author sketches a proof of equivalence of "computability" in his sense and "effective calculability" in the sense of the present reviewer (***American Journal of Mathematics***, vol. 58 (1936), pp. 345–363, see review in this Journal, vol. 1, pp. 73–74). The author's result concerning the existence of uncomputable sequences was also anticipated, in terms of effective calculability, in the cited paper. His work was, however, done independently, being nearly complete and known in substance to a number of people at the time that the paper appeared.

As a matter of fact, there is involved here the equivalence of three different notions: computability by a Turing machine, general recursiveness in the sense of Herbrand–Gödel–Kleene, and the λ-definability in the sense of Kleene and the present reviewer. Of these, the first has the advantage of making the identification with effectiveness in the ordinary (not explicitly defined) sense evident immediately—i.e., without the necessity of proving preliminary theorems. The second and third have the advantage of suitability for embodiment in a system of symbolic logic.

So, Turing's notion is presumed to make the identification with effectiveness in the ordinary sense "evident immediately". How this is to be understood is a little clearer from the first paragraph of the review, where it is claimed to be immediately clear "that computability, so defined, can be identified with ... the

notion of effectiveness as it appears in certain mathematical problems ... ".
This claim is connected to previous sentences by "thus": the premises of this "inference" are: (1) computability is defined via computing machines (that occupy a finite space and have working parts of finite size), and (2) human calculators, "provided with pencil and paper and explicit instructions", can be regarded as Turing machines.

The review of Turing's paper is immediately followed by Church's review of (Post 1936); the latter is reprinted in Appendix C. Church is sharply critical of Post; this is surprising, perhaps, as Church notices the equivalence of Post's and Turing's notions. The reason for the criticism is methodological: Post does not "identify" his formulation of a finite 1-process with effectiveness in the ordinary sense, but rather considers it as a "working hypothesis". The program of reducing wider and wider formulations to this basic one is seen as a way of changing "this hypothesis not so much to a definition or an axiom but to a *natural law*". Church objects "that effectiveness in the ordinary sense has not been given an exact definition, and hence the working hypothesis in question has not an exact meaning". The need for a working hypothesis disappears, so Church argues, if effectiveness is defined as "computability by an arbitrary machine, subject to restrictions of finiteness". The question here is, why does that seem "to be an adequate representation of the ordinary notion"? Referring back to the "inference" isolated in the review of Turing's paper, we may ask, why do the two premises support the identification of Turing computability with the informal notion of effectiveness as used for example in the formulation of the decision problem? Thus we are driven to ask the more general question, what is the real character of Turing's analysis?[27]

Let me emphasize that Turing's analysis is neither concerned with *machine computations* nor with general *human mental processes*. Rather, it is *human mechanical computability* that is being analyzed, and the special character of this intended notion motivates the restrictive conditions that are brought to bear by Turing.[28] Turing exploits in a radical way that a *human computor* is performing *mechanical procedures* on *symbolic configurations*: the immediate recognizability of symbolic configurations is demanded so that basic (computation) steps cannot be further subdivided. This demand and the evident limitation of the computor's sensory apparatus lead to the formulation of boundedness and locality conditions. Turing requires also a *determinacy condition* (**D**), i.e., the computor carries out deterministic computations, as his internal state together with the observed configuration fixes uniquely the next computation step. The boundedness conditions can be formulated as follows:

---

[27] The following analysis was given in (Sieg 1994); it is also presented in the synoptic (Sieg & Byrnes 1997).

[28] This is detailed in (Sieg 1994).

(**B.1**) *there is a fixed bound for the number of symbolic configurations a computor can immediately recognize; and*
(**B.2**) *there is a fixed bound for the number of a computor's internal states that need to be taken into account.*[29]

For a given computor there are consequently only boundedly many different combinations of symbolic configurations and internal states. Since his behavior is, according to (**D**), uniquely determined by such combinations and associated operations, the computor can carry out at most finitely many different operations. These operations are restricted by the following *locality conditions*:

(**L.1**) *only elements of observed configurations can be changed; and*
(**L.2**) *the distribution of observed squares can be changed, but each of the new observed squares must be within a bounded distance of an immediately previously observed square.*[30]

Turing's computor proceeds deterministically, must satisfy the boundedness conditions, and the elementary operations he can carry out must be restricted as the locality conditions require. Every number-theoretic function such a computor can calculate, Turing argues, is actually computable by a Turing machine over a two-letter alphabet. Thus, on closer inspection, Turing's Thesis that the concept "mechanical procedure" can be identified with machine computability is seen as the result of a two-part analysis. The first part yields axioms expressing boundedness conditions for symbolic configurations and locality conditions for mechanical operations on them, together with the *central thesis* that any mechanical procedure can be carried out by a computor satisfying the axioms. The second part argues for the *claim* that every number-theoretic function calculable by such a computor is computable by a Turing machine. In Turing's presentation these quite distinct aspects are intertwined and important steps in arguments are only hinted at.[31] Indeed, the claim that is actually established in Turing's paper is the more modest one that Turing machines operating on single letters can simulate Turing machines operating on strings.

In the historical context in which Turing found himself, he asked exactly the right question: What are the elementary processes a computor carries out (when calculating a number)? Turing was concerned with *symbolic* processes, not—as the other proposed explications—with processes directly related to the

---

[29]This condition (and the reference to internal states) can actually be removed and was removed by Turing; nevertheless, it has been a focus of critical attention.

[30]This is almost literally Turing's formulation. Obviously, it takes for granted particular features of the precise model of computation, namely, to express that the computor's attention can be shifted only to symbolic configurations that are not "too far away" from the currently observed configuration.

[31]Turing's considerations are sharpened and generalized in (Sieg & Byrnes 1996).

evaluation of (number theoretic) functions. Indeed, the general "problematic" required an analysis of the idealized capabilities of a computor, and it is precisely this feature that makes the analysis epistemologically significant. The separation of conceptual analysis and rigorous proof is essential for clarifying on what the correctness of Turing's central thesis rests, namely, on recognizing that the boundedness and locality conditions are true for a computor and also for the particular precise, analyzed notion.

§6. Conceptual analyses: a brief comparison. Church's way of approaching the problem was at first deeply affected by quasi-empirical considerations. That is true also for his attitude to the consistency problem for the systems in (Church 1932 and 1933); his letter of July 27, 1932 to Gödel is revealing. His review of Turing's 1936 paper shows, however, that he moved away from that position; how far is perhaps even better indicated by the very critical review of (Post 1936). In any event, Turing's approach provides immediately a detailed conceptual analysis realizing, it seems, what Gödel had suggested in conversation with Church, namely "to state a set of axioms which would embody the generally accepted properties of this notion [effective calculability, WS], and to do something on that basis". The analysis leads convincingly to the conclusion that Turing machines (over a two letter alphabet) can compute "effectively calculable" functions. The former mathematical notion, appropriately, serves as the starting point for *Computability Theory*; cf. (Soare 1996).

Turing's analysis divides, as I argued in the last section, into conceptual analysis and rigorous proof. The conceptual analysis leads *first* to a careful and sharper formulation of the intended informal concept, here, "mechanical procedures carried out by a human computor", and *second* to the axiomatic formulation of determinacy, boundedness, and locality conditions. Turing's central thesis connects the informal notion and the axiomatically restricted one. Rigorous proof allows us then, third, to recognize that a Turing machine can simulate all the actions of an axiomatically restricted computor. Thus, the analysis together with the proof allows us to "replace" the boldly claimed thesis, all effectively calculable functions are Turing computable, by a carefully articulated argument that includes a sharpened informal notion and an axiomatically characterized one.

Once such a "meta-analysis" of Turing's ways is given, one can try and see whether there are other mathematical concepts that have been analyzed in a similar way.[32] It seems to me that Dedekind's recasting of "geometric" continuity in "arithmetic" terms provides a convincing second example; the

---

[32] Mendelson and Soare, for example, draw in their papers parallels between Turing's or Church's Thesis and other mathematical "theses". G. Kreisel has reflected on "informal rigor" generally and on its application to Church's Thesis in particular; a good presentation of Kreisel's views and a detailed list of his relevant papers can be found in (Odifreddi 1996).

steps I will describe now are explicitly in (Dedekind 1872). The intended informal concept, "continuity of the geometric line", is *first* sharpened by the requirement that the line must not contain "gaps". The latter requirement is characterized, *second*, by the axiomatic condition that any "cut" of the line determines a unique geometric point. This "completeness" of the line is taken by Dedekind to be the "essence of continuity" and corresponds, as a normative demand, to Turing's central thesis. What corresponds to the third element in Turing's analysis, namely the rigorous proof? — Dedekind's argument, that the continuous geometric line and the system of rational cuts are isomorphic, does: the rationals can be associated with geometric points by fixing an origin on the line and a unit; the geometric cuts can then be transferred to the arithmetic realm. (To be sure, that requires the consideration of arbitrary partitions of the rationals satisfying the cut conditions and the proof that the system of rational cuts is indeed complete.) It is in this way that Dedekind's Thesis, or rather Dirichlet's demand that Dedekind tried to satisfy, is now supported: every statement of algebra and higher analysis can be viewed as a statement concerning natural numbers (and sets of such).[33]

Hilbert presented considerations concerning the continuum in his lectures from the winter term 1919, entitled *Natur und mathematisches Erkennen*; he wanted to support the claim that the formation of concepts in mathematics is constantly guided by intuition and experience, so that on the whole mathematics is a non-arbitrary, unified structure.[34] Having presented Dedekind's construction and his own investigation on non-Archimedean extensions of the rationals, he formulated the general point as follows:

> The different existing mathematical disciplines are consequently necessary parts in the construction of a systematic development of thought; this development begins with simple, natural questions and proceeds on a path that is essentially traced out by compelling internal reasons. There is no question of arbitrariness. Mathematics is not like a game that determines the tasks by arbitrarily invented rules, but rather a conceptual system of internal necessity that can only be thus and not otherwise.[35]

---

[33] Let me add to the above analogy two further remarks: (i) both concepts are highly idealized—in Dedekind's case, he is clear about the fact that not all cuts are needed to have a model of Euclidean geometry, i.e., the constructibility of points is not a concern; for Turing, feasibility of computations is not a concern; (ii) both concepts are viewed by me as "abstract" mathematical concepts in the sense of (Sieg 1996).

[34] l.c., p. 8; .... vielmehr zeigt sich, daß die Begriffsbildungen in der Mathematik beständig durch Anschauung und Erfahrung geleitet werden, so daß im großen und ganzen die Mathematik ein willkürfreies, geschlossenes Gebilde darstellt.

[35] l.c., p. 19: Es bilden also die verschiedenen vorliegenden mathematischen Disziplinen notwendige Glieder im Aufbau einer systematischen Gedankenentwicklung, welche von einfachen, naturgemäß sich bietenden Fragen anhebend, auf einem durch den Zwang innerer Gründe

Hilbert's remarks are fitting not only for the theory of the continuum, but also for the theory of computability.

Appendix

A. Church's letter of June 8, 1937, to Pepis was enclosed with a letter to Bernays sent on June 14, 1937. Other material, also enclosed, were the "card" from Pepis to which Church's letter is a reply and the manuscript of Pepis' paper, *Ein Verfahren der mathematischen Logik*; Church asked Bernays to referee the paper. Church added, "Not because they are relevant to the question of acceptance of this particular paper, but because you may be interested in seeing the discussion of another point, I am sending you also a card received from Pepis and a copy of my reply. Please return Pepis' card when you write." In his letter of July 3, 1937, Bernays supported the publication of the paper which appeared in the 1938 edition of ***The Journal of Symbolic Logic***; he also returned the card (which may very well be in the Church *Nachlass*).

Dear Mgr. [Magister] Pepis:

This is to acknowledge receipt of your manuscript, <u>Ein Verfahren der mathematischen Logik</u>, offered for publication in the Journal of Symbolic Logic. In accordance with our usual procedure we are submitting this to a referee to determine the question of acceptance for publication, and I will write you further about the matter as soon as I have the referee's report.

In reply to your postal I will say that I am very much interested in your results on general recursiveness, and hope that I may soon be able to see them in detail. In regard to your project to construct an example of a numerical function which is effectively calculable but not general recursive I must confess myself extremely skeptical—although this attitude is of course subject to the reservation that I may be induced to change my opinion after seeing your work.

I would say at the present time, however, that I have the impression that you do not fully appreciate the consequences which would follow from the construction of an effectively calculable non-recursive function.

For instance, I think I may assume that we are agreed that if a numerical function f is effectively calculable then for every positive integer a there must exist a positive integer b such that a valid proof can be given of the proposition $f(a) = b$ (at least if we are not agreed on this then our ideas of effective calculability are

---

im wesentlichen vorgezeichneten Wege fortschreitet. Von Willkür ist hier keine Rede. Die Mathematik ist nicht wie ein Spiel, bei dem die Aufgaben durch willkürlich erdachte Regeln bestimmt werden, sondern ein begriffliches System von innerer Notwendigkeit, das nur so und nicht anders sein kann.

so different as to leave no common ground for discussion). But it is proved in my paper in the ***American Journal of Mathematics*** that if the system of *Principia Mathematica* is omega-consistent, and if the numerical function f is not general recursive, then, whatever permissible choice is made of a formal definition of f within the system of *Principia*, there must exist a positive integer a such that for no positive integer b is the proposition $f(a) = b$ provable within the system of *Principia*. Moreover this remains true if instead of the system of *Principia* we substitute any one of the extensions of *Principia* which have been proposed (e.g., allowing transfinite types), or any one of the forms of the Zermelo set theory, or indeed any system of symbolic logic whatsoever which to my knowledge has ever been proposed.

Therefore to discover a function which was effectively calculable but not general recursive would imply discovery of an utterly new principle of logic, not only never before formulated, but never before actually used in a mathematical proof—since all extant mathematics is formalizable within the system of *Principia*, or at least within one of its known extensions. Moreover this new principle of logic must be of so strange, and presumably complicated, a kind that its metamathematical expression as a rule of inference was not general recursive (for this reason, if such a proposal of a new principle of logic were ever actually made, I should be inclined to scrutinize the alleged effective applicability of the principle with considerable care).

Sincerely yours,
Alonzo Church

B. This is the part of Bernays' letter to Church of July 3, 1937, that deals with the latter's reply to Pepis.

Your correspondence with Mr. Pepis on his claimed discovery has much interested me. As to the consequence you draw from your result p. 357 Amer. Journ. Math., it seems to me that you have to use for it the principle of excluded middle. Without this there would remain the possibility that for the expression f it can <u>neither</u> be proved that to every μ standing for a positive integer m, there is a ν standing for a positive integer n such that the formula $\{f\}(\mu) = \nu$ is deducible within the logic, <u>nor</u> there can be denoted a positive integer m for which it can be proved that for no positive integer n the formula $\{f\}(\mu) = \nu$, where μ stands for m and ν for n, is deducible within the logic.

C. Church's review of Turing's paper in the Journal of Symbolic Logic is followed directly by his review of (Post 1936):

> The author proposes a definition of "finite 1-process" which is similar in formulation, and in fact equivalent, to computation by a Turing machine (see the preceding review). He does not, however, regard his formulation as certainly to be identified with effectiveness in the ordinary sense, but takes this identification as a "working hypothesis" in need of continual verification. To this the reviewer would object that effectiveness in the ordinary sense has not been given an exact definition, and hence the working hypothesis in question has not an exact meaning. To define effectiveness as computability by an arbitrary machine, subject to restrictions of finiteness, would seem to be an adequate representation of the ordinary notion, and if this is done the need for a working hypothesis disappears.
>
> The present paper was written independently of Turing's, which was at the time in press but had not yet appeared.

D. On July 25, 1983 Church wrote a letter to John W. Dawson responding to the latter's inquiry, whether he (Church) had been "among those who thought that the Gödel incompleteness theorem might be found to depend on peculiarities of type theory". Church's letter is a rather touching (and informative) reflection on his work in the early thirties.

> Dear Dr. Dawson:
>
> In reply to your letter of June eighth, yes I was among those who thought that the Gödel incompleteness theorem might be found to depend on peculiarities of type theory (or, as I might later have added, of set theory) in a way that would show his results to have less universal significance than he was claiming for them. There was a historical reason for this, and that is that even before the Gödel results were published I was working on a project for a radically different formulation of logic which would (as I saw it at the time) escape some of the unfortunate restrictiveness of type theory. In a way I was seeking to do the very thing that Gödel proved impossible, and of course it's unfortunate that I was slow to recognize that the failure of Gödel's first proof to apply quite exactly to the sort of formulation of logic I had in mind was of no great significance.
>
> The one thing of lasting importance that came out of my work in the thirties is the calculus of λ-conversion. And indeed this might be claimed as a system of logic to which the Gödel incompleteness theorem does not apply. To ask in what sense this claim is sound

and in what sense it is not altogether pointless, as it may give some insight into the question where the boundary lies for applicability of the incompleteness theorem.

In my monograph on the calculus of λ-conversion (Annals of Mathematics Studies), in the section of the calculus of λ-δ-conversion (a minor variation of the λ-calculus) it is pointed out how, after identifying the positive integer 1 with the truth-value falsehood and the positive integer 2 with the truth-value truth, it is possible to introduce by definition, first the connectives of propositional calculus, and then an existential quantifier, but the latter only in the sense that: $(\exists x)M$ reduces to truth whenever there is some positive integer such that M reduces to truth after replacing x by the standard name of that positive integer, and in the contrary case $(\exists x)M$ has no normal form. The system is complete within its power of expression. But an attempt to introduce a universal quantifier, whether by definition or by added axioms, will give rise to some form of the Gödel incompleteness.—I'll not try to say more, as I am writing from recollection and haven't the monograph itself before me.

Sincerely,
Alonzo Church

BIBLIOGRAPHY

(Some starred items were added for this reprinting: they are referred to only in the Postscriptum.)

Bernays, Paul

1935 –1936	*Logical Calculus*; Notes by Prof. Bernays with assistance of Mr. F. A. Ficken, The Institute for Advanced Studies, Princeton, 125 pp.

Church, Alonzo

1927	*Alternatives to Zermelo's assumption*; **Transactions of the American Mathematical Society**, vol. 29, pp. 178–208. (Ph.D. thesis; Princeton.)
1928	*On the law of the excluded middle*; **American Mathematical Society. Bulletin**, vol. 34, pp. 75–78.
1932	*A set of postulates for the foundation of logic, part I*; **Annals of Mathematics**, vol. 33, pp. 346–366.
1933	*A set of postulates for the foundation of logic, part II*; **Annals of Mathematics**, vol. 34, pp. 839–864.
1934	*The Richard Paradox*; **The American Mathematical Monthly**, vol. 41, pp. 356–361.
1935	*A proof of freedom from contradiction*; **Proceedings of the National Academy of Sciences of the United States of America**, vol. 21, pp. 275–81; reviewed in (Gödel 1936b).

1935a	*An unsolvable problem of elementary number theory*; **American Mathematical Society. Bulletin**, vol. 41, pp. 332–333.
1936	*An unsolvable problem of elementary number theory*; **American Journal of Mathematics**, vol. 58, pp. 345–363.
1936a	*A note on the Entscheidungsproblem*; **The Journal of Symbolic Logic**, vol. 1, pp. 40–41, and Corrections, ibid., pp. 101–102.
1937	*Review of* (Turing 1936), **The Journal of Symbolic Logic**, vol. 2, pp. 42–43.
1937a	*Review of* (Post 1936), **The Journal of Symbolic Logic**, vol. 2, p. 43.
1956	**Introduction to mathematical logic, volume I**; Princeton University Press, Princeton.

Church, Alonzo and J. Barkley Rosser

1935	*Some properties of conversion*; **American Mathematical Society. Bulletin**, vol. 41, p. 332.
1936	*Some properties of conversion*; **Transactions of the American Mathematical Society**, vol. 39, pp. 472–482.

Church, Alonzo and Stephen C. Kleene

1935	*Formal definitions in the theory of ordinal numbers*; **American Mathematical Society. Bulletin**, vol. 41, p. 627.
1936	*Formal definitions in the theory of ordinal numbers*; **Fundamenta Mathematicae**, vol. 28, pp. 11–21.

Davis, Martin

*1958	**Computability and unsolvability**; McGraw-Hill.
1965	**The Undecidable**; Raven Press, Hewlett, New York.
1982	*Why Gödel didn't have Church's Thesis*; **Information and Control**, vol. 54, pp. 3–24.
*1994	*Emil L. Post: His life and his work*; in **Solvability, Provability, Definability: The Collected Works of Emil L. Post**, (M. Davis, editor), Birkhäuser, pp. xi–xxviii.
1995	*American logic in the 1920s*; **The Bulletin of Symbolic Logic**, vol. 1, pp. 273–278.

Dawson, John W.

1997	**Logical Dilemmas—The life and work of Kurt Gödel**; A. K. Peters, Wellesley.

Dedekind, Richard

1872	**Stetigkeit und irrationale Zahlen**; Vieweg, Braunschweig.
1888	**Was sind und was sollen die Zahlen?**; Vieweg, Braunschweig.

Enderton, Herb

1995	*In Memoriam: Alonzo Church*; **The Bulletin of Symbolic Logic**, vol. 1, pp. 486–488.

Feferman, Solomon

1988	*Turing in the land of O(z)*; in **The Universal Turing Machine**, (Rolf Herken, editor), Oxford University Press, Oxford, 1988, pp. 113–147.

Gandy, Robin

1988   *The confluence of ideas*; in **The Universal Turing Machine**, (Rolf Herken, editor), Oxford University Press, Oxford, 1988, pp. 55–111.

Gödel, Kurt

1931   *On formally undecidable propositions of* Principia Mathematica *and related systems I*; in **Collected Works I**, pp. 145–195.

1933o   *The present situation in the foundations of mathematics*; in **Collected Works III**, pp. 45–53.

1934   *On undecidable propositions of formal mathematical systems*; in **Collected Works I**, pp. 346–371.

1934e   *Review of Church 1933*; in **Collected Works I**, pp. 381–383.

1936a   *On the length of proofs*; in **Collected Works I**, pp. 397–399.

1936b   *Review of Church 1935*; in **Collected Works I**, pp. 399–401.

*193?   [[*Undecidable Diophantine propositions*]]; in **Collected Works III**, pp. 164–175.

1946   *Remarks before the Princeton bicentennial conference on problems in mathematics*; in **Collected Works II**, pp. 150–153.

1986   **Collected Works I**, Oxford University Press, Oxford.

1990   **Collected Works II**, Oxford University Press, Oxford.

1995   **Collected Works III**, Oxford University Press, Oxford.

*2003a   **Collected Works VI**, Oxford University Press, Oxford.

*2003b   **Collected Works V**, Oxford University Press, Oxford.

Hilbert, David

1919   *Natur und mathematisches Erkennen*; Lecture given in the Winter term 1919, the notes were written by Paul Bernays; reprint by the Mathematical Institute of the University Göttingen, 1989.

Hilbert, David and Paul Bernays

1934   **Grundlagen der Mathematik I**, Springer Verlag, Berlin.

1939   **Grundlagen der Mathematik II**, Springer Verlag, Berlin.

Kleene, Stephen C.

1934   *Proof by cases in formal logic*; **Annals of Mathematics**, vol. 35, pp. 529–544.

1935   *A theory of positive integers in formal logic*; **American Journal of Mathematics**, vol. 57, pp. 153–173 and 219–244.

1935a   *General recursive functions of natural numbers*; **American Mathematical Society. Bulletin**, vol. 41, p. 489.

1935b   *λ-definability and recursiveness*; **American Mathematical Society. Bulletin**, vol. 41, p. 490.

1936   *General recursive functions of natural numbers*; **Mathematische Annalen**, vol. 112, 727–742.

1936a   *λ-definability and recursiveness*; **Duke Mathematical Journal**, vol. 2, pp. 340–353. (1935a and b are abstracts of these two papers and were received by the AMS on July 1, 1935.)

1936b   *A note on recursive functions*; **American Mathematical Society. Bulletin**, vol. 42, pp. 544–546.
1952   ***Introduction to metamathematics***; Wolters-Noordhoff Publishing, Groningen.
1981   *Origins of recursive function theory*; **Annals of the History of Computing**, vol. 3, pp. 52–66.

Kleene, Stephen C. and J. Barkley Rosser

1935   *The inconsistency of certain formal logics*; **American Mathematical Society. Bulletin**, vol. 41, p. 24.
1935   *The inconsistency of certain formal logics*; **Annals of Mathematics**, vol. 36, no. 2, pp. 630–636.

Mendelson, Elliott

1990   *Second thoughts about Church's Thesis and mathematical proofs*; **The Journal of Philosophy**, vol. 87, no. 5, 225–233.

Odifreddi, Piergiorgio

1996   *Kreisel's Church*; in ***Kreiseliana*** (P. Odifreddi, editor), A. K. Peters, Wellesley, pp. 389–415.

Pepis, József

1938   *Ein Verfahren der mathematischen Logik*; **The Journal of Symbolic Logic**, vol. 3, pp. 61–76.

Post, Emil

1936   *Finite combinatory processes. Formulation I*; **The Journal of Symbolic Logic**, vol. 1, pp. 103–105.
*1941   *Absolutely unsolvable problems and relatively undecidable propositions—Account of an anticipation*; submitted to the **American Journal of Mathematics** in 1941, but published only in Davis' "The Undecidable" in 1965. (The mathematical core was published in (Post 1943).)
*1943   *Formal reductions of the general combinatorial decision problem*; **American Journal of Mathematics**, vol. 55, no. 2, pp. 197–215.
*1947   *Recursive unsolvability of a problem of Thue*; **The Journal of Symbolic Logic**, vol. 12, pp. 1–11.

Rosser, J. Barkley

1935   *A mathematical logic without variables*; **Annals of Mathematics**, vol. 36, no. 2, pp. 127–150 and **Duke Mathematical Journal**, vol. 1, pp. 328–355.
1984   *Highlights of the history of the lambda-calculus*; **Annals of the History of Computing**, vol. 6, no. 4, pp. 337–349.

Shagrir, Oron

*2006   *Gödel on Turing on computability*; in **Church's Thesis after 70 years**, (A. Olszewski, J. Wolenski and R. Janusz, editors), Ontos Verlag, pp. 393–419.

Shoenfield, Joseph R.

1967   **Mathematical Logic**; Addison-Wesley, Reading (Massachusetts).

Sieg, Wilfried

1994   *Mechanical procedures and mathematical experience*; in **Mathematics and Mind**, (A. George, editor), Oxford University Press, Oxford, pp. 71–117.

1996   *Aspects of mathematical experience*; in **Philosophy of Mathematics Today**, (E. Agazzi and G. Darvas, editors), Kluwer, pp. 195–217.

*2002   *Calculations by man and machine: conceptual analysis*; in **Reflections on the Foundations of Mathematics** (W. Sieg, R. Sommer and C. Talcott, editors), Lecture Notes in Logic 15, Association for Symbolic Logic, pp. 390–409.

*2005   *Only two letters: The correspondence between Herbrand and Gödel*; **The Bulletin of Symbolic Logic**, vol. 11, no. 2, pp. 172–184.

*2006   *Gödel on computability*; **Philosophia Mathematica**, vol. 14, pp. 189–207.

*2009   *On computability*; in **Philosophy of Mathematics**, (Andrew Irvine, editor), pp. 535–630.

Sieg, Wilfried and John Byrnes

1996   *K-graph machines: generalizing Turing's machines and arguments*; in **Gödel '96**, (Petr Hájek, editor), Lecture Notes in Logic 6, Springer Verlag, pp. 98–119.

1997   *Gödel, Turing, and K-graph machines*; in **Logic and Foundations of Mathematics**, (Andrea Cantini et al. editors), Kluwer Academic Publisher, pp. 57–66.

Soare, Robert

1996   *Computability and recursion*; **The Bulletin of Symbolic Logic**, vol. 2, pp. 284–321.

Turing, Alan

1936   *On computable numbers, with an application to the Entscheidungsproblem*; **Proceedings of the London Mathematical Society, Second Series**, vol. 42, pp. 230–265.

*1950   *The word problem in semi-groups with cancellation*; **Annals of Mathematics**, vol. 52, pp. 491–505.

*1954   *Solvable and unsolvable problem*; **Science News**, vol. 31, pp. 7–23.

Postscriptum (added on July 20, 2011)

This essay was published in *The Bulletin of Symbolic Logic*, vol. 3 (1997), no. 2, pp. 154–180, and is reprinted here with the permission of the Association for Symbolic Logic.—The essay is unchanged except for some minor stylistic improvements, a modified description of Post's (1936) in section 5, and a correction: Polish colleagues told me that "Mgr." in Church's address "Dear Mgr. Pepis" should be expanded to "Magister". The rationale for republication is direct: the historical discussions and conceptual analyses presented here are still most informative for the ongoing lively debate of the "Church–Turing Thesis". They depict the genuine difficulties Church and other pioneers faced, when attempting to articulate a conceptual framework, which would allow the conclusive formulation of central undecidability and incompleteness results. The complexities of interactions and the currents of influences are clear, but could be deepened now by examining, in particular, the archives of Church and Post; they have become accessible. The evolution of Gödel's thinking on computability has already become clearer, in particular, through the publication of his brief correspondence with Herbrand (see (Dawson 1997, pp. 100–103) and (Sieg 2005)) and the manuscript (Gödel 193?). My essay Gödel *On computability* traces that evolution.

On a number of issues directly connected to the content of this essay, I have gained a better understanding of both the relevant historical sources and the important mathematical analyses. I have modified my views on some issues, but not on one of the central insights: there was exactly *one core concept* that served in the 1930s to explicate effective calculability of number theoretic functions, namely, computability of their values in "logical calculi" or "deductive formalisms"; i.e., calculations were viewed as deductions proceeding by elementary, formal steps. That explication underlies not only Church's definition of effectively calculable functions and Hilbert & Bernays' characterization of reckonable functions, but provides also the basis for Gödel's absoluteness claims; cf. sections 3 and 4. I will make, in two groups, supplementary remarks on issues where I have modified my views; they are preceded by a discussion of Gödel's notion of a computable function in his (193?). The first group of remarks concerns the concept of "abstract deductive step" or "formal syntactic operation" as analyzed in the work of Turing and Post. Already in 1936, that work was conceptually much more closely related than I had previously realized. In the second group of remarks, I restructure Turing's analysis presented in section 5 and discuss briefly a way of characterizing "computability by man, respectively by machine" as abstract concepts that allow suitable representation theorems.

1. *Computable functions*. The elementary steps in Church's definition are ultimately taken to be recursive, but it also seems that they are taken as mathematically meaningful; just consider Church's emphatic insistence in his letter

to Pepis that if a function f is effectively calculable, then "for every positive integer a there must exist a positive integer b such that a valid proof can be given of the proposition $f(a) = b$ (at least if we are not agreed on this then our ideas of effective calculability are so different as to leave no common ground for discussion)." In Gödel's equational calculus, on which Church's more general logic is modeled, the individual steps are of course arithmetical ones.

The equational calculus presented in (Gödel 193?) has a simpler and more concise form than the original version of (Gödel 1934). Recursion equations of particular functions are, as before, the starting-points for calculating their values for specific arguments. Gödel asserts then that, "by analyzing in which manner this calculation proceeds", one finds "that it makes use only of the following two rules". The first rule, R1, allows the substitution of variables by numerals, whereas the second rule, R2, permits the replacement of equals for equals. I.e., if one has arrived at an equation of the form $T_1 = T_2$, then R2 permits the replacement of T_1 by T_2 in any other derived equation. "And now", Gödel continues, "it turns out that these two characteristics [recursion equations plus rules R1 and R2, WS] *are exactly those* that give *the correct definition of a computable function*." [My emphasis, WS.] This remark is followed by a proper mathematical definition of "computable" function.[36] It is perfectly clear that the equational calculus is an extremely natural framework for calculating the values of recursively given functions, and Gödel seems to endorse here implicitly Church's definitional identification of effectively calculable (computable) functions with general recursive ones.

2. *Machines & workers*. Two-letter Turing machines and Post's workers, carrying out finite combinatory processes, are at the heart of the same conception of

---

[36] Having defined "admissible postulate" and "true (wrong) elementary equation", Gödel writes: "And now we shall call a function of integers 'computable' if there exists a finite number of admissible postulates in f and perhaps some other auxiliary functions $g_1, \ldots, g_n$ such that every true elementary equation for f can be derived from these postulates by a finite number of applications of the rules R1 and R2 and no false elementary equation for f can be thus derived." (Gödel 193?, p. 168) Gödel, in the very next paragraph, asserts that Turing has shown "beyond any doubt" that this is not only the correct definition of computability, but of *mechanical* computability. Turing did that, according to Gödel, by establishing that the computable functions defined as above are exactly those whose values can be determined "by a machine with a finite number of parts".

Some, e.g., Shagrir in his (2006), have claimed that Gödel, through this remark, fully endorsed Turing's notion of machine computability as fundamental. However, that interpretation does not seem to be warranted: Gödel appeals to a theorem (Turing did not prove in his classical paper) and does not indicate even briefly Turing's analysis leading to his notion of machine computability; Gödel describes the machines that determine the values of number theoretic functions as simple office calculators with a crank and does not even allude to the need for a potentially infinite computing space. In addition, Shagrir does not take into account Gödel's actual definition of computable function and his detailed reasons, as just described, why deducibility in the equational calculus yields "the correct definition of a computable function".

computability. Indeed, when introducing his paper (1950), Turing asserts that the method for showing the unsolvability of the word problem for semi-groups with cancellation "depends on reducing the unsolvability of the problem in question to a known unsolvable problem connected with the logical computing machines introduced by Post (1936) and the author (1936)." Thus, Turing simply states that Post and he himself introduced the logical computing machines we call Turing machines; that is a rather striking statement. The obvious similarity of these conceptions based on machines and workers can be contrasted with their equally obvious dissimilarity with the effective calculability conceptions described above. For one, the direct arithmetic meaning of operations is given up: marking (writing) and unmarking (erasing) of boxes (of individual symbols on tape squares) are not conceived of as arithmetically meaningful. More importantly, the starting-points for Post and Turing's considerations are not only quite different from those for Gödel, Church and Hilbert & Bernays discussed above, but actually allow us to recognize a deeper connection between their conceptions.

That deeper connection does not just rest on the observation that Turing's machines are codifications of human computing agents, computors, but on the shared goal of determining the basic steps underlying mechanical operations on finite syntactic configurations. For Turing, that is of course clear from section 9 of his (1936) and as described in the paper above. For Post, that goal is not so directly apparent from his (1936). It is hinted at (on p. 104), when he discusses the applicability of a set of directions (for his worker) to a particular problem, possibly concerning symbolic logics; it is implicit in the clearly formulated program of considering wider and wider formulations and reducing them to "formulation I" (on p. 105). For a proper understanding, the context of Post's earlier work has to be considered. He attempted, starting in the early 1920s, to analyze the generating syntactic operations needed for the formalization of mathematics in Whitehead and Russell's *Principia Mathematica*. He conjectured, as Davis puts it in his (1994, p. xvi) "that whatever methods one could imagine for systematically 'generating' a set of strings on a finite alphabet could be formalized in *Principia*" and that *Principia* itself could be reduced to one of his "canonical systems". Thus, it was not too far-fetched to think that all systematically generated sets of strings can be generated by canonical systems.[37] Ultimately, as described in the last footnote of (Post 1943) and in greater detail in (Post 1941), he was striving for a characterization of all finite generating operations independent of their formalizability in *Principia*! Such a characterization would require, as

---

[37]This could be appropriately called *Post's central thesis*; the assertion that all such sets can be generated by normal systems, Martin Davis calls *Post's thesis*; see (Davis 1994, p. xvi) and (Davis 1982). Thus, Post's thesis is "decomposable" into Post's central thesis and the mathematical fact that canonical systems can be reduced to normal ones. This is structurally the "same" decomposition as for Turing's thesis.

Post remarked in (1941), "a complete analysis ... of all the possible ways in which the human mind could set up finite processes for generating sequences." (Davis 1994, p. xvii) That analysis proved to be elusive, but the core of the early mathematical work presented in (Post 1943) was most important; there, Post describes *canonical systems* and proves their reducibility to *normal* ones of a strikingly simple form.

In contrast, Turing's analysis *leads* from general mechanical operations on syntactic, even two-dimensional configurations to operations of string machines, i.e., to Post's production systems; Turing's letter machines can carry out the latter operations. How close his perspective was to that of Post, Turing recognized in his (1950), when extending Post's 1947—result of the unsolvability of the word-problem from semi-groups to semi-groups with cancellation. How deep the conceptual links between their conceptions were is perhaps best seen from (Turing 1954). In that remarkable essay, published in a popular science magazine, Turing formulated the foundational issues and some paradigmatic mathematical results all in terms of what he calls "substitution puzzles", i.e., Post's production systems. Machines are not alluded to.

3. *Turing's argument*. That is analyzed in my essay towards the end of section 5 and is discussed further in section 6. The needed modifications concern (i) the "axiomatic" formulation of determinacy, boundedness, and locality conditions, and (ii) the place of Turing's central thesis. Turing's analysis leads to boundedness and locality constraints, but in the form they are given here they are not mathematically rigorous. Thus the assertion (in the second paragraph of section 6) that "Turing's central thesis connects the informal notion and the axiomatically restricted one" cannot be maintained. Rather, I recognized later, the central thesis connects the informally constrained notion of *computability for computors* to the sharp concept of computability for string machines. This restructuring was an initial step towards a mathematically rigorous formulation of axioms for discrete dynamical systems that characterize *Turing computors* and *Gandy parallel machines*; see (Sieg 2002) and (Sieg 2009). The need for a central thesis disappears, as its claim is replaced by the recognition that the axioms are correct for the intended notions of human mechanical, respectively parallel machine computability. The connection with ordinary Turing computability is secured through *representation theorems* of the following form: Turing machines can carry out the computations of any model of the axioms.

DEPARTMENT OF PHILOSOPHY
CARNEGIE MELLON UNIVERSITY
PITTSBURGH, PENNSYLVANIA, USA

TURING AND THE DISCOVERY OF COMPUTABILITY

ROBERT IRVING SOARE

Contents

1. **A very brief overview of computability**   468
   1.1. Hilbert's programs 468
   1.2. Gödel, Church, and recursive functions 469
   1.3. Turing's analysis 470
   1.4. Gödel accepts Turing's analysis 470
   1.5. Art and mathematics 471
   1.6. Church's influence at Princeton 471
2. **Analyses of themes in computability theory**   472
3. **Computability and recursion**   472
   3.1. Gödel rejects the term "Recursive Function Theory" 473
   3.2. Computer scientists Hopcroft and Ullman [1979] 475
   3.3. Changing "recursive" back to "inductive" 475
   3.4. Computability in Europe 475
4. **Wittgenstein and wrong turnings**   476
5. **Turing's oracle machine in [1939]**   476
   5.1. An almost incidental definition 476
   5.2. An oracle machine with the halting problem K 477
   5.3. Producing a non Π_2^0 set 477
   5.4. Post [1944] defined relative computability 477
   5.5. Online or interactive computing 478
   5.6. Trial and error computing 478
6. **Why Turing's Thesis is not a thesis**   479
   6.1. The use of terms 479

    Robert Irving Soare is the Paul Snowden Russell Distinguished Service Professor of Mathematics and Computer Science at the University of Chicago. He was the founding Chairman of the Department of Computer Science in 1983–1987 and is the author of numerous articles and two books on computability theory and Alan Turing. His 1974 *Annals of Mathematics* paper on automorphisms was chosen by Gerald Sacks of Harvard University for the Sacks book [2003] on the most important papers in mathematical logic in the 20th century. As an undergraduate at Princeton University in 1959–1963, Soare studied mathematical logic and computability theory with Alonzo Church.

Turing's Legacy: Developments from Turing's Ideas in Logic
Edited by Rod Downey
Lecture Notes in Logic, 42

6.2. Turing [1936] proves a theorem........................... 480
6.3. What is a thesis?... 480
6.4. Effectively calculable functions 481
6.5. It was never a thesis..................................... 481
6.6. The Turing completeness theorem 482
7. **Formalism and informalism in computability** 483
   7.1. Kleene's formalism 483
   7.2. The Turing–Post informalism 483
   7.3. The renaissance by Rogers and Lachlan.................. 485
8. **Darwinian evolution and natural selection** 485
   8.1. Survival of the fittest.................................... 486
   8.2. The λ-calculus in 1931-1934............................. 486
   8.3. Herbrand–Gödel recursive function in 1934............... 486
   8.4. Kleene T-predicate and μ-recursive functions in 1936 486
   8.5. The computability renaissance of 1967–1970 487
   8.6. The renaming in 1996.................................... 487
   8.7. Church, Turing, Michelangelo, and the marble 487
9. **The fathers of the subject** 488
   9.1. The father of recursive function theory.................... 488
   9.2. The fathers of computability theory 488
10. **A full bibliography** 488

Abstract. In §1 we give a short overview[1] for a general audience of Gödel, Church, Turing, and the discovery of computability in the 1930s. In the later sections we mention a series of our previous papers where a more detailed analysis of computability, Turing's work, and extensive lists of references can be found. The sections from §2–§9 challenge the conventional wisdom and traditional ideas found in many books and papers on computability theory. They are based on a half century of my study of the subject beginning with Church at Princeton in the 1960s, and on a careful rethinking of these traditional ideas.

The references in all my papers and books are given in the format, author [year], as in Turing [1936], in order that the references are easily identified without consulting the bibliography and are uniform over all papers. A complete bibliography of historical articles from all my books and papers on computability is given on the page as explained in §10.

§1. A very brief overview of computability.

1.1. Hilbert's programs. Around 1880 Georg Cantor, a German mathematician, invented naive set theory. A small fraction of this is sometimes taught to elementary school children. It was soon discovered that this naive set theory was inconsistent because it allowed unbounded set formation, such

---

[1]This work was partially supported by grant number 204186 *Computability Theory and Applications* from the Simons Foundation Program for Mathematics and the Physical Sciences. The work was also partially supported by the Isaac Newton Institute of the University of Cambridge where Soare spent four months as a visiting fellow in 2012.

as the set of all sets. David Hilbert, the world's foremost mathematician from 1900 to 1930, defended Cantor's set theory but suggested a formal axiomatic approach to eliminate the inconsistencies. He proposed two programs. *First*, Hilbert wanted an axiomatization for mathematics, beginning with arithmetic, and a finitary consistency proof of that system. *Second*, Hilbert suggested that the statements about mathematics be regarded as formal sequences of symbols, and his famous *Entscheidungsproblem* (decision problem) was to find an algorithm to decide whether a statement was valid or not. Hilbert characterized this as the fundamental problem of mathematical logic.

Hilbert retired and gave a special address in 1930 in Königsberg, the city of his birth. Hilbert spoke on the importance of mathematics in science and the importance of logic in mathematics. He asserted that there are no unsolvable problems and stressed, "We must know. We will know." At a mathematical conference preceding Hilbert's address, a quiet, obscure young man, Kurt Gödel, only a year beyond his PhD, refuted Hilbert's consistency program with his famous incompleteness theorem [1931] and changed forever the foundations of mathematics. Gödel soon joined other leading figures, Albert Einstein and John von Neumann, at the Institute for Advanced Study in Princeton.

1.2. Gödel, Church, and recursive functions. The refutation of Hilbert's first program on consistency gave hope for refuting his second program on the *Entscheidungsproblem*. However, this was no ordinary problem in number theory or analysis. To prove the unsolvability of a certain problem, such as Hilbert's famous Tenth Problem on Diophantine equations of 1900, one must:
(1) find a precise mathematical definition for the intuitive idea of algorithm;
(2) demonstrate beyond doubt that every algorithm has been captured; and
(3) prove that no algorithm on the list can be the solution of the Diophantine equation problem.

Work began independently at Princeton and Cambridge. Alonzo Church completed an A.B. degree at Princeton in 1924 and his PhD degree there under Oswald Veblen in 1927. Church joined the mathematics department at Princeton from 1929 until his retirement in 1967 when he moved to UCLA. Church worked from 1931 through 1934 with his graduate student, Stephen Cole Kleene, on the formal system of λ-definable functions. They had such success that in 1934 Church proposed privately to Gödel that a function is effectively calculable (intuitively computable) if and only if it is λ-definable. Gödel rejected this as "thoroughly unsatisfactory." In addition, Kleene reported "chilly receptions from audiences around 1933–35 to disquisitions on λ-definability." By 1935 Church and Kleene had moved to the formalism of Gödel's *recursive functions*. Gödel [1931] had used the *primitive* recursive functions, those where one computes a value $f(n)$ by using previously computed values $f(m)$, for $m < n$, such as the factorial function $f(0) = 1$ and $f(n+1) = (n+1) \cdot f(n)$.

In his 1934 lectures at Princeton, Gödel extended this to the (Herbrand–Gödel) (*general*) recursive functions. Church eagerly embraced them and formulated his famous *Church's Thesis* in 1935–36 that the effectively calculable functions coincide with the recursive functions. Again, Gödel failed to accept this thesis even though he was the author of the recursive functions. Gödel noted that recursive functions are clearly effectively calculable, but the converse, "cannot be proved, since the notion of finite computation is not defined, but it serves as a heuristic principle."

1.3. Turing's analysis. Independently, Turing attended lectures in 1935 at the University of Cambridge by topologist M.H.A. (Max) Newman on Gödel's paper [1931] and Hilbert's *Engscheidungsproblem*. A year later Turing submitted his solution to the incredulous Newman on April 15, 1936. Turing's monumental paper [1936] was distinguished because: (1) Turing analyzed an idealized *human* computing agent, call it a *"computor"*, which brought together the intuitive conceptions of a "function produced by a mechanical procedure" that had been evolving for more than two millenia from Euclid to Leibniz to Babbage and Hilbert; (2) Turing specified a remarkably simple formal device (*Turing machine*) and demonstrated the equivalence of (1) and (2); (3) Turing proved the unsolvability of Hilbert's *Entscheidungsproblem*, which prominent mathematicians had been studying intently for some time; (4) Turing proposed a *universal* Turing machine, one which carried within it the capacity to duplicate any other, an idea which was later to have great impact on the development of high speed digital computers and considerable theoretical importance. As a boy, Turing had been fascinated by his mother's typewriter. He devised his Turing machine as a kind of idealized typewriter with a reading head moving over a fixed unbounded tape or platen[2] on which the head writes. Turing's model was by far the most convincing then and now. From 1936–1938 Turing completed his PhD at Princeton under Church. His PhD thesis was on a different topic but contained a crucial idea (5), that of a local machine communicating with a data base, the same mechanism we use today when a laptop communicates with the Internet.

1.4. Gödel accepts Turing's analysis. Gödel enthusiastically accepted Turing's analysis and always thereafter gave Turing credit for the definition of mechanical computability. For the Princeton Bicentennial in 1946 Gödel wrote, "one [Turing] has for the first time succeeded in giving an absolute definition of an interesting epistemological notion, *i.e.*, one not depending on the formalism chosen." Gödel also wrote, "That this really is the correct definition

---

[2]The platen is the cylindrical roller in a typewriter against which the paper is held. In 1930 the typing head was fixed in the center and the platen and carriage moved back and forth under it as the keys struck the platen. By 1980 the IBM selectric tpewriter had a fixed carriage and a movable writing ball which passed back and forth across the platen. This was Turing's design [1936]. I do not know whether IBM paid royalties to Turing's estate.

of mechanical computability was established beyond any doubt by Turing." Church wrote that of the three notions: computability by a Turing machine, general recursiveness of Herbrand–Gödel–Kleene, and lambda-definability, "The first has the advantage of making the identification with effectiveness in the ordinary (not explicitly defined) sense evident immediately—*i.e.*, without the necessity of proving preliminary theorems."

By 1937, the three definitions of computable functions had been proved mathematically equivalent. In retrospect, Church got it right and got it first. Why should Church not get the credit? However, the problem was not purely a mathematical one but one of art in first selecting an appealing model, and then convincingly demonstrating that the model captures the informal notion of computability. Why Turing and not Church? The answer in part is, why Michelangelo and not Donatello or Verrochio?

1.5. Art and mathematics. David slaying the giant Goliath was a very popular theme in Florence after 1400. Florence, like David, was militarily weak, but David served as a reminder of courage and a warning against aggressors. Donatello was the greatest sculptor of the early Renaissance. His bronze statue of David around 1430 was the first free-standing statue since ancient times. It showed a graceful David with helmet and sword standing over the head of Goliath. Verrocchio completed a similar bronze statue of David around 1475. David stands over the head of Goliath in battle costume holding a sword.

Michelangelo's David, the most famous statue in the world, was cut from the pure Cararra marble in 1501–1505. David is portrayed as manly and ready to fight, but he carries no sword or armor, only a slingshot, and is shown *before* the battle. The tension on his face shows him just after he has made the decision to face Goliath. For Michelangelo's David this was a *mental* achievement. Michelangelo wrote,

> "In every block of marble I see a statue as plain as though it stood before me, shaped and perfect in attitude and action. I have only to hew away the rough walls that imprison the lovely apparition to reveal it to the other eyes as mine see it."

Turing saw the figure of computability in the marble more clearly than anyone else, and skillfully revealed it. There are similarities between David, Michelangelo, and Turing. All were very young at the time of their achievement: Michelangelo was 24 and Turing 23. All made a stunning accomplishment, far beyond their contemporaries. All the achievements were mental, not physical. All required a completely new creative vision which no one else could even imagine.

1.6. Church's influence at Princeton. Church excelled in another area. Church and Princeton were a mecca for mathematical logic for nearly 40 years. The 30 graduate students produced by Church at Princeton were involved in the solution or advancement of some of most important open problems. As an undergraduate at Princeton, I took courses from Church in 1961–1963, including a theoretical computer science course and his graduate course in mathematical logic. I also wrote a senior thesis with him and his associate on solvable cases of the decision problem. Church was a scholar of great insight. He was the first to recognize that the λ-definable functions and recursive functions comprise the effectively calculable ones. Church was eventually proved correct because both have been shown equivalent to the now most accepted model of Turing computable functions.

§2. **Analyses of themes in computability theory.** Over the past twenty-five years, and particularly over the last two years with the Turing Centennial celebration, I have written (or am writing) concerning various themes on Church, Turing, Gödel, and computability theory, two books, [1987] and [2014], and ten papers: [1996], [1999], [2007], [2009], [2012a], [2012b], [2013a], [2013b], [2013c], and [2013d] which is the present paper, and which will be listed as Soare [2013d] in future Soare bibliographies and lists of publications. Rather than repeat those analyses here, I shall simply provide in the next several sections a guide to some of these themes and where I have written about them. The bibliographic references not listed here can be found in Soare [1987], [1996], [2009], and [2013b], and the full bibliography will appear on my web page as described in §10 for easy reference.

§3. **Computability and recursion.** This section is a brief summary of the arguments in Soare [1996], [1999], and especially [2013b] section 10 about displacements in computability theory. The original term "computable" in the 1930s as used by Gödel and Turing was displaced by the term "recursive" to mean "calculable" by Church and Kleene from 1936 to 1995. Then in 1996 the term "recursive" was returned to its original meaning of "inductively defined," and no longer was used to mean "effectively calculable."

After seeing Gödel's lectures in 1934, Church and Kleene dropped the λ-definable functions and adopted the recursive functions as the formal definition of the intuitive notion "effectively calculable" functions before they had seen Turing's analysis. Church was very eager for mathematicians to accept his thesis and he knew that the recursive functions were more familiar to a mathematical audience than λ-definable ones.

By 1936 Kleene and Church had begun thinking of the word "recursive" to mean "effectively calculable" (intuitively computable). Church had seen his first thesis rejected by Gödel and was heavily invested in the acceptance of his 1936 thesis in terms of recursive functions. Without the acceptance of this thesis Church had no unsolvable problem. Church wrote in [1936: 96] printed in Davis [1965] that a *"recursively enumerable set"* is one which is the range of a recursive function. This is apparently the first appearance of the term "recursively enumerable" in the literature and the first appearance of "recursively" as an adverb meaning "effectively" or "computably."

In the same year Kleene [1936: 238] cited in Davis [1965: 238] mentioned a *"recursive enumeration"* and noted that there is no recursive enumeration of Herbrand–Gödel recursive systems of equations which gives only the systems which define the (total) recursive functions. By a "recursive enumeration" Kleene states that he means "a recursive sequence (*i.e.*, the successive values of a recursive function of one variable)." Post [1944], under the influence of Church and Kleene, adopted this terminology of "recursive" and "recursively enumerable" over his own earlier terminology [1943], [1944] of "effectively generated set," "normal set," and "generated set." Thereafter, it was firmly established.

Martin Davis entitled his book *Computability and Unsolvability* [1958] but did not challenge the prevailing Kleene terminology at the time that calculable functions should be called "recursive functions." Indeed, Davis wrote on page vii of his preface, "This book is an introduction to the theory of computability and noncomputability, usually referred to as the theory of recursive functions." Davis refers several times to "the theory of computability" as including "purely mechanical procedures for solving various problems."

This is very typical of usage from Kleene [1936] through Davis [1958] and Hopcroft and Ullman [1979], Rogers [1967] and Soare [1987] and many more, where the term computability theory was often used for the concept, especially to a nontechnical general audience. However, a computable function was always called a "recursive function," and a computably enumerable set was always called a recursively eumerable (r.e.) set. In contrast, Turing's epochal paper [1936] uses only the terminology "computable function" and calculable function, and never mentions recursive in the sense of calculable. In his paper [1982] *Why Gödel did not have Church's Thesis* Martin Davis referred to "workers in the field of recursive function theory," as a formal *field* but later wrote about the "algorithmic unsolvability of important problems" when referring to the *concept* of unsolvability.

3.1. Gödel rejects the term "Recursive Function Theory". The first renaissance in computability theory of 1967–1970 had replaced the Kleene T-predicate by the more intuitive Turing–Post informal methods of Turing machines, but the *ancien régime* of "Recursive Function Theory" based on the concept of recursion (induction) rather than the direct concept of "calculability" remained a serious obstacle to advancement in the subject. Students at major universities like Cornell interested in computability theory failed to take a course called "recursive function theory" because they had no idea what that meant. This archaic terminology helped build a wall around the subject and its researchers separating them from interaction with computability in broader areas of mathematics and science.

By 1995 the situation had become very difficult. Most people had access to a personal computer on their desks and the terms of computing were familiar to the general population but "recursive" was limited to very small number who mainly associated it with a first year programming course or a definition by induction in mathematics and almost never with computability. So few people understood the meaning of "recursive" that by 1990 I had to begin my papers with,

> "Let f be a recursive function (that is, a computable function),"

as if I were writing in Chinese and translating back into English. By 1996 it was time for a second renaissance in computability theory in Soare [1996] returning "recursive function theory" to "computability theory" in the sense of Gödel and Turing with their broad interest in the interaction of computability with other areas of mathematics and seicnce, and most important basing everything on the concept of computability not recursion (induction)).

Neither Turing nor Gödel ever used the word "recursive" to mean "computable." Gödel *never* used the term "recursive function theory" to name the subject; when others did Gödel reacted sharply negatively, as related by Martin Davis.

> In a discussion with Gödel at the Institute for Advanced Study in Princeton about 1952–54, Martin Davis casually used the term "recursive function theory" as it was used then. Davis related, "To my surprise, Gödel reacted sharply, saying that the term in question should be used with reference to the kind of work Rosza Peter did."

(See Peter's work on recursions in [1934] and [1951].)

The traditional meaning of "recursive" as "inductive" led to ambiguity. Kleene often wrote of calculations and algorithms dating back to the Babylonians, the Greeks and other early civilizations. However, Kleene [1981b: 44] wrote,

> "I think we can say that recursive function theory was born there ninety-two years ago with Dedekind's Theorem 126 ('Satz der Definition durch Induktion') that functions can be defined by primitive recursion."

Did he mean that recursion and inductive definition began with Dedekind or that computability and algorithms began there? The latter would contradict his several other statements, such as Kleene [1988: 19] where he wrote, "The recognition of algorithms goes back at least to Euclid (c. 330 B.C.)." When one uses a term like "recursive" to also mean "computable" or "algorithmic" as Kleene did, then one is never sure whether a particular instance means "calculable" or "inductive," and our language has become imprecise. Returning "recursive" to its original meaning of "inductive" has made its use much clearer. We do not need another word to mean "computable." We already have one.

3.2. Computer scientists Hopcroft and Ullman [1979]. The same pattern occurred in computer science. Hopcroft and Ullman [1979] wrote a very popular book which was studied by most computer science students, and which appeared in several editions. They began with some of the terminology of "computable" on page 9, but then they reverted in the text to at least two dozen pages which use the Kleene terminology of *recursive function, recursively enumerable set* (and r.e. language), and *recursive function theory*. Worse still, they use the Kleene ambiguity of defining on page 151 a function using a Turing machine, but then *naming* it a *recursive function* even though they know that "recursive" means "inductive" and there is no induction here. On p. 189 they speak of "recursively enumerable" sets not "computably enumerable sets" because the latter term was invented only in Soare [1996] to eliminate this kind of Kleene ambiguity. On p. 207 they speak of *"Recursive Function Theory"* a term rejected by Gödel and Turing as quoted above. Now that term has been nearly completely replaced by by "Computability Theory," even though changing a name which has been established for over a half century is very difficult in science or mathematics.

3.3. Changing "recursive" back to "inductive". By 1996 the confusion had become intolerable. I wrote an article on *Computability and Recursion* for the *The Bulletin of Symbolic Logic* [1996] on the history and scientific reasons for why we should use "computable" and not "recursive" to mean "calculable." "Recursive" should mean "inductive" as it had for Dedekind and Hilbert. At first few were willing to make such a dramatic change, overturning a sixty year old tradition of Kleene, considered by many the father of the subject. Furthermore, the terms "computability theory" and "computably enumerable (c.e.) set" did not come trippingly on the tongue.[3] However, in a few months more people were convinced by the undeniable logic of the situation. Three years later at the A.M.S. conference in Boulder, Colorado referenced in Soare [2000], most researchers, especially those under forty years old, had adopted the new terminology and conventions. Changing back from "recursive" to "computable" during 1996–1999 has had a number of advantages. This usage is more historically and scientifically accurate and has much greater name recognition for students taking courses.

3.4. Computability in Europe. The change to a focus on computability rather than recursion helped open the field to new connections with other areas. Around 2002 Barry Cooper, Benedikt Löwe, Arnold Beckman, Elvira Mayordomo, and others founded the organization *Computability in Europe*

---

[3]Shakespeare, Hamlet, Act 3, scene 2, 1–4: "Speak the speech, I pray you, as I prounonc'd it to you, trippingly on the tounge, but if you mouth it, as many of our players do, I had as lief the town-crier spoke my lines."

which has a well-attended meeting every summer and has been a great success. It and the Isaac Newton Institute (INI) year at Cambridge University in 2012 have helped establish many connections between computability theory and other areas.

§4. Wittgenstein and wrong turnings. (On October 21, 2011, Barry Cooper wrote to me the following section §4 about the significance of the name change in computability theory.)

"By the Way, there is no doubt that the subject could not have changed without the name-change. I thought—ah, Wittgenstein will have said things about this, and came across (from 1931):

> 'Language sets everyone the same traps; it is an immense network of easily accessible wrong turnings. And so we watch one man after another walking down the same paths and we know in advance where he will branch off, where walk straight on without noticing the side turning, etc. etc. What I have to do then is erect signposts at all the junctions where there are wrong turnings so as to help people past the danger points.'

I think your reconstructed terminology played the role of useful signposts to how we might think about the subject. Nowadays I get out my notes to lecture to my students on 'Computability and Unsolvability,' and find I say things with a far surer grasp of the world I'm describing than even a couple of years ago."

§5. Turing's oracle machine in [1939].

5.1. An almost incidental definition. In the Royal Society paper Soare [2012b: §6.1] we describe Turing's extraordinary but almost incidental discovery of his oracle machine. It appeared very briefly in Turing [1939: §4] as an aside and was unnecessary there. Turing's oracle machine was developed by Post [1944] into Turing reducibility and other reducibilites. Turing reducibility allows us to measure the information content and complexity of structures and sets. It is the most important concept in computability theory.

After Turing's a-machine discovery in April, 1936, his adviser, Max Newman, at Cambridge suggested that Turing go to Princeton to study with Church. Turing completed his PhD at Princeton under Church from 1936–1938. Many mathematicians found Gödel's Incompleteness Theorem unsettling. Turing's dissertation [1939] was on ordinal logics, apparently a suggestion of Church, and was an attempt to get around Gödel's incompleteness theorem by adding new axioms. If T_1 is a consistent extension of Peano arithmetic, then the arithmetical sentence σ_1 asserting the unprovability of itself is independent of T_1, but we can form a new theory $T_2 = T_1 \cup \{\sigma_1\}$ which strictly extends T_1. One can continue this sequence through all the computable (constructive) ordinals. Solomon Feferman [2006] gives an excellent analysis of Turing's paper and of ordinal logics.

5.2. An oracle machine with the halting problem K. Turing [1939: §4] introduced oracle machines for a very specific purpose. In the preceding section [1939: §3] Turing had considered Π_2 predicates ($\forall \exists$-predicates over a computable matrix), and had shown that the Riemann Hypothesis and other common problems were Π_2 which Turing called "number theoretic." For example, the question whether a Turing a-machine computes a partial function with infinite domain is Π_2. More precisely, let φ_e be the partial computable function computed by the Turing program P_e with Gödel number e and let W_e be the domain of φ_e. Now W_e is a Σ_1-set. Define

$$\mathrm{Inf} = \{e : W_e \text{ is infinite}\}.$$

Now Inf is Π_2, and in fact Π_2-complete, although the latter concept arose only later. See Soare [1987] Theorem 3.2 that Inf is Π_2-complete, *i.e.*, that for every Π_2 set V there is a computable function f such that $x \in V$ iff $f(x) \in \mathrm{Inf}$.

Turing invented oracle machines to construct a set which was not Π_2. This could easily have been accomplished by a diagonal argument without oracle machines. Turing put the oracle Inf on the oracle tape. By the same diagonal argument as in Turing [1936] for a noncomputable function he used the oracle machine to construct a non Π_2 set. Turing wrote in [1939: 173],

> "Given any one of these machines we may ask the question whether it prints an infinity of figures 0 or 1; I assert that this class of problem is not number-theoretic [*i.e.*, Π_2]."

5.3. Producing a non Π_2^0 set. For a fixed set $A \subseteq \omega$, the set of positive integers, effectively number all Turing oracle programs. Let $\Phi_e^A(x)$ denote the partial function computed by the o-machine with Gödel number e and with A on the oracle tape. Define the relative halting set

$$K^A := \{e : \Phi_e^A(e) \text{ halts}\}. \tag{1}$$

THEOREM 5.1. *Turing,* [1939: 173] *Given A the set K^A is not computable in A.*

Turing [1936] had shown that there is a diagonal set not computable by a Turing machine. The same proof on o-machines relativized to A establishes the theorem. His specific application is that if $A = \mathrm{Inf}$, then K^A is not Π_2. In 1939 Turing left the topic of oracle machines, never to return. It mostly lay dormant for five years until it was developed in a beautiful form by Post [1944], [1948], Kleene–Post [1954], and other papers.

5.4. Post [1944] defined relative computability. Turing never developed the idea of relative computability but Post [1944] developed it beautifully as we describe in the Royal Society paper Soare [2012b: §7.3]. Turing's oracle machine idea lay dormant for five years until Post [1944] studied computably enumerable (c.e.) sets and their decision problems. Post believed that along with decidable and undecidable problems the relative reducibility (solvability) or nonreducibility of one problem to another is a crucial issue. Post studied not only the structure of the computably enumerable sets, those which could be effectively listed, but he initiated a series of reducibilities of one set to another culminating in the most general reducibility which he generously named "Turing" reducibility. Post's aim was to use these reducibilities to study the information content of one set relative to another.

5.5. Online or interactive computing. The original implementations of computing devices were generally offline devices such as calculators. However, in recent years the implementations have been increasingly online computing devices which can access or interact with some external database or other device. The Turing oracle-machine [1939] section 4 is a better model to study them because the original Turing machine [1936] lacks this online capacity. We explore this theme in our MIT Press paper [2013b].

An *online* or *interactive* computing process is one which interacts with its environment, for example a computer communicating with an external data base such as the Internet. An *offline* computing process is one which begins with a program and input data, and proceeds internally, not interacting with any external device. This includes a calculator, and *batch processing* devices where a user handed a deck of punched IBM cards to an operator, who fed them to the computer and produced paper output later. Even the universal Turing machine [1936] is offline because given inputs e and x is first decodes e into a Turing program P_e and then uses P_e to process the input x.

The Turing oracle machine is an excellent theoretical model to analyze an interactive process because there is a fixed procedure at the core, which by Turing's thesis we can identify with a Turing oracle machine, and there is a mechanism for the process of communicating with its environment, which when coded into integers may be regarded as a Turing type oracle A. Under the Post–Turing Thesis explained in Soare [2013b] Thesis 5.2 these real world online or interactive processes can be described by a Turing oracle machine.

In real world computing the oracle may be called a *database* or an environment. A laptop obtaining data from the Internet is a good example. In the real world the database would not be literally infinite but may appear so (like the Internet) and is usually too large to be conveniently downloaded into the local computer. Sometimes the local computer is called the "client" and the remote device the "server."

5.6. Trial and error computing. We expect Turing machines and oracle machines to be absolutely correct. However, there are many computing processes in the real world which give a sequence of approximations converging to the final answer. Turing considered machines which make mistakes. In his talk to the London Mathematical Society, February 20, 1947, quoted in Hodges, pp. 360–361, Turing said,

"I would say that fair play must be given to the machine. Instead of it sometimes giving no answer we could arrange that it gives occasional wrong answers. But the human mathematician would likewise make blunders when trying out new techniques. It is easy for us to regard these blunders as not counting and give him another chance, but the machine would probably be allowed no mercy. In other words if a machine is expected to be infallible, it cannot also be intelligent. There are several theorems which say exactly that. But these theorems say nothing about how much intelligence may be displayed if a machine makes no pretence at infallibility."

Hillary Putnam [1965] described *trial and error* predicates as ones for which there is a computable function which arrives at the correct answer after a finite number of mistakes. In modern terminology this is called a *limit computable* function as described in Soare [1987] Chapter 3. This is a model for many processes in the real world which allow finitely many mistakes but gradually converge to the correct answer. A function f is defined to be *limit computable* if there is a computable sequence $\{\widehat{f}_s(x)\}_{s \in \omega}$ such that for all x,

$$f(x) = \lim_s \widehat{f}_s(x).$$

By the Shoenfield Limit Lemma in Soare [1987] Chapter 3 we know that f is limit computable iff f is computable in K ($f \leq_T K$) where K is the halting problem. The point is that we can use an oracle machine with K on the oracle tape to analyze limit computable functions.

§6. Why Turing's Thesis is not a thesis.
In the Birkhäuser volume Soare [2013c] presents the argument for why Turing's Thesis is not a thesis. This paper depends upon the careful and correct use of language.

6.1. The use of terms.
Philosopher Charles Sanders Peirce argued that advancement in science depends on precision of thought and precision of thought depends on precision of language. Peirce [1960] described the importance of language for science this way in *The Ethics of Terminology* Volume II *Elements of Logic* p. 129.

"the woof and warp of all thought and all research is symbols, and the life of thought and science is the life inherent in symbols; so that it is wrong to say that a good language is *important* to good thought, merely; for it is of the essence of it."

In §6.3 we shall take pains to explain the meaning of the term "thesis" as a proposition put forward for debate not a fact or theorem. The article Soare [1996] *Computability and Recursion* explored the *intensional* meaning of the terms "recursive" and "computable" and argued that "recursive" should be used in the meaning of "defined by induction" but never in the sense of calculable as had been done from 1935 to 1995. After 1995 most of the community has followed this suggestion.

6.2. Turing [1936] proves a theorem. In 1936 Alan Turing showed that any effectively calculable function is computable by a Turing machine. Scholars at the time, such as Kurt Gödel and Alonzo Church, regarded this as a convincing *demonstration (proof)* of this claim, not as a mere hypothesis in need of continual reexamination and justification. In 1988 Robin Gandy said that Turing's analysis "proves a theorem." However, Stephen C. Kleene [1943, 1952] introduced the term "thesis" in 1943 and in his book in 1952. The phrase "Turing's thesis" has been used ever since and might suggest that Turing made a mere hypothesis. Many agree with Gandy that Turing proved a completeness result analogous to Gödel's Completeness Theorem for the predicate calculus which states that a semantical notion can be exactly captured by a syntactical one, thereby providing a a crucial connection between semantics and syntax. In §6.6 we call this the "Turing Completeness Theorem" and explain why.

6.3. What is a thesis? The English term "thesis" comes from the Greek word $\theta\acute{\epsilon}\alpha\sigma\iota\varsigma$, meaning "something put forth." In logic and rhetoric it refers to a "proposition laid down or stated, esp. as a theme to be discussed and proved, or to be maintained against attack."[4] It can be a hypothesis presented *without* proof, or it can be an assertion put forward with the intention of defending and debating it. For example, a PhD thesis is a dissertation prepared by the candidate and defended by him before the faculty in order to earn a PhD degree.

The Harvard online dictionary says that a thesis is "not a topic; nor is it a fact; nor is it an opinion." A theorem such as the Gödel Completeness Theorem is not a thesis. It is a fact with a proof in a formal axiomatic system which cannot be refuted.

The conclusion is that a *theorem or fact* is a proposition which is laid down and which may possibly need to be demonstrated, but about which there will be no debate. It will be accepted in the future without further question. A *thesis* is a weaker proposition which invites debate, discussion, and possibly repeated verification. Attaching the term "thesis" to such a proposition invites continual reexaminationn. It signals that to the reader that the proposition may not be completely valid, but rather it should continually be examined more critically.

This is not a question of mere semantics, but about what Turing actually *achieved*. If we use the term "thesis" in connection with Turing's work, then we are continually suggesting some doubt about whether he really gave an authentic characterization of the intuitively calculable functions. The central question of this paper is to consider whether Turing proved his assertion beyond any reasonable doubt or whether it is merely a thesis, in need of continual verification.

---

[4]Oxford English Dictionary.

6.4. Effectively calculable functions. The work of Turing [1936] rests on the meaning of *effectively calculable* function the term in the early 1930s for one which was intuitively computable. Today the term "algorithm" has acquired a more general meaning as explored in Gurevich [2013] and many other papers.

However, here we shall use the term "effectively calculable" (abbreviated here as "calculable") and "procedure" in a more restricted way understood by Church, Turing and Gödel in the 1930s and arising from an analysis of the *Entscheidungsproblem* of Hilbert. This will better clarify Turing's accomplishment of the Turing machine characterization in [1936]. Turing [1936: §9] defined for the first time the intuitive notion of effectively calculable function.

In the Nachlass printed in Gödel Volume III [1995: 166] Gödel wrote,

"When I first published my paper about undecidable propositions the result could not be pronounced in this generality, because for the notions of mechanical procedure and of formal system no mathematically satisfactory definition had been given at that time. . . .

The essential point is to define what a procedure is."

Gödel believed that Turing had done so but he was not convinced by Church's argument.

Turing [1936: §9] defined *procedure* and simultaneously *Turing machine* its formal mechanical equivalent. Turing's analysis of a procedure is roughly the following.

- Turing proposed a number of **simple operations** "so elementary that it is not easy to imagine them further subdivided."
- Divide the work space into squares. May assume one dimensional.
- Finitely many symbols. Each square contains one symbol.
- Finitely many *states* (of mind).
- Action of the machine determined by the present state and the squares observed.
- Squares **observed** are bounded by B, say $B = 1$.
- Reading head examines one symbol in one square.
- May assume the machine moves to only squares within a **radius** of L (of current square). May assume $L = 1$.
- Machine may print a symbol in the current square, change state, and move to adjacent square.

Kurt Gödel wrote,

"That this really is the correct definition of mechanical computability was established beyond any doubt by Turing."

"But I was completely convinced only by Turing's paper."

6.5. It was never a thesis. None of the experts in the 1930s, Turing, Church, Gödel, thought of Turing's achievement as a thesis but rather as a definition or characterization of effectively computable. When Post [1936] suggested it may be a "working hypothesis in need of continual verification," Church [1936b] in his review of Post [1936] wrote,

> "The author ... [Post] proposes a definition of "finite 1-process" which is equivalent, to computation by a Turing machine. He does not, however, regard his formulation as certainly to be identified with effectiveness in the ordinary sense, but takes this identification as a "working hypothesis" in need of continual verification. To define effectiveness as computability by an arbitrary machine, subject to restrictions of finiteness, would seem to be an adequate representation of the ordinary notion, and if this is done the need for a working hypothesis disappears."

In contrast, Church [1937a] wrote the following in his review of Turing [1936].

> " ... in particular, a human calculator, provided with pencil and paper and explicit instructions, can be regarded as a kind of Turing machine. It is thus immediately clear that computability, so defined, can be identified with ... the notion of effectiveness as it appears in certain mathematical problems (various forms of the *Entscheidungsproblem*, various problems to find complete sets of invariants in topology, group theory, etc., and in general any problem which concerns the discovery of an algorithm)."

Church clearly saw Turing's [1936] characterization as a definition and not as a hypothesis, conjecture, or a thesis. The term "thesis" first appeared in Kleene [1943] when he used the term "Thesis I" for what we now call "Church's Thesis." In his very influential book [1952] Kleene introduced the terms "Church's Thesis" and "Turing's Thesis" and the former was used thereafter.

6.6. The Turing completeness theorem. Turing's monumental achievement [1936] was:

1. to give a precise but informal analysis in [1936: §9] of the intuitive notion of an effectively calculable function;
2. to give a formal mathematical definition of an automatic machine (Turing machine) in [1936: §92]; and
3. to prove them equivalent (although this became obvious by the end of his paper).

Gödel's completeness theorem [1930] links a certain semantical notion to a syntactic notion by proving (as a theorem in ZFC set theory) that a formula of the predicate calculus is valid if and only if it is derivable from the axioms and rules of inference of the predicate calculus. We may view Turing's [1936] work as likewise a theorem in ZFC. First, Turing analyzed in §9 just what an effectively calculable function is. This semantical analysis gave rise naturally to the corresponding notion of an automatic machine, even though these are presented in the reverse order in Turing [1936].

§7. Formalism and informalism in computability.

7.1. Kleene's formalism. Kleene's Fixed Point Theorem (Recursion Theorem) explained in Soare [1987: 36] is one of the most elegant and important theorems in the subject with one of the shortest proofs. Kleene himself was the fixed point of recursion theory, recursive function theory as he called it. His career began at Princeton in 1931 as a graduate student of Church and continued through the 1970s with some of the most influential work. In addition, he wrote several historical and retrospective articles in the 1980s from [1981] to [1988]. Kleene enthusiastically supported Church's Thesis. He also defined the Kleene T-predicate $T(e, x, y)$ for enumerating the Herbrand–Gödel recursive functions and added to the five primitive schemata a sixth one Scheme (VI) the unbounded search, as discussed in Soare [1987: 10]. The resulting class constitutes the Kleene μ-recursive functions, equivalent to the Herbrand–Gödel recursive functions or to the Turing computable functions.

The Kleene μ-recursive functions give a precise and mathematically correct model but one which is very formal, unwieldy, and lacking in intuition. The Royal Society paper Soare [2012b] explores this dual influence of very formal models such as Kleene's versus the informal and intuitive approaches of Turing [1936] and Post [1944].

7.2. The Turing–Post informalism. As Turing left the subject of pure computability theory in 1939, his mantle fell on Post. Post continued the tradition of clear, intuitive proofs, of exploring the most basic objects such as computability and computably enumerable sets, and most of all, exploring relative computability and Turing reducibility. During the next decade and a half, from 1940 until his death in 1954, Post played an extraordinary role in shaping the subject.

Post called attention to the importance of stripping away the formalism and presenting computability as one would present group theory. In his introduction to [1944] Post wrote,

> "Recent developments of symbolic logic have considerable importance for mathematics both with respect to its philosophy and practice. That mathematicians generally are oblivious to the importance of this work of Gödel, Church, Turing, Kleene, Rosser and others as it affects the subject of their own interest is in part due to the forbidding, diverse and alien formalisms in which this work is embodied. Yet, without such formalism, this pioneering work would lose most of its cogency. But apart from the question of

importance, these formalisms bring to mathematics a new and precise mathematical concept, that of the general recursive function of Herbrand–Gödel–Kleene, or its proved equivalents in the developments of Church and Turing. It is the purpose of this lecture to demonstrate by example that this concept admits of development into a mathematical theory much as the group concept has been developed into a theory of groups. Moreover, that stripped of its formalism, such a theory admits of an intuitive development which can be followed, if not indeed pursued, by a mathematician, layman though he be in this formal field. It is this intuitive development of a very limited portion of a sub-theory of the hoped for general theory that we present in this lecture. We must emphasize that, with a few exceptions explicitly so noted, we have obtained formal proofs of all the consequently mathematical theorems here developed informally. Yet the real mathematics involved must lie in the informal development. For in every instance the informal "proof" was first obtained; and once gotten, transforming it into the formal proof turned out to be a routine chore."

Researchers in the 1930s concentrated on a formal definition of an effectively calculable function, not a definition of a function computable in an oracle. The exception was Turing [1939: §4]. Turing defined oracle machines as in §5. The idea lay dormant for five years until Post [1944] studied computably enumerable (c.e.) sets and their decision problems. Post believed that along with decidable and undecidable problems the *relative* reducibility (solvability) or nonreducibility of one problem to another is a crucial issue. Post studied not only the structure of the computably enumerable sets, those which could be effectively listed, but he initiated a series of reducibilities of one set to another culminating in the most general reducibility which he generously named "Turing" reducibility. Post's aim was to use these reducibilities to study the information content of one set relative to another.

Here Post is not merely introducing reducibility of one problem to another for the sake of demonstrating solvability or unsolvability. He takes it further by introducing for the first time in the subject the term "degree of unsolvability" to mean that two sets code the same information, *i.e.*, have the same information content. For seventy years since then researchers have classified objects in algebra, model theory, complexity, and computability theory according to their degree of unsolvability or information content.

Post wrote an intuitive section §11 in Post [1944] on the general case of Turing reducibility, but it was not well understood at that time. Post continued to study it for a decade and introduced [1948] the concept of *degree of unsolvability*, now called "Turing degree." Before his death in 1954, Post gave his notes to Kleene, who published the joint paper Kleene–Post [1954] which laid the foundation of all the subsequent results on Turing reducibility and Turing degree. Post became quite ill and died in 1954 before the Kleene–Post paper was written, but he gave his notes to Kleene. Kleene added some parts, and unfortunately put it all into his μ-recursive formalism making it very hard to read, including Friedberg's [1957] solution to Post's problem of the existence of an incomplete noncomputable c.e. set. With both Turing and Post dead in 1954, the Kleene formalism dominated for another decade until the mid 1960s and caused the papers to become nearly unreadable.

7.3. The renaissance by Rogers and Lachlan. By 1965 the influence of formalism, the Kleene Normal Form, the μ-recursive functions, and lack of interest in explaining intuition in proofs led to a completely unacceptable state where researchers could no longer read papers. The field was rescued from this excessive formalism of 1965 by two events: the book by Hartley Rogers [1967] which presented the subject in the style of Turing and Post, and several remarkable and groudbreaking papers by Alistair Lachlan [1970], [1973], and [1975] on Lachlan games, tree proofs, and other devices for explaining the intuition behind complicated proofs.

Hartley Rogers' book [1967] was a significant advance in intuition and understanding of computability theory. Rogers based his book on the ideas of Post's [1944] epochal paper. Rogers wrote [1967: vii] in his preface,

"Although Post's paper concerns only one part of a larger subject, it marks an epoch in that subject, not only for the specific problems and methods that it presents, but also for the emphasis that it places on intuitive naturalness in basic concepts."

Rogers [1967: vii] continued about Post [1944] which was the basis for Rogers's own book which he describes this way.

"The book presents its subject matter in semiformal terms, with an intuitive emphasis that is, hopefully, appropriate and instructive. The use of semiformal procedures in recursive function theory is analogous to the use, in other parts of mathematics, of set-theoretical arguments that have not been fully formalized. It is possible for one who possesses a good grasp of the simple, primitive ideas of our theory to do research just as it is possible for a student of elementary algebra in school to do research in the theory of natural numbers."

From 1965 to 1980 a number of new researchers had entered the field, influenced by Rogers and Lachlan, and with fresh ideas expressed in the language of Turing, Post, Rogers, and Lachlan, and abandoning the Kleene μ-recursive functions and T-predicate notation. New areas grew up around computability, including computable model theory, computable algebra, higher computability, and more.

§8. Darwinian evolution and natural selection.

8.1. Survival of the fittest. One can think of the development of computability theory since 1931 in terms of Darwinian evolution, natural selection, survival of the fittest, and adaptation to the environment. Many ideas are generated, the strongest and most useful survive and are modified to adapt to the environment, and they remain in the center. The lesser ideas move to one of the side branches of the subject. This is closely related to the wrong turning described by Wittgenstein in §4.

8.2. The λ-calculus in 1931-1934. In a remarkable discovery, Alonzo Church was the first to notice that the λ-calculus provided a formal definition of the informal notion of effectively calculable function, even though unlike Turing, he did not define the latter notion precisely or convince Gödel that this was a correct formalization. Not even Church's own student, Kleene, was convinced until he tried unsuccessfully to diagonalize out of the class of partial recursive functions, and discovered instead the Kleene fixed point theorem as discussed in the book by Soare [1987: 36].

However, Church's first thesis delivered privately to Gödel in 1934 was replaced for reasons of natural selection in presenting computability to a general audience, because it lacked popular appeal. Kleene himself [1981] admitted,

> "I myself, perhaps unduly influenced by rather chilly receptions from audiences around 1933–35 to disquisitions on λ-definability, chose, after general recursiveness had appeared, to put my work in that format...."

Nevertheless, the λ-calsulus has been very useful both as a theoretical tool and in software design such as programming languages where machines can more easily read the code than human beings. Turing himself wrote his thesis [1939] in it, probably at the suggestion of his thesis adviser, Church.

8.3. Herbrand–Gödel recursive function in 1934. Beginning in Church's abstract [1935] and his paper [1936] attention shifted to the next evolutionary branch of recursion theory, the Herbrand–Gödel recursive functions defined by Gödel in [1934]. This model was more recognizable to mathematicians with its generalization of primitive recursive functions but was unwieldly for presenting proofs, and was rarely used directly in proofs.

8.4. Kleene T-predicate and μ-recursive functions in 1936. Kleene refined this model by defining the Kleene T-predicate $T(e, x, y)$, making it more competitive in the evolutionary process. This did not give a new formalism because it simply numbered the Herbrand–Gödel recursive functions of [1934] using the Gödel coding method of [1931]. However, it gave a precise, if very formal and unintuitive, model in which to present proofs. This lasted from about Kleene [1936] until the mid 1960s when it largely collapsed under its own weight, and moved to a side branch of the subject. However, the general idea of recursion (induction) still plays a very important role in the subject at the theoretical and practical level.

The fact that the Kleene μ-recursive model lasted as long as it did into the mid 1960s was probably due to the fact that the competing and more intuitive models of Turing machines [1936], Turing oracle machine [1939] and Post's informalism of [1944] and [1948] lay dormant for a decade. Both Turing and Post died in 1954 leaving Kleene as the last active researcher from the 1930s, and the by far the dominant one.

8.5. The computability renaissance of 1967–1970. Again natural selection played a key role in the late 1960s. By 1965 the Kleene μ-recursive function model was no longer suitable for presenting clear, convincing proofs. Under natural selection and survival of the fittest, attention returned overwhelmingly to the previous models of Turing and Post, which have remained in the center of the subject ever since. As soon as the Rogers book [1967] appeared, there was a very rapid expansion of researchers and results in the subject. Lachlan gave the new tools necessary to understand the existing results and to prove new deep results.

8.6. The renaming in 1996. The renaming of the subject from recursion theory to computability theory in 1996 was not really a new development but a return to an old one because both Turing and Gödel had used the term "computable" and neither had used "recursive" to mean "calculable." The return to their original terminology was a natural evolutionary step in the subject to bring it closer to its roots and to other branches of mathematics and science. Soon there appeared other books and papers such as Cooper's book [2004] and books linking computability to other areas such as algorithmic complexity such as Nies [2009] and Downey and Hirschfeldt [2010] as well as many books and papers relating other areas to computability.

8.7. Church, Turing, Michelangelo, and the marble. Consider the quote by Michelangelo in §1.5 where he saw the statue plainly in the block of marble and simply hewed the marble to release it. Church was the first to see the figure of computability in the block of marble, but in two attempts, λ-definable functions, and Herbrand–Gödel functions, was unsuccessful in releasing it, partly because Church was tied to the formal deductive systems of the previous decades in which he has been raised. He could not convey his vision convincingly to Gödel or to modern scholars. Turing, with a completely fresh view, saw the figure of computability while reclining in a meadow outside Cambridge after running, and saw it as a mechanical extension of his mother's typewriter. Just as Michelangelo had captured the male figure of David in the pure Cararra marble, Turing captured the centuries old intuitive idea of an effectively calculable function. The achievement by Michelangelo was the next natural evolution in sculpture; the achievement by Turing the next natural evolution in calculability.

§9. **The fathers of the subject.**

9.1. The father of recursive function theory. Kleene's great contributions to the founding and development of recursive function theory were recognized by two of the leaders in the field. In his book [1971] *Degrees of Unsolvability*, Joseph Shoenfield dedicated his book to "S.C. Kleene who made recursive function theory into a theory." Likewise, Sacks [2003] described Kleene as the father of recursion theory. Because of Kleene's great influence from the 1930s through the 1970s I agree that Kleene was *the father of Recursive Function Theory*, particularly with respect to the Kleene T-predicate, μ-recursive functions, and for his steady, if formal, leadership.

9.2. The fathers of computability theory. However, Turing and Post were *the fathers of Computability Theory*, as we understand it today, particularly as it has emerged since 1967, three decades after the original definition of recursive function and Turing machine, and after 1996, three decades later, when the subject returned to its original terminology and concept of computablity rather than recursion. Turing [1936] introduced the intuitive model of a Turing machine as a device to capture effective calculability and the oracle machine [1939: §4] for a local machine to communicate with a remote server. Post [1944] developed the brief oracle machine description by Turing into a full theory of relative computability, and called importance to the study and classification of computably enumerable (c.e.) set. These contributions are fundamental to modern computability theory. Rogers' book [1967] abandoned the Kleene formalism and proofs with μ-recursive functions and the Kleene T-predicate. Rogers returned to the original ideas or Turing [1936], [1939: §4] and Post [1944]. After that the Kleene style formalism and proofs largely disappeared.

§10. **A full bibliography.** To keep the bibliography here short we have omitted many primary references from the following short list of references which contains all the Soare papers and books. The remaining references mentioned here are included in the full bibliography which may be found on the following web page with the following url which contains ten pages, including a hundred references covering most of those mentioned in the Soare books or papers below.

`www.people.cs.uchicago.edu/~soare/`

There the file name risBib.pdf will contain the continually updated list of Soare papers and books on Turing and computability theory, and the file name `computBib.pdf` will contain most of the cited historical and analytical papers on computability theory, but not technical research articles.

REFERENCES

[Church, 1936]
A. CHURCH, *An unsolvable problem of elementary number theory*, **The American Journal of Mathematics**, vol. 58 (1936), pp. 345–363.

[Church, 1937a]
——, *Review of Turing 1936*, **The Journal of Symbolic Logic**, vol. 2 (1937), no. 1, pp. 42–43.

[Church, 1937b]
——, *Review of Post 1936*, **The Journal of Symbolic Logic**, vol. 2 (1937), no. 1, pp. 43.

[Church, 1956]
——, **Introduction to mathematical logic**, vol. I, Princeton University Press, Princeton, New Jersey, 1956.

[Cooper, 2004]
S. B. COOPER, **Computablility theory**, Chapman & Hall/CRC Mathematics, London, New York, 2004.

[Davis, 1958]
M. DAVIS, **Computability and unsolvability**, Mc-Graw-Hill, New York, 1958, reprinted in 1982 by Dover Publications.

[Davis, 1965]
—— (editor), **The undecidable. Basic papers on undecidable propositions, unsolvable problems, and computable functions**, Raven Press, Hewlett, New York, 1965.

[Davis, 1982]
——, *Why Gödel did not have Church's Thesis*, **Information and Control**, vol. 54 (1982), pp. 3–24.

[Davis, 1988]
——, *Mathematical logic and the origin of modern computers*, in Herken, [1988], pp. 149–174.

[Downey–Hirschfeldt, 2010]
R. DOWNEY AND D. R. HIRSCHFELDT, **Algorithmic randomness and complexity**, Springer-Verlag, New York, 2010.

[Feferman, 2006]
S. FEFERMAN, *Turing's thesis*, **Notices of the American Mathematical Society**, vol. 53, (2006), pp. 1200–1206.

[Gödel, 1931]
K. GÖDEL, *Über formal unentscheidbare Sätze der Principia Mathematica und verwandter Systeme. I*, **Monatshefte für Mathematik und Physik**, vol. 38, (1931), pp. 173–178, (English translation in Davis [1965], pp. 4–38 and in van Heijenoort [1967], pp. 592–616).

[Gödel, 1934]
——, *On undecidable propositions of formal mathematical systems*, Notes by S. C. Kleene and J. B. Rosser on lectures at the Institute for Advanced Study, Princeton, New Jersey, 1934, 30 pp. (reprinted in Davis [1965: 39–74]).

[Gödel, 1946]
——, *Remarks before the Princeton bicentennial conference of problems in mathematics*, 1946, reprinted in Davis [1965], pp. 84–88.

[Gödel, 1986]
——, *Collected works Volume I: Publications* 1929–1936, (S. Feferman et al., editors), Oxford University Press, Oxford, 1986.

[Gödel, 1990]
——, *Collected works Volume II: Publications* 1938–1974, (S. Feferman et al., editors), Oxford University Press, Oxford, 1990.

[Gödel, 1995]
——, *Collected works Volume III: Unpublished essays and lectures*, (S. Feferman et al., editors), Oxford University Press, Oxford, 1995.

[Gurevich, 2013]
Y. GUREVICH, *Analyzable and nonanalyzable computations*, **Turing centenary volume**, (Thomas Strahm and Giovanni Sommaruga,editors), Birkhäuser/Springer, Basel, 2013, (to appear).

[Herken, 1988]
R. Herken (editor), **The universal Turing machine: A half-century survey**, Oxford University Press, 1988.

[Hopcroft–Ullman]
J. E. HOPCROFT AND J. D. ULLMAN, **Introduction to automata, languages and computation**, Addison-Wesley, 1979.

[Kleene, 1936]
S. C. KLEENE, *General recursive functions of natural numbers*, **Mathematische Annalen**, vol. 112, (1936), pp. 727–742.

[Kleene, 1936b]
——, *λ-definability and recursiveness*, **Duke Mathematical Journal**, vol. 2, (1936), pp. 340–353.

[Kleene, 1943]
——, *Recursive predicates and quantifiers*, **Transactions of the American Mathematical Society**, vol. 53, (1943), pp. 41–73.

[Kleene, 1952]
——, **Introduction to metamathematics**, Van Nostrand, New York, 1952, ninth reprint 1988, Walters-Noordhoff Publishing Company, Groningën and North-Holland, Amsterdam.

[Nies, 2009]
A. NIES, **Computability and randomness**, Oxford University Press, Oxford, UK, 2009.

[Post, 1943]
E. L. POST, *Formal reductions of the general combinatorial decision problem*, **American Journal of Mathematics**, vol. 65, (1943), pp. 197–215.

[Post, 1944]
——, *Recursively enumerable sets of positive integers and their decision problems*, **American Mathematical Society. Bulletin**, vol. 50, (1944), pp. 284–316.

[Sacks, 2003]
G. E. SACKS, **Mathematical logic in the 20th century**, World Scientific Series in 20th Century Mathematics, vol. 6, Singapore University Press and World Scientific Publishing Co., Singapore, New Jersey, London, Hong Kong, 2003.

[Sieg, 1994]
W. SIEG, *Mechanical procedures and mathematical experience*, **Mathematics and mind** (A. George, editor), Oxford University Press, 1994, pp. 71–117.

[Soare, 1987]
R. I. SOARE, **Recursively enumerable sets and degrees: A study of computable functions and computably generated sets**, Springer-Verlag, Heidelberg, 1987.

[Soare, 1996]
——, *Computability and recursion*, **The Bulletin of Symbolic Logic**, vol. 2, (1996), pp. 284–321.

[Soare, 1999]
——, *The history and concept of computability*, **Handbook of computability theory** (E. R. Griffor, editor), North-Holland, Amsterdam, 1999, pp. 3–36.

[Soare, 2007]
——, *Computability and incomputability, computation and logic in the real world*, **Proceedings of the third conference on Computability in Europe, CiE 2007, Siena, Italy, June 18–23, 2007** (S. B. Cooper, B. Löwe, and Andrea Sorbi, editors), Lecture Notes in Computer Science, vol. 4497, Springer-Verlag, Berlin, Heidelberg, 2007.

[Soare, 2009]
——, *Turing oracle machines, online computing, and three displacements in computability theory*, **Annals of Pure and Applied Logic**, vol. 160, (2009), pp. 368–399.

[Soare, 2012a]
——, *An interview with Robert Soare: Reflections on Alan Turing*, published in **Crossroads, The ACM Magazine for Students**, Association of Computing Machinery, XRDS, Spring 2012, vol. 18 (2012), no. 3, a telephone interview with Robert Soare, transcribed by Arefin Huq, a computer science graduate student at Northwestern University, on December 15, 2011.

[Soare, 2012b]
——, *Formalism and intuition in computability theory*, **Philosophical Transactions of the Royal Society A**, vol. 370, (2012), pp. 3277–3304.

[Soare, 2013a]
——, *Turing and the art of classical computability*, **Alan Turing—his work and impact** (Barry Cooper and Jan van Leeuwen, editors), Elsevier, 2012, to appear.

[Soare, 2013b]
——, *Interactive computing and Turing–Post relativized computability*, **Computability: Gödel, Church, Turing, and beyond** (Jack Copeland, Carl Posy, and Oron Shagrir, editors), MIT Press, 2012, to appear.

[Soare, 2013c]
——, *Why Turing's thesis is not a thesis*, **Turing centenary volume** (Thomas Strahm and Giovanni Sommaruga, editors), Birkhäuser/Springer, Basel, 2013, to appear.

[Soare, 2014]
——, *Turing and the art of classical computability: Theory and applications*, Computability in Europe Series, Springer-Verlag, Heidelberg, 2012, to appear.

[Turing, 1936]
A. M. TURING, *On computable numbers, with an application to the Entscheidungsproblem*, **Proceedings of the London Mathematical Society. Second Series**, vol. 42, (1936), no. 3 and 4, pp. 230–265, reprinted in Davis [1965], pp. 116–154.

[Turing, 1937a]
——, *A correction*, **Proceedings of the London Mathematical Society. Second Series**, vol. 43, (1937), pp. 544–546.

[Turing, 1937]
——, *Computability and λ-definability*, **The Journal of Symbolic Logic**, vol. 2, (1937), pp. 153–163.

[Turing, 1939]
——, *Systems of logic based on ordinals*, **Proceedings of the London Mathematical Society**, vol. 45, (1939), no. 3, pp. 161–228, reprinted in Davis [1965], pp. 154–222.

DEPARTMENT OF MATHEMATICS
THE UNIVERSITY OF CHICAGO
5734 UNIVERSITY AVENUE
CHICAGO, IL 60637-1546, USA
E-mail: soare@uchicago.edu
URL: www.people.cs.uchicago.edu/~soare/

TRANSFINITE MACHINE MODELS

P. D. WELCH

§1. Introduction. In recent years there has emerged the study of discrete computational models which are allowed to act *transfinitely*. By 'discrete' we mean that the machine models considered are not analogue machines, but compute by means of distinct *stages* or in *units of time*. The paradigm of such models is, of course, Turing's original machine model. If we concentrate on this for a moment, the machine is considered to be running a program P perhaps on some natural number input $n \in \mathbb{N}$ and is calculating $P(n)$. Normally we say this is a successful computation if the machine halts after a finite number of stages and we may read off some designated form of output: '$P(n)\downarrow$'. However if the machine fails to halt after a finite time it may be exhibiting a variety of behaviours on its tape. Mathematically we may ask what happens 'in the limit' as the number of stages approaches ω. The machine may of course go haywire, and simply be rewriting a particular cell infinitely often, or else the Read/Write head may go 'off to infinity' as it moves inexorably down the tape. These kind of considerations are behind the notion of 'computation in the limit' which we consider below.

Or, it may only rewrite finitely often to any cell on the tape, and leave something meaningful behind: an infinite string of 0, 1s and thus an element of Cantor space $2^{\mathbb{N}}$. What kind of elements could be there? Considerations of what may lay on an output tape at an infinite stage first surface in the notion of 'computation in the limit' or 'limit decidable'. Whilst the first publication on the matter seems to be two papers coincidentally appearing in the same year, 1965, as Martin Davis has commented, surely this was already known to Post?

DEFINITION 1 (Putnam [59]). *P is a* trial and error predicate *if and only if there is a general recursive function f such that for every $x_1, \ldots, x_n$:*

$$P(x_1, \ldots, x_n) \equiv \lim_{y \to \infty} f(x_1, \ldots, x_n, y) = 1,$$

$$\neg P(x_1, \ldots, x_n) \equiv \lim_{y \to \infty} f(x_1, \ldots, x_n, y) = 0.$$

Then it is possible to write out conditions for P holding that are Δ_2^0. Moreover any property that can be expressed in a Δ_2^0 way, can be expressed also in this form. However there can be no decidable function $g(m_1,\ldots,m_n)$ which tells us how long one must wait for the value to have settled down: if such existed then one may easily see that the relation P becomes decidable.

By allowing ourselves to survey the whole tape 'at stage ω' we are performing a so-called *'super-task'* which is generally conceived to be a process which is brought to completion, but involves infinitely many stages of activity. If we allow supertasks into computational models there might be some interesting phenomena to discover. For the Turing machine model the whole of the tape is then usable, both for input strings as well as output strings. We then have essentially a computation at a *higher type* and such were much studied following Kleene's papers [35], [38], [37], [36] and [39] in the 1960s and early 1970s; the work here of Aczel, Gandy, Moschovakis and others, would eventually lead to the *theory of inductive definitions*, and the abstract theory of *Spector classes*, and to the theory of *admissible sets* in the work of Kripke, Platek, Barwise. Another strand of thought, under the influence of Kreisel and Sacks, led to *meta-recursion theory*, and ultimately *α-recursion theory* where the objects of computation were not numbers but ordinals (see [63] for an account of this latter development).

We should like to see how the theory of transfinitely running discrete computational machine models fits into this broader, older, more abstract theory. The theory of Infinite Time Turing Machines (ITTM) of Hamkins and Kidder (originating in the 90s but first published in [22] in 2000) lays down an attractive formulation for this, in particular for defining actions at limit stages of time. Previously one might have considered standard computational machines 'stacked up' or computations repeated using one infinite stage output recycled as the next input *etc*. This might be done either conceptually, or else thought of as inhabiting a particular spacetime (such as in the work of Etesi–Neméti and Hogarth) that allow—in a particular sense—for supertasks. The ITTMs provide a suitable yardstick for measuring or modelling the capabilities of other machines or arrangements.

Perhaps surprisingly *Infinite Time Register Machines* (ITRMs) which again keep the usual hardware of a finite number of registers, with numerical contents, lead to surprisingly strong computational models. In contradistinction to the finite case where the capabilities of Turing machines and register machines are the same, in the transfinite realm they markedly diverge.

Omitted in this chapter is any discussion of TMs or other machines in particular physical settings or implementations. Thus we have not discussed the models of Beggs and Tucker [3] that compute sets of numbers by some kinematic based device, or the models of Davies [12] that attempt to run a supertask within a fragment of Newtonian mechanics by means of accelerating machine components of exponentially decreasing size *etc.*, *etc*. These are essentially only standard Turing machines in a particular format or physical setting, so we have decided to exclude these. Also, as we have discussed the capabilities of machines in Malament–Hogarth spacetimes elsewhere [81], this is not included here. More seriously we omit any discussion of finite automata acting on, say infinite graphs, or infinite binary trees.

Rather we emphasise here the logico-mathematical aspects rather than the physical. Since we are often considering runs of machines along countable wellorders, a certain amount of analysis or second order number theory is needed to ensure such ordinals exist. We have to hand the methodology of reverse mathematics with which to discuss the strengths of the various assertions that certain machines loop or halt in certain ordinal times *etc.* This we do where possible. In Section 2 we very briefly sketch connections with Kleene's seminal papers in higher type recursion theory. There the theory of *Kleene Recursion* on sets of reals is outlined, with its connections to hyperarithmetic sets, Borel sets and so forth. This then is expanded to include definitions of Spector Class, and hyperdegrees. This we believe is the right picture to have before one, when thinking about the Infinite Time Turing Machine model (Section 4) of Hamkins and Kidder. Here we explore this formalism in quite some detail, and express the classes defined within Δ_2^1 as Spector classes.

Since these machine models require transfinite ordinals to compute along the question arises as to what ordinals, and how strong a theory is needed to justify their existence. At a couple of points below we illustrate in terms of traditional theories such as Π_1^1-CA$_0$ and subsystems of Π_3^1-CA$_0$ that play a role. This is in the spirit of the reverse mathematics of [70].

We set $\mathcal{L}_{\dot{\in}}$ to be the language of set theory. By ZF$^-$ we mean the axiom system of Zermelo–Fraenkel with the Axiom Scheme of Collection, but the Axiom of Power Sets removed. By Σ_n-KP we mean ZF$^-$ but with Collection restricted to Σ_n, and Separation to Δ_n, formulae respectively. (Σ_1-KP is then just KP.) When notions are used from ordinary Turing computability theory and there is danger of confusing them with some new broader notion under discussion we refer to them as 'standard'. By '$A \leq_1 B$' for $A, B \subseteq \mathbb{N}$ we mean that $A = f^{-1}$"B for a total (1–1) computable (*i.e.*, standard) function f. By '$A \equiv_1 B$' we mean that $A = f^{-1}$"B, for a (standard) computable bijection $f \colon \mathbb{N} \to \mathbb{N}$, in other words that A is computably isomorphic to B. $A \leq_T B$ is the standard Turing reducibility. We shall use notions of 'Γ *being a pointclass*' of either a set of integers, or sets of reals—the latter will be mostly thought of as coextensive with $2^\mathbb{N}$ (but sometimes $\mathbb{N}^\mathbb{N}$, however it will be clear which), in the sense of Moschovakis [56]. We shall not have any occasion to consider Polish spaces other than products of these two, and the latter reference will contain all the descriptive set-theoretic definitions needed for Spector classes, Uniformization, Scales and the like. Everything we do will take place within the pointclass of Δ_2^1. For an account of recursive-theoretic hierarchies within this pointclass, see [32]. We let WO denote the set of real numbers coding wellorderings. For λ a limit ordinal and $\langle l(\alpha) \mid \alpha < \lambda \rangle$ a sequence of natural numbers, by $\mathrm{Liminf}^*\langle l(\alpha) \mid \alpha < \lambda \rangle$ we mean the usual $\mathrm{Liminf}\langle l(\alpha) \mid \alpha < \lambda \rangle = \sup_{\beta < \lambda} \left(\inf\{l(\alpha) \mid \alpha > \beta\} \right)$, but only if this is less than ω. If the latter equals ω, then the Liminf^* operation returns 0.

§2. Kleene's equational calculus for higher types.

In a series of papers ([35]–[38]) Kleene developed a generalisation of his earlier equational calculus for the Gödel–Herbrand general recursive functions from the 30s. In this generalisation *higher-type recursion* was developed. This theory saw numbers as objects of *type*—0 a function $f \colon \mathbb{N}^k \to \mathbb{N}$ as an object of *type* 1, and $F \colon (2^{\mathbb{N}})^k \times \mathbb{N}^l \to \mathbb{N}$ as of *type*-2, and so forth. (See the account of Normann [57] in this volume.) This was continued into further types. It is not the purpose of this article to describe in any great details those theories but concentrating only on the type-1 and type-2 cases. One notion of computation can be given a mechanistic flavour, and illustrated by what Rogers [61] (attributing the phrase to Spector and Addison) called the "$\aleph_0$-mind". One imagines a Turing machine with a genuinely infinite tape of cells $\langle C_i \mid i < \omega \rangle$, but with the capability of surveying and manipulating a countable amount of information in a single step. The machine is deemed capable of consulting an oracle that gives yes/no answers to queries of the form *Is* $y \in A$? where $y \in 2^{\mathbb{N}}$ is the tape's contents at any stage of computation, and $A \subset \mathbb{N} \cup 2^{\mathbb{N}}$. The machine must thus both be able to calculate the values of $y(k)$ to complete y, and be thought of as being able to present y to the oracle for query. Computations are thus truly infinitary objects and indeed a successful computation is best thought of as represented by a wellfounded tree of computation steps and subroutine calls.

Such a tree has *infinite branching* below any node (perhaps representing the calculations $y(0), y(1) \ldots$ *etc.*). The output of such a machine could then be either integer or real form.

Generalising to several variables $\vec{n}$ of number type and $\vec{x}$ of real type, we may index programs of such machines by $e \in \omega$ and write a typical such calculation with oracle $\mathcal{I}$ as $\{e\}(\vec{n}, \vec{x}, \mathcal{I})$. Recursion in ^2E is essential for much of the development of relative recursiveness for the computation theory at this type, where, for $\alpha \in 2^{\mathbb{N}}$:

$$\mathcal{E}(x) = \begin{cases} 0 & \text{if } \exists n\, x(n) = 0, \\ 1 & \text{otherwise.} \end{cases}$$

For any oracle $\mathcal{T}$ the class of relations semi-decidable in $\mathcal{T}$ is closed under universal number quantification; by requiring that computations be relative to $\mathcal{E}$ we also ensure their closure under existential number quantification (as the form of $\mathcal{E}$ shows). This then implies that any arithmetical sets of reals is Kleene-computable. However requiring computations to be relative to $\mathcal{E}$ also ensures that we have an *Ordinal Comparison Theorem*; this is needed to develop the theory of relations semi-computable in a type-2 functional. Kleene then showed that the decidable relations were precisely the hyperarithmetic, and the semi-decidable are those 'Kleene reducible' to a complete Π_1^1 set of reals, say WO, that of the codes of countable wellorders.

THEOREM 1 (Kleene). *The hyperarithmetic relations $R(\vec{n}, \vec{x}) \subseteq \mathbb{N}^k \times (\mathbb{N}^\mathbb{N})^l$ for any $k, l \in \mathbb{N}$.are precisely those computable in $\mathcal{E}$.*

The Π_1^1 relations are precisely those semi-computable in $\mathcal{E}$.

One may define the notion of *Kleene degree* from the notion of *Kleene reducibility*:

DEFINITION 2 (Kleene reducibility). *Let $A, B \subseteq \mathbb{R}$; we say that $A \leq_K B$ iff there is an index e and $y \in \mathbb{R}$ so that*

$$\forall x \in \mathbb{R} \, (x \in A \longleftrightarrow \{e\}(x, y, B, \mathcal{E}) \downarrow 1);$$
$$\forall x \in \mathbb{R} \, (x \notin A \longleftrightarrow \{e\}(x, y, B, \mathcal{E}) \downarrow 0).$$

A is Kleene-semi-computable in B iff there is an index e and $y \in \mathbb{R}$ so that

$$\forall x \in \mathbb{R} \, (x \in A \longleftrightarrow \{e\}(x, y, B, \mathcal{E}) \downarrow 1).$$

Notice that the obvious degree notion has become a boldface one: we are allowed a real parameter y into the definition. We may if we wish think of y as a variable adding some countable amount of information to the computation, but in any case the bottom-most degree $\mathbf{0}_K$, say, containing $\varnothing, \mathbb{R}$, in fact now consists of the *Borel sets*. The semi-computable sets are now those of degree precisely the co-analytic Π_1^1 sets, In terms of a jump notation the Kleene degree of a complete Π_1^1 such as WO, may be written $\mathbf{0}'_K$. Then the degree of $A \subseteq \mathbb{R}$ contains continuum many sets $B \subseteq \mathbb{R}$. The presence of the parameters y ensures that the degree is closed under continuous pre-images, and hence is a *Wadge degree*.

The notion of Kleene degree here is then tied in very much with hyperarithmeticity and WO. It is thus a finer degree notion than that of Δ_2^1-degree. We may formulate a general slice through the Δ_2^1-pointclass as follows. Suppose $f : \mathbb{R} \longrightarrow \omega_1$ is such that $x \leq_T y \longrightarrow f(x) \leq f(y)$. The function f is then *Turing invariant*. Further let us suppose f is Σ_1 definable over $(HC, \in)$ without parameters, by a formula in $\mathcal{L}_{\dot{\in}}$.

DEFINITION 3. *Let f be as described; let Φ be a class of formulae of $\mathcal{L}_{\dot{\in}}$. Then $\Gamma = \Gamma_{f,\Phi}$ is the pointclass of sets of reals A so that $A \in \Gamma$ if and only if there is $\varphi \in \Phi$ with:*

$$\forall x \in \mathbb{R} \, (x \in A \leftrightarrow L_{f(x)}[x] \models \varphi[x]).$$

The idea is that we are allowed, for example if $\Phi = \Sigma_1$, to only search through the $L[x]$ hierarchy as far as $f(x)$ for a witness to show that $\varphi[x]$ holds. Allowing f and Φ to vary then carves out a slice through the Δ_2^1 sets.

As an example, with $\Phi = \Sigma_1$ and $f(x) = \omega_{1,ck}^x$ (the first ordinal that is not x-recursive) this yields the Π_1^1 sets; allowing an arbitrary $y \in \mathbb{R}$ as a fixed parameter to go with a choice of $\varphi \in \Sigma_1$ would yields the $\Pi_1^1(y)$ sets. Hence Definition 3 can be relativised uniformly to any parameter.

As is well known $L_{\omega_{1,ck}^z}[z]$ is the least transitive model of Kripke–Platek set theory containing ω and z; it is thus the "least z-admissible" ordinal (cf. [1], V 5.11). We may extend this form and define $\omega_{1,ck}^{B,y}$ to be the ordinal rank of the least B-y-admissible set, now in a language extended with a predicate for the set of reals B. The relation $A \leq_K B$ of Definition 2 can be usefully reformulated.

LEMMA 1. *$A \leq_K B$ iff there are Σ_1-formulae in $\mathcal{L}_{\in,\dot{X}}$ $\varphi_1(\dot{X}, v_0, v_1)$, $\varphi_2(\dot{X}, v_0, v_1)$, and there is $y \in \mathbb{R}$, so that*

$$\forall x \in \mathbb{R} \left(x \in A \iff L_{\omega_1^{B,y,x}}[B, y, x] \models \varphi_1[B, y, x] \right.$$
$$\left. \iff L_{\omega_1^{B,y,x}}[B, y, x] \models \neg \varphi_2[B, y, x] \right).$$

Similarly A would be Kleene-semi-computable in B if only if there is a formula φ_1 as above with just the first line holding (see [33]).

Solovay showed that, assuming Axiom of Determinacy, the ordering of the Kleene degrees under strict Kleene-reducibility was wellordered.

The structure of the Kleene degrees of sets of reals is very much dependent on the ambient set-theoretical universe: Solovay observed ([71]) that if the Axiom of Determinacy holds then $\leq_K$ is ordered; if $V = L$ (or set forcing extensions thereof) then there are intermediate degrees $\mathbf{0}_K <_K B <_K \mathbf{0}'_K$ (indeed 2^c many mutually incompatible such) (Hrbacek–Simpson [33]); however if $\mathrm{Det}(\Pi_1^1)$ holds then there are none (Harrington [30]).

DEFINITION 4 (Moschovakis [56]). *A* Spector class *of pointsets* $\Gamma \subseteq \mathbb{N}^k \times (\mathbb{N}^\mathbb{N})^l$ *for any k, l, is a collection that is closed under $\cap, \cup$, number quantification $\exists^\mathbb{N}, \forall^\mathbb{N}$; closed under (standard) recursive substitutions, has a universal set U indexing by $\mathbb{N}$ all members of Γ, and lastly has the Prewellordering property*:

PW: *For any $P \in \Gamma$ there is $\sigma : P \longrightarrow \lambda$ for some ordinal λ with the property that there are relations: $x \leq_0^\sigma y \in \Gamma$, $x \leq_1^\sigma y \in \check{\Gamma}$ so that*:

$$P(y) \Rightarrow \forall x \left([P(x) \wedge \sigma(x) \leq \sigma(y)] \iff x \leq_0^\sigma y \iff x \leq_1^\sigma y \right).$$

The paradigm here is the class of Π_1^1 sets itself: as any such $P \in \Pi_1^1$ has the form: $P = \{ y \mid L_{\omega_{1,ck}^y}[y] \models \varphi[y] \}$ for some $\Sigma_1 \varphi$, we may define $\sigma(y) \simeq \mu\alpha.L_\alpha[y] \models \varphi[y]$, and this is a suitable prewellordering. We then may set

$$x \leq_0^\sigma y \iff \exists \eta < \omega_1^{x,y} \left(L_\eta[x, y] \models \sigma(x) \leq \sigma(y) \right);$$
$$x \leq_1^\sigma y \iff \forall \eta < \omega_1^{x,y} \left(L_\eta[y] \models \varphi[y] \longrightarrow L_\eta[x] \models \varphi[x] \right).$$

As relations $\Pi_1^1(x, y)$ are precisely those $\Sigma_1([L_{\omega_1^{x,y}}[x,y])$ (essentially by the Spector–Gandy Theorem, [72], [21]) we have relations of the right complexity. (The existence of a universal set U can be justified by the existence of a universal Σ_1 formula and Skolem function in $\mathcal{L}_{\dot{\in}}$ for models of "$V = L[\dot{z}]$".) Other Spector classes are the projective classes Σ_k^1 ($k \geq 2$) assuming $V = L$, or the classes Π_{2n+1}^1, and Σ_{2n+2}^1 ($n \geq 0$) under Projective Determinacy (PD cf. [56]), but there are many others. We shall be deriving some from the transfinite machine model, analogously to that obtained from Kleene reducibility, which we shall describe below.

The notions above for sets of integers are clearly closely related to that of hyperarithmeticity between sets of integers, and hyperdegrees where '$x \leq_h y$' is the reducibility of the set x is hyperarithmetic in the set y. For such we have a so-called *Spector Criterion for hyperdegrees*:

$$x \leq_h y \longrightarrow (x' \leq_h y \longleftrightarrow \omega_{1,\mathrm{ck}}^x < \omega_{1,\mathrm{ck}}^y)$$

where x' is the complete $\Sigma_1^1(x)$ set of integers—the *hyperjump* of x. We shall see that other reducibilities related to transfinite computational models have an associated jump operation, and can have ordinal assignments $x \longmapsto \lambda^x$ which satisfy the analogous form of criterion.

§3. Infinite Time Turing Machines.

We turn now to considering the behaviour of ordinary machines, those designed by Turing for the analysis of computability in the ordinary sense, but allowing them to have transfinite amounts of time or even space in which to run.

3.1. Basic properties.

We shall give a somewhat historically ordered account of these over the last decade, starting with the description of the Infinite Time Turing Machines (ITTMs) due to Hamkins and Kidder. These were considered in the 1990s but only appeared in print in an extensive account by Hamkins and Lewis in 2000, [22]. (Hamkins tells the story that he and Kidder invented the ITTM in their office directly after hearing Leonore Blum speak on the Blum–Shub–Smale machines at the Berkeley Logic Colloquium.) This paper awakened interest in such transfinitely running models. Later others turned to Register machines, or Turing machines with ordinal length tapes, thus increasing the space component of such machines.

We shall look at the mathematics of the ITTM model in detail first, before turning to these others. Such a machine is essentially a standard Turing machine, with a one way infinite tape of cells $\langle C_i \mid i \in \mathbb{N} \rangle$, which we consider divided up into three infinite tapes for *Input*, *Scratch*, and *Output* tapes, say as $\langle C_{3i+j} \mid i \in \mathbb{N} \rangle$ for $j < 3$ for definiteness (Figure 1). In [22] a special *limit state*, q_L, was added to the already finite number of states of the machine $q_0, \ldots, q_N$ say. This was the state the machine entered into at a limit stage of time, but which we shall dispense with here. The program software of the

	\multicolumn{2}{c}{}	R/W							
Input:	1	1	0	1	1	0	0	0	0	...
Scratch:	0	1	1	1	1	1	1	0	0	...
Output:	1	0	0	0	1	1	0	1	0	...

FIGURE 1. A 3-Tape Infinite Time Turing Machine.

machine with its finite transition state table is thus unchanged: programs of the standard machine $\langle P_e \mid e \in \mathbb{N}\rangle$ can be carried over to this transfinite context. All we have to do is specify a behaviour at limit stages λ of time. Firstly we by decree define that the cells' values, $C_i(\lambda)$, are determined by a final segment of the values $C_i(\alpha)$ at ordinal times $\alpha < \lambda$:

$$C_i(\lambda) = k \quad \text{if } \exists \alpha < \lambda \, \forall \beta < \lambda \, (\alpha < \beta \longrightarrow C_i(\beta) = k) \text{ for } k \in \{0,1\};$$
$$= 0 \quad \text{otherwise.}$$

This 'limit rule' is thus a Liminf, and is of an $\exists \forall$ nature. (The original paper used *Limsup*s instead but this is an inessential difference.) Having specified the cell values, if we denote the state of the machine at time α as $q(\alpha)$, and the position the R/W head is at as $l(\alpha)$, then we need only to specify the values of $q(\lambda)$ and $l(\lambda)$. We again use a liminf rule and specify:

$$q(\lambda) = \text{Liminf}\langle q(\alpha) \mid \alpha < \lambda \rangle,$$
$$l(\lambda) = \text{Liminf}^*\langle l(\alpha) \mid \alpha < \lambda \rangle.$$

The reader may like to verify that the upshot is that if we set out our Turing machine transition table as lines of a program with suitably nested subroutine calls and so forth in a sensible fashion, then matters can be arranged so that then the machine is placed back in the state/instruction at the beginning of the outermost subroutine in the program that was called unboundedly often before time λ; moreover the head is then also placed at that cell from which it entered into that subroutine, again unboundedly often before λ. This is a pleasing but inessential detail in the arrangement. (The original paper had the machine enter the special limit time state q_L and the R/W head always went back to the beginning of the tapes.)

We now have defined the behaviour of the machine for any ordinal stage of time, for any sequence $x \in 2^{\mathbb{N}}$ that may be written on the input tape at the outset. Note that computations with oracles $Z \subseteq \omega$ are rendered superfluous, since Z may be written to the input tape, and the whole tape can be scanned in ω many steps. We can thus think now of oracles as being sets $A \subseteq 2^{\mathbb{N}}$, and we may quiz the oracle as to whether, *e.g.*, the contents of the scratch tape is in A. We thus shall have a model of computations in a higher type. Hamkins and Lewis explore the possibilities of this model, mostly emphasising the analogy with Turing computations, forming notions of relative computability, computation degrees and the like. Given the affinity with the standard model one sees that one has many of those features; they prove a wealth of features which we cannot list all here, but they include the existance of a *universal machine*, there are S_m^n and *Recursion Theorems*. To give a flavour of several of the proofs in [22] we prove:

THEOREM 2 (Hamkins–Lewis—[22]). (i) '$x \in $ WO' *is decidable by an* ITTM, *where* WO *is the set of reals that are codes of wellorders*. (ii) *Moreover let* α *be any recursive ordinal and* $A \subseteq 2^{\mathbb{N}}$ *an* α-Π_1^1 *set, that is, at the* αth *level of the difference hierarchy on* Π_1^1 *sets. Then* '$x \in A$' *is decidable by an* ITTM.

PROOF. (i) We do this just for the single Π_1^1 set. The proof for (ii) at the higher levels is just complication. It can be quickly established that such machines will be able to decide whether an input real codes a wellorder: note first that any arithmetic predicate is decidable: one may even decide a Σ_2^0 predicate of integers in ω steps: to ask if $x \in A$ with $A \in \Sigma_2^0$, defined as $\exists u \forall v \Upsilon(u, v, x)$ with Υ in Δ_0, one computes of $\Upsilon(0, 0, x), \Upsilon(0, 1, x) \ldots$ in a standard Turing manner, until, if ever, a k is found so that $\neg\Upsilon(0, k, x)$, if so then a 'flag cell', say C_1 has its value switched $1 \to 0 \to 1$, and then one proceeds with $\Upsilon(1, 0, x), \Upsilon(1, 1, x) \ldots$ etc. Then $x \in A \leftrightarrow C_1(\omega) = 1$. (In a further ω many steps we may decide Σ_4^0 predicates, and thus by time ω^ω any arithmetic predicate.) As being a linear order is Π_2^0, the machine may decide if x codes such a linear order, and thereafter seeks for the least element in the ordering coded by x. It makes a guess by taking an element from the field of x, k_0 say, and starts searching for a lesser element k_1, if such exists, then it is found in finitely many steps; repeating a search for k_2 etc; if each time a lesser element is found we can flash a flag cell $1 \to 0 \to 1$, then after ω many stages if the flag is zero, then we found an infinite descending chain, and the ordering is illfounded; if it is 1, then there is some value k_m that was $<_x$-minimal. In ω further steps we go through the field of the wellorder, erasing all mention of k_m; we then pick another guess as a new k_0' and repeat. Eventually we either erase the whole of Field($<_x$) or we find an infinite descending chain, and in the former case we know we have a wellorder. ⊣

They show that even further levels of the difference hierarchy on Π_1^1 can be decided by such programs, in fact for α that are *writable*, in a sense to be made precise below. We may denote by '$P_e(x) \downarrow y$' that y is the output of a halting computation of a machine on program P_e with input x. We may input an integer n by a string of $n+1$ 1s followed by 0s, but we abbreviate this by writing '$P_e(n) \downarrow y$'. If the computation fails to halt then we write '$P_e(x)\uparrow$'.

DEFINITION 5. (i) *We write* '$P_e(n) \downarrow^\alpha y$' *if* $P_e(n) \downarrow y$ *in exactly* α *steps. We call* α clockable *if* $\exists e \exists n \in \omega \exists y \, P_e(n) \downarrow^\alpha y$.

(ii) *A real* $y \in 2^{\mathbb{N}}$ *is writable if there are* $n, e \in \omega$ *with* $P_e(n) \downarrow y$; *an ordinal* β *is called writable, if* β *has a writable code* y.

An ITTM calculation such as that of $P_e(n)$ can of course now halt in a transfinite number of steps, and it is amusing to think of programs that halt *precisely* in $\omega, \omega^2, \omega^\omega.2$ steps. An elementary argument shows that if we regard a *snapshot* of such a machine at time α as consisting of $\langle l(\alpha), q(\alpha), \langle C_i(\alpha) \rangle_i \rangle$ then a non-halting program must eventually after a countable number of steps enter into some permanently repeating loop. (One should note that it is not sufficient to merely have a pair of identical snapshots in order to declare that a computation is in a permanent loop: it is necessary to have a closed increasing $\omega + 1$-sequence of ordinals γ_i ($i \leq \omega$) with identical snapshots for this: even an ω-sequence $\langle \delta_i \rangle_i < \omega$ may result in $C_k(\delta_i) = 1$ whilst having $C_k(\sup\{\delta_i\}) = 0$ by the *liminf* formulation. It can be shown, however, for the universal machine that any repeating pair of snapshots will be permanently repeating.) We thus see that the property of being a 'well-ordered sequence of snapshots in the computation $P_e(x)$' is Π_1^1 as a relation of e and x; thus '$P_e(x) \downarrow y$' is Δ_2^1:

$\exists w$ (w *codes a halting computation of* $P_e(x)$, *with* y *written on the output tape at the final stage*) $\iff$

$\forall w$ (w *codes a computation of* $P_e(x)$ *that is either halting or performs a repeating infinite loop* $\longrightarrow w$ *codes a halting computation with* y *on the output tape*).

Similarly $P_e(x)\uparrow$ is also Δ_2^1. In [22] the authors conduct an analysis of *gaps* in the clockable ordinals and prove a number of results, including: the first maximal gap in which no computation of the form $P_e(n)$ halts is $[\omega_1^{ck}, \omega_1^{ck}+\omega)$; further no calculation halts in precisely an admissible β number of steps; no gap is shorter than ω but there are arbitrarily long gaps appearing cofinally below the supremum γ of clockable ordinals.

DEFINITION 6. (i) $x^\nabla = \{e \mid P_e(x)\downarrow\}$ (*The halting set on integers*).

(ii) $X^\blacktriangledown = \{(e, y) \mid P_e^X(y)\downarrow\}$ (*The halting set on reals relativised to* $X \subseteq 2^{\mathbb{N}}$).

It is natural to ask for characterisations of, e.g., 0^∇, or to ask: what are the clockable ordinals? Before turning to this we define:

DEFINITION 7. (i) $R(x)$ *is an* ITTM-semi-decidable *predicate if there is an index* e *so that*:

$$\forall x \, (R(x) \leftrightarrow P_e(x) \downarrow 1).$$

(ii) *A predicate* R *is* ITTM-decidable *if both* R *and* $\neg R$ *are* ITTM-semi-decidable.

Clearly one would like to know what the ITTM-(semi-)decidable sets are. To get a grasp on an upper bound note, at the risk of stating the obvious, that we have one clear difference between standard Turing computations and that of an ITTM: for the former a single integer suffices to code a whole *course of computation*, whereas to code a course of an ITTM-computation is to code a wellordered sequence of snapshots the form $\langle q(\alpha), l(\alpha) \langle C_i(\alpha) \rangle_{i<\omega} \rangle$ each of which may be coded as real. We might try to get an analogue of Kleene's canonical *Normal Form Theorem* arising from his T-predicate. For any halting standard Turing computation there is an integer K that will code the whole course of the computation $P_e(n)$, and we may effectively find an e' so that $\forall n \, (P_e(n)\!\downarrow \longrightarrow P_{e'}(n) \downarrow K)$. For this to work in the ITTM arena even for computations on integers, we must have that the *ordinal lengths* of the computations are writable by computations. In short: *is every clockable ordinal writable?*

Before answering this we remark that non-halting behaviour has two aspects: there can be a computation that, whilst not formally halting nevertheless continues without making any further changes to its output tape—it may merely alter entries to the scratch tape for ever; in one sense then the output has *stabilized* even if the machine has not formally halted. The other kind of non-halting behaviour results in the output tape being infinitely modified. This leads to:

DEFINITION 8. (i) *Suppose for the computation $P_e(x)$ the machine does not halt then we write $P_e(x)\!\uparrow$; if eventually the output tape does have a stable value $y \in 2^{\mathbb{N}}$ then we write: $P_e(x) \uparrow y$ and we say that y is* eventually writable.

(ii) $R(x)$ *is an* eventually ITTM-semi-decidable *predicate if there is an index e so that*:

$$\forall x (R(x) \leftrightarrow P_e(x) \uparrow 1).$$

(iii) *A predicate R is* eventually ITTM-decidable *if both R and $\neg R$ are eventually* ITTM-*semi-decidable*.

There are concomitant questions as to what these predicates are, how long it takes for a computation to start looping, and if the output is eventually writable, after how many stages *etc.*, and indeed about halting or stabilizing behaviour in general. Resolving the question of whether all clockable ordinals were all writable (in short *halting* behaviour) turned out to require an analysis of the *stabilizing* behaviour of individual cells C_i during the course of a universal machine computation (*cf.* [78]). In this sense stabilization is prior or anterior to halting. Once stabilizing behaviour is fully analysed one has:

THEOREM 3 (The λ, ζ, Σ-Theorem (Welch *cf.* [77], [83])).

(i) *Any* ITTM *computation $P_e(x)$ which halts, does so by time λ^x, the latter being defined as the supremum of the x-writable ordinals*;

(ii) *any computation $P_e(x)$ with eventually stable output tape, will stabilize before the time ζ^x defined as the supremum of the eventually x-writable ordinals*;

(iii) *moreover ζ^x is the least ordinal so that there exists $\Sigma^x > \zeta^x$ with the property that*

$$L_{\zeta^x}[x] \prec_{\Sigma_2} L_{\Sigma^x}[x];$$

(iv) *then λ^x is also characterised as the least ordinal satisfying*:
$$L_{\lambda^x}[x] \prec_{\Sigma_1} L_{\zeta^x}[x].$$

Considering just the integer computational version, by taking $x = \emptyset$ in the above, since the ITTM operations are simply defined and absolute, one sees that running the computation inside L, that if $P_e(n)$ is non-halting, then, with hindsight, the Σ_2 liminf rules will ensure that the snapshots at times $\zeta (= \zeta^\emptyset)$ and $\Sigma (= \Sigma^\emptyset)$ (taken as defined by the characterisation of (iii) above) are the same. It is then easy to argue that if the computation has not halted by time ζ then it will never halt but continue with periodicity Σ. Thus the lexicographically least pair $(\bar{\zeta}, \bar{\Sigma})$ standing in the relationship of (iii) of the theorem, forms an upperbound to the least pair of permanently repeating snapshots of the universal computation. (The reader may be concerned about the warning issued earlier that having merely a pair of ordinals (ζ, Σ) with identical snapshots is not sufficient to guarantee that we are in a permanently looping cycle. This is true, but note here no cell C_j that is stable on a tail interval $(\delta, \zeta]$ will change its value in $[\zeta, \Sigma)$: the first ordinal $\delta' > \delta$ where C_j changed its value (were it to exist) would be $\Sigma_2(L_\Sigma)$ definable, and hence (by the characterisation of (iii)) would be less than ζ. So actually any pair of snapshots of any computation on integers, at this particular pair of stages must be permanently cycling.) As a lower bound one can show that if one has a code for α on a tape one can use this to build a code for L_α on another.

Since $\lambda (= \lambda^\emptyset)$ has the Σ_1-elementarity property above, and since '$P_e(n)\downarrow$' is a Σ_1 statement, we have immediately that any clockable ordinal is below λ. This, together with the observation from [22] that the 'counting through' algorithm of any code for a wellordering of order type τ takes at least τ many steps before halting, yields that any writable ordinal is also clockable. The corollary is that the two classes are the same. Hence, following Kleene:

THEOREM 4 (Normal Form Theorem (Welch)). (a) *For any* ITTM *computable function φ_e we can effectively find another* ITTM *computable function $\varphi_{e'}$ so that on any input x from $2^{\mathbb{N}}$ if $\varphi_e(x)\downarrow$ then $\varphi_{e'}(x) \downarrow y \in 2^{\mathbb{N}}$, where y codes a wellordered computation sequence for $\varphi_e(x)$.*

(b) *There is a universal predicate $\mathfrak{T}_1$ which satisfies $\forall e \, \forall x$*:
$$P_e(x) \downarrow z \leftrightarrow \exists y \in 2^{\mathbb{N}} \left[\mathfrak{T}_1(e, x, y) \wedge Last(y) = z \right].$$

Further, as a corollary (to Theorem 3):

COROLLARY 1. (i) $x^\nabla \equiv_1 \Sigma_1\text{-Th}(\langle L_{\lambda^x}[x], \in, x \rangle)$—*the latter the Σ_1-theory of the structure.*

(ii) *Let $x^\infty =_{df} \{e \mid \exists y \, P_e(x) \uparrow y\}$ be the set of x-stable indices. Then*
$$x^\infty \equiv_1 \Sigma_2\text{-Th}(\langle L_{\zeta^x}[x], \in, x \rangle).$$

The ITTM-decidable sets of integers consist precisely of those in L_λ, and the ITTM-semidecidable sets are those $\Sigma_1(L_\lambda)$; both of these with the obvious uniform relativised counterparts. The ITTM-eventually decidable sets turn out then, to be those which are $\Delta_2(L_\zeta)$, and as L_ζ is a Σ_2-admissible set, are elements of L_ζ; and the ITTM-eventually-semi-decidable those that are $\Sigma_2(L_\zeta)$. Again their relativisations to real inputs are uniform. To see these latter claims from Cor. 1(ii), note that if the n'th cell of the output tape of the machine running $P_e(0)$ is stable from some point on, then this fact is expressed by a Σ_2 sentence about e and n over L_ζ; hence if all the cells of the output tape stabilize a simple application of Σ_2-admissibility of ζ shows that this will have happened for them all by a stage $\bar{\zeta} < \zeta$. Hence if $P_e(0) \uparrow y$ then $y \in L_\zeta$.

The last corollary should be compared with Kleene's result that the complete Π_1^1-set of integers coding indices of wellfounded recursive trees, that is the notation system $\mathcal{O}$ (see [61]), is recursively isomorphic to the Σ_1 theory of $\langle L_{\omega_1^{ck}}, \in \rangle$. Indeed it is easy to see that one can literally extend Kleene's $\mathcal{O}$ to an $\mathcal{O}^+$ and also to an $\mathcal{O}^\infty$ by simply allowing indices into $\mathcal{O}^+$ of those programs that halt, and into $\mathcal{O}^\infty$ those of stable output: we use precisely the same formalism as for $\mathcal{O}$, and simply widen the definition of 'computable'. Thus $\mathcal{O}$ is to $L_{\omega_1^{ck}}$ as $\mathcal{O}^+$ is to L_λ as $\mathcal{O}^\infty$ is to L_ζ (the details of this are in [40]). It is natural to regard both x^∇ and x^∞ as jump operators and later we shall refer to them as such.

DEFINITION 9. (i) *A set of integers x is* semi-decidable *in a set y if and only if*:
$$\exists e \, \forall n \in x \, \left[P_e^y(n) \downarrow 1 \longleftrightarrow n \in x\right].$$

(ii) *A set of integers x is* decidable *in a set y if and only if*:
$$\exists e \, \forall n \in x \, \left[(P_e^y(n) \downarrow 1 \leftrightarrow n \in x) \wedge (P_e^y(n) \downarrow 0 \leftrightarrow n \notin x)\right].$$

We may write $x \leq_\infty y$ for the reducibility ordering.

(iii) *A set of integers x is* eventually-(semi)-decidable *in a set y if and only if the above holds with $\uparrow$ replacing $\downarrow$. For this reducibility ordering we write $x \leq^\infty y$.*

We thus compactly express the two reducibilities using the notations $\leq_\infty$ and $\leq^\infty$ with the first denoting computability, and the second eventual computability.

LEMMA 2. (i) *The assignment $x \longmapsto \lambda^x$ satisfies the Spector criterion*:
$$x \leq_\infty y \longrightarrow (x^\nabla \leq_\infty y \leftrightarrow \lambda^x < \lambda^y).$$

(ii) *Similarly for the assignment $x \longmapsto \zeta^x$*:
$$x \leq^\infty y \longrightarrow (x^\infty \leq^\infty y \leftrightarrow \zeta^x < \zeta^y).$$

The above shows that degree structure induced by $\leq_\infty$ is more akin to that of hyperdegrees than Turing degrees; indeed the proper analogy is probable more that with Δ_2^1-degrees; in terms of fineness, ITTM-reducibility and eventual-ITTM-reducibility are strictly in between these two. The following is the natural version for real computation.

DEFINITION 10. *A set of reals A is* semi-decidable *in a set of reals B if and only if*:

$$\exists e \, \forall x \in 2^{\mathbb{N}} \left[P_e^B(x) \downarrow 1 \leftrightarrow x \in A \right].$$

(ii) *A set of reals A is* decidable *in a set of reals B if and only if*:

$$\exists e \, \forall x \in 2^{\mathbb{N}} \left[P_e^B(x) \downarrow 1 \leftrightarrow x \in A \land P_e^B(x) \downarrow 0 \leftrightarrow x \notin A \right].$$

(iii) *If in the above we replace $\downarrow$ everywhere by $\uparrow$ then we obtain the notion in* (i) *of A is* eventually decidable *in B and in* (ii) *of A is* eventually semi-decidable *in B*.

DEFINITION 11.
(i) $A \leq_\infty B$ *iff for some* $e \in \omega$, *for some* $y \in 2^{\mathbb{N}}$: *A is decidable in* (y, B).
(ii) $A \leq^\infty B$ *iff for some* $e \in \omega$, *for some* $y \in 2^{\mathbb{N}}$: *A is eventually decidable in* (y, B).

Again two reducibilities, one for each notion. Also a real parameter has been inserted into the last definition; this brings it into line with the previous notion of Kleene reducibility between sets of reals, and is closed under continuous pre-images, and thus again will ensure that each degree is a Wadge pointclass. Just as for Kleene degrees the structure of either of these induced degree orderings will depend on the ambient set theory: whether $V = L$ (and this implies there are many degrees, even incomparable degrees below the complete ITTM-(eventually)-semi-decidable set) or there is "sufficient determinacy". In the latter case the degrees will be wellordered and without any intermediate degrees at all. This is discussed further below.

By analogy with Kleene recursion we have:

LEMMA 3.
(i) $A \leq_\infty B$ *iff there are Σ_1-formulae in $\mathcal{L}_{\in, \dot{X}}$ $\varphi_1(\dot{X}, v_0, v_1), \varphi_2(\dot{X}, v_0, v_1)$, and $y \in \mathbb{R}$, so that*

$$\forall x \in \mathbb{R} \left(x \in A \iff L_{\zeta^{B,y,x}}[B, y, x] \models \varphi_1[B, y, x] \right.$$
$$\left. \iff L_{\zeta^{B,y,x}}[B, y, x] \models \neg \varphi_2[B, y, x] \right).$$

(ii) $A \leq^\infty B$ *iff there are Σ_2-formulae in $\mathcal{L}_{\in, \dot{X}}$ $\varphi_1(\dot{X}, v_0, v_1), \varphi_2(\dot{X}, v_0, v_1)$, and $y \in \mathbb{R}$, so that*

$$\forall x \in \mathbb{R} \left(x \in A \iff L_{\zeta^{B,y,x}}[B, y, x] \models \varphi_1[B, y, x] \right.$$
$$\left. \iff L_{\zeta^{B,y,x}}[B, y, x] \models \neg \varphi_2[B, y, x] \right).$$

The lemma then identifies structures in which we can look to ascertain the outcomes of our ITTM computations relative to a set of reals B say. By way of analogy with ζ, the ordinal $\zeta^{B,y,x}$ is the least that is not ITTM-(B, x, y)-eventually-semi-decidable. It is thus also least such that $L_{\zeta^{B,y,x}}[B, y, x]$ has a proper Σ_2-elementary end-extension in the $L[B, y, x]$ hierarchy. If $\lambda^{B,y,x}$ is the least λ which is the height of a transitive elementary Σ_1-substructure of $L_{\zeta^{B,y,x}}[B, y, x]$, we could replace $\zeta^{B,y,x}$ in (i) with $\lambda^{B,y,x}$ of course.

3.2. Degree theory and Post's problem for ITTM-degrees. In [22] the degree theory of the ITTM-reducibility together with the jump operator $x \longmapsto x^\nabla$ is explored, pursuing the analogy with Turing degrees under Turing jump. We shall consider for the moment this simply on sets of integers. In [23] the same authors continue this investigation. In particular they consider the analogue of Post's problem in both settings. With the definition of degree coming naturally from Definition 9(i) they establish that in fact there are no degrees at all between 0 and 0^∇.

In [77] degree theory was also explored but it pursued instead the analogy with hyperjump and hyperdegrees. Cor. 1(i) establishes that the jump x^∇ is essentially a mastercode, or equivalently at this level, the Σ_1-truth set, for a particular admissible set, namely that of $L_{\lambda^x}[x]$ where the natural association of $x \longmapsto \lambda^x$ satisfies the Spector criterion from Lemma 2(i). This is just as the hyperjump of x is the complete $\Sigma_1^1(x)$ set of integers, and is then recursively isomorphic to the Σ_1-Th$(L_{\omega_1^x}[x])$. Just as there are no reals of hyperdegree between that of the hyperarithmetic sets and $\mathcal{O}$, we do not expect there to be any such intermediate degrees, which is just as the argument from [23] shows. As we iterate the jump operation (starting say from $0 = \varnothing$) we obtain iterates $0^{\nabla\alpha}$ in a linear fashion. At limit stages $\nu < \lambda$ we want to take $0^{\nabla\nu}$ as some minimally chosen upper bound $\{0^{\nabla\alpha} : \alpha < \nu\}$. For $\alpha < \zeta$ these are characterised as follows. We set $S = S_\zeta^1 =_{df} \{\mu < \zeta : L_\mu \prec_{\Sigma_1} L_\zeta\}$ the ordinals Σ_1-stable in ζ. By Theorem 3(iv) the least element of S is λ. Let μ_ι for $\iota < \zeta$ enumerate $S \cup \{0\}$. By standard arguments a non-projectible element of S must be a limit point μ_ν of S. (Note that easily L_ζ is a Σ_2-admissible set, and hence S has this maximal order-type.) Recall that ν is non-projectible if L_ν is a model of Σ_1-Separation. We let T_μ^k be the Σ_k-Th(L_μ). We already have seen that $0^\nabla \equiv_1 T_\lambda^1$ and $0^\infty \equiv_1 T_\zeta^2$. The following completes this picture in L below ζ.

Let $a_\alpha =_{df} T_{\mu_\alpha}^1$ unless μ_α is non-projectible, in which case let $a_\alpha =_{df} T_{\mu_\alpha}^2$. Define by recursion: $0^{\nabla 0} = \emptyset$; $0^{\nabla\alpha+1} = (0^{\nabla\alpha})^\nabla$; and $0^{\nabla\eta} = a_\eta$ for $\text{Lim}(\eta)$. Then we have:

THEOREM 5 (Welch [77]). (i) *For $\alpha < \zeta$:*
  (a) $\mu_{\alpha+1} = \lambda^{0^{\nabla\alpha}}$;
  (b) $0^{\nabla\alpha} \equiv_1 a_{\mu_\alpha}$; *and*
  (c) $[0^{\nabla\alpha}]_\infty = \{z : z \in L_{\mu_{\alpha+1}} \backslash L_{\mu_\alpha}\}$.
Hence it follows that:
(ii) $L_\zeta \models$ "*The $=_\infty$-degrees are linearly ordered in type ζ, and* $\text{Lim}(\eta) \longrightarrow [0^{\nabla\eta}]_\infty$ *is the l.u.b. of* $\{[0^{\nabla\alpha}]_\infty : \alpha < \eta\}$".
(iii) $L_\Sigma \models$ "$\neg \exists y \, (\forall \alpha < \zeta)(0^{\nabla\alpha} <_\infty y <_\infty 0^\infty)$".

The fact that the jump may be iterated taking least upper bounds at limit stages *inside* L_ζ is no guarantee that l.u.b.s exist outside of L and indeed it can be shown the natural hierarchy defined in this way stops at ω: there is no least upper bound of the $\{0^{\nabla^n}: n < \omega\}$ in V but rather, continuum many minimal upper bounds (see [76]).

Q: *If $D = \{d_n: n < \omega\}$ is a countable set of $=_\infty$-degrees, does D have a minimal upper bound?*

Although much is known and there are positive answers for Δ^1_{2n} degrees (assuming PD), for Δ^1_{2n+1}-degrees this remains open, even for hyperdegrees. *Minimal degrees* (compare minimal hyperdegrees) are known to exist by similar arguments for minimal hyperdegrees using Sacks style perfect set arguments (again see [76]).

If one turns to the other notion of jump $x \longmapsto x^\infty$ from Definition 1(ii) the picture is somewhat similar, but there are significant differences: call an ordinal τ Σ_2-*extendible* if there is $\sigma > \tau$ with $L_\tau \prec_{\Sigma_2} L_\sigma$. Let Z be the least Σ_2-extendible that is a limit of such. Then the natural hierarchy of $=^\infty$ degrees which takes l.u.b.s at limit stages (where they exist) has length Z (rather than ω as for $=_\infty$-degrees).

DEFINITION 12. (i) [22] *Let $A \subseteq 2^\mathbb{N}$, then $A^\blacktriangledown =_{df} \{(e,x): P_e^A(x)\!\downarrow\}$.*
(ii) $\Gamma_0 =_{df} \{A \subseteq 2^\mathbb{N} \mid A \text{ is semi-decidable from a real}\}$;
(iii) $\Delta_0 =_{df} \Gamma_0 \cap \neg\Gamma_0$.

Hamkins and Lewis in [22] showed that the decidable sets of reals, Δ_0, form a σ-algebra, closed under the Suslin $\mathcal{A}$-operator and properly containing the Selivanovski C-sets, and are within Δ^1_2. By a result of Solovay they are absolutely measurable, and have the property of Baire. Lemma 3(i) then describes abstractly the place of the class Γ_0 within the Wadge hierarchy. The structure of sets of reals under either of the orderings $\leq_\infty$ or $\leq^\infty$ is dependent on one's set theory: if $V = L$ (or set generic extensions thereof) then following [33] there are 2^c mutually incomparable sets A, B below the complete Γ set. Following an argument of Steel [73], if we let Boolean(Γ_0) denote Boolean combinations of sets from Γ_0, we have:

THEOREM 6. Det(Boolean(Γ_0)) $\implies$ *any complete Γ_0-set is reducible to any $A \in \Gamma_0 \backslash \Delta_0$. Consequently there are no ∞-degrees between those of sets in Δ_0 and those of sets in Γ_0.*

The solution to Post's problem then in this context depends on the ambient set theoretical universe. (See [79] for a further discussion of this, and [84] where it is shown that Det(Γ_0) proves the existence of inner models with a proper class of strong cardinals at the very least. However it is not known how sharp this bound is. One should contrast this strength with the corresponding result of Martin as it affects the Kleene degrees: Det(Boolean(Π^1_1))—which implies there are no Kleene degrees between Borel and complete analytic, a result of Harrington [30]—is equivalent to $\forall x\, (x^\sharp \text{exists})$.) There are entirely similar results and effects for the eventually decidable reduction $\leq^\infty$. However if Γ is defined to be the class of sets eventually-semi-decidable in a real, then it is not known how to separate in terms of inner models $\text{Det}(\Gamma_0)$ from $\text{Det}(\Gamma)$.

3.3. Complexity of ITTM**-computation.** Given the fruitfulness of complexity theory in the theory of (standard) Turing computation, it may seem also tempting to try and define complexity classes for their transfinite computations. In one sense Hamkins and Lewis initiated this with their attempts to pin down the clockable ordinals (think of 'time' here), and by showing, eg, that any arithmetic set could be decided in time less than ω^2 (Thm. 2.6 [22]). They had shown that no admissible ordinal is clockable: this can be shown directly as they do, and is in essence also an application of Π_2-reflection. Schindler [65] considered 'polynomial time'. He gave a more encompassing definition.

DEFINITION 13. (i) *Let $A \subseteq 2^{\mathbb{N}}$, and $\alpha \leq \omega_1$, then we say that $A \in P^\alpha$ if there is $\beta < \alpha$ and an index e so that $\forall x\, \varphi_e(x)\downarrow^{<\beta}$ and $x \in A \leftrightarrow \varphi_e(x) \downarrow 1$.*

(ii) *More generally if $\mathcal{D}$ is the class of (standard) Turing degrees and $f: \mathcal{D} \to \omega_1$ we say that $A \in P^f$, if there is an index e so that $\forall x\, \varphi_e(x)\downarrow^{<f([x]_T)}$ and $x \in A \leftrightarrow \varphi_e(x) \downarrow 1$.*

We thus think of output 1 as 'acceptance'. Then (i) with $\alpha = \omega^\omega$ is Schindler's 'polynomial time' and was named P. He further observed that $P^{\omega_1^{\text{ck}}} = \text{HYP}$ and (both he and Hamkins) that P^{ω_1} coincided with those sets A that were $\Delta_1^1(\beta)$ for some $\beta < \omega_1$. Schindler defined $P^+ = P^{f_0}$ where $f_0(x) = \omega_{1\,\text{ck}}^x + 1$. [14] (Thm. 4) shows directly $P^{f_0} = P^{\omega_1^{\text{ck}}}$. Underlying this is perhaps the following Bounding Lemma:

LEMMA 4 (Bounding Lemma [80]). (i) *Let φ_e be a total computable function and suppose $\forall x\, (\varphi_e(x)\downarrow^{<\omega_1^x})$ then there is $\gamma < \omega_1^{\text{ck}}$ so that $\forall x\, (\varphi_e(x)\downarrow^{<\gamma})$.*

(ii) *Let β be admissible, and φ_e a total computable function and suppose $\forall x\, (\varphi_e(x)\downarrow^{\leq\beta})$ then there is $\gamma < \beta$ so that $\forall x\, (\varphi_e(x)\downarrow^{<\gamma})$.*

These are straightforward applications of, for (i), Σ_1^1-Bounding, and for (ii), Barwise Compactness. Schindler also defined some 'non-deterministic' classes:

DEFINITION 14. (i) *Let $A \subseteq 2^{\mathbb{N}}$, and $\alpha \leq \omega_1$, then we say that $A \in NP^\alpha$ if there is $\beta < \alpha$ and an index e so that $\forall z\, \varphi_e(z)\downarrow^{<\beta}$ and $x \in A \leftrightarrow \exists y\, \varphi_e(x \oplus y) \downarrow 1$.*

(ii) *More generally if $f: \mathcal{D} \to \omega_1$ we say that $A \in NP^f$, if there is an index e so that $\forall z\, \varphi_e(z)\downarrow^{<f([z]_T)}$ and $x \in A \leftrightarrow \exists y\, \varphi_e(x \oplus y) \downarrow 1$.*

Then $NP = NP^{\omega^\omega}$. From the definition it is clear that $P \neq NP$ as the latter is clearly the class of Σ_1^1. Löwe defines various complexity classes and surveys their relationships in [51]. As he points out, Schindler does not explicitly give the connection between his NP classes and an actual non-deterministic ITTM. Löwe makes the appropriate definition and he suggests possible notions of *space* required for a computation $P_e(x)$; one such could be the supremum of the $L[x]$-constructible ordinal ranks of the snapshots of the scratch tape arising in the computation. We refer the reader to this paper for the discussion and question of the $PSPACE^f$ classes that arise.

Hamkins and Welch [29] showed that for almost all f, $NP^f \neq P^f$ and later [14] showed that for many α $P^\alpha = NP^\alpha \cap \text{co-}NP^\alpha$ including those α that begin a gap in the clockable ordinals. (Such are admissible by [83] Thm. 50). However, for many f, P^f is properly contained in $NP^f \cap \text{co-}NP^f$. If one restricts to sets of integers rather than sets of reals, a somewhat different picture emerges: for many β, for example β that are admissible and limits of gaps in the clockables then $P^\beta \cap \mathcal{P}(\mathbb{N}) = NP^\beta \cap \mathcal{P}(\mathbb{N})$ ([80]). For other β they are different.

It is arguably best to think of these results as epiphenomena on the constructible hierarchies that the ITTMs inhabit, with their relevant input; there is any case some danger with this nomenclature that casual readers will be misled into thinking that these classes 'P^f','NP^f' *etc.* have something in common with the classical $P = NP?$ problem, which really they do not.

3.4. Decidable *versus* eventually decidable. A feature that has not been much remarked on so far, is that there are two notions of ITTM-reducibility: that of semi-decidability and that of eventual semi-decidability, which, to take the simpler case of integers, are respectively decided by a Σ_1 relation over L_λ, or a Σ_2 relation over L_ζ. After all, even after all halting computations on integer input have halted the machine still carries on and may produce 'eventually' interesting data on the output tape. Furthermore the universal machine behaviour really does get its character from the Σ_2-extendible pair of ordinals (ζ, Σ). Is there then a case for singling out one of these reducibilities as somehow the 'primary' underlying relation? It is not part of the this survey's aims to examine applications of ITTMs, hence we shall not go in to the proposals for such applications that have been made for equivalence relation theory (such as in [8], [9], [10]) or in ITTM-computable model theory ([27]). Clearly when discussing such reducibilities, either a choice has to be made, or both reducibilities are carried along.

One might argue (as this author has on occasion) that in one sense the notion of eventual semi-decidability is the more general and encompassing, and more fully captures the essence of the machine action (whilst most of the papers cited tend to concentrate on the semi-decidable). That may seem satisfying, but on the other side there are arguments that the Spector pointclass of semi-decidable sets Γ_0 has better closure properties than that of the eventually semi-decidable Γ. For example, Theorem 1 of [20] shows that the uniformization and scale properties hold for Γ_0 but both fail for Γ. This also for the generalisations of the ITTMs to those of Σ_n limit rules: there again uniformization and scale holds for the pointclasses Γ_n defined there for the

Σ_1-semi-decidable notion. (It is open whether the weaker property holds that Γ relations can be uniformized by a Σ_3-function defined over L_Σ.) We leave this discussion at this point.

3.5. Correspondences with other theories. We discuss the relations between ITTM-reducibility and other theories. As is well known Π_1^1-monotone inductive definitions over, for example the structure $\mathbb{N}$ of the natural numbers (see, *e.g.*, [63]), results in sets of integers that are themselves Π_1^1. It is also well known that strategies for Gale–Stewart games with open pay-off sets contained in $\mathbb{N}^\mathbb{N}$ are definable over $\langle L_{\omega_1^{ck}}, \in \rangle$. We explore the connections and generalisations here.

Let $\Phi \colon \mathcal{P}(\mathbb{N}) \to \mathcal{P}(\mathbb{N})$ be any map, which in this context we call an *operator*. We call it 'recursive', 'arithmetical', 'Π_1^1', 'Γ', *etc.*, if the relation "$n \in \Phi(X)$" is recursive, ... Γ, and so on. It is *monotone* if $A \subseteq B \longrightarrow \Phi(A) \subseteq \Phi(B)$.

DEFINITION 15. *Let* $\Phi \colon \mathcal{P}(\mathbb{N}) \to \mathcal{P}(\mathbb{N})$ *be a Γ-operator. We define the Γ-quasi-inductive operator using iterates of* Φ *as*:

$$\Phi_0(X) = X; \quad \Phi_{\alpha+1}(X) = \Phi(\Phi_\alpha(X)); \text{ and}$$
$$\Phi_\lambda(X) = \liminf_{\alpha \to \lambda} \Phi_\alpha(X) =_{df} \cup_{\alpha<\lambda} \cap_{\lambda>\beta>\alpha} \Phi_\beta(X).$$

We set the stability set *to be* $\Phi_{\text{On}}(X)$.

It is easy to recast the ITTM computation steps on the tapes' contents as an example of a recursive quasi-inductive operator on $\mathbb{N}$, and in this case, *e.g.*, $\Phi_{\text{On}}(\varnothing)$ for the universal machine is $\Phi_\zeta(\varnothing)$. We define $A \subseteq \omega$ to be a Γ-*quasi-inductive set* if there is a Γ-operator Φ with A (1–1) reducible to the stability set $\Phi_{\text{On}}(\varnothing)$: $A \leq_1 \Phi_{\text{On}}(\varnothing)$. (Again this should be compared to a set being *inductive* over $\mathbb{N}$ if it is (1–1) reducible to a fixed point of a Π_1^1-operator.) Recall the following characterisation of infinite two person games into sets $A \subseteq \mathbb{N}^\mathbb{N}$:

• Every Σ_1^0-game which is a winner for player I has a Π_1^1-monotone-inductive winning strategy (which in this context is equivalent to it being the stability set for a Π_1^1-monotone operator Φ; similar comments apply to the next bullet point).

• (Solovay—unpublished but see Kechris [34]) Every Σ_2^0-game which is a winner for player I has a Σ_1^1-monotone-inductive winning strategy.

One might tempted be to ask the question as to whether for Σ_3^0-games the winning strategies for I are ITTM-decidable, or putting it another way: recursively-quasi-inductive? It is easy to see that since an ITTM can emulate arithmetic operations, and even Π_1^1 ones, that any arithmetic or Π_1^1-quasi-inductive set is also recursively quasi-inductive. The answer turns out to be negative, but we shall see it is a close race.

Clearly a certain amount of analysis is needed to prove that there are wellorderings of sufficient length along which to run an ITTM and see that it either halts or loops. How much of second order number theory is needed for this? Burgess in [6] first defined, to our knowledge, *arithmetic-quasi inductive* (in a very different context) so in deference to that we formulate in second order number theory the existence of sufficiently long wellorderings for arithmetic operators, although by the comment in the last paragraph, this would have been equally valid using recursive, or Π_1^1-operators. By a 'repeat pair' of ordinals for a quasi-inductive operator Φ, we mean the least pair $(\zeta, \Sigma) = (\zeta(\Phi, x), \Sigma(\Phi, x))$ with $\Phi_\zeta(x) = \Phi_\Sigma(x) = \Phi_{\mathrm{On}}(x)$.

DEFINITION 16. AQI *is the sentence*: "*for every arithmetic operator* Φ, *for every* $x \subseteq \mathbb{N}$, *there is a wellordering* W *with a repeat pair* $(\zeta(\Phi, x), \Sigma(\Phi, x))$ *in* Field(W)". *If an arithmetic operator* Φ *acting on* x *has a repeat pair, we say that* Φ *converges* (*with input* x).

AQI thus asserts precisely that we have sufficiently long wellorders to discover the behaviour of quasi-inductions along them, or equivalently, the looping or otherwise behaviour of an ITTM machine. Typically then one may ask:

Q: *What is the strength of* ACA$_0$ + AQI?
We then have:

THEOREM 7 (Welch [85]). *The theories*:

$$\Pi_3^1\text{-CA}_0, \quad \Delta_3^1\text{-CA}_0 + \Sigma_3^0\text{-Det}, \quad \Delta_3^1\text{-CA}_0 + \mathsf{AQI}, \text{ and } \Delta_3^1\text{-CA}_0$$

are in strictly descending order of strength, meaning that each theory proves the existence of a β-model of the next.

We remark here that in determining the proof-theoretic ordinal for Π_2^1-CA$_0$, Rathjen ([60]) was lead to analyse Σ_1-chains of arbitrary but finite length of models in the L-hierarchy of the form $L_{\gamma_0} \prec_{\Sigma_1} L_{\gamma_1} \prec_{\Sigma_1} \cdots \prec_{\Sigma_1} L_{\gamma_n}$. (Recall also that the least β-model of Π_2^1-CA$_0$ occurs as $\mathbb{R} \cap L_{\gamma_\omega}$ where γ_ω is the least γ so that L_γ is a union of an infinite such chain of Σ_1-substructures.) Speculatively, to analyse Π_3^1-CA$_0$ in the same manner will need a theory of such chains but with Σ_2-elementarity replacing Σ_1. Hence analysing the pair $L_\zeta \prec_{\Sigma_2} L_\Sigma$ derived from ITTM-theory is the first step along this way. The "closeness" of Σ_3^0-Det to AQI is indicated by the fact that a chain of length 2: $L_{\zeta_0} \prec_{\Sigma_2} L_{\zeta_1} \prec_{\Sigma_2} L_{\zeta_2}$ is sufficient to establish that $(\Sigma_3^0\text{-Det})^{L_{\zeta_0}}$. (The fact of the matter is that the theories are even closer than just a chain of length 2, but part of the theorem above asserts that a chain of length 1 is insufficient—see [85] for a discussion of this.)

One conjecture we mention here since it concerns this level of determinacy, is a possible connection with R. Lubarsky's "*Feedback*-ITTMs": one envisages an ITTM which has the capability to call as a subroutine some other ITTM-computation for input, handing data over to the called machine and awaiting 'feedback'. This seems uncontroversial, but there is the possibility for infinite descending chains of calls. Lubarsky analyses the situation in [52]. One may prohibit descending chains by various devices: he does this by assigning ordinals to calls, and requiring naturally that further subcalls must attach lower ordinals. However he also discusses the situation where the machines are purposefully free to make such infinite chains of calls on certain inputs (such a computation he calls 'freezing'). The level of L where strategies for Σ_3^0-games first are definable, is precisely that β so that L_β has an illfounded end-extension $\mathcal{M}$ with infinite depth nesting of Σ_2 extendible pairs $L_{\gamma_i} \prec_{\Sigma_2} L_{b_i}$ with, for all $i < \omega$, $\gamma_i \leq \gamma_{i+1} <_\mathcal{M} b_{i+1} <_\mathcal{M} b_i$. Conjecturally there appears *prima facie* a connection between this level of determinacy and such a machine model.

Theories of Truth—a historical note. We remarked above that *arithmetically quasi-inductive* had been coined by Burgess. In [6] he analysed a *Herzberger Revision Sequence*. Herzberger ([31]) had defined in effect a revision operator just beyond arithmetic by defining, in a language for arithmetic with an additional predicate symbol $\dot{T}$ (for 'Truth'), for any $X = H_0 \subseteq \mathbb{N}$:

$$H_{\alpha+1} = \{\ulcorner \sigma \urcorner : \langle \mathbb{N}, +, \times, \ldots, H_\alpha \rangle \models \sigma\}; \text{ with } H_\alpha \text{ interpreting } T;$$
$$H_\lambda = \cup_{\alpha < \lambda} \cap_{\lambda > \beta > \alpha} H_\beta.$$

Benedikt Löwe was the first to point out the similarity between this formalism of a revision sequence of truth sets and ITTMs, and in [50] also wrote a program that could produce any Herzberger sequence as long as it possessed ITTM computable length. This left open the nature of the relationship between the two formalisms, but as may be gleaned from the above, essentially they are similar as in fact any H-revision sequence starting with $H_0 = x$ can be mimicked on an ITTM with similar input, irrespective of length, and indeed this is how one sees that essentially a 'repeat pair' for a Herzberger sequence must exist (as Herzberger had asserted). Burgess had calculated that repeat pair for $X = \emptyset$ as being precisely our (ζ, Σ) here and established these as least so that $L_\zeta \prec_{\Sigma_2} L_\Sigma$. This much earlier work on theories of truth was unknown to those working on ITTMs in 2000 when the significance of this pair of ordinals for ITTM theory was established in [77].

Burgess also then had defined arithmetical quasi-inductive sets as above. More recently Field (*cf.* [17] and subsequently) has started to develop a theory of truth which uses revision theoretic models given by a Π_1^1-quasi-inductive operator. That all three formalisms are essentially mathematically similar—in that they all produce recursively isomorphic stable sets—is established in [82].

3.6. Varying the limit rules—hypermachines. One might ask whether other limit rules are conceivable. It was shown in [77] that the Liminf rule is in one sense 'complete' for Σ_2 limit rules. This is entirely unsurprising in view of the fact that the actions of any machine whose operations are recursive at successor steps, and further obeys a definable rule at limit stages that sets a cell's value according only to what has happened at previous stages, will be absolute to L: consequently one may argue that $C_i(\alpha)$ is always $\Sigma_2(L_\alpha)$ (assuming say integer input). Conversely given a wellorder w (either as input or self-computed) a subroutine may inductively define codes along w for L_βs for β less than the order type of w. The conclusion is that for many stages α a code for L_α will be uniformly arithmetic in the snapshot of a machine (and so in particular the universal machine) at precisely stage α. This is probably explicit and implicit in [22] and the earlier papers to varying degrees, but it is explicitly shown in [19] that one program computes on the one hand codes for J_α (levels of the Jensen hierarchy, for $\alpha < \Sigma$), whilst simultaneously on the other their Σ_2-truth sets, T_α^2 say, (under some recursive enumeration of sentences). *Inter alia* a code for J_α is written by stage $\omega^2.(\alpha+1)$. For limit stages μ this has to be done in such a fashion that the Σ_2-truth sets are (standardly) uniformly c.e. in the machine's snapshots at stages $\omega^2 \cdot \mu$. This uses the fact that for $\alpha < \beta_0$ (where $L_{\beta_0} = J_{\beta_0}$ is the least model of ZF^-) there are *uniform* Σ_n-skolem functions for each $n < \omega$. This requires one to show at limit stages μ that T_μ^2, although not necessarily equal to $\text{Liminf}_{\beta \to \mu} T_\beta^2$, is nevertheless (standardly) uniformly c.e. in it.

Can we do this for more complicated machines with say a Σ_3 or a Σ_n 'limit rule' for cell values? This is investigated in [20] where such rules are supplied and an analysis of programs that, as above, produce again simultaneously codes for J_α and truth sets T_α^n for $\alpha < \Sigma(n)$, where now $\Sigma(n)$ is least so that $L_{\Sigma(n)}$ has a Σ_n-proper elementary substructure $L_{\zeta(n)}$. 'Σ_n-Machines' operating under such limit rules exhibit entirely analogous behaviours as the ITTMs: a universal such machine will produce entries on its tapes until $\Sigma(n)$ but the snapshot at $\Sigma(n)$ will be equal to that at $\zeta(n)$, and thus it will enter a permanent loop of periodicity $\Sigma(n)$. The rules are defined inductively: those for $\Sigma(n)$ are defined assuming the rules for $\Sigma(n-1)$-machines. It is probably fair to say that as n increases the rules become increasingly distant from any machine intuition. Even for Σ_3 the limit rule is somewhat complicated, and depends on taking a liminf for a limit stage λ, not on all ordinals $\alpha < \lambda$ but roughly, as a first approximation, those of 'stable informational content'. More precisely, we might say that, relative to λ, an ordinal $\alpha < \lambda$ is of such stable informational content, if for any real $y \in 2^\mathbb{N}$ that has appeared on a tape before stage α, then if by stage λ something 'new about y has been written to the scratch tape' then actually that new fact had already been written at some stage less than α. The intuition is that at stage α, anything that is going to be known about such a y by stage λ is already known. One then determines cell values $C_i(\lambda)$ by taking a liminf of only those $C_i(\alpha)$ where the ordinals below α that are of stable content relative to α are those below α of stable content relative to λ.

3.7. Alternative ITTM machine models. One might ask: are three tapes necessary? what if one enlarges the alphabet? Enlarging the alphabet perhaps unsurprisingly does not change the class of functions that are computable, nor adding extra tapes. Perhaps more surprisingly reducing the number of tapes does change the class of computable functions. In [28] Hamkins and Seabold investigate what happens if the triple tape arrangement is reduced to a single tape of cells $\langle C_i \rangle_i$. Except for one small, but vital, difference, the class of computable functions remains the same: for $f \colon \mathbb{N} \to \mathbb{N}$, or $f \colon 2^\mathbb{N} \to \mathbb{N}$ there is absolutely no difference. However for $f \colon 2^\mathbb{N} \to 2^\mathbb{N}$ there is a difference. It is easy to arrange a pre-run of the program that expands the input so that it lies only on the cells $\langle C_{3i+1} \rangle_i$ now, with 0s on the other cells. One then may successfully mimick the run of a 3-tape-ITTM machine on the single tape using $\langle C_{3i} \rangle_i$ and $\langle C_{3i+2} \rangle_i$ as the two other 'tapes' for scratch work and output. If this successfully comes to a conclusion there is the natural output y written on $\langle C_{3i+2} \rangle_i$, and we need to squash this back down on to the whole tape $\langle C_i \rangle_i$ erasing the other values. How does one know when this has been completed? In [28] it is shown that this problem cannot be surmounted. Different solutions can be proposed: a special tab cell, not on the tape $\langle C_i \rangle_i$ and reserved precisely for this purpose is one. Another is to allow a special symbol $*$ say onto the tape to mark the progress of this squashing process.

One solution advocated in [79] is that of allowing the alphabet of a 1-tape machine to consist of 0, 1 and 'B' the latter for a blank or empty cell. One then changes the limit rule so that if at limit stage μ C_i has changed value cofinally often in μ, then $C_i(\mu)$ is set to B—thus B represents ambiguity or non-determination to some extent. (In the other cases the cell values are set to that fixed value on a tail below μ of course.) This then remarkably also suffices to compute the full complement of 3-tape-ITTM-computable functions $f \colon 2^\mathbb{N} \to 2^\mathbb{N}$.

§4. **Longer tapes.** Once one has freed Turing's machine from the confines of time, it is tempting to free up some extra space too. One can allow for cell sequences of arbitrary ordinal length and even consider a machine acting on a proper class sequence of cells C_i for each ordinal $i \in \mathrm{On}$. The idea is that one keeps the programs fixed and the lim inf limit rule as for an ITTM. We must devise some specific rule to handle the limit cells $C_\omega, C_{\omega+\omega}, \ldots, C_\lambda, \ldots$ when the R/W head is over a limit cell but then receives an instruction to move left. One formalism is to just to return the head to the start of the tape C_0. The position function for the head at time α, $l(\alpha)$, is then governed by the lim inf rule, and thus can now ascend to higher limit level cells.

4.1. Ordinal length tapes. What can such machines now compute? Essentially we are allowing an infinite sequence of 0, 1s to come into play and there is the possibility of coding arbitrary sets by such sequences. Dawson and Koepke came up independently with this concept. Dawson in [13] formulated an *Axiom of Computability* that said that every set could appear coded on the output tape of such a machine whilst it was running; thus at some point for some program P_e a machine (not necessarily halting) would have a code for a set y on the output tape for any set y. He then proves that the class of such sets form a transitive class satisfying ZFC. Studying the grid of snapshots produced yields a form of Löwenheim–Skolem theorem that allows for a proof of the Condensation Lemma and so the Generalised Continuum Hypothesis. The "computable sets" of this construction are of course the same as the constructible sets of the Gödel hierarchy, since as remarked earlier the machine theoretic operations are absolute to L. Koepke in [41] and with Koerwien in [42] considered halting computations starting from an On-length tape with input only finitely many 1s representing a finite set of ordinal parameters. Remarkably one has:

THEOREM 8 (Koepke [41]). *A set $x \subseteq$ On is On-ITTM-computable from a finite set of ordinal parameters if and only if it is a member of the constructible hierarchy.*

The left to right direction is just the observation that the machine operations are absolute to L. It is the converse that takes some work. Koepke goes into considerable programming detail to show that the *bounded truth predicate*:

$$\{(\alpha, \varphi, \vec{x}) \mid \varphi \text{ an } \in \text{-formula, } \alpha \in \text{On}, \vec{x} \in L_\alpha, L_\alpha \models \varphi(\vec{x})\}$$

is computable. The reduction of sets in L to finite sequences of ordinals is achieved by noting (by an induction on α) that any set $x \in L_\alpha$ can be defined by one of countably many terms of $\mathcal{L}_\in$, evaluated over L_α, whilst using lower levels as parameters: $t(L_{\alpha_0}, L_{\alpha_1}, \ldots, L_{\alpha_k})$. This allows a bounded truth function of the form: $W(a, \varphi) = 1 \leftrightarrow \varphi(a)$ to be recursively defined for bounded formulae φ. Once the theorem is proven, then GCH and $\Diamond$ arguments follow by considering Löwenheim–Skolem arguments applied to computations defined inside transitive ZF$^-$ models.

An unrealised hope was that computational models, which slow down the construction of the L hierarchy, might, rather like the *Silver machine* precursors for constructing L, yield up alternatives to, or at least insights into the more delicate fine structural arguments that go into proving $\square$ or morasses. Silver machines were indeed invented to avoid the Jensen hierarchy of J_αs for this very purpose (see [4] and [15]). Both the Silver machines, and the *Hyperfine Structure* of Friedman and Koepke [18] however make use of a Finiteness Property that works against the ITTM formalism (as indeed does the Σ_2-nature of the limit rules.)

4.2. Admissible length tapes. Another natural candidate is to compare computable reducibilities of a machine model with length of tape an admissible ordinal $\alpha > \omega$ with α-recursion theory.

Takeuti [74] seems to have been the first to consider (at least in print) computation involving ordinals themselves. He considered a scheme equivalent to

Σ_1-definability on ordinals to replace the notion of 'recursive enumerability'. There were a number of other attempts to extrapolate from Kleene's equational calculus to ordinal valued functions. Such calculi had been developed by Machover [53], Levy [49], Tugué [75], Kripke [48], Platek [58]. Kripke–Platek set theory emphasised the utility of Σ_1-Replacement and the notion of an *admissible set* and concomitant weakening of ZF as an axiom system emerged.

Although some of the above schemes involved forms of an equational calculus (such as Kleene's) and others imagined ordinals *per se* as the objects of computation (see, *e.g.*, the discussion in [63] Part V for an account of some of this development), we shall first examine ITTMs with lengths of tape an arbitrary admissible ordinal, and in a later section consider machine models dealing with ordinals as objects directly. Clearly an α-ITTM with tape an admissible α is just a cut-down version of the On-ITTM just described. The sets imagined as being described by such a version of computation would be those that could be enumerated, or traversed as input, in α many steps.

In terms of α-recursion theory a set $A \subseteq \alpha$ which is $\Sigma_1(L_\alpha)$ is called α-*recursively enumerable* (α-r.e.). It is α-*recursive* if both it and its complement are α-r.e. and thus it is $\Delta_1(L_\alpha)$. For the notion of relative α-recursion a notion of reduction procedure was defined; originally a notion now called *weak α-recursive* $A \leq_{w\alpha} B$ was employed, but this was discovered by Driscoll [16] to be intransitive. The notion of $A \leq_\alpha B$, A is α-*recursive in* B is more stringent and $\leq_\alpha \subsetneq \leq_{w\alpha}$ in general. Koepke and Seyfferth in [44] use the restriction of Koepke's ordinal tape machines to an α, and define A *is computable in* B to mean that the characteristic function of A can be computed by a machine in α many stages from an oracle for B. For them this is exactly the relation that $A \in \Delta_1(L_\alpha[B])$. This is again a strictly proper weakening of the relation $\leq_{w\alpha}$. However this has the advantage that the notion of α-computability and α-computable enumerability tie up exactly with the notions of α-recursiveness and α-r.e. With this modelling of α-r.e. they proceed to give a reworking of the argument of Sacks–Simpson Theorem ([64]) for solving Post's problem:

THEOREM 9 (Sacks–Simpson [64]). *There are α-computable enumerable sets A, B with both $A \not\leq_{w\alpha} B$ and $B \not\leq_{w\alpha} A$.*

This approach although it ties up neatly with the notions of α-recursion, and is a very natural extension of the Turing model to α-length tapes, has again the disadvantage that the relation of 'computable in' is intransitive.

Dawson addressed this problem in [13]. He considered various modes of behaviour: a 'punch-hole tape' of length α that can only be written to once; and 'α-*sequential computation*' whereby the cells of the output tape are written out in sequential order. The term 'punch-hole' refers to the fact that the output tape may be written to once only. (Such ω-tape versions were also initially considered by Hamkins and Kidder as prototypes for an ITTM. It is an easy observation (S.-D. Friedman and the author) to check that such ω-tape machines decide all and only arithmetic predicates and, when halting on integer input, do so by ω^2.) The effect of the restriction to α-sequential computation is that any α-length computation has all the initial segments of its output as elements of L_α. In terms of traditional α-recursion theory we should say that the initial segments of the computation are thus α-*finite*. An α-r.e. set with all initial segments α-finite was called *regular*. This is in contradistinction to a general α-recusively enumerable set (for example there are ω_1^{ck}-r.e. subsets of ω that are not in $L_{\omega_1^{ck}}$). α-recursion theorists recognized the utility of regular α-r.e. sets: proofs are more tractable when sets can be assumed to be regular. Usefully it was shown by Sacks ([62]) that any α-degree of α-r.e. sets contains a regular set. This points the way to attempting to find a notion of computation which restricts attention to regular sets.

Dawson defines:

DEFINITION 17. (i) *A is uniformly-α-computably-enumerable in B ("A is α-c.e. in B") if there is a program index e and ordinal parameter $\eta < \alpha$ for which, when given B as input on one tape, there is a function $\delta: \alpha \to \alpha$ so that it will write to the punch-hole output tape in such a way that $A \upharpoonright \gamma$ will be written by time $\delta(\gamma)$.*

(ii) *A is uniformly-α-computable in B ($A \leq_c B$) if both A and cA are α-c.e. in B.*

The presence of the function δ thus simply enforces regularity: $A \cap \gamma \in L_\alpha[B]$ for any $\gamma < \alpha$. He then has:

THEOREM 10 (Dawson). *A is α-c.e. if and only if it is α-r.e. and regular. A is α-computable if and only if it is α-recursive.*

Although Dawson does not discuss the other useful notion for classical α-recursion theory, *hyper-regularity*, it is easy to argue that an α-c.e. set A is hyper-regular if there is pair (e, η) and an α-*computable* function δ all as in Definition 17(i) outputting A.

The upshot of Dawson's definition is that he is considering α-degrees, but restricting to where all α-c.e. sets are α-regular, and any set that is A-α-c.e. will be A-α-regular *etc.* He then has that the structure of the α-degrees of the α-r.e. sets in the classical sense, is isomorphic to that of the α-computable-degrees of the α-c.e. sets. Hence those theorems of the classical α-recursion theory about α-r.e. sets whose proofs rely on, or use regular α-r.e. sets will carry over to his theory. These are such as the Sacks–Simpson theorem above: there are α-c.e. sets A, B that are α-computably, $\leq_c$-incomparable. However the Shore Splitting Theorem (that any regulat α-r.e. sets A may be split into two disjoint α-r.e. sets B_0, B_1 with $A \not\leq_\alpha B_i$, [68]) is less amenable to this argument as in general taking unions and intersections does not preserve the isomorphism between the α-r.e. degrees and the α-c.e. degrees.

The proof of the Shore Density theorem (that between any two α-r.e. sets $A <_\alpha B$ there lies a third α-r.e. C: $A <_\alpha C <_\alpha B$, [69]) is more complex than the other arguments, and seemingly relies on more of the fine structure of L for its proof. This can however be generalised, and for predicates $\mathbb{B}$ that result in so-called *acceptable* and *sound* hierarchies (see, *e.g.*, [66] Defs. 1.20 & 5.7), Dawson lifts the notion of α-computation to $\mathbb{B}$-α-computation where now $\mathbb{B} =_{df} \langle J_\alpha^\mathbb{B}, \in \mathbb{B}\rangle$ is a transitive, admissible, acceptable, and sound structure. These assumptions make the structure $\mathbb{B}$ sufficiently L-like for Shore's arguments to go through. He then has (working just in the case that $\mathbb{B} \subseteq \alpha$):

THEOREM 11 (Dawson—The α-c.e. Density Theorem). *Let $\mathbb{B}$ be as above. Let A, B be two $\mathbb{B}$-α-c.e. sets, with $A <_{\mathbb{B},\alpha} B$. Then there is C also $\mathbb{B}$-α-c.e. with $A <_{\mathbb{B},\alpha} C <_{\mathbb{B},\alpha} B$.*

§5. Infinite time register machines.

5.1. Number register machines.
Koepke has been instrumental in looking at the behaviour of *register machines* such as those of Shepherdson and Sturgis [67], as described in Cutland [11], or those of or Minsky [54]. Such machines have a finite number of natural number registers R_i for $i < N$, acting under a program consisting of a finite list of instructions $\vec{I} = I_0, \ldots, I_q$. These instructions come from a standard set for such machines of incrementing or setting to zero, a register, transferring of register contents and a conditional jump on comparison of two registers to another instruction. We shall say that the machine at time τ is about to perform instruction $I(\tau)$, and the register contents are $R_i(\tau) \in \mathbb{N}$ for $i < N$. We let $\vec{R}(\tau)$ be the N-vector of the registers' contents at time τ. We specify that at limit times λ the next instruction to be performed $I(\lambda) =_{df} \liminf_{\alpha \to \lambda} I(\alpha)$. The reader may like to verify that result of this is that the machine is at the beginning of the outermost subroutine in the program that was called unboundedly often before time λ; register values at limit stages are also defined according to $\liminf^*$ values:

$$R_i(\lambda) =_{df} \liminf_{\alpha \to \lambda} R_i(\alpha) \text{ if this is finite; otherwise we set } R_i(\lambda) = 0.$$

It is this resetting of unbounded values to zero (rather than allowing the machine to crash) that gives the model its surprising strength. A function $F: \mathbb{N}^N \to \mathbb{N}$ is then ITRM-*computable* if there is some program P of the above sort with $P(\vec{k}) \downarrow F(\vec{k})$ for every $\vec{k} \in \mathbb{N}^N$. In order to handle reals, or sets of integers, we allow the machine to have an oracle query as instruction. Thus a register R_i may be reset to zero if its contents is in an oracle set $Z \subseteq \mathbb{N}$. This machine was investigated in [7], [43].

In the former paper the clockable ordinals are again defined as the lengths of times taken for a halting computation, and in general the relations between clockable ordinals and computable ordinals are investigated in the same spirit as Hamkins and Lewis did for ITTMs. Here however the clockables form an initial segment of the countable ordinals, and every *computable ordinal* (that is an ordinal which has real code whose characteristic function is ITRM-computable) is clockable. They show that the ITRMs are (perhaps unsurprisingly) weaker than the ITTMs as the latter can simulate the former and calculate their halting sets.

DEFINITION 18 (*n-register halting set*).

$$H_N =_{df} \{\langle e, r_0, \ldots, r_{N-1}\rangle \mid P_e(r_0, \ldots, r_{N-1})\downarrow\}.$$

They give a criterion for when a machine has started looping after θ many steps. Note that the two functions I and $\vec{R}$ give the course of computation for θ steps.

LEMMA 5. *Let* $I: \theta \longrightarrow \omega$, $R: \theta \longrightarrow {}^N\omega$ *be a computation of the N register machine with program P and with oracle Z for order type θ many stages. Then if this computation has not halted by stage θ, then it will never do so if θ is sufficiently large so that there is some constellation (I', R') so that*

$$\text{otp}\,\{\beta \mid I(\beta) = I' \wedge R(\beta) = R'\} \geq \omega^\omega.$$

They present an algorithm using a stack that when programmed on such a machine gives a yes/no output as to whether the oracle Z contains a set of integers coding a wellfounded relation. This is a back-tracking algorithm that searches for the leftmost descending path through Z's coded ordering, if it exists. Thus Π_1^1-complete sets are decidable by such ITRMs. It is also easy to see that Boolean combinations of Π_1^1 sets are also decidable. They also prove

THEOREM 12 (Koepke–Miller [43]). *For any N the N-halting problem: '$\langle e, \vec{r}\rangle \in H_N$' is decidable by an* ITRM. *Similarly for any oracle Z, the (N, Z)-halting problem '$\langle e, \vec{r}\rangle \in H_N^Z$' is decidable by a Z-*ITRM *with an oracle for Z.*

Crucially in the above as N increases the number of registers needed to run a program deciding H_N also increases. As they remark this shows (in contradistinction to ITTMs, or indeed finitary register machines) that there can be no universal ITRM.

In order to ascertain the exact strength of ITRMs it is argued in [47] that a simple one register machine (!) must halt or enter an infinite loop by the *second* admissible ordinal. Note that ω_1^{ck} is not *a priori* a candidate here: if $\text{Liminf}_{\beta \to \omega_1} R_0(\beta) = p < \omega$ then indeed a Π_2-reflection argument shows that on a closed and unbounded in ω_1^{ck} set of ordinals, the same instruction number and this liminf value of p are revisited, and so by the Lemma 5 the machine is looping; however there is the possibility that this liminf value is ω for the first time and then $R_0(\omega_1^{ck})$ is reset to 0. At $\omega_1^{ck} + \omega_1^{ck}$ this could reoccur, but now the next instruction number could differ! However this pattern cannot go on for ever and by ω_2^{ck} the machine must halt or loop. An induction on N shows that each time a register is added another admissible ordinal is needed for this argument. Hence by adding registers we can find programs that clock any ordinal below ω_ω^{ck}—the first limit of admissibles. The argument above can be used to show that given sufficient admissible ordinals we can prove that the halting sets H_n exist. More formally in the case of number register machines, a well known subsystem of second order number theory measures their strength.

THEOREM 13 (Koepke–Welch [47]).

(i) *Let* ITRM_N *be the assertion*: *"The N-register halting set H_N exists."*
Then: KP + *"there exist $N + 1$ admissible ordinals $> \omega$"* $\vdash \text{ITRM}_N$.

(ii) *Let* ITRM *be the similar relativized statement that "For any $Z \subseteq \omega$, for any $N < \omega$ the N-register halting set H_N^Z exists." Then*: $\Pi_1^1\text{-CA}_0 \vdash \text{ITRM}$.

Now for the converse (where $\text{HJ}(N, y)$ denotes the n'th hyperjump of y):

THEOREM 14 (Koepke–Welch [47]). (i) $\text{ATR}_0 + \text{ITRM} \vdash \Pi_1^1\text{-CA}_0$.
(ii) *There is a fixed $k < \omega$ so that for any $N < \omega$*

$$\text{ATR}_0 + \text{ITRM}_{N \cdot k} \vdash \text{"HJ}(N, \varnothing) \text{ exists."}$$

5.2. Ordinal register machines (ORM). Koepke also considered register machines that contained ordinals. Platek (in private correspondence) indicated that he had thought of his equational calculus on recursive ordinals as being also implementable as some kind of ordinal register machine. Ryan Siders had also been considering such machines and in a series of papers they considered what could be done with unbounded ordinals. The format is the same as for ITRMs except now that we allow ordinals as register entries; there is no longer any need for a register's contents to be reset to 0 if a liminf becomes unbounded in ω (or any larger limit ordinal λ) at a limit stage. Then the ordinal arithmetic operations of addition, mulitplication and exponentiation can be shown to be ORM-computable, as well as the Gödel pairing function.

THEOREM 15 (Koepke–Siders [46]). *A set $x \subseteq \text{On}$ is ORM-computable from a finite set of ordinals parameters if and only if it is a member of the constructible hierarchy.*

They do this as indicated for the OTM model in Section 4.1 above by following that plan: as G and the ordinal arithmetic is ORM computable, they may show that the recursively given truth predicate $T \subseteq \text{On}$ is ORM-computable. They thus can characterize ordinal computability by the theorem above.

They remark that their program is implementable on a 12 register machine, and assert that this can even be lowered to 4—so conceptually this is a rather simple machine.

Hamkins and Miller at roughly the same time took Koepke and Siders ORM-model and proved a number of facts about them. First a definition of jump.

DEFINITION 19. *Let P_e be the e'th ORM program.* (i) *The* (weak) jump *is the set*

$$0^\diamond = \{e \in \mathbb{N} \mid P_e(0)\downarrow\};$$

(ii) *The* strong jump *is the set*

$$0^\blacklozenge = \{(e, \alpha) \in \mathbb{N} \times \text{On} \mid P_e(\alpha)\downarrow\}.$$

For finite time register machines the sets of which these are analogues are recursively isomorphic, but in the infinite case they differ. Indeed if we let $<_{\text{ORM}}$ be the natural relative computability relation, we have $0 <_{\text{ORM}} 0^\diamond <_{\text{ORM}} 0^\blacklozenge$. Hamkins and Miller show:

THEOREM 16 (Hamkins–Miller [26]). *Let $F: \text{On}^n \to \text{On}$ be any partial function. Then F is* ORM-*computable if and only if it is* OTM-*computable*.

They further show:

THEOREM 17 (Hamkins–Miller [26] as abstracted in [25]). *There exist* ORM-*computable enumerable sets A, B sets of ordinals both below $0^\diamond$ which are $\leq_{\text{ORM}}$-incomparable*.

They give an explicit proof of this fact, with a Friedberg–Muchnik style priority argument. As they also remark, a softer proof of this fact can be deduced from the observation due to Koepke, that, defining γ-ORM machines where ordinals parameters are restricted to γ, and run-times are less than γ, then the γ-ORM-computable sets coincide with the γ-recursive sets (in the sense of α-recursion theory). Here γ is the supremum of the clockable ordinals (in the ORM-sense, which they remark in this case are easily argued to be all ORM-writable). Hence the result actually follows by the Sacks–Simpson Theorem 9 (and the proof could indeed be translated over). They also prove that A, B must be unbounded subsets of γ: shorter sets will not work.

In [24] the authors take up the observation that an ITTM can essentially in ω-steps decide Σ_2^0 truths (note again the Σ_2-nature of the limit rule) and by repeating this can decide $\Sigma_{2 \cdot n}^0$ truth within $\omega \cdot n$ steps. However an ORM needs to take times unboundedly times in ω^ω to decide arithmetic truths. Both styles of machine catch up with each other at stage ω_1^{ck}, with membership in hyperarithmetic sets of integers being so decidable.

§6. Infinite time Blum–Shub–Smale machines and polynomial time set recursion. It is natural to consider what other models of computation might be capable of running transfinitely. The computational model ([5]) of Blum–Shub–Smale is a natural candidate. Such machines act on the real continuum directly rather than on Cantor or Baire space. A study of an infinite time version has been initiated by Koepke and Seyfferth ([45]). Such machines can be thought of as having a finite number of registers $R_1, \ldots, R_N$ containing reals $r_1, \ldots, r_n$. A finite program, which can best be thought of as a flow diagram with either conditional branching nodes, or function nodes where a rational function computation takes place. (A test is made to avoid division by zero.) At limit stages of time λ the current instruction number is set to be the Liminf of the previous instructions, but the register values $R_i(\lambda)$ must be the continuous limit of the values $R_i(\alpha)$ for $\alpha < \lambda$. If for any register R_i, the previous values in it fail to converge to a unique limit then the whole computation is deemed to crash. This requirement of continuity on real values is quite stringent and as we shall see limits the machine's capabilities. However, and usefully, such computations as $sin(x)$, e^x ... etc. can now be calculated with the machine halting after ω steps with the correct values (they cannot on the finite time model).

The requirement on continuity at limit points of time force some coding tricks on the programmer in order to get the machine to recognise limit ordinals: typically if one real register is thought of as having the function of a "scratch tape" then at limit stages one does not normally expect a continuous limit. However by continually dividing by 2, at a limit stage the tape will have the real 0.0. If a register is in danger of becoming unboundedly large at a limit time then it is best to calculate $\frac{1}{x}$ than x. With such devices as these, together with coding ω sequences of 0, 1 bits into decimals, the authors of [45] can ensure that a machine with n nodes in its flow diagram, can halt (on rational number input) at ordinal times up to ω^{n+1}, but this is the correct upper bound. Hence such machines, as a class, will on rational input either crash, halt, or be in an infinite loop by stage ω^ω.

As they ask, the question arises as to what is the computational scope of such machines. One observation is that the construction of such machines, as always, is absolute to the constructible L-hierarchy. Hence methods we have discussed above ensure that imagining the machine running on rational input with a final output can be done inside L_{ω^ω}. Hence the machine can only produce reals constructible by exactly this level. However, they ask if it can compute all reals in L_{ω^ω}? We may answer this as follows: a program can be written to compute successive iterates of the following continuously decreasing hierarchy, we set:

$$h_0 = \mathbb{N}; \quad h_{\alpha+1} = (h_\alpha)' \cap h_\alpha; \quad \text{Lim}(\lambda) \to h_\lambda = \cap_{\alpha<\lambda} h_\alpha.$$

(Here $(h_\alpha)'$ denotes the standard Turing jump.) By a result of Shoenfield this hierarchy continues throughout the recursive ordinals, and every hyperarithmetic is (ordinary Turing) reducible to a member of this sequence. We can restrict this to just ω^ω stages and have an ITBSS-machine compute a code for any element of this initial segment. Hence we have an exact characterisation of the ITBSS computable reals as those recursive in some h_α for an $\alpha < \omega^\omega$. One may then argue that these coincide with the reals of L_{ω^ω}.

One might consider such machines with the ability to take Liminf's at limit stages rather than just continuous limits; however now the notion of Liminf is in the topology of the reals. An upper bound for the reals such a model could produce is that of the reals in $L_{\pi(3)}$ where $\pi(3)$ is the least Π_3-*reflecting ordinal*. However it is not known if this the best possible.

Finally we mention another connection of the ω^ω'th level of constructibility with the *Safe Recursive Set Functions* of Beckmann, Buss and S. Friedman ([2]). They develop a notion of set recursion using the notion of safe recursion developed on integers by Bellantoni and Cook. The key idea is to divide variables in a many-place function into *safe* and *normal*: such as $f(\vec{a}/\vec{b})$. Recursion is only allowed on safe variables—here $\vec{b}$. However input may now be sets, whether integers, ordinals, strings or otherwise. This typing of the variables ensures that ranks of sets are kept low by any SRSF-function, indeed they prove the following:

THEOREM 18. *Let f be any* SRSF. *Then there is a ordinal polynomial q_f so that*

$$rk(f(\vec{a}/\vec{b})) \leq \max_i rk(a_i) + q_f(rk(\vec{a})).$$

Using an adaptation of Arai, such functions, whilst restricting to finite strings, may decide problems computed by (standard) polynomial time Turing machines. On the other hand it is shown that the least SRSF-closed set which contains ω as an element is again L_{ω^ω} which arose as above for ITBSS-machines. Hence the SRSF functions computing on 2^ω to 2^ω will compute the same functions as an ITTM machine bounded by that ordinal number of stages. This coincides with Schindler's definition of 'polynomial time' functions for ITTMs mentioned above. Thus we have three different notions of polynomial time for sequences on ω (or reals) from three different computational or recursion notions coinciding. This leads dangerously close to formulating a thesis that the continuous limit ITBSS machines (or is it the ITTMs up to stage ω^ω?) capture the informal notion of 'polynomial time infinitary computation'.

§7. Conclusions.

On reflection, what the above all illustrate is that there were always undiscovered avenues of concrete machine models that can contribute to our knowledge or assessment of what is 'generalised computation'. By way of being wonderfully simple examples of higher type recursion they throw a light on the earlier abstract (and at times difficult) theory. The set theorist might conclude that what the models do (at least in their oracle free versions) is mostly construct sets at the bottom of the Gödel constructible hierarchy L. The examples of Koepke and his co-workers of ITTMs and ITRMs, show that we can give a presentation of L very much in a computational spirit, and that presentation is yet another to lay besides the Gödel, Jensen, Silver, and the Friedman–Koepke hyperfine structural versions. The computability theorist may have a different view of matters and will see that there is a wealth of inventiveness when it comes to generalising the actual model constructs to allow for transfinite resources, and may calculate precisely how rapidly the computational strength of such machines grows. Then the reverse mathematician can gauge this strength in terms of axiomatic theories. Thus the theory of such machine models constitutes a very pleasing nexus between various branches of logic and computability theory.

Acknowledgements. The author would like to express his thanks to many people for their time and energy on this subject, in particular to Peter Koepke, and to Benedikt Löwe for, *inter alia*, pointing out to me the formal similarity between ITTMs and the Herzberger version of Revision Theory, but most particularly to Joel Hamkins who introduced me to the delightful ITTMs whilst in Kobe.

REFERENCES

[1] K. J. BARWISE, *Admissible sets and structures*, Perspectives in Mathematical Logic, Springer Verlag, 1975.

[2] A. BECKMANN, S. BUSS, and S.-D. FRIEDMAN, *Safe recursive set functions*, submitted, 2012.

[3] E. BEGGS and J. V. TUCKER, *Can Newtonian systems, bounded in space, time, mass and energy compute all functions?*, **Theoretical Computer Science**, vol. 371 (2007), pp. 4–19.

[4] A. BELLER and A. LITMAN, *A strengthening of Jensen's □ principles*, **The Journal of Symbolic Logic**, vol. 45 (1980), no. 2, pp. 251–264.

[5] L. BLUM, M. SHUB, and S. SMALE, *On a theory of computation and complexity over the real numbers*, **Notices of the American Mathematical Society. New Series**, vol. 21 (1989), no. 1, pp. 1–46.

[6] J. P. BURGESS, *The truth is never simple*, **The Journal of Symbolic Logic**, vol. 51 (1986), no. 3, pp. 663–681.

[7] M. CARL, T. FISCHBACH, P. KOEPKE, R. MILLER, M. NASFI, and G. WECKBECKER, *The basic theory of infinite time register machines*, **Archive for Mathematical Logic**, vol. 49 (2010), no. 2, pp. 249–273.

[8] S. COSKEY, *Infinite time Turing machines and Borel reducibility*, **Mathematical theory and computational practice**, Lecture Notes in Computer Science, vol. 5635, Springer, 2009, pp. 129–133.

[9] S. COSKEY and J. D. HAMKINS, *Infinite time decidable equivalence relation theory*, **Notre Dame Journal for Formal Logic**, vol. 52 (2011), no. 2, pp. 203–228.

[10] ———, Effective Mathematics of the Uncountable, ch. Infinite time Turing machines and an application to the hierarchy of equivalence relations on the reals, to appear.

[11] N. CUTLAND, *Computability: an introduction to recursive function theory*, Cambridge University Press, 1980.

[12] E. B. DAVIES, *Building infinite machines*, **British Journal for the Philosophy of Science**, vol. 52 (2001), no. 4, pp. 671–682.

[13] B. DAWSON, *Ordinal time Turing computation*, Ph.D. thesis, Bristol, 2009.

[14] V. DEOLALIKAR, J. D. HAMKINS, and R-D. SCHINDLER, $P \neq NP \cap co\text{-}NP$ *for the infinite time Turing machines*, **Journal of Logic and Computation**, vol. 15 (2005), pp. 577–592.

[15] K. DEVLIN, *Constructibility*, Perspectives in Mathematical Logic, Springer Verlag, Berlin, Heidelberg, 1984.

[16] G. H. C. DRISCOLL JR., *Metarecursively enumerable sets and their metadegrees*, **The Journal of Symbolic Logic**, vol. 33 (1968), pp. 389–411.

[17] H. FIELD, *A revenge-immune solution to the semantic paradoxes*, **Journal of Philosophical Logic**, vol. 32 (2003), no. 3, pp. 139–177.

[18] S.-D. FRIEDMAN and P. KOEPKE, *An elementary approach to the fine structure of L*, **The Bulletin of Symbolic Logic**, vol. 3 (1997), no. 4, pp. 453–468.

[19] S.-D. FRIEDMAN and P. D. WELCH, *Two observations concerning infinite time Turing machines*, **BIWOC 2007 Report** (I. Dimitriou, editor), Bonn, January 2007, Hausdorff Centre for Mathematics, also at http://www.logic.univie.ac.at/sdf/papers/joint.philip.ps, pp. 44–47.

[20] S.-D. FRIEDMAN and P. D. WELCH, *Hypermachines*, **The Journal of Symbolic Logic**, vol. 76 (2011), no. 2, pp. 620–636.

[21] R. O. GANDY, *On a proof of Mostowski's conjecture*, **Bulletin de l'Academie Polonaise des Sciences. Série des Sciences Mathématiques, Astronomiques et Physiques**, vol. 8 (1960), pp. 571–575.

[22] J. D. HAMKINS and A. LEWIS, *Infinite time Turing machines*, **The Journal of Symbolic Logic**, vol. 65 (2000), no. 2, pp. 567–604.

[23] ——— , *Post's problem for supertasks has both positive and negative solutions*, **Archive for Mathematical Logic**, vol. 41 (2002), pp. 507–523.

[24] J. D. HAMKINS, D. LINETSKY, and R. MILLER, *The complexity of qickly ORM-decidable sets*, **Computation and logic in the real world: Proceedings of CiE2007, Siena** (S. B. Cooper and A. Sorbi, editors), Lecture Notes in Computer Science, vol. 4497, Springer-Verlag, Berlin, 2007, pp. 488–496.

[25] J. D. HAMKINS and R. MILLER, *Post's problem for ordinal register machines*, **Computation and logic in the real world: Proceedings of CiE2007, Siena** (S. B. Cooper and A. Sorbi, editors), Lecture Notes in Computer Science, vol. 4497, Springer-Verlag, Berlin, 2007, pp. 358–367.

[26] ——— , *Post's problem for ordinal register machines: an explicit approach*, **Annals of Pure and Applied Logic**, vol. 160 (2009), no. 3, pp. 302–309.

[27] J. D. HAMKINS, R. MILLER, D. SEABOLD, and S. WARNER, *Infinite time computable model theory*, **New computational paradigms** (S. B. Cooper et al., editors), Springer-Verlag, Berlin, 2008, pp. 521–557.

[28] J. D. HAMKINS and D. SEABOLD, *Infinite time Turing machines with only one tape*, **Mathematical Logic Quarterly**, vol. 47 (2001), no. 2, pp. 271–287.

[29] J. D. HAMKINS and P. D. WELCH, $P^f \neq NP^f$ *almost everywhere*, **Mathematical Logic Quarterly**, vol. 49 (2003), no. 5, pp. 536–540.

[30] L. HARRINGTON, *Analytic determinacy and* $0^{\#}$, **The Journal of Symbolic Logic**, vol. 43 (1978), no. 4, pp. 684–693.

[31] H. G. HERZBERGER, *Notes on naive semantics*, **Journal of Philosophical Logic**, vol. 11 (1982), pp. 61–102.

[32] P. HINMAN, **Recursion-theoretic hierarchies**, Ω Series in Mathematical Logic, Springer, Berlin, 1978.

[33] K. HRBACEK and S. SIMPSON, *On Kleene degrees of analytic sets*, **Proceedings of the Kleene symposium** (H. J. Keisler, J. Barwise, and K. Kunen, editors), Studies in Logic, North-Holland, 1980, pp. 347–352.

[34] A. S. KECHRIS, *On Spector classes*, **Cabal seminar 76–77** (A. S. Kechris and Y. N. Moschovakis, editors), Lecture Notes in Mathematics Series, vol. 689, Springer, 1978, pp. 245–278.

[35] S. C. KLEENE, *Recursive quantifiers and functionals of finite type I*, **Transactions of the American Mathematical Society**, vol. 91 (1959), pp. 1–52.

[36] ——— , *Turing-machine computable functionals of finite type I*, **Proceedings of the 1960 conference on Logic, Methodology and Philosophy of Science**, Stanford University Press, 1962,

pp. 38–45.

[37] ———, *Turing-machine computable functionals of finite type II*, **Proceedings of the London Mathematical Society**, vol. 12 (1962), pp. 245–258.

[38] ———, *Recursive quantifiers and functionals of finite type II*, **Transactions of the American Mathematical Society**, vol. 108 (1963), pp. 106–142.

[39] S. C. KLEENE, *Recursive functionals and quantifiers of finite types revisited*, **Generalized Recursion Theory II, Proceedings 2nd Scandinavian Logic Symposium, Oslo, 1977**, Studies in Logic and Foundations of Mathematics, vol. 94, North-Holland, Amsterdam, New York, 1978, pp. 185–222.

[40] A. KLEV, Magister thesis, ILLC Amsterdam, 2007.

[41] P. KOEPKE, *Turing computation on ordinals*, **The Bulletin of Symbolic Logic**, vol. 11 (2005), pp. 377–397.

[42] P. KOEPKE and M. KOERWIEN, *Ordinal computations*, **Mathematical Structures in Computer Science**, vol. 16 (2006), no. 5, pp. 867–884.

[43] P. KOEPKE and R. MILLER, *An enhanced theory of infinite time register machines*, **Logic and the theory of algorithms** (A. Beckmann et al., editors), Lecture Notes in Computer Science, vol. 5028, Springer, 2008, Swansea, pp. 306–315.

[44] P. KOEPKE and B. SEYFFERTH, *Ordinal machines and admissible recursion theory*, **Annals of Pure and Applied Logic**, vol. 160 (2009), no. 3, pp. 310–318.

[45] ———, *Towards a theory of infinite time Blum–Shub–Smale machines*, **CiE2012 Proceedings**, 2012.

[46] P. KOEPKE and R. SIDERS, *Computing the recursive truth predicate on ordinal register machines*, **Logical approaches to computational barriers** (A. Beckmann et al., editors), Computer Science Report Series, Swansea, 2006, p. 21.

[47] P. KOEPKE and P. D. WELCH, *A generalised dynamical system, infinite time register machines, and Π_1^1-CA_0*, **Proceedings of CiE 2011, Sofia** (B. Löwe, editor), 2011.

[48] S. KRIPKE, *Transfinite recursion on admissible ordinals I, II*, **The Journal of Symbolic Logic**, vol. 29 (1964), pp. 161–162.

[49] A. LEVY, *Transfinite computability (abstract)*, **Notices of the American Mathematical Society**, vol. 10 (1963), p. 286.

[50] B. LÖWE, *Revision sequences and computers with an infinite amount of time*, **Journal of Logic and Computation**, vol. 11 (2001), pp. 25–40.

[51] ———, *Space bounds for infinitary computations*, **Logical approaches to computational barriers, CiE 2006** (A. Beckmann et al., editors), Lecture Notes in Computer Science, vol. 3988, Springer-Verlag, 2006, pp. 319–329.

[52] R. LUBARSKY, *Well founded iterations of infinite time turing machines*, **Ways of proof theory** (R.-D. Schindler, editor), Ontos, 2010.

[53] M. MACHOVER, *The theory of transfinite recursion*, **Bulletin of the American Mathematical Society**, vol. 67 (1961), pp. 575–578.

[54] M. MINSKY, **Computation: Finite and infinite machines**, Prentice-Hall, 1967.

[55] A. MONTALBÁN and R. SHORE, *The limits of determinacy in second order number theory*, **Proceedings of the London Mathematical Society**, (to appear).

[56] Y. N. MOSCHOVAKIS, **Descriptive set theory**, Studies in Logic and the Foundations of Mathematics, North-Holland, Amsterdam, 1980.

[57] D. NORMANN, *Turing's models*, in *this volume*, 2012.

[58] R. PLATEK, **Foundations of recursion theory**, Ph.D. thesis, Stanford, 1966.

[59] H. PUTNAM, *Trial and error predicates and the solution to a problem of Mostowski*, **The Journal of Symbolic Logic**, vol. 30 (1965), pp. 49–57.

[60] M. RATHJEN, *An ordinal analysis of parameter-free Π_2^1 comprehension*, **Archive for Mathematical Logic**, vol. 44 (2005), no. 3, pp. 263–362.

[61] H. ROGERS, *Recursive function theory*, Higher Mathematics, McGraw, 1967.

[62] G. E. SACKS, *Post's problem, admissible ordinals and regularity*, **Transactions of the American Mathematical Society**, vol. 124 (1966), pp. 1–23.

[63] ———, *Higher recursion theory*, Perspectives in Mathematical Logic, Springer Verlag, 1990.

[64] G. E. SACKS and S. G. SIMPSON, *The α-finite injury method*, **Annals of Mathematical Logic**, vol. 4 (1972), pp. 343–367.

[65] R.-D. SCHINDLER, *$P \neq NP$ for infinite time Turing machines*, **Monatshefte für Mathematik**, vol. 139 (2003), no. 4, pp. 335–340.

[66] R.-D. SCHINDLER and M. ZEMAN, *Fine structure theory*, **Handbook of set theory** (M. Magidor, M. Foreman, and A. Kanamori, editors), vol. 1, Springer Verlag, Heidelberg, New York, 2010, pp. 605–656.

[67] J. SHEPHERDSON and H. STURGIS, *Computability of recursive functionals*, **Journal of the Association of Computing Machinery**, vol. 10 (1963), pp. 217–255.

[68] R. A. SHORE, *Splitting an α recursively enumerable set*, **Transactions of the American Mathematical Society**, vol. 204 (1975), pp. 65–78.

[69] ———, *The recursively enumerable α-degrees are dense*, **Annals of Mathematical Logic**, vol. 9 (1976), pp. 123–155.

[70] S. SIMPSON, **Subsystems of second order arithmetic**, Perspectives in Mathematical Logic, Springer, January 1999.

[71] R. M. SOLOVAY, *Determinacy and type-2 recursion*, **The Journal of Symbolic Logic**, vol. 36 (1971), p. 374.

[72] C. SPECTOR, *Hyperarithmetic quantifiers*, **Fundamenta Mathematicae**, vol. 48 (1959), pp. 313–320.

[73] J. R. STEEL, *Analytic sets and Borel isomorphisms*, **Fundamenta Mathematicae**, vol. 108 (1980), pp. 83–88.

[74] G. TAKEUTI, *On the recursive functions of ordinal numbers*, **Journal of the Mathematical Society of Japan**, vol. 12 (1960), pp. 119–128.

[75] T. TUGUÉ, *On the partial recursive functions of ordinal numbers*, **Journal of the Mathematical Society of Japan**, vol. 16 (1964), pp. 1–31.

[76] P. D. WELCH, *Minimality arguments in the infinite time Turing degrees*, **Sets and proofs: Proceedings of the Logic Colloquium 1997, Leeds** (S. B. Cooper and J. K. Truss, editors), London Mathematical Society Lecture Note Series, vol. 258, Cambridge University Press, 1999.

[77] ———, *Eventually Infinite Time Turing degrees: infinite time decidable reals*, **The Journal of Symbolic Logic**, vol. 65 (2000), no. 3, pp. 1193–1203.

[78] ———, *The length of infinite time Turing machine computations*, **Bulletin of the London Mathematical Society**, vol. 32 (2000), pp. 129–136.

[79] ———, *Post's and other problems in higher type supertasks*, **Classical and new paradigms of computation and their complexity hierarchies, papers of the conference Foundations of the Formal Sciences III** (B. Löwe, B. Piwinger, and T. Räsch, editors), Trends in Logic, vol. 23, Kluwer, October 2004, pp. 223–237.

[80] ———, *Bounding lemmata for non-deterministic halting times of transfinite Turing machines*, **Theoretical Computer Science**, vol. 394 (2008), pp. 223–228.

[81] ———, *Turing unbound: The extent of computations in Malament–Hogarth spacetimes*, **British Journal for the Philosophy of Science**, vol. 15 (2008), no. 4, pp. 659–674.

[82] ———, *Ultimate truth vis à vis stable truth*, **The Review of Symbolic Logic**, vol. 1 (2008), no. 1, pp. 126–142.

[83] ———, *Characteristics of discrete transfinite Turing machine models: halting times, stabilization times, and normal form theorems*, **Theoretical Computer Science**, vol. 410 (2009), pp. 426–442.

[84] ———, *Determinacy in strong cardinal models*, **The Journal of Symbolic Logic**, vol. 76 (2011), no. 2, pp. 719–728.

[85] ——, *Weak systems of determinacy and arithmetical quasi-inductive definitions*, **The Journal of Symbolic Logic**, vol. 76 (2011), no. 2, pp. 418–436.

SCHOOL OF MATHEMATICS
UNIVERSITY OF BRISTOL
BS8 1TW, BRISTOL, ENGLAND
URL: http://www.maths.bris.ac.uk/~mapdw/